T0255030

An Introduction to Complex Analysis and the Laplace Transform

Textbooks in Mathematics

Series editors:
Al Boggess, Kenneth H. Rosen

Mathematical Modeling in the Age of the Pandemic
William P. Fox

Games, Gambling, and Probability
An Introduction to Mathematics
David G. Taylor

Linear Algebra and Its Applications with R
Ruriko Yoshida

Maple™ Projects of Differential Equations
Robert P. Gilbert, George C. Hsiao, Robert J. Ronkese

Practical Linear Algebra
A Geometry Toolbox, Fourth Edition
Gerald Farin, Dianne Hansford

An Introduction to Analysis, Third Edition
James R. Kirkwood

Student Solutions Manual for Gallian's Contemporary Abstract Algebra, Tenth Edition
Joseph A. Gallian

Elementary Number Theory
Gove Effinger, Gary L. Mullen

Philosophy of Mathematics
Classic and Contemporary Studies
Ahmet Cevik

An Introduction to Complex Analysis and the Laplace Transform
Vladimir Eiderman

An Invitation to Abstract Algebra
Steven J. Rosenberg

Numerical Analysis and Scientific Computation
Jeffery J. Leader

Introduction to Linear Algebra
Computation, Application and Theory
Mark J. DeBonis

https://www.routledge.com/Textbooks-in-Mathematics/book-series/CANDHTEXBOOMTH

An Introduction to Complex Analysis and the Laplace Transform

Vladimir Eiderman

CRC Press is an imprint of the
Taylor & Francis Group, an **informa** business
A CHAPMAN & HALL BOOK

First edition published 2022
by CRC Press
6000 Broken Sound Parkway NW, Suite 300, Boca Raton, FL 33487-2742

and by CRC Press
2 Park Square, Milton Park, Abingdon, Oxon, OX14 4RN

CRC Press is an imprint of Taylor & Francis Group, LLC

ISBN: 978-0-367-40978-4 (hbk)
ISBN: 978-1-032-16203-4 (pbk)
ISBN: 978-0-367-81028-3 (ebk)

DOI: 10.1201/9780367810283

Publisher's note: This book has been prepared from camera-ready copy provided by the authors.

Contents

Preface

The present textbook is designed for university students studying science and engineering, and can be used in one-semester or two-semester courses. I have aimed for a comparatively short book that would give a sufficiently full exposition of the fundamentals of the theory of functions of a complex variable to prepare the student for various applications. Several important applications in physics and engineering are considered in the book.

At the same time I have not skimped in the thoroughness of the presentation: all theorems (with a few exceptions) are presented with proofs. No previous exposure to complex numbers is assumed. In one respect this book is larger than usual, namely in the number of detailed solutions of typical problems. I hope this makes the book useful for self-study as well.

A specific point of the book is the inclusion of the Laplace transform in a textbook on Complex Analysis. As a rule the Laplace transform is studied in textbooks on differential equations, separately from Complex Analysis. But such a presentation is incomplete because these two topics are closely related to one another. Concepts in Complex Analysis are needed to formulate and prove some of the basic theorems in the theory of Laplace transform, such as the inverse Laplace transform formula. Moreover, methods of Complex Analysis provide an alternative approach for solving typical problems involving Laplace transforms. For instance, we provide a second method of calculating the inverse Laplace transform.

An essential part of the book is a translation from the Russian edition [3]. Work on the original version of the textbook was begun at the suggestion of Professor B.P. Osilenker, to whom I am grateful for discussions about the structure of the book and other valuable advice. I also owe a debt of thanks to Professor S.Y. Khavinson, who carefully read the manuscript of the Russian edition and made numerous comments and suggestions contributing to its improvement.

I am grateful to the translator of the book Dr. Andrew Dabrowski, cooperation with whom contributed to the deep revision of the original text. Moreover, he provided me with invaluable technical support.

On the advice of reviewers, a large chapter has been added to the book containing applications of Complex Analysis. I want to express my deepest gratitude to the reviewers Professors James Brennan and Al Boggess for their advice, which contributed to the improvement of the book.

Preface

Author

Vladimir Eiderman holds a Ph.D. from Mathematical Institute of Academy of Sciences, Armenian SSR. He is Rothrock Lecturer at Indiana University. He has been Professor, Moscow State University of Civil Engineering, Visiting Professor of University of Kentucky, University of Wisconsin-Madison, and Indiana University. Dr. Eiderman has more than 30 research publications.

Author

[illegible]

Introduction

Over the course of human history the introduction of increasingly broader classes of numbers arose from practical needs. Thus the need to count objects led to the emergence of *natural numbers,* the nonnegative whole numbers. Even this concept did not arise at once; its formulation took centuries. For the measurement of magnitudes allowing arbitrary subdivisions, fractions were required. With the development of algebra the need to solve linear and quadratic equations led to the introduction of negative numbers, which Diophantus employed freely already in the 3rd century C.E. In this way were introduced the *rational numbers*, which can be represented as ratios of integers, or equivalently all integers and fractions, whether positive or negative, and zero.

Already the ancient Greeks had encountered the need to expand the class of fractions, having discovered that not all line segment lengths can be expressed rationally in terms of a given unit length segment—for example, the diagonal of a square cannot be expressed rationally if the square's side is taken as the unit. Numbers which are not rational, i.e. which cannot be expressed as a ratio of two whole numbers, are called *irrational.* The union of the rational and irrational numbers forms the set of *real numbers.* Each real number corresponds to a point on the number line, and conversely each point on the number line corresponds to a real number. In this way the real numbers fill and exhaust the number line.

Operations with "new" numbers, the squares of which are less than zero, were begun by the Italian mathematicians Cardano and Bombelli in the 16th century. They arrived at the need for these strange numbers in developing methods for the solution of the cubic equation. Of course the square of any real number is greater than or equal to zero; so these new numbers, and the wider class of complex numbers in which they were included, were treated with suspicion for a long time—in fact as late as the 16th and 17th centuries many European mathematicians did not even recognize negative numbers, calling them false or impossible.

The distrust of complex numbers dissipated only at the end of the 18th century, after the celebrated work of Euler and Gauss and the geometric interpretation of complex numbers as points in the plane. This geometric interpretation shows that there is no fundamental distinction between real and complex numbers: they both can be drawn in the plane, the real numbers corresponding to just the x-axis while the complex numbers correspond to the whole plane. Operations on real and complex numbers have the same basic properties and are performed according to the same rules. Thus the set of real numbers forms part of the set of complex numbers.

Complex numbers lend clarity and completion to some areas of classical analysis. They permit the solution of a number of problems to which complex numbers might seem irrelevant; one of these, the evaluation of improper integrals, will be considered later. Note that complex numbers, unlike the real ones, arose not from the practical needs, but from the demands of mathematical theory. Subsequently, these numbers found important applications in the mathematical descriptions of processes in physics and engineering, for example in hydrodynamics, aeronautics, electro-magnetism, elasticity, etc. Some of these applications will be considered in this book. The former suspicion lingers only in a few vestigial terms like "imaginary"; the development of the theory and its diverse applications have long since established equal rights for real and complex numbers.

1

Complex Numbers and Their Arithmetic

1.1 Complex Numbers

A *complex number* z is an ordered pair (x, y) of real numbers x and y. That (x, y) is an *ordered pair* means that if $x \neq y$ then (x, y) is distinct from (y, x); in other words not just the values of x and y are important in identifying the pair, but also the order in which they are listed.

Complex numbers have a simple geometric meaning. Introduce a system of rectangular Cartesian coordinates into the plane (Fig. 1). Then every pair $z = (x, y)$ of real numbers corresponds to a point in the plane with coordinates (x, y), and conversely every point in the plane corresponds to a complex number $z = (x, y)$.

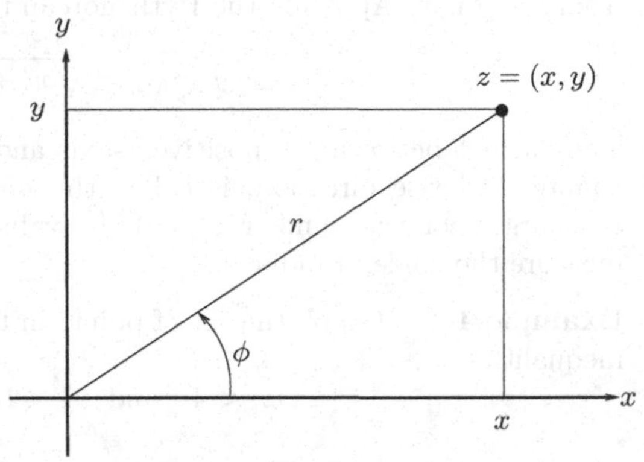

Fig. 1

The first component x of the complex number $z = (x, y)$ is called the *real part* of z and is denoted by $x = \operatorname{Re} z$; the second component y is called the *imaginary part* of z and is denoted by $y = \operatorname{Im} z$. The terms "real" and "imaginary" harken back to the uncomfortable early history of the theory of complex numbers; it is plain to us nowadays that the first coordinate of the point z is no more real than the second.

When the points are thought of as complex numbers we refer to the plane as the *complex plane*, and denote it by $\mathbb{C}$. Two complex numbers $z_1 = (x_1, y_1)$ and $z_2 = (x_2, y_2)$ are considered equal if $x_1 = x_2$ and $y_1 = y_2$, or in other words if z_1 and z_2 correspond to the same point in the plane.

Complex numbers of the form $z = (x, 0)$ fall on the x-axis, which is identified with the real number line. The x-axis is therefore called the *real axis*, and the number $z = (x, 0)$ is usually written simply as x:

$$z = (x, 0) = x.$$

DOI: 10.1201/9780367810283-1

In particular, $z = (0,0)$ is denoted simply as 0.

Complex numbers of the form $z = (0, y)$ fall on the y-axis, which is therefore called the *imaginary axis*, honoring historical practice, and $z = (0, y)$ is called a *pure imaginary* number. The particular number $(0, 1)$ is called the *imaginary unit*, and is denoted by i (Fig. 2):

$$z = (0, 1) = i.$$

y

1 $\quad i = (0, 1)$

x

Fig. 2

Sometimes it is convenient to think of a complex number $z = (x, y)$ as the vector $\mathbf{0z}$ from the origin to the point (x, y), in other words as the position vector of the point z. The length of this vector is called the *modulus* of the number z and is denoted by $|z|$. So $|z|$ is equal to the distance of the point z from the origin. Applying the Pythagorean theorem to Fig. 1,

$$r = |z| = \sqrt{x^2 + y^2} \tag{1.1}$$

The angle ϕ between the positive x-axis and the vector $\mathbf{0z}$, measured in the counterclockwise direction, is called the *argument* of z, and is denoted by $\phi = \arg z$; for the number $z = 0$ the value of $\arg z$ is undefined. We will measure the angle in radians.

Example 1.1 Graph the set of points in the complex plane defined by the inequalities

$$|z| < 4 \quad \text{and} \quad 1 < \operatorname{Im} z < 3.$$

Solution The inequality $|z| < 4$ says that the distance from z to $z_0 = 0$ is less than 4; this set of points is the interior of the circle centered at the origin with radius 4.
The condition $1 < \operatorname{Im} z < 3$ says that the y-coordinate of the point z is between 1 and 3; the set of such z is the strip lying between the horizontal lines $y = 1$ and $y = 3$.
The desired set is the intersection of the disk and the strip, shown in Fig. 3; note that the boundary of the shaded region is not included in the set.

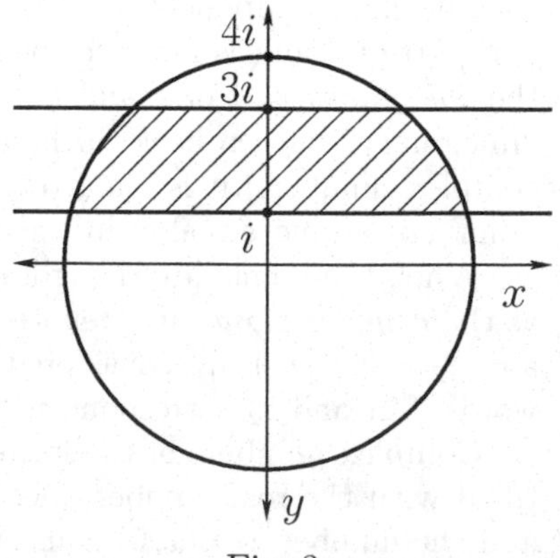

Fig. 3

It is clear from Fig. 1 that

$$x = r\cos\phi, \qquad y = r\sin\phi \tag{1.2}$$

from which we get

$$\cos\phi = \frac{x}{r} = \frac{x}{\sqrt{x^2+y^2}};$$
$$\sin\phi = \frac{y}{r} = \frac{y}{\sqrt{x^2+y^2}}. \tag{1.3}$$

The value of $\arg z$ is not determined uniquely, but only up to a term $2\pi n$, where n is an integer. It is useful to define the *principal value of the argument* of z to be the value of the argument that satisfies the inequality

$$-\pi < \phi \le \pi.$$

The principal value of the argument of z is denoted by $\operatorname{Arg} z$. In the future the vector $\mathbf{0z}$ will usually be denoted simply by z.

1.2 Operations with Complex Numbers

1. The sum and difference of complex numbers as well as the multiplication of a complex number by a real one are determined in exactly the same way as the corresponding operations on vectors. Let $z_1 = (x_1, y_1)$ and $z_2 = (x_2, y_2)$ (Fig. 4). Then the *sum* of z_1 and z_2 is defined as the complex number

$$z = z_1 + z_2 = (x_1 + x_2,\ y_1 + y_2).$$

The *difference* is defined as the complex number

$$z = z_1 - z_2 = (x_1 - x_2,\ y_1 - y_2).$$

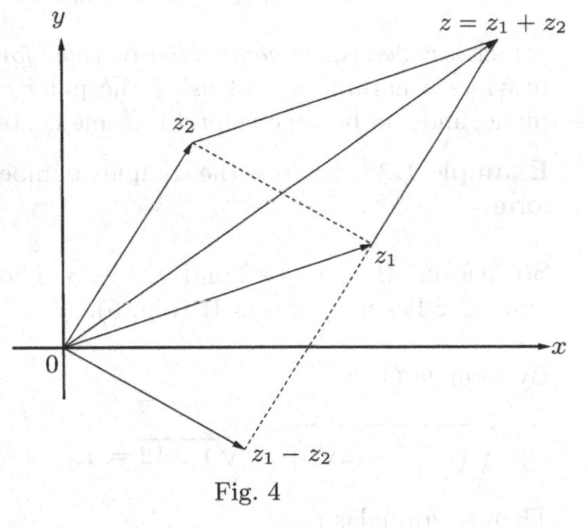

Fig. 4

So the sum or difference of two complex numbers corresponds to the sum or difference of their real and imaginary parts. From Fig. 4 we see that the modulus $|z_1 - z_2|$ is equal to the distance between the points z_1 and z_2. Therefore, the distance $\rho(z_1, z_2)$ between two points $z_1 = x_1 + iy_1$ and $z_2 = x_2 + iy_2$ is found by the formula

$$\rho(z_1, z_2) = \sqrt{(x_1 - x_2)^2 + (y_1 - y_2)^2} = |z_1 - z_2|.$$

By this, the equation of the circle of radius R with center at the point z_0 looks like $|z - z_0| = R$, and the set of points z lying within the same circle is given by the inequality $|z - z_0| < R$.

The *product* of the complex number $z = (x, y)$ with the real number λ is defined as

$$\lambda z = (\lambda x, \lambda y).$$

Since (see Fig. 4) side $0z$ of the triangle $0zz_1$ cannot be longer than the sum of the lengths of the other two sides, we get the *triangle inequality*

$$|z_1 + z_2| \leq |z_1| + |z_2|;$$

equality occurs if and only if the vectors z_1 and z_2 are *co-directed*, that is pointing in exactly the same direction.

Now we introduce another way of expressing the complex number $z = (x, y)$. Writing $(x, 0) = x$ and $(0, y) = (0, 1)y = iy$, we get

$$z = (x, y) = (x, 0) + (0, y) = x + iy.$$

The expression $z = x + iy$ is called the *algebraic* or *Cartesian form* of the complex number z. Substituting the formulas from equations (1.2) gives the expression

$$z = r\cos\phi + i\,r\sin\phi = r(\cos\phi + i\sin\phi), \quad r = |z|, \tag{1.4}$$

which is called the *trigonometric* or *polar form* of the complex number z. This provides a natural way of using the polar coordinates (r, ϕ) in the complex plane, and can be very helpful in some contexts.

Example 1.2 Express the complex number $z = -2+2\sqrt{3}i$ in trigonometric form.

Solution Here $x = -2$ and $y = 2\sqrt{3}$. Therefore the point representing the number z lies in quadrant II (Fig. 5).

By formula (1.1)

$$r = \sqrt{(-2)^2 + (2\sqrt{3})^2} = \sqrt{4 + 12} = 4.$$

Then by formulas (1.3)

$$\cos\phi = -\frac{2}{4} = -\frac{1}{2},$$
$$\sin\phi = \frac{\sqrt{3}}{2}.$$

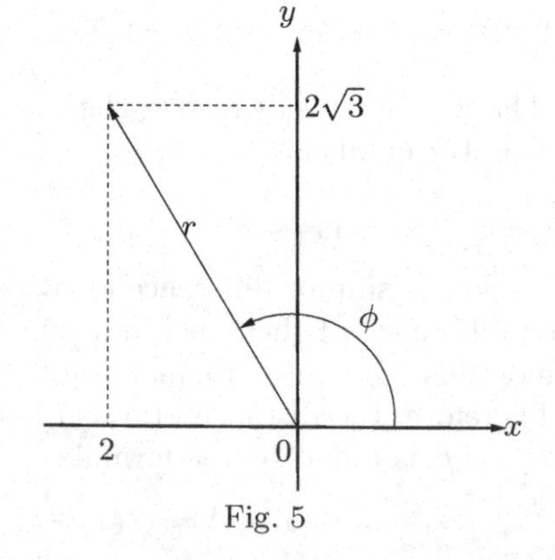

Fig. 5

Since $\cos\phi = -\frac{1}{2}$ we have

$$\phi = \pm\cos^{-1}\left(-\frac{1}{2}\right) + 2\pi k = \pm\frac{2}{3}\pi + 2\pi k,$$

with $k \in \mathbb{Z}$, where $\mathbb{Z}$ denotes the set of integers $\{0, \pm 1, \pm 2, \dots\}$. Since ϕ lies in quadrant II, we must have

$$\phi = \arg z = \frac{2}{3}\pi + 2\pi k,$$

and the principal value of the argument is $\phi = \frac{2}{3}\pi$. Therefore in trigonometric form the number z becomes

$$z = 4\left(\cos\frac{2\pi}{3} + i\sin\frac{2\pi}{3}\right).$$

Of course one may use other values of the argument:

$$z = 4\cos\left(\frac{2\pi}{3} + 2\pi k\right) + 4i\sin\left(\frac{2\pi}{3} + 2\pi k\right),$$

for $k = 0, \pm 1, \pm 2, \dots$.

Example 1.3 Plot the set of points z in the complex plane which satisfy the equation

$$|z - 2 + i| = |z + 5 - 2i|.$$

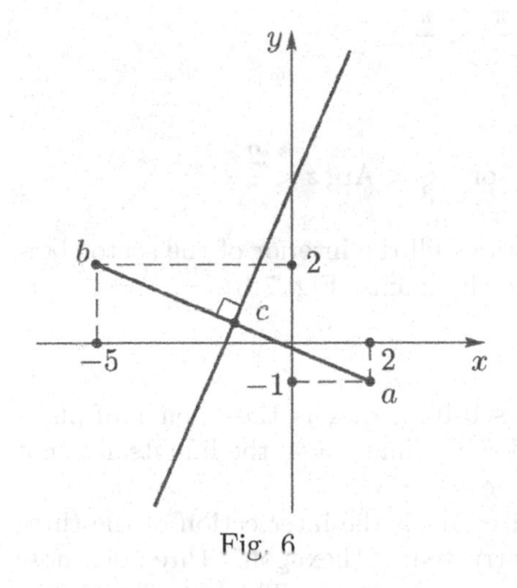

Fig. 6

Solution The value of $|z - z_0|$ is the distance between the points z and z_0. So the given equation says that z is equidistant to the points $a = 2 - i$ and $b = -5 + 2i$. From geometry we know that the set of such points is precisely the perpendicular bisector of the segment $\overline{ab}$ as shown as Fig. 6. Here the midpoint of a and b is

$$c = \frac{a + b}{2} = \frac{2 - i - 5 + 2i}{2} = -\frac{3}{2} + i\frac{1}{2}.$$

So the desired set is the line through c which is perpendicular to $\overline{ab}$.

Example 1.4 Graph the set of points in the complex plane defined by the system of inequalities

$$|z - 3i| < 2, \quad \left|\operatorname{Arg} z - \frac{\pi}{2}\right| < \frac{\pi}{6}, \quad \operatorname{Im} z < 4.$$

Solution (a) First we graph the points satisfying $|z - 3i| < 2$, which is the set of points of distance less than 2 from $3i$: this is the interior of the disk centered at $3i$ with radius 2 (Fig. 7 a); the boundary of the disk is not included in the set.

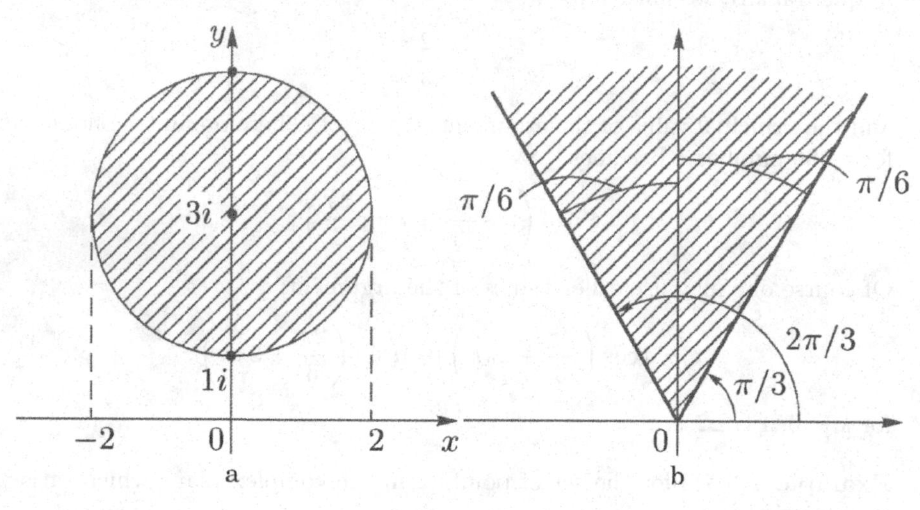

Fig. 7

(b) The inequality

$$\left|\operatorname{Arg} z - \frac{\pi}{2}\right| < \frac{\pi}{6}$$

is equivalent to the relations

$$-\frac{\pi}{6} < \operatorname{Arg} z - \frac{\pi}{2} < \frac{\pi}{6}, \quad \text{or} \quad \frac{\pi}{3} < \operatorname{Arg} z < \frac{2\pi}{3}.$$

The points z, which satisfy these conditions fill the interior of the sector bordered by the rays at angles $\frac{\pi}{3}$ and $\frac{2\pi}{3}$ to the x-axis (Fig. 7 b).

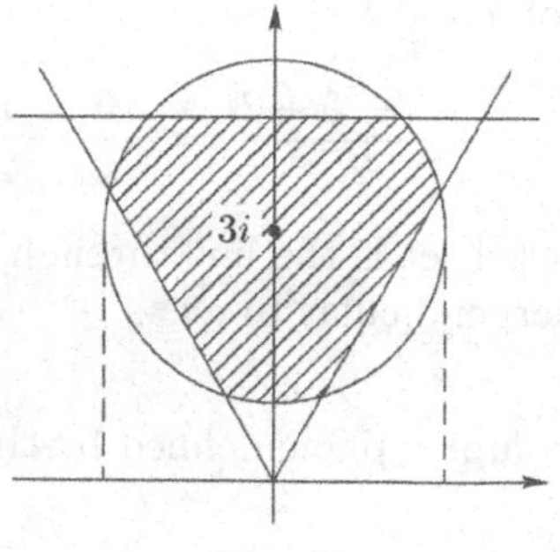

Fig. 8

(c) The set $\operatorname{Im} z < 4$ is the open half-plane lying below the line $y = 4$; the line itself is not part of the set.
The desired set is the intersection of the three sets constructed: a "hexagon", three of whose sides are curved (Fig. 8). The boundary is not part of the set.

2. Now we consider the multiplication operation on two complex numbers. Let us start with the simple product $i \cdot i = i^2$. We *define*

$$i^2 = -1. \tag{1.5}$$

This definition is of fundamental importance. We know that for any real number, represented by a point on the x-axis, the square is nonnegative. But among this wider class, the complex numbers, we find some numbers whose square is negative. Such numbers must correspond to points in the plane that are not on the x-axis.

On first seeing complex numbers students often ask, Why is i^2 equal to -1 rather than some other value? From a formal point of view, equation (1.5) is a definition which does not require proof (beyond its consistency with existing facts). But that answer is not considered convincing. Let us try to find a reason for setting the value of i^2 equal to precisely -1. Given that we naturally want to preserve the customary rules of algebra in our multiplication of complex numbers, one particular rule we would like to keep is the distributive law:

$$z_1 z_2 = (x_1 + i\,y_1)(x_2 + i\,y_2) = x_1 x_2 + i\,y_1 x_1 + i\,x_1 y_2 + i^2\,y_1 y_2,$$

while also maintaining the property of the modulus that

$$|z_1 z_2| = |z_1|\,|z_2|,$$

which holds for all real numbers.

We will show that these two properties can be preserved only by assuming equation (1.5). Take $z_1 = 1 + i$ and $z_2 = 1 - i$. If these two properties hold then

$$z_1 z_2 = (1+i)(1-i) = 1 + i - i - i^2 = 1 - i^2;$$
$$|1 - i^2| = |z_1 z_2| = |z_1|\,|z_2| = \sqrt{1^2 + 1^2}\sqrt{1^2 + (-1)^2} = 2.$$

Setting $w = -i^2$, we will show that $w = 1$. In fact,

$$1 - i^2 = 1 + (-i^2) = 1 + w,$$

from which

$$|1 + w| = |1 - i^2| = 2.$$

On the other hand $w = i(-i)$, so by the modulus property

$$|w| = |i|\,|-i| = 1.$$

This means that

$$|1 + w| = 2 = |1| + |w|.$$

The equation $|1+w| = |1| + |w|$ implies (by the triangle inequality) that the vectors corresponding to 1 and w are in the same direction. Since $|w| = 1$, that means that w and 1 are identical, i.e. $w = 1$. Hence, $-i^2 = 1$, and equation (1.5) follows.

With that established, the product of two arbitrary complex numbers given in algebraic form can now be calculated as follows. Let $z_1 = x_1 + i\,y_1$ and

$z_2 = x_2 + i\,y_2$. We expand the product using the standard distributive law, and then simplify using the identity $i^2 = -1$.

$$\begin{aligned} z_1 z_2 &= (x_1 + i\,y_1)(x_2 + i\,y_2) \\ &= x_1 x_2 + i\,y_1 x_2 + i\,x_1 y_2 + i^2 y_1 y_2 \\ &= (x_1 x_2 - y_1 y_2) + i(y_1 x_2 + x_1 y_2). \end{aligned}$$

Therefore,

$$z_1 z_2 = (x_1 x_2 - y_1 y_2) + i(y_1 x_2 + x_1 y_2).$$

But when multiplying complex numbers it is preferable to carry out this procedure directly rather than memorizing the formula.

Example 1.5 Find the product of the numbers

$$z_1 = 2 - 3i \quad \text{and} \quad z_2 = -4 + i.$$

Solution

$$\begin{aligned} z_1 z_2 &= (2 - 3i)(-4 + i) = -2 \cdot 4 + 3 \cdot 4i + 2 \cdot i - 3 \cdot i^2 \\ &= -8 + 12i + 2i + 3 = -5 + 14i. \end{aligned}$$

From the definitions of addition and multiplication on complex numbers it follows that they have the same properties as the corresponding operations on real numbers:

$$\begin{aligned} z_1 + z_2 &= z_2 + z_1; \\ z_1 z_2 &= z_2 z_1; \\ z_1(z_2 + z_3) &= z_1 z_2 + z_1 z_3; \\ z_1(z_2 z_3) &= (z_1 z_2) z_3. \end{aligned} \tag{1.6}$$

This in turn means that all the usual multiplication formulas apply also to complex arithmetic, for example:

$$\begin{aligned} (z_1 + z_2)^2 &= z_1^2 + 2 z_1 z_2 + z_2^2, \\ (z_1 + z_2)(z_1 - z_2) &= z_1^2 - z_2^2, \end{aligned}$$

etc. In particular note that

$$(x + i\,y)(x - i\,y) = x^2 - (iy)^2 = x^2 + y^2. \tag{1.7}$$

Two complex numbers whose real parts are equal and whose imaginary parts differ in sign are said to be *mutually conjugate*. If $z = x + i\,y$ then its conjugate is denoted by

$$\overline{z} = x - i\,y.$$

Points representing conjugate numbers are symmetric about the real axis (Fig. 9). It is easy to see that

$$|z| = |\overline{z}| = \sqrt{x^2 + y^2}$$
$$\operatorname{Arg} z = -\operatorname{Arg} \overline{z},$$

with the exception in the latter identity of negative real numbers, for when $z = x < 0$ then $\operatorname{Arg} z = \operatorname{Arg} \overline{z} = \pi$. By virtue of equation (1.7),

$$z \cdot \overline{z} = x^2 + y^2 = |z|^2. \qquad (1.8)$$

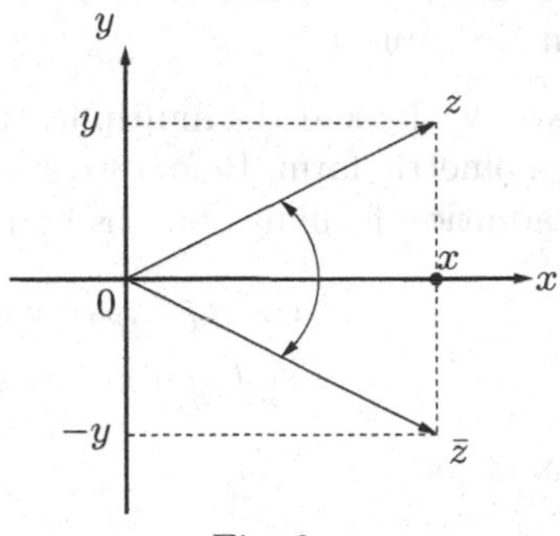

Fig. 9

3. Division of complex numbers, as with real numbers, is defined as the inverse operation to multiplication. Division is possible by any complex number except 0. To find the *quotient* $\frac{z_1}{z_2}$ of two complex numbers given in algebraic form we multiply the numerator and denominator of the fraction by $\overline{z_2}$ and then simplify using equation (1.8):

$$\frac{z_1}{z_2} = \frac{x_1 + i\,y_1}{x_2 + i\,y_2} = \frac{(x_1 + i\,y_1)(x_2 - i\,y_2)}{(x_2 + i\,y_2)(x_2 - i\,y_2)} = \frac{(x_1 + i\,y_1)(x_2 - i\,y_2)}{x_2^2 + y_2^2}.$$

After this we must carry out the multiplication in the numerator, and then divide the real and imaginary parts of the result by the denominator:

$$\begin{aligned}\frac{z_1}{z_2} &= \frac{(x_1 + i\,y_1)(x_2 - i\,y_2)}{x_2^2 + y_2^2} \\ &= \frac{(x_1x_2 + y_1y_2) + i(y_1x_2 - x_1y_2)}{x_2^2 + y_2^2} \\ &= \frac{x_1x_2 + y_1y_2}{x_2^2 + y_2^2} + i\frac{y_1x_2 - x_1y_2}{x_2^2 + y_2^2}.\end{aligned}$$

As in the case of multiplication we recommend that, instead of memorizing this formula, the student should become comfortable with this method for evaluating quotients.

Example 1.6 Find the quotient of the numbers $z_1 = -5 + 14i$ and $z_2 = -4 + i$.

Solution

$$\begin{aligned}\frac{z_1}{z_2} &= \frac{-5 + 14i}{-4 + i} = \frac{(-5 + 14i)(-4 - i)}{(-4 + i)(-4 - i)} \\ &= \frac{20 - 56i + 5i + 14}{(-4)^2 + 1^2} = \frac{34 - 51i}{17} = 2 - 3i.\end{aligned}$$

Note that here we have carried out the inverse operation of the multiplication from Example 1.5.

4. Now we look at the multiplication and division of complex numbers given in trigonometric form. Before we start, you will need to recall from trigonometry the addition formulas for cos and sin,

$$\begin{aligned}\cos(\phi_1+\phi_2) &= \cos\phi_1\cos\phi_2 - \sin\phi_1\sin\phi_2\\ \sin(\phi_1+\phi_2) &= \sin\phi_1\cos\phi_2 + \cos\phi_1\sin\phi_2.\end{aligned} \tag{1.9}$$

Now let

$$\begin{aligned}z_1 &= r_1(\cos\phi_1 + i\sin\phi_1)\\ z_2 &= r_2(\cos\phi_2 + i\sin\phi_2)\end{aligned}$$

where $r_1 = |z_1|$ and $r_2 = |z_2|$. Then

$$\begin{aligned}z_1z_2 &= r_1r_2((\cos\phi_1\cos\phi_2 - \sin\phi_1\sin\phi_2) + i(\sin\phi_1\cos\phi_2 + \cos\phi_1\sin\phi_2))\\ &= r_1r_2((\cos(\phi_1+\phi_2) + i\sin(\phi_1+\phi_2)).\end{aligned} \tag{1.10}$$

So we see that *in the multiplication of two complex numbers, their moduli are multiplied and their arguments are added:*

$$\begin{aligned}|z_1z_2| &= |z_1|\,|z_2|\\ \arg(z_1z_2) &= \arg z_1 + \arg z_2.\end{aligned} \tag{1.11}$$

These formulas mean that the vector z_1z_2 is obtained from the vector z_1 by rotating it through the angle ϕ_2 and multiplying its length by $|z_2|$. For example, the product of a number z with i is obtained from z by rotating it counterclockwise through the angle $\frac{\pi}{2}$, since $\operatorname{Arg} i = \frac{\pi}{2}$ and $|i| = 1$.

From formula (1.11) it follows that an analogous property holds for the product of an arbitrary finite number of complex numbers. For example,

$$|z_1z_2z_3| = |(z_1z_2)z_3| = |z_1z_2|\,|z_3| = |z_1|\,|z_2|\,|z_3|$$

and

$$\arg(z_1z_2z_3) = \arg((z_1z_2)z_3) = \arg(z_1z_2) + \arg z_3 = \arg z_1 + \arg z_2 + \arg z_3.$$

In particular,

$$\begin{aligned}|z^n| &= |\overbrace{z\cdot z\cdot z\cdot \ldots \cdot z}^{n\text{ times}}| = |z|^n = r^n;\\ \arg(z^n) &= n\arg z.\end{aligned}$$

Therefore

$$z^n = r^n(\cos n\phi + i\sin n\phi). \tag{1.12}$$

We turn now to the division of numbers z_1 and z_2 given in trigonometric form. Note that if

$$z_2 = r_2(\cos\phi_2 + i\sin\phi_2),$$

then

$$\overline{z_2} = r_2(\cos(-\phi_2) + i\sin(-\phi_2)).$$

Therefore

$$\frac{z_1}{z_2} = \frac{z_1\overline{z_2}}{z_2\overline{z_2}} = \frac{r_1r_2(\cos(\phi_1-\phi_2) + i\sin(\phi_1-\phi_2))}{r_2^2},$$

from which we get

$$\frac{z_1}{z_2} = \frac{r_1}{r_2}(\cos(\phi_1-\phi_2) + i\sin(\phi_1-\phi_2)). \tag{1.13}$$

So *the modulus of the quotient of complex numbers is equal to the quotient of their moduli, and the argument of the quotient is equal to the difference of the arguments of the dividend and the divisor:*

$$\left|\frac{z_1}{z_2}\right| = \frac{|z_1|}{|z_2|}, \qquad \arg\frac{z_1}{z_2} = \arg z_1 - \arg z_2, \qquad \text{when } z_2 \neq 0.$$

In particular, when

$$z_1 = 1 = 1(\cos 0 + i\sin 0)$$

and

$$z_2 = z^n = r^n(\cos n\phi + i\sin n\phi),$$

then we get

$$z^{-n} = \frac{1}{r^n}(\cos(-n\phi) + i\sin(-n\phi)). \tag{1.14}$$

We set $z^0 = 1$ by definition when $z \neq 0$. Then, by virtue of formulas (1.12) and (1.14), for any integer m (positive, negative, or zero) the following equation, called *de Moivre's formula*[1], holds:

$$z^m = r^m(\cos m\phi + i\sin m\phi). \tag{1.15}$$

Example 1.7 Express the number $(\sqrt{2} - \sqrt{6}\,i)^4$ in algebraic form.

Solution 1) Let us plot the number $\sqrt{2} - \sqrt{6}\,i$ (Fig. 10) and write it in trigonometric form:

---

[1]Abraham de Moivre (1667–1754) was a French mathematician.

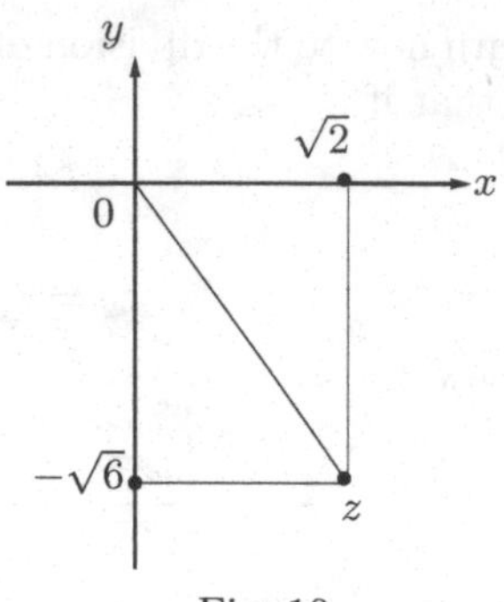

Fig. 10

$$r = |z| = \sqrt{(\sqrt{2})^2 + (\sqrt{6})^2} = \sqrt{2+6} = \sqrt{8};$$

$$\cos\phi = \frac{x}{r} = \frac{\sqrt{2}}{\sqrt{8}} = \frac{1}{2},$$

from which we get

$$\phi = \pm\frac{\pi}{3} + 2\pi k, \quad k \in \mathbb{Z}.$$

Since the angle ϕ lies in quadrant IV, in fact

$$\phi = -\frac{\pi}{3} + 2\pi k,$$

so may use the principal value $-\frac{\pi}{3}$ to evaluate z:

$$z = \sqrt{8}\left(\cos\left(-\frac{\pi}{3}\right) + i\sin\left(-\frac{\pi}{3}\right)\right).$$

2) We use de Moivre's formula (1.15) to find z^4:

$$z^4 = (\sqrt{8})^4\left(\cos\left(-\frac{4\pi}{3}\right) + i\sin\left(-\frac{4\pi}{3}\right)\right)$$
$$= 64\left(\cos\frac{2\pi}{3} + i\sin\frac{2\pi}{3}\right) = 64\left(-\frac{1}{2} + \frac{\sqrt{3}}{2}i\right) = 32(-1+\sqrt{3}\,i).$$

(Note that we have made use of $-\frac{4\pi}{3} + 2\pi = \frac{2\pi}{3}$.)

5. By the *nth root* of a complex number z we mean a number w, which when raised to the nth power equals z, i.e. $w^n = z$. An nth root of z is denoted by $\sqrt[n]{z}$.

Let $z = r(\cos\phi + i\sin\phi)$ be given, and let $w = \rho(\cos\theta + i\sin\theta)$ be the nth root of z which we wish to find. Since by definition $w^n = z$, then

$$|w^n| = |z|, \quad \arg w^n = \arg z,$$

and therefore

$$\rho^n = r, \quad n\theta = \phi + 2\pi k, \quad k \in \mathbb{Z}. \tag{1.16}$$

Because r and ρ are nonnegative real numbers, the equation $\rho^n = r$ implies $\rho = \sqrt[n]{r}$, that is, the nonnegative real number whose nth power equals r. The second equation in (1.16) gives

$$\theta = \frac{\phi + 2\pi k}{n}, \quad k = 0, \pm1, \pm2, \ldots$$

In this way we get

$$w = \sqrt[n]{z} = \sqrt[n]{r}\left(\cos\frac{\phi + 2\pi k}{n} + i\sin\frac{\phi + 2\pi k}{n}\right), \tag{1.17}$$

where $k = 0, \pm 1, \pm 2, \ldots$. Substituting in (1.17) the values $k = 0, 1, 2, \ldots, n-1$, we obtain n distinct values for the nth root; let us call them $w_0, w_1, \ldots, w_{n-1}$. For each of them $|w_k| = \sqrt[n]{r}$, so that the corresponding points lie on a circle of radius $\sqrt[n]{r}$ centered at the origin (Fig. 11). The arguments of the w_k

$$\theta_k = \frac{\phi + 2\pi k}{n} = \frac{\phi}{n} + \frac{2\pi}{n}k$$

increase by $\frac{2\pi}{n}$ as k increases by 1. When $k = n$ we get

$$\theta_k = \frac{\phi}{n} + 2\pi = \theta_0 + 2\pi.$$

This means that the points w_0 and w_n are identical. When $k = n+1, n+2, \ldots$ we will again get the points w_1, w_2, etc. Analogously, when $k = -1, -2, \ldots$, the corresponding points still give us the same $w_0, w_1, \ldots, w_{n-1}$ (in the opposite order). So we are led to the following conclusion.

Every complex number $z \neq 0$ has exactly n distinct nth roots. All these roots are found from formula (1.16) with $k = 0, 1, 2, \ldots, n-1$. The corresponding points form the vertices of a regular n-gon centered at the origin.

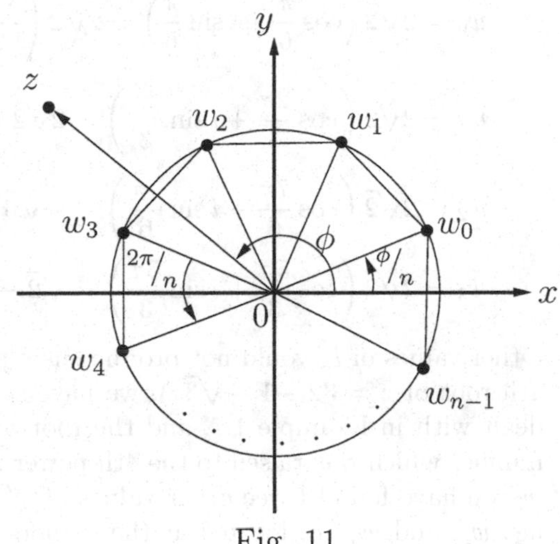

Fig. 11

Formula (1.17) is also called de Moivre's formula; it essentially corresponds to substituting $m = \frac{1}{n}$ into formula (1.15).

Example 1.8 Find all values of $\sqrt[4]{z}$ when $z = 32(-1 + \sqrt{3}\,i)$.

Solution First we find the modulus and argument of the number z:

$$r = \sqrt{32^2((-1)^2 + (\sqrt{3})^2)} = 32 \cdot 2 = 64;$$

$$\cos\phi = \frac{-32}{64} = -\frac{1}{2}, \quad \phi = \pm\frac{2\pi}{3} + 2\pi k, \quad k \in \mathbb{Z}.$$

Since $\operatorname{Re} z < 0$ and $\operatorname{Im} z > 0$, the angle ϕ lies in quadrant II, therefore

$$\phi = \frac{2\pi}{3} + 2\pi k, \quad k \in \mathbb{Z}.$$

In formula (1.17) we can use any value of the argument ϕ, so let us take the simplest, $\phi = \frac{2\pi}{3}$. Substituting these values of r and ϕ into formula (1.17) we get

$$\begin{aligned}\sqrt[4]{32(-1+\sqrt{3}\,i)} &= \sqrt[4]{64}\left(\cos\frac{\frac{2\pi}{3}+2\pi k}{4} + i\sin\frac{\frac{2\pi}{3}+2\pi k}{4}\right)\\ &= 2\sqrt{2}\left(\cos\left(\frac{\pi}{6}+\frac{\pi k}{2}\right) + i\sin\left(\frac{\pi}{6}+\frac{\pi k}{2}\right)\right)\end{aligned}$$

(using $\sqrt[4]{64} = \sqrt{8} = 2\sqrt{2}$). Substituting the values $k = 0, 1, 2, 3$ we obtain four distinct values for $\sqrt[4]{z}$:

$$\begin{aligned} w_0 &= 2\sqrt{2}\left(\cos\frac{\pi}{6} + i\sin\frac{\pi}{6}\right) = 2\sqrt{2}\left(\frac{\sqrt{3}}{2} + i\frac{1}{2}\right) = \sqrt{6} + \sqrt{2}\,i;\\ w_1 &= 2\sqrt{2}\left(\cos\frac{2\pi}{3} + i\sin\frac{2\pi}{3}\right) = 2\sqrt{2}\left(-\frac{1}{2} + i\frac{\sqrt{3}}{2}\right) = -\sqrt{2} + \sqrt{6}\,i;\\ w_2 &= 2\sqrt{2}\left(\cos\frac{7\pi}{6} + i\sin\frac{7\pi}{6}\right) = -\sqrt{6} - \sqrt{2}\,i;\\ w_3 &= 2\sqrt{2}\left(\cos\frac{5\pi}{3} + i\sin\frac{5\pi}{3}\right) = \sqrt{2} - \sqrt{6}\,i.\end{aligned}$$

Other values of k would not produce new points. Note that in extracting the 4th root of $z = 32(-1+\sqrt{3}\,i)$, we have solved a problem inverse to the one dealt with in Example 1.7, and the root $w_3 = \sqrt{2} - \sqrt{6}\,i$ happens to be the number which was raised to the 4th power in the earlier example. But besides w_3 we have found three other values of $\sqrt[4]{z}$: the points corresponding to w_0, w_1, w_2, and w_3 are located at the corners of a square inscribed in the circle of radius $2\sqrt{2}$ and centered at the origin.

The ability to extract roots of any given number allows us to solve the quadratic equation

$$az^2 + bz + c = 0$$

with arbitrary (complex) coefficients a, b, and c. The roots of the equation are given by the quadratic formula

$$z_{1,2} = \frac{-b \pm \sqrt{b^2 - 4ac}}{2a}, \tag{1.18}$$

which is derived in the same way as in the case of real numbers a, b, c, and z, by completing the square of the quadratic polynomial $az^2 + bz + c$. For the term $\sqrt{b^2 - 4ac}$ we can take either of the two values of the roots; these are related by the equality $w_0 = -w_1$.

Example 1.9 Solve the equation $z^2 + 2z + 2 = 0$.

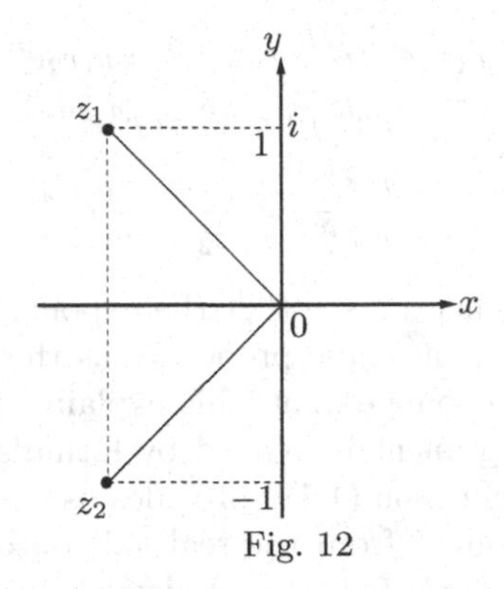

Fig. 12

Solution Let $D = b^2 - 4ac = 2^2 - 4 \cdot 1 \cdot 2 = -4$. Then

$$\sqrt{D} = \sqrt{-4} = \sqrt{-1 \cdot 4} = 2\sqrt{-1} = \pm 2i.$$

By formula (1.18)

$$z_{1,2} = \frac{-2 \pm 2i}{2} = -1 \pm i,$$

so that $z_1 = -1 + i$, $z_2 = -1 - i$. The roots z_1 and z_2 are shown on Fig. 12.

In a high school math course, students are usually taught that if the discriminant $D < 0$ then the equation has no solutions. In fact there are no solutions if we are looking only for *real* solutions, that is points lying on the x-axis. But in the wider class of complex numbers solutions can indeed be found, but the corresponding points lie off of the real axis.

So *every quadratic equation has exactly two, possibly equal, roots.* Later we will show that for every natural number n the equation

$$a_0 z^n + a_1 z^{n-1} + \cdots + a_n = 0$$

has exactly n roots—generally speaking, complex valued. This means that to find solutions for algebraic equations of higher degree, it is not necessary to further expand the set of numbers, e.g. by looking at points in 3-space. The present set of complex numbers is already enough for finding a solution to any algebraic equation.

6 . To conclude Chapter 1 we introduce one more form for representing a complex number. We define the *exponential* function on imaginary numbers $i\phi$ by the following equation:

$$e^{i\phi} = \cos\phi + i \sin\phi. \tag{1.19}$$

Recall that a number z can be expressed in the trigonometric form

$$z = r(\cos\phi + i \sin\phi)$$

where $r = |z|$. This can now be shortened to

$$z = re^{i\phi}, \quad r = |z|,$$

called the *exponential form* of a complex number.

Let

$$z_1 = r_1 e^{i\phi_1}, \quad z_2 = r_2 e^{i\phi_2}.$$

In view of formulas (1.10), (1.12), and (1.13), the following equalities hold:

$$\begin{aligned} r_1r_2e^{i\phi_1}e^{i\phi_2} &= z_1z_2 = r_1r_2e^{i(\phi_1+\phi_2)} && \text{so that} \quad e^{i\phi_1}e^{i\phi_2} = e^{i(\phi_1+\phi_2)}, \\ (re^{i\phi})^n &= z^n = r^ne^{in\phi} && \text{so that} \quad (e^{i\phi})^n = e^{in\phi}, \\ \frac{r_1e^{i\phi_1}}{r_2e^{i\phi_2}} &= \frac{z_1}{z_2} = \frac{r_1}{r_2}e^{i(\phi_1-\phi_2)} && \text{so that} \quad \frac{e^{i\phi_1}}{e^{i\phi_2}} = e^{i(\phi_1-\phi_2)}. \end{aligned} \tag{1.20}$$

Hence we see that the exponential function defined in equation (1.19) above has the same properties as the usual exponential function of a real variable. To some extent, this explains why the exponential function of an imaginary argument is defined by formula (1.19). In Section 4.3.1 we will see that the definition (1.19) provides us with a natural extension of the exponential function e^x from the real axis onto the complex plane. Formula (1.19) is called *Euler's formula*[2] A deeper account of the origin of Euler's formula will be given in Section 6.4.

Problem 1.10 Write the number

$$z = \frac{6e^{i\pi/3}}{(3-2i)^2}$$

in algebraic form.

Solution By Euler's formula (1.19)

$$6e^{i\pi/3} = 6\left(\cos\frac{\pi}{3} + i\sin\frac{\pi}{3}\right) = 6\left(\frac{1}{2} + i\frac{\sqrt{3}}{2}\right) = 3(1+i\sqrt{3});$$

$$(3-2i)^2 = 9 - 2\cdot 6i + (2i)^2 = 9 - 12i - 4 = 5 - 12i;$$

$$\begin{aligned} z = \frac{3(1+i\sqrt{3})}{5-12i} &= 3\frac{(1+i\sqrt{3})(5+12i)}{(5-12i)(5+12i)} = 3\frac{(5-12\sqrt{3}) + i(12+5\sqrt{3})}{25+144} \\ &= \frac{3(5-12\sqrt{3})}{169} + i\frac{(12+5\sqrt{3})}{169}. \end{aligned}$$

Problems

1. Use the triangle inequality to prove that

$$|z_1 - z_2| \geq |z_1| - |z_2|,$$
$$|z_1 + z_2| \geq |z_1| - |z_2|.$$

2. Using formula (1.18), show that a quadratic equation with *real* coefficients whose discriminant is negative will have solutions that are mutually conjugate.

[2]Leonhard Euler (1707–1783) was a great Swiss mathematician, physicist, astronomer, geographer, and engineer. He spent most of his adult life in Saint Petersburg, Russia.

3. Prove the identities and interpret them geometrically:

$$(a) \quad |z_1 - z_2|^2 = |z_1|^2 + |z_2|^2 - 2|z_1|\,|z_2|\cos\theta,$$

where θ, $0 \le \theta \le \pi$, is the angle between vectors z_1 and z_2.

$$(b) \quad |z_1 + z_2|^2 + |z_1 - z_2|^2 = 2(|z_1|^2 + |z_2|^2).$$

4. Let z_1, z_2, z_3 be consecutive vertices of a parallelogram. Find the fourth vertex z_4 which is opposite to z_2.

5. Prove properties (1.6).

6. Prove that (a) $\overline{z_1 + z_2} = \overline{z_1} + \overline{z_2}$; (b) $\overline{z_1 z_2} = \overline{z_1}\,\overline{z_2}$.

7. Express the number $(\sqrt{3} - i)^7$ in algebraic form.

8. Express the following as numbers in algebraic form.

a) $(-1 - i\sqrt{3})^6$; b) $\dfrac{e^{i\frac{\pi}{4}}}{(1-3i)^2}$.

9. Find all values of the root $\sqrt[4]{\dfrac{-1 - i\sqrt{3}}{2}}$.

10. Find all values of the roots.

a) $\sqrt[3]{-1}$; b) $\sqrt[4]{\dfrac{-1 + i\sqrt{3}}{2}}$; c) $\sqrt[4]{-16}$; d) $\sqrt{i}$.

11. Calculate the values of

$$\sqrt[3]{\frac{(2-2i)^4 + 72 + 4i}{(1-2i)^2 + 5i}},$$

and plot the values found in the complex plane.

12. Carry out the operations on the complex numbers and plot the values found in the complex plane.

a) $\sqrt[3]{\dfrac{3(2+2i)^5}{(1+3i)^2+14}}$; *b)* $\sqrt[6]{\dfrac{(2+3i)^2 - (2-3i)^2}{48(1+i)^{10}}}$; *c)* $\sqrt[2]{\dfrac{(2-i)^2 - 3}{(-1+i)^8}}$;

d) $\sqrt[3]{\dfrac{(1-2i)^3 + (1+2i)^3}{44(1+i)}}$; *e)* $\sqrt[4]{\dfrac{\left(1+i\sqrt{3}\right)^6 - 60 + 2i}{(2-i)^3 - 6 + 9i}}$.

13. Find all solutions of the equation $z^6 + 28z^3 + 27 = 0$.

14. Find all roots of the equation $z^4 - 2z^2 + 4 = 0$.

15. Find all roots of the equations.

a) $z^6 - 9z^3 + 8 = 0$; b) $z^4 + 8z^2 + 16 = 0$; c) $z^2 + 2z + 4 = 0$.

16. Graph in the complex plane the set of points given by the system of inequalities.

a) $|z - i| < 2,\ \operatorname{Im} z > 2$

b) $|z - 4| < 4,\ |z - 2| > 2,\ |\operatorname{Arg} z| < \frac{\pi}{4}$

c) $|z - 1 + i| < 1,\ |\operatorname{Arg} z + \frac{\pi}{4}| < \frac{\pi}{6}$

d) $|z - 3 - 3i| < 2,\ 2 < \operatorname{Re} z < 4,\ \operatorname{Im} z > 3$

e) $|z - 1 + i| < 2,\ |\operatorname{Arg} z| < \frac{\pi}{3},\ \operatorname{Im} z > -1$

f) $|z - 2i| < 3,\ |z + 2i| < 3,\ |\operatorname{Im} z| < \frac{1}{2}$

17. Find systems of inequalities that describe the set of points graphed in Fig. 13. The boundaries are included into the sets.

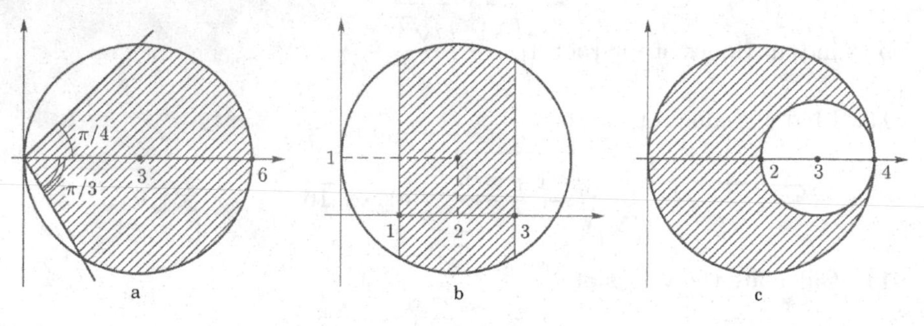

Fig. 13

18. Graph in the complex plane the set of points z satisfying the equation

$$|z - 1 - i| = |z + 3 - 3i|.$$

2

Functions of a Complex Variable

2.1 The Complex Plane

If x and y represent variable (real) quantities, then $z = x + iy$ is called a *complex variable.* By changing the values of x and y, the corresponding point $z = x + iy$ runs through some set of points in the complex plane $\mathbb{C}$. The entire complex plane therefore is calledplane of the complex variable *the plane of the complex variable* z.

2.1.1 Curves in the complex plane

First we recall some facts about curves. Any curve in the plane can be written in the form of parametric equations

$$x = x(t), \quad y = y(t), \quad \alpha \leq t \leq \beta, \tag{2.1}$$

where $x(t)$ and $y(t)$ are real-valued functions of the real variable t. Since each point (x, y) in the plane can also be thought of as a complex number $z = x+iy$, the equations (2.1) can be rewritten as

$$z(t) = x(t) + iy(t), \quad \alpha \leq t \leq \beta.$$

When t runs from α to β, the corresponding point $z(t)$ runs along the curve. The curve is called a *Jordan curve* if the functions $x(t)$, $y(t)$ are continuous on the closed interval $[\alpha, \beta]$ and $z(t_1) \neq z(t_2)$ as $\alpha \leq t_1 < t_2 < \beta$. It means that the curve does not intersect itself when t is in this half-open interval $[\alpha, \beta)$. A Jordan curve is called *closed Jordan curve* or *contour* if in addition $z(\alpha) = z(\beta)$.

A curve is said to be *smooth* on the interval $[\alpha, \beta]$ if the functions $x(t)$ and $y(t)$ have continuous derivatives on $[\alpha, \beta]$ and $x'(t)$ and $y'(t)$ are never zero at the same t; the derivatives at the endpoints α and β are taken from the right and left, respectively. A curve is *piecewise smooth* if it is continuous (that is the functions $x(t)$, $y(t)$ are continuous) on $[\alpha, \beta]$, and consists of a finite number of smooth curves.

DOI: 10.1201/9780367810283-2

2.1.2 Domains

Recall that $|z_1 - z_2|$ is the distance between two points z_1, and z_2. Therefore, the equation of the circle of radius R with center at the point z_0 looks like $|z - z_0| = R$, and the set of points z lying within the same circle is given by the inequality $|z - z_0| < R$.

For a fixed point z_0 in the plane of the complex variable and a positive value δ, the set of points satisfying the inequality $|z - z_0| < \delta$ is the interior of the circle of radius δ with center at z_0. This set is called the *δ-neighborhood of the point* z_0, or, if the value of δ is unimportant, just *a neighborhood of* z_0 (Fig. 14).

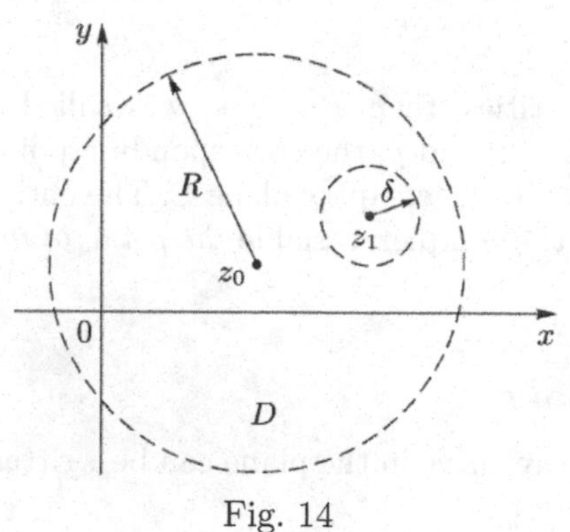

Fig. 14

A set D of points is said to be *open* if all of its points have a neighborhood contained in D, (i.e. entirely consisting of points from the given set D). For example, the set $D = \{z : |z - z_0| < R\}$ (see Fig. 14) is open. Indeed, take any point $z_1 \in D$. Then $|z_1 - z_0| < R$, and $d = R - |z_1 - z_0| > 0$ will be the distance from z_1 to the circle $|z - z_0| = R$. So if δ is chosen so that $0 < \delta < d$, then the set $|z - z_0| < \delta$ lies in D; this shows that D is open.

A set D is said to be *connected* if for any two points of D there exists a continuous curve, lying entirely in D, which contains the two given points. A set which is both open and connected is called a *domain.* Here are some examples of domains:

1. the disk $|z - z_0| < R$;
2. the ring $r < |z - z_0| < R$, where $0 \le r < R$;
3. the entire plane $\mathbb{C}$;
4. the half-plane $\operatorname{Re} z > a$, where a is a real number.

However the similar disk $|z - z_0| \le R$ is *not* a domain, as it is not an open set: for points z for which $|z - z_0| = R$, there is no neighborhood lying completely within the disk.

A point z_1 is called a *boundary point of the set D* if every neighborhood of z_1 contains both some points belonging to D as well as some points *not* belonging to it. The set of boundary points is referred to as the *boundary of the set D*. The set containing its boundary is called *a closed set.* The set composed of a domain D and all of the boundary points of D is called a *closed domain* and is denoted $\overline{D}$. For example, the disk $|z - z_0| \le R$ and the ring $r \le |z - z_0| \le R$ are closed domains; but the ring $r < |z - z_0| \le R$ is not a closed domain (as the boundary points lying on the circle $r = |z - z_0|$ do not belong to the set), neither is it a domain (it contains boundary points on the circle $|z - z_0| = R$, and thus is not open).

A domain is said to be *bounded*, if it lies inside some circle of sufficiently large radius.

Example 2.1 Give a system of inequalities which describe the closed bounded domain shown in Fig. 15.

Solution The points in this region lie within the closed disk centered at $-2i$ with radius 2; therefore they satisfy the inequality

$$|z + 2i| \leq 2.$$

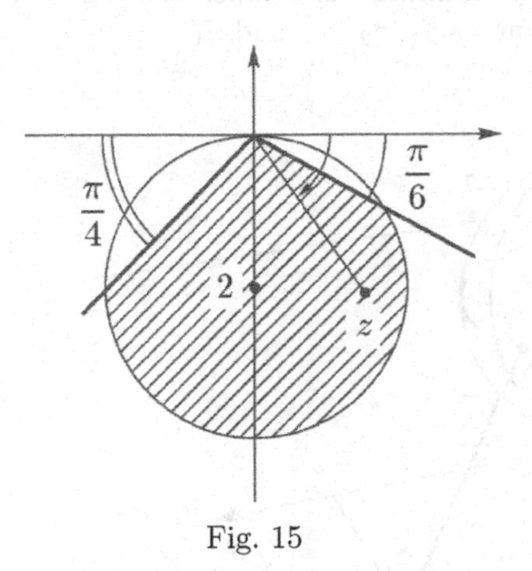

Fig. 15

In addition, these points lie within a sector whose vertex is at the origin, putting constraints on their arguments. It is easiest to state these constraints in terms of the principal value of the argument, for which $-\pi < \operatorname{Arg} z \leq \pi$. For the points of the given region, the angle $\operatorname{Arg} z$ is measured clockwise from the positive x-axis, and so the values are negative:

$$-\frac{3\pi}{4} \leq \operatorname{Arg} z \leq -\frac{\pi}{6}.$$

Thus the desired system of inequalities is

$$|z + 2i| \leq 2, \quad -\frac{3\pi}{4} \leq \operatorname{Arg} z \leq -\frac{\pi}{6}.$$

Domains in the plane are divided into two categories: simply connected and multiply connected. A bounded domain D is said to be *simply connected* if its boundary is connected (for example, consists of a single Jordan curve). It means that for any closed Jordan curve lying in D, all points inside the contour belong to D. Informally, D is simply connected if it does not have holes.[1] In particular, the disk $|z - z_0| < R$ and half-plane are simply-connected domains.

[1] For unbounded domains in $\mathbb{C}$ this definition requires refinement. For example, the exterior of the unit disk has a hole, but its boundary is a single Jordan curve, namely the unit circumference. Later on we will extend the definition of simply connected domains to unbounded domains.

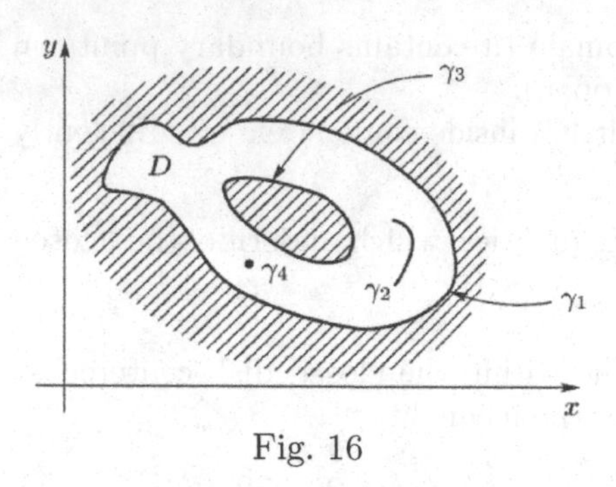

Fig. 16

A domain which is not simply connected is called *multiply connected.* More specifically, a bounded multiply connected domain is said to be n-*connected* if its boundary consists of n ($n > 1$) connected components; some of the components may be degenerate, i.e. just points. For example, in Fig. 16 a 4-connected domain D is depicted; its boundary is composed of the four curves γ_1, γ_2, γ_3, and γ_4.

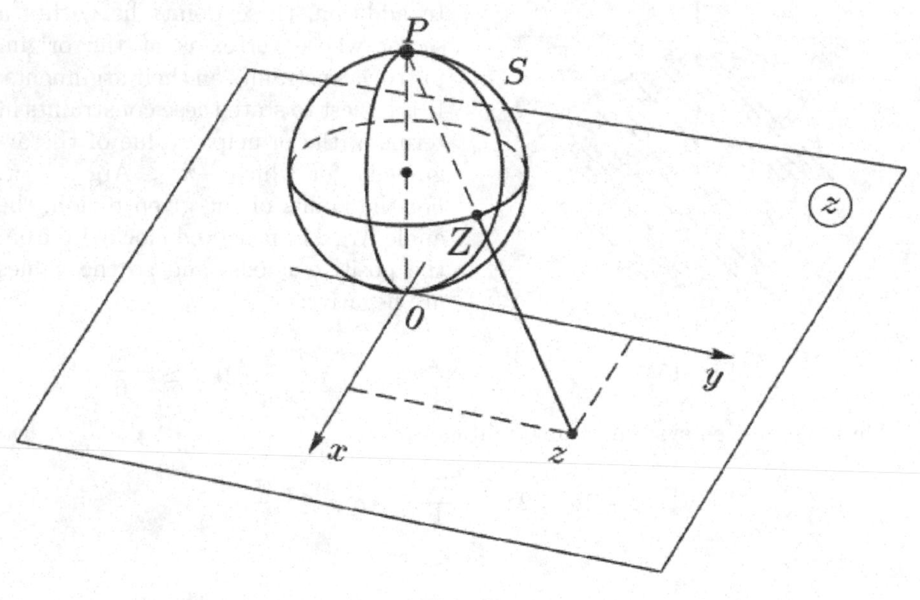

Fig. 17

Next we will look at one additional geometric interpretation of complex numbers. Let S be a sphere of radius $\frac{1}{2}$, touching the complex plane $\mathbb{C}$ at the point $z = 0$ (Fig. 17), and let P be a point on the sphere diametrically opposite to the point 0; P is the north pole of the sphere and 0 is the south pole. Take an arbitrary point $z \in \mathbb{C}$ and construct the ray Pz. This ray has a unique intersection Z with the sphere S; obviously, $Z \neq P$. In this way each point $z \in \mathbb{C}$ can be put into correspondence with a point $Z \in S$, $Z \neq P$. Conversely, if we are given a point $Z \in S$, $Z \neq P$, an analogous construction gives us a point $z \in \mathbb{C}$. In this way we have defined a 1-1 correspondence between the points of the complex plane $\mathbb{C}$ and the points of the sphere S excluding P. This correspondence is called the *stereographic projection* of $\mathbb{C}$ onto $S \setminus P$. It's not hard to see that if the points z_n in the plane $\mathbb{C}$ grow infinitely far away from the origin, then the corresponding points Z_n on the sphere approach P; the point P does not correspond to any point in $\mathbb{C}$.

Now we bring into consideration an additional, special, point not shown in the diagram: this is called *the point at infinity* and is denoted by $z = \infty$. This imaginable point $z = \infty$ will be added to the plane $\mathbb{C}$, and we place it in correspondence with the point $P \in S$. This combination of $\mathbb{C}$ with the point at infinity is called the *extended complex plane* and is denoted by $\overline{\mathbb{C}}$ (this notation is used because in fact $\overline{\mathbb{C}}$ has some of the properties of a closed domain). Each point $z \in \overline{\mathbb{C}}$ corresponds to a unique point $Z \in S$, and conversely. The sphere S is called *the Riemann sphere.*[2]

The Riemann sphere shows that the point $z = \infty$ is, in some sense, just as valid as the other, *finite*, points of $\overline{\mathbb{C}}$: both finite and infinite points represent points of the sphere S. The Riemann sphere is often a convenient place to work in situations that require consideration of the point at infinity, or points approaching it. It is possible to show that the stereographic projection maps lines and circles in $\mathbb{C}$ to circles in S, and that angles between intersecting curves are preserved.

In Fig. 18 we see lines γ_1 and γ_2 in $\mathbb{C}$ which meet at the point $z_0 \in \mathbb{C}$; the circles Γ_1 and Γ_2 are the images in the Riemann sphere of the lines under the stereographic projection, which maps them to circles. Also shown is the image Z_0 of z_0; and strips of the planes formed by P and γ_1 and by P and γ_2; the line containing P and z_0 is the line of intersection of these strips, and the top edges of these strips are parallel to γ_1, γ_2, and are tangent to the Riemann sphere at P, as well as to Γ_1 and Γ_2 (respectively). The angle between γ_1 and γ_2 will be the same as the angle between Γ_1 and Γ_2 at P; the proof of this is omitted, but it is visually obvious. When two circles like Γ_1 and Γ_2 intersect at distinct points, the angle between them is the same at both intersections; so the angle between Γ_1 and Γ_2 at Z_0 in the Riemann sphere is the same as the original angle between γ_1 and γ_2 at z_0 in $\mathbb{C}$.

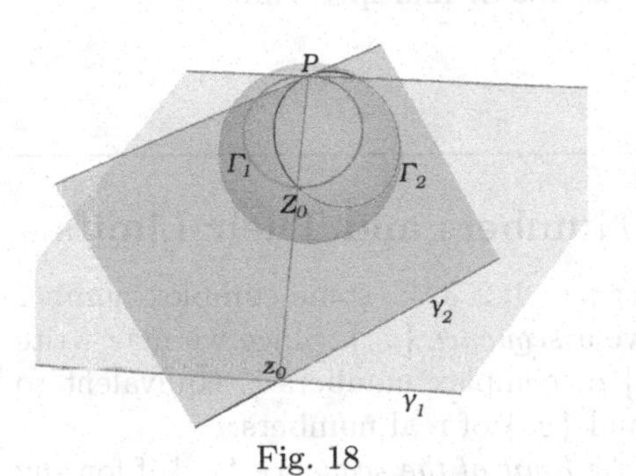

Fig. 18

Let us take some neighborhood in S of the point P, that is, a spherical cap centered at P. A point $Z \neq P$ of this neighborhood will correspond to a point $z \in \mathbb{C}$, lying outside some disk centered at the origin. A *neighborhood of the point at infinity* refers to a set consisting of points in $\mathbb{C}$ for which $|z| > R$, and also the point $z = \infty$; that is, the exterior of the disk of radius R, centered at the origin, with the addition of the point $z = \infty$ (usually R will be large).

Now we define n-connected domains in $\overline{\mathbb{C}}$ exactly in the same way as for bounded domains. Namely, a domain D in $\overline{\mathbb{C}}$ is said to be *n-connected* if its boundary consists of n connected components. If $n = 1$, we say that a domain is simply connected. For example, the domain $D = \overline{\mathbb{C}} \setminus \{|z| \leq$

[2]G. Riemann (1826–1866) was the great German mathematician.

$1\} = \{|z| > 1\} \cup \infty$ is *simply connected.* Note that the set of points on the Riemann sphere corresponding to D is a cap "without holes". But the domain $\mathbb{C} \setminus \{|z| \le 1\} = \{|z| > 1\}$, $z \in \mathbb{C}$, is 2-connected, since its boundary consists of the circumference $|z| = 1$ and the point $z = \infty$.

Problems

1. Prove that the strip $0 < \operatorname{Im} z < 1$ is a simply connected domain. What is the image of the strip under the stereographic projection onto the Riemann sphere?

2. Graph the set of points in the complex plane defined by the inequalities

a) $|z - 2i| < 2$ and $1 < \operatorname{Im} z < 3$; *b)* $|z - 2i| < 2$ and $1 < \operatorname{Im} z \le 3$.

Is this set a domain? If so, is it simply connected or multiply connected?

3. Determine whether the sets of points defined in problem 16 of Section 1.2 are domains? If so, are they simply connected or multiply connected?

4. Determine whether the sets of points defined in problem 17 of Section 1.2 are domains? If so, are they simply connected or multiply connected?

2.2 Sequences of Complex Numbers and Their Limits

Suppose that for each counting number $n = 1, 2, 3, \ldots$ some complex number z_n is defined. Then we say that we have a *sequence* $\{z_n\}$. Since we may write $z_n = x_n + iy_n$, a given sequence $\{z_n\}$ of complex numbers is equivalent to being given a pair of sequences $\{x_n\}$ and $\{y_n\}$ of real numbers.

A complex number A is said to be *the limit of the sequence* $\{z_n\}$ if for any positive number ϵ there exists some counting number N (which depends on ϵ) such that for all $n > N$ the inequality $|z_n - A| < \epsilon$ is satisfied.

Satisfying these conditions means that however small the ϵ-neighborhood of the point A may be, all points z_n with $n > N$ fall into that neighborhood, and only finitely many z_n remain outside it. In fact an equivalent definition is: A is the limit of the sequence $\{z_n\}$ if any neighborhood of A contains all but finitely many of the z_n.

A sequence which has a limit is said to be *convergent.* The statement that A is the limit for a sequence $\{z_n\}$ can be written as

$$\lim_{n \to \infty} z_n = A$$

or as $z_n \to A$ as $n \to \infty$, or just $z_n \to A$ if the context is clear. Our definition of limit matches that given for limits of sequences of real numbers.

Theorem 2.2. *A sequence of complex numbers $\{z_n = x_n + iy_n\}$ has the limit $A = a + ib$ if and only if the sequences of real numbers $\{x_n\}$ and $\{y_n\}$ have limits and specifically*

$$\lim_{n\to\infty} x_n = a \quad \text{and} \quad \lim_{n\to\infty} y_n = b.$$

Proof. **Only if (Necessity):** Let us assume that

$$\lim_{n\to\infty} z_n = A.$$

We must show that $x_n \to a_n$ and $y_n \to b_n$. Note that

$$|z_n - A| = \sqrt{(x_n - a)^2 + (y_n - b)^2}. \tag{2.2}$$

It follows that

$$|x_n - a| \le |z_n - A| \quad \text{and} \quad |y_n - b| \le |z_n - A|. \tag{2.3}$$

Take an arbitrary $\epsilon > 0$. Since $z_n \to A$ as $n \to \infty$, we can find some number N such that when $n > N$ the inequality $|z_n - A| < \epsilon$ is satisfied. Combined with (2.3) we get

$$|x_n - a| < \epsilon \quad \text{and} \quad |y_n - b| < \epsilon \quad \text{when} \quad n > N.$$

From the definition of limit of a sequence of real numbers, we then have $x_n \to a_n$ and $y_n \to b_n$, as required.

If (Sufficiency): Now assume that

$$\lim_{n\to\infty} x_n = a \quad \text{and} \quad \lim_{n\to\infty} y_n = b.$$

We will show that $z_n \to A = a + ib$. Take an arbitrary $\epsilon > 0$; since $x_n \to a$ we can find some number N_1 such that for any $n > N_1$ the inequality

$$|x_n - a| < \frac{\epsilon}{\sqrt{2}} \tag{2.4}$$

is satisfied (making use of the definition of the limit of a sequence of real numbers). Analogously, from the assumption $y_n \to b$ there follows the existence of some number N_2 such that for any $n > N_2$

$$|y_n - b| < \frac{\epsilon}{\sqrt{2}}. \tag{2.5}$$

Set $N = \max(N_1, N_2)$. Then whenever $n > N$ both inequalities (2.4) and (2.5) will be satisfied. Combined with equation (2.2) we get

$$|z_n - A| < \sqrt{\frac{\epsilon^2}{2} + \frac{\epsilon^2}{2}} = \epsilon \quad \text{when} \quad n > N.$$

Thus, for an arbitrary $\epsilon > 0$ a number N can be found such that when $n > N$ the inequality $|z_n - A| < \epsilon$ is satisfied. This means that $z_n \to A$ as $n \to \infty$.

This completes the proof of Theorem 2.2. □

Using Theorem 2.2 it is easy to show that convergent sequences of complex numbers have the same properties as convergent sequences of real numbers:

$$\begin{aligned} \lim_{n\to\infty}(z_n+w_n) &= \lim_{n\to\infty} z_n + \lim_{n\to\infty} w_n \\ \lim_{n\to\infty}(z_n w_n) &= \lim_{n\to\infty} z_n \cdot \lim_{n\to\infty} w_n \\ \lim_{n\to\infty}\frac{z_n}{w_n} &= \frac{\lim\limits_{n\to\infty} z_n}{\lim\limits_{n\to\infty} w_n}, \quad \text{if} \quad \lim_{n\to\infty} w_n \neq 0. \end{aligned} \tag{2.6}$$

The derivation of these formulas (2.6), from the corresponding properties of sequences of real numbers, is left to the reader.

The notion of limit introduced above deals with the case when the limit $A \neq \infty$. Now we will consider the case of sequences tending toward infinity, i.e. $A = \infty$.

The limit of the sequence $\{z_n\}$ equals infinity if, for any value $R > 0$ however large, there can be found some number N (which depends on R) such that for any $n > N$ the inequality $|z_n| > R$ is satisfied. In this case we write

$$\lim_{n\to\infty} z_n = \infty.$$

Using the notions of the point at infinity and its neighborhoods, which were introduced at the end of Section 2.1, we can reformulate the new definition in the following form:

$\lim\limits_{n\to\infty} z_n = \infty$ if, for any neighborhood of the point $A = \infty$, there is a positive integer N such that all points z_n with $n > N$ fall in that neighborhood.

In this form the definitions of finite and infinite limits are analogous to each other.

Problems

1. Using Theorem 2.2 and the properties of convergent sequences of real numbers, prove the formulas (2.6).

2. Prove the following statements.

(a) If $\lim\limits_{n\to\infty} z_n = \infty$, then $\lim\limits_{n\to\infty} |z_n| = \infty$ and $\lim\limits_{n\to\infty} \dfrac{1}{z_n} = 0$.

(b) If $\lim\limits_{n\to\infty} w_n = 0$ and all $w_n \neq 0$, then $\lim\limits_{n\to\infty} \dfrac{1}{w_n} = \infty$.

2.3 Functions of a Complex Variable; Limits and Continuity

Let D be some set of complex numbers. *A single-valued function of a complex variable* is a rule that assigns, to every complex number z in the set D, a unique complex number w. This association is denoted by $w = f(z)$, or $f : z \mapsto w$; here f refers to the rule.

The set D is called the domain of the function f. For example, the function $w = z^2$ assigns to each complex number $z = x + iy$ the complex number $w = (x + iy)^2 = x^2 - y^2 + 2ixy$; this function is defined on the entire plane of the complex variable z, and if we set $f(\infty) = \infty$, then it will be defined on the entire extended complex plane.

If we let $z = x + iy$ and $w = u + iv$, then being given a function $w = f(z)$ of a complex variable is equivalent to being given, on the same domain, two real-valued functions of the real variables x and y: $u = u(x, y)$ and $v = v(x, y)$. For example, for the function $w = z^2$ we have $u = x^2 + y^2$ and $v = 2xy$.

Along with the plane of the complex variable $z = x+iy$, let us also consider the plane of the complex variable $w = u + iv$. Under the function $w = f(z)$, each point $z = x + iy$ of the domain D is associated with a well-defined point $w = u + iv$ in the (u, v)- or w-plane. As the point z runs through the set D in the z plane, the associated point w runs through some other set E in the w-plane. In this way, the single-valued function f *maps* the set D *onto* the set E, i.e. each point $z \in D$ is associated with a point $w \in E$, and for every point $w \in E$ there is at least one point z in D such that $w = f(z)$. The point w is called the *image* of the point z, and the point z is called the *preimage* of the point w, under the mapping $w = f(z)$. A point w may have several (or even infinitely many) preimages. For example, under the mapping $w = z^n$ each point $w \neq 0$ has n preimages—the nth degree roots of w.

From this it is evident that the behavior of a function of a complex variable cannot be illustrated with the help of two- or three-dimensional Cartesian graphs; four-dimensional graphs would be required, but they are rather hard to work with. So in order to understand visually the geometric properties of a function $w = f(z)$, we must explore how this or that region or curve in the z-plane is mapped under f to some type of set in the w-plane.

Example 2.3 Consider the mapping $w = f(z) = z^2$, and the region D of the quarter disk, in the first quadrant, with radius R centered at the origin. To what region in the w-plane is D mapped under f?

Solution Let us write the variables z and w in exponential form: $z = re^{i\phi}$, $w = \rho e^{i\theta}$. Since $w = z^2$, then $\rho e^{i\theta} = (re^{i\phi})^2 = r^2 e^{i2\phi}$, from which it follows that

$$\rho = r^2 \quad \text{and} \quad \theta = 2\phi. \tag{2.7}$$

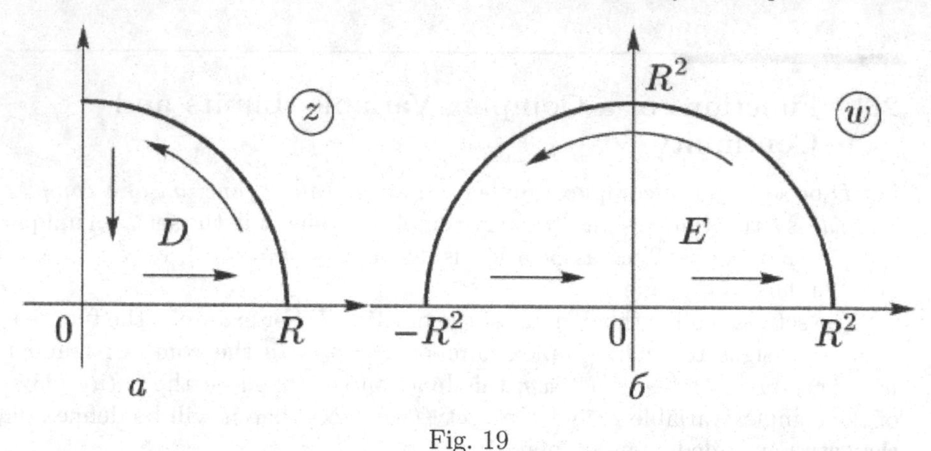

Fig. 19

Let us find out what kind of curve it is that the boundary of the region D maps to under $w = z^2$. Let us traverse this boundary starting at the point $z = 0$ and moving in the positive direction (i.e. keeping the region D on the left side as we move—Fig. 19, a). From formula (2.7) it follows that the section of the real axis $0 \leq r \leq R$, $\phi = 0$ in the z-plane goes to the section $0 \leq \rho \leq R^2$, $\theta = 0$ of the real axis in the w-plane; the quarter circle $r = R$, $0 \leq \phi \leq \pi/2$ goes to the half circle $\rho = R^2$, $0 \leq \theta \leq \pi$ (Fig. 19, b); and, last, the section of the imaginary axis $0 \leq r \leq R$, $\phi = \pi/2$ goes to the section $0 \leq \rho \leq R^2$, $\theta = \pi$, i.e. to the interval $[-R, 0]$ in the real axis. Every interior point z in the quarter circle D goes to an interior point of the half circle E in the w-plane, and moreover the entire half disk E will be completely filled by images of the points z, without any "holes".

Example 2.4 What region is the image of the disk $|z| < R$ under the function $w = \frac{1}{z}$?

Solution As in the previous example, we write $z = re^{i\phi}$ and $w = \rho e^{i\theta}$ so that $w = \frac{1}{z} = \frac{1}{r}e^{-i\phi}$. Therefore

$$\rho = \frac{1}{r} \quad \text{and} \quad \theta = -\phi.$$

Now consider the circle $|z| = r$, $r \leq R$ (Fig. 20, a); it goes to the circle $|w| = \frac{1}{r} \geq \frac{1}{R}$ (Fig. 20, b). Therefore the disk $|z| < R$ is mapped to the exterior of the circle of radius $\frac{1}{R}$, or in other words to the set $|w| > \frac{1}{R}$. Also, if we traverse the circle $|z| = r$ in the counterclockwise direction, then by virtue of the equality $\theta = -\phi$ the direction is reversed while traversing the circle $|w| = \frac{1}{r}$.

Other elementary functions of a complex variable will be considered in Chapter 4.

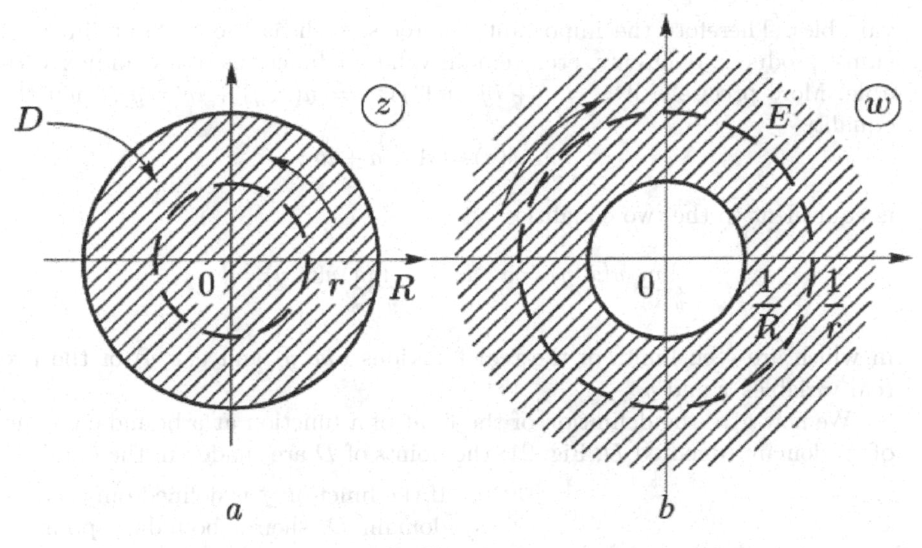

Fig. 20

Now we introduce the important notion of the limit of a function at a point. Let a point $z_0 \in \mathbb{C}$ and a positive number δ be given. *A punctured neighborhood of a point* z_0 is a δ-neighborhood of that point, which however excludes the point z_0 itself (i.e. it is the disk of radius δ centered at z_0 from which the center has been removed). This set can be written in the form $0 < |z - z_0| < \delta$.

Now let the function $w = f(x)$ be defined in some punctured neighborhood of the point z_0. A number A is called *the limit of the function* $w = f(z)$ *at the point* z_0 if, for any $\epsilon > 0$, there is a number $\delta > 0$ (dependent on ϵ), such that for all points of the punctured δ-neighborhood of the point z_0 (that is for all points z for which $0 < |z - z_0| < \delta$), the inequality

$$|f(z) - A| < \epsilon \tag{2.8}$$

is satisfied.

The statement that a limit A exists for the function $w = f(z)$ at the point z_0 can be written as

$$\lim_{z \to z_0} f(z) = A$$

and can be paraphrased as: for any neighborhood U_A of the point A, however small, we can find some punctured neighborhood of the point z_0, such that for all points z from the punctured neighborhood, the associated value $w = f(z)$ lies in U_A. In this form the definition of limit also covers the case when $z = \infty$ and/or when $A = \infty$; a *punctured neighborhood of the point* $z = \infty$ is understood as a set of the form $|z| > R$, $z \in \mathbb{C}$.

The definition of limit for a function of a complex variable is closely analogous to the definition given for the limit of functions of one and of two real variables. Therefore the important theorems, such as those about limits of sums, products, quotients, etc., remain valid for functions of a complex variable. More precisely, if $z = x + iy$ and $f(z) = u(x, y) + iv(x, y)$, then the equality

$$\lim_{z \to z_0} f(z) = A = a + ib$$

is equivalent to the two equalities

$$\lim_{\substack{x \to x_0 \\ y \to y_0}} u(x, y) = a \quad \text{and} \quad \lim_{\substack{x \to x_0 \\ y \to y_0}} v(x, y) = b,$$

in which are seen limits of the real functions $u(x, y)$ and $v(x, y)$ of the two real variables x and y.

We now give the definition of the limit of a function at a boundary point of its domain (denoted D, Fig. 21; the points of D are shaded in the detail).

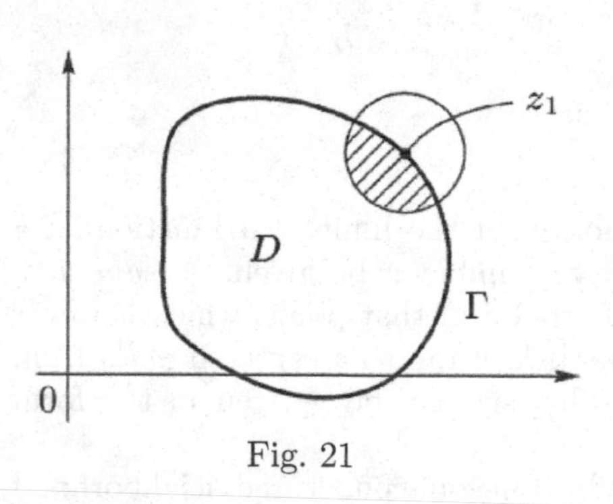

Fig. 21

If the function f is defined only on the domain D, then a boundary point z_1 will have no punctured neighborhood in which $f(z)$ always has a value; this is the difference with the preceding case. But this is easily remedied. A number A is called *the limit of the function* $w = f(z)$ at the boundary point z_0 if, for any $\epsilon > 0$, there is $\delta > 0$, such that the inequality $|f(z) - A| < \epsilon$ holds when z is both in the punctured δ-neighborhood of z_0 *and also* in D.

Now we move to the definition of continuity for functions of a complex variable. A function $w = f(z)$, defined in a neighborhood (non-punctured!) of z_0, is said to be *continuous at the point* z_0 if

$$\lim_{z \to z_0} f(z) = f(z_0).$$

The continuity of the function $w = f(z) = u(x, y) + iv(x, y)$ at the point z_0, is equivalent to the continuity of the two real-valued functions $u(x, y)$ and $v(x, y)$ of the variables x and y at the point (x_0, y_0).

A function $w = f(z)$, defined on a domain D, is said to be *continuous on* D if f is continuous at each point of D. A function $w = f(z)$ is said to be *continuous on the closed domain* $\overline{D}$ if it is defined on $\overline{D}$ and for each point $z_0 \in \overline{D}$ (including boundary points) the equality

$$\lim_{z \to z_0} f(z) = f(z_0)$$

is satisfied.

Let us fix a point z_0 in D, and compare it to another point $z \in D$. The change in value $\Delta z = z - z_0 = \Delta x + i\Delta y$, is called *the increment of* z. The corresponding change in the function

$$\Delta w = f(z) - f(z_0) = f(z_0 + \Delta z) - f(z_0)$$

is called *the increment of the function.* It is not difficult to show that if $z_0 \neq \infty$ and $f(z_0) \neq \infty$, then a function $w = f(z)$ is continuous at the point z_0 if and only if

$$\lim_{\Delta z \to 0} \Delta w = 0$$

— the proof is left to the reader as problem 1 below.

Problems

1. Prove that if $z_0 \neq \infty$ and $f(z_0) \neq \infty$, then a function $w = f(z)$ is continuous at the point z_0 if and only if $\lim_{\Delta z \to 0} \Delta w = 0$.

2. Let $f(z)$, $z \in \overline{\mathbb{C}}$, be such that $f(z) = \frac{1}{z}$ as $z \neq 0$, $z \neq \infty$, and $f(0) = \infty$, $f(\infty) = 0$. Prove that $f(z)$ is continuous on $\overline{\mathbb{C}}$.

3. Use the definition of limit to prove that

$$a)\ \lim_{z \to 1+i} (2z + i) = 2 + 3i; \quad b)\ \lim_{z \to i} \frac{1}{z - i} = \infty;$$

$$c)\ \lim_{z \to \infty} (2z + i) = \infty; \quad d)\ \lim_{z \to \infty} \frac{1}{z} = 0.$$

3

Differentiation of Functions of a Complex Variable

3.1 The Derivative. Cauchy-Riemann Conditions

3.1.1 The derivative and the differential

The definitions of derivative and differential of a function of a complex variable match exactly the corresponding definitions given for a function of a real variable.

Let the function $w = f(z) = u + iv$ be defined on some neighborhood U of the point $z_0 \in \mathbb{C}$. We give to the independent variable $z = x + iy$ an increment $\Delta z = \Delta x + i\Delta y$, not taking z outside the neighborhood U. Then the function $w = f(z)$ gets a corresponding increment $\Delta w = f(z + \Delta z) - f(z_0)$.

The derivative of the function $w = f(z)$ *at the point* z_0 is the limit of the ratio of the increment of the function, Δw, to that of the argument, Δz, as Δz approaches zero—from any direction.[1]

The derivative is denoted in several ways: $f'(z)$, w', $\dfrac{dw}{dz}$, or $\dfrac{df}{dz}$. The definition of the derivative may be written as

$$f'(z) = \lim_{\Delta z \to 0} \frac{\Delta w}{\Delta z} = \lim_{\Delta z \to 0} \frac{f(z_0 + \Delta z) - f(z_0)}{\Delta z}. \tag{3.1}$$

The finite limit in (3.1) might not exist; then we say that the function $w = f(z)$ does not have a derivative at the point z_0.

A function $w = f(z)$ is said to be *differentiable at the point* z_0 if it is defined on some neighborhood of z_0 and its increment Δw can be expressed as

$$\Delta w = A\Delta z + \alpha(\Delta z) \cdot \Delta z, \tag{3.2}$$

where the complex number A does not depend on Δz, while the function $\alpha(\Delta z)$ is vanishingly small when Δz is near zero, i.e. $\alpha(\Delta z) \to 0$ as $\Delta z \to 0$.

As also is true for real-valued functions, it can be shown that a function $f(z)$ is differentiable at the point z_0 if and only if it has a derivative at z_0,

[1] We will see that this "omnidirectionality" of the derivative imposes severe restrictions on functions for which the limit exists; and that in turn means that such functions of a complex variable have many surprising special properties.

DOI: 10.1201/9780367810283-3

in which case $A = f'(z_0)$—see problem 1. The term $f'(z_0)\Delta z$ is called the *differential of the function f at the point* z_0 and can be referred to by dw or $df(z_0)$. Similarly the increment Δz of the independent variable z is also called the differential of z and can be referred to by dz. In this way we have

$$dw = df(z_0) = f'(z_0)dz.$$

The differential is the *principal linear part* of the increment of the function, or in other words, the best linear approximation to the increment $\Delta f(x_0)$ in terms of dz.

Example 3.1 Determine whether the function $w = f(z) = \operatorname{Re} z$ has a derivative at an arbitrary point z_0.

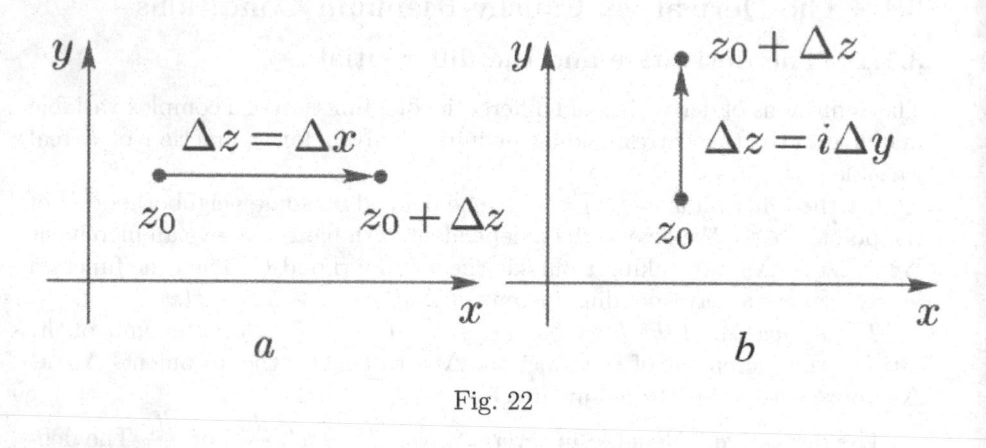

Fig. 22

Solution According to the definition of derivative, the limit (3.1) must not depend on the way in which $z = z_0 + \Delta z$ approaches z_0 as $\Delta z \to 0$. In this case $w = f(z) = \operatorname{Re} z = x$, and we first take $\Delta z = \Delta x$ (Fig. 22, *a*). So $\Delta w = \Delta x$, and

$$\frac{\Delta w}{\Delta z} = \frac{\Delta x}{\Delta x} = 1.$$

If we now take $\Delta z = i\Delta y$ (Fig. 22, *b*), then $\Delta x = 0$, and consequently $\Delta w = 0$, which means

$$\frac{\Delta w}{\Delta z} = 0.$$

Therefore the limit of the ratio $\frac{\Delta w}{\Delta z}$ as $\Delta z \to 0$ does not exist, and thus the function $w = \operatorname{Re} z = x$ does not have a derivative at any point.

On the other hand the function $w = z = x + iy$ obviously has a derivative at every point, and $f'(z) = 1$ (why?). From these examples it is clear that the real and imaginary parts of a differentiable function f may not be arbitrary—they must be connected by some additional relation. This relationship arises from the fact, mentioned earlier, that the existence of $f'(z)$ is substantially more restrictive than the existence of the derivative of a function of one real variable, or partial derivatives in the case of several real variables: the limit in (3.1) must not depend on the path by which the point $z = z_0 + \Delta z$ approaches z_0 as $\Delta z \to 0$. To arrive at the necessary relationship we first recall the definition of differentiability for functions of two real variables.

A real-valued function $u = u(x, y)$ of real variables x and y is said to be differentiable at the point $P_0 = (x_0, y_0)$ if it is defined on some neighborhood of the point P_0 and its total increment $\Delta u = u(x_0 + \Delta x, y_0 + \Delta y) - u(x_0, y_0)$ can be expressed in the form

$$\Delta u = B\Delta x + C\Delta y + \beta(\Delta x, \Delta y) \cdot \Delta x + \gamma(\Delta x, \Delta y) \cdot \Delta y, \tag{3.3}$$

where B and C are real numbers independent of Δx and Δy, while β and γ are real-valued functions of the variables Δx and Δy, tending to zero as $\Delta x \to 0$ and $\Delta y \to 0$.

If the function u is differentiable at the point P_0, then it has partial derivatives at P_0, and

$$B = \frac{\partial u}{\partial x}(P_0) \quad \text{and} \quad C = \frac{\partial u}{\partial y}(P_0).$$

But, in contrast to the case of functions of a single variable, the existence of partial derivatives of $u(x, y)$ does *not* imply that it is differentiable.

3.1.2 Cauchy-Riemann conditions

Theorem 3.2. *Let the function $w = f(z)$ of the complex variable $z = x+iy = (x, y)$ be defined on a neighborhood of the point $z_0 = (x_0, y_0)$, and let u and v be the real and imaginary parts of f, i.e. $f(z) = u(x, y) + iv(x, y)$. Then f is differentiable at the point z_0 if and only if the functions u and v are differentiable at the point (x_0, y_0), and the conditions*

$$\frac{\partial u}{\partial x} = \frac{\partial v}{\partial y} \quad \text{and} \quad \frac{\partial u}{\partial y} = -\frac{\partial v}{\partial x} \tag{3.4}$$

hold at this point.

The equations (3.4) are called *the Cauchy*[2]*-Riemann conditions.*[3]

Notice that the necessity of the conditions (3.4) follows easily from the requirement that the derivative of $w = f(z)$ must be same in all directions; so the derivative in the real direction

$$\frac{dw}{dx} = \frac{du}{dx} + i\frac{dv}{dx}$$

must be equal to the derivative in the imaginary direction

$$\frac{dw}{d(iy)} = \frac{du}{d(iy)} + i\frac{dv}{d(iy)} = \frac{dv}{dy} - i\frac{du}{dy},$$

[2] Augustin-Louis Cauchy (1789–1857) was the great French mathematician.

[3] The conditions (3.4) were already investigated in the 18th century by D'Alembert and Euler, so they are sometimes called the D'Alembert-Euler conditions, which is more historically accurate.

(using $diy = idy$, since i is a constant). Then setting the real and imaginary parts equal yields (3.4). But these arguments do not yield differentiability of u and v.

Proof. **Only if (necessity):** Suppose the function $w = f(z)$ is differentiable at the point z_0, i.e.

$$\Delta w = \Delta u + i\Delta v = f'(z_0)\Delta z + \alpha(\Delta z)\cdot\Delta z. \tag{3.5}$$

We will write

$$\begin{aligned} f'(z_0) &= a + ib, \\ \alpha(\Delta z) &= \beta(\Delta x, \Delta y) + i\gamma(\Delta x, \Delta y), \\ \Delta z &= \Delta x + i\Delta y, \end{aligned}$$

where β and γ are real functions of the variables Δx and Δy tending to zero as $\Delta x \to 0$ and $\Delta y \to 0$. Substituting these expressions into (3.5) and separating the real and imaginary parts we get

$$\begin{aligned} \Delta u + i\Delta v &= (a + ib)(\Delta x + i\Delta y) + (\beta + i\gamma)(\Delta x + i\Delta y) \\ &= (a\Delta x - b\Delta y + \beta\Delta x - \gamma\Delta y) + i(b\Delta x + a\Delta y + \gamma\Delta x + \beta\Delta y). \end{aligned} \tag{3.6}$$

Since an equality between complex numbers is equivalent to the equalities of the real and imaginary parts, then (3.6) is equivalent to the system of equations

$$\begin{aligned} \Delta u &= a\Delta x - b\Delta y + \beta\Delta x - \gamma\Delta y, \\ \Delta v &= b\Delta x + a\Delta y + \gamma\Delta x + \beta\Delta y. \end{aligned} \tag{3.7}$$

The equations (3.7) mean that the functions u and v satisfy the conditions (3.3), and therefore are differentiable. Since the coefficients of Δx and Δy are equal to the partial derivatives by x and y respectively, we get

$$\begin{aligned} a &= \frac{\partial u}{\partial x}, & -b &= \frac{\partial u}{\partial y}; \\ b &= \frac{\partial v}{\partial x}, & a &= \frac{\partial v}{\partial y}, \end{aligned} \tag{3.8}$$

from which the conditions (3.4) follow.

If (sufficiency): Now we assume that the functions $u(x, y)$ and $v(x, y)$ are differentiable at the point (x_0, y_0), and the conditions (3.4) are satisfied. If we denote

$$a = \frac{\partial u}{\partial x} \quad \text{and} \quad b = -\frac{\partial u}{\partial y}$$

then applying (3.4) gives the equations (3.8). From that and the differentiability conditions for u and v, we have

$$\begin{aligned} \Delta u &= a\Delta x - b\Delta y + \beta_1\Delta x + \gamma_1\Delta y, \\ \Delta v &= b\Delta x + a\Delta y + \beta_2\Delta x + \gamma_2\Delta y, \end{aligned}$$

where β_1, γ_1, β_2, and γ_2 are functions tending to zero as $\Delta x \to 0$ and $\Delta y \to 0$. Thus

$$\Delta u + i\Delta v = (a + ib)(\Delta x + i\Delta y) + (\beta_1 + i\beta_2)\Delta x + (\gamma_1 + i\gamma_2)\Delta y. \tag{3.9}$$

We define the function α by

$$\alpha(\Delta z) = \frac{(\beta_1 + i\beta_2)\Delta x + (\gamma_1 + i\gamma_2)\Delta y}{\Delta z}$$

and we set $A = a + ib$. Then (3.9) can be rewritten in the form

$$\Delta w = \Delta u + i\Delta v = A\Delta z + \alpha(\Delta z) \cdot \Delta z,$$

which agrees with (3.2). So, to prove the differentiability of the function $f(z)$ it remains only to show that

$$\lim_{\Delta x \to 0} \alpha(\Delta z) = 0.$$

From the equation

$$|\Delta z| = \sqrt{(\Delta x)^2 + (\Delta y)^2}$$

it follows that $|\Delta x| \leq |\Delta z|$ and $|\Delta y| \leq |\Delta z|$. Therefore

$$|\alpha(\Delta z)| \leq \frac{|\beta_1 + i\beta_2|\,|\Delta z| + |\gamma_1 + i\gamma_2|\,|\Delta z|}{|\Delta z|} = |\beta_1 + i\beta_2| + |\gamma_1 + i\gamma_2|.$$

If $\Delta z \to 0$ then $\Delta x \to 0$ and $\Delta y \to 0$, which means that the functions β_1, γ_1, β_2, and γ_2 also tend to zero as $\Delta z \to 0$. Therefore $\alpha(\Delta z) \to 0$ as $\Delta z \to 0$, and the proof of Theorem 3.2 is complete. □

Example 3.3 Check whether the function $w = z^2$ is differentiable. If so, at which points?

Solution Writing the function as

$$w = u + iv = (x + iy)^2 = x^2 - y^2 + 2ixy,$$

we get

$$u = x^2 - y^2 \quad \text{and} \quad v = 2xy.$$

Clearly u and v are differentiable, and

$$\frac{\partial u}{\partial x} = 2x = \frac{\partial v}{\partial y} \quad \text{and} \quad \frac{\partial u}{\partial y} = -2y = -\frac{\partial v}{\partial x}.$$

Therefore the Cauchy-Riemann conditions (3.4) are satisfied at all points, and the function $w = z^2$ is differentiable on $\mathbb{C}$.

Example 3.4 Check the differentiability of the function $w = \overline{z} = x - iy$.

Solution Writing the function as $w = u + iv = x - iy$, we get $u = x$ and $v = -y$, so that

$$\frac{\partial u}{\partial x} = 1 \quad \text{and} \quad \frac{\partial v}{\partial y} = -1.$$

So the Cauchy-Riemann conditions are not satisfied anywhere, and the function $w = \overline{z}$ is not differentiable.

Checking a function for differentiability and finding its derivative can also be carried out directly from formula (3.1).

Example 3.5 Use formula (3.1) to check the function $w = z^2$ for differentiability.

Solution Here

$$\Delta w = (z_0 + \Delta z)^2 - z_0^2 = 2z_0 \Delta z + (\Delta z)^2,$$

from which

$$\lim_{\Delta z \to 0} \frac{\Delta w}{\Delta z} = \lim_{\Delta z \to 0} \frac{2z_0 \Delta z + (\Delta z)^2}{\Delta z} = \lim_{\Delta z \to 0} (2z_0 + \Delta z) = 2z_0.$$

Consequently the function $w = z^2$ is differentiable at any point z_0, and its derivative is $f'(z_0) = 2z_0$.

Remark 3.6 Seeing as the basic theorems about limits hold for functions of a complex variable, and the definition of the derivative as a limit also does not differ from the corresponding definition for functions of a real variable, then the well known rules for differentiating sums, differences, products, quotients, and compositions of functions remain valid for functions of a complex variable. Also analogously it can be shown that if a function f is differentiable at a point z_0, then it is continuous at that point (the converse is not true!).

3.1.3 Analytic functions

A function $w = f(z)$ which is differentiable not only at the point z_0, but also on a neighborhood of that point, is said to be *analytic at the point* z_0. If it is analytic at each point in a domain D, then it it is said to be *analytic on the domain* D. (If D is open, this is no different from saying that f is differentiable at each point in D.)

From the properties of the derivative it follows immediately that if $f(z)$ and $g(z)$ are analytic functions on the domain D, then so are the functions $f(z)+g(z)$, $f(z)-g(z)$, and $f(z)\cdot g(z)$; also the quotient $f(z)/g(z)$ is analytic at all points in D where $g(z) \neq 0$. For example, the function

$$f(z) = \frac{z}{(z-1)(z-i)}$$

is analytic on the plane $\mathbb{C}$ excluding the points $z = 1$ and $z = i$.

From the theorem about derivatives of compositions of functions we get the following statement: if the function $u = u(z)$ is analytic on the domain D, and maps D to the domain D' of the variable u, and the function $f(u)$ is analytic on the domain D', then the composed function $f(u(z))$ of the variable z is analytic on D.

Now we introduce the notion of a function which is analytic on a closed domain $\overline{D}$. Unlike the case of open domains, here there are also points on the boundary, which do not have neighborhoods lying within $\overline{D}$; therefore the derivative at these points is not defined. The function $f(z)$ is said to be *analytic on the closed domain* $\overline{D}$ if this function can be extended to some wider open domain D_1, containing $\overline{D}$, on which it is analytic.

The following property of analytic functions is analogous to the corresponding property of differentiable functions of one real variable. But the proof is essentially different.

Theorem 3.7. *Suppose $f(z)$ is analytic on a domain D, and its derivative is zero everywhere in D. Then $f(z)$ is constant.*

Proof. Writing $f(z) = u(z) + iv(z)$, we see that

$$f'(z) = \frac{\partial u}{\partial x} = 0 \quad \text{and} \quad \frac{\partial v}{\partial x} = 0,$$

so u and v do not depend on x. And by the Cauchy-Riemann conditions, we also have

$$\frac{\partial u}{\partial y} = 0 \quad \text{and} \quad \frac{\partial v}{\partial y} = 0,$$

so u and v do not depend on y either. Thus those functions, and hence f, are constants. □

Problems

1. Prove that a function f is differentiable at the point z_0 if and only if it has a derivative at z_0 and $A = f'(z_0)$.

2. Find the points at which the function

$$w = z(\overline{z} - 3 \cdot \operatorname{Im} z)$$

is differentiable, and if any exist, compute the derivative at those points.

3. Determine the points at which the given functions $w = f(z)$ are differentiable, and if there are any, calculate the derivative at those points.

a) $w = z \operatorname{Im} z$ b) $w = z \operatorname{Re} z$

c) $w = z(z + 2 \operatorname{Re} z)$ d) $w = (\overline{z} + 1)z$

e) $w = (z + 1)\overline{z}$

3.2 The Connection between Analytic and Harmonic Functions

A real-valued function $u = u(x, y)$ of two variables x and y is said to be *harmonic on a domain* D, if it is defined on D, has continuous first and second partial derivatives everywhere in D, and satisfies Laplace's equation

$$\frac{\partial^2 u}{\partial x^2} + \frac{\partial^2 u}{\partial y^2} = 0 \tag{3.10}$$

at each point in D. Laplace's equation and harmonic functions play an important role in physics and technology. For example, the static temperature distribution on a domain D, and electric potential fields on charge-free domains, are harmonic functions. The connection between analytic and harmonic functions, which we will study in this section, is exploited in various applications of analytic functions.

Theorem 3.8. *The real and imaginary parts of an analytic function are harmonic functions.*

Proof. Let the function $f(z) = u(x, y) + iv(x, y)$ be analytic on the domain D; we must show that functions $u(x, y)$ and $v(x, y)$ are harmonic on D. We need the following fact, which will be proved (in greater generality) in Section 5.4 (see Theorem 5.18): the real and imaginary parts of analytic functions have continuous first and second derivatives.

Since f is analytic on D, then at each point in the domain D the Cauchy-Riemann conditions (3.4) are satisfied. Differentiating the first identity with respect to x, and the second with respect to y, we get

$$\frac{\partial^2 u}{\partial x^2} = \frac{\partial^2 v}{\partial y \partial x} \quad \text{and} \quad \frac{\partial^2 u}{\partial y^2} = -\frac{\partial^2 v}{\partial x \partial y}. \tag{3.11}$$

In a course on multivariable calculus (see for example [9], Section 14.3) the reader will have learned that if a real-valued function $v(x, y)$ has continuous first and second derivatives, then the mixed second derivatives are equal, i.e.

$$\frac{\partial^2 v}{\partial y \partial x} = \frac{\partial^2 v}{\partial x \partial y}.$$

Adding up the equalities (3.11), we get Laplace's equation (3.10), as required. The harmonicity of the function v is proved in a similar way. □

So if $f(z) = u + iv$ is an analytic function, then u and v are harmonic. But the converse is false: if u and v are arbitrarily chosen harmonic functions, then the function $f(z) = u + iv$ will not necessarily be analytic. For example, the function $f(z) = \operatorname{Re} z = x + i0$ is not analytic (see Example 3.1), although the functions $u = x$ and $v = 0$ are harmonic. In order for the function $f(z) = u+iv$ to be analytic, the functions u and v must not only be harmonic, but must also satisfy the Cauchy-Riemann conditions (3.4). A harmonic function v which is related to the harmonic function u by the Cauchy-Riemann conditions, is said to be *(harmonically) conjugate to* u. From Theorems 3.2 and 3.8 we get the following proposition.

Theorem 3.9. *Given two harmonic functions u and v, the function $f(z) = u + iv$ is analytic if and only if v is conjugate to u.*

Suppose that $u = u(x, y)$ is a harmonic function on a simply connected domain D, and it is known that it forms the real part of some analytic function $f(z) = u + iv$. Then for the imaginary part $v = v(x, y)$ we find from the Cauchy-Riemann conditions (3.4) that

$$\frac{\partial v}{\partial x} = -\frac{\partial u}{\partial y} \quad \text{and} \quad \frac{\partial v}{\partial y} = \frac{\partial u}{\partial x}.$$

In this way, whenever we are given such a function u, we can find the partial derivatives of the function v. A theorem of multivariate analysis says that a function of several variables on a simply connected domain can be determined from its partial derivatives, and uniquely so, up to an additive constant (one method for doing this calculation is shown below, Example 3.10). Thus, given a harmonic function u on a simply connected domain D, we can uniquely determine (up to an additive constant) the function v which is conjugate to it, and thereby also determine the analytic function $f(z) = u+iv$. Similarly, f can also be determined uniquely (up to an additive constant) from its imaginary part v.

Example 3.10 Find an analytic function whose imaginary part is

$$v = 4xy + y.$$

Note that it's easy to check that this function v is harmonic, since

$$\frac{\partial^2 v}{\partial x^2} = 0 = \frac{\partial^2 v}{\partial y^2}.$$

Solution From the Cauchy-Riemann conditions (3.4) we find the partial derivatives of the as yet unknown function u:

$$\frac{\partial u}{\partial x} = \frac{\partial v}{\partial y} = 4x + 1, \tag{3.12}$$

$$\frac{\partial u}{\partial y} = -\frac{\partial v}{\partial x} = -4y. \tag{3.13}$$

We will integrate the equation $\frac{\partial u}{\partial x} = 4x + 1$ with respect to x; the constant of integration $C = C(y)$ does not depend on x, but may depend on y:

$$u = \int (4x + 1)\, dx = 2x^2 + x + C(y).$$

We substitute this expression for u in equation (3.13) to determine the function $C(y)$:

$$\frac{\partial u}{\partial y} = C'(y) = -4y,$$

so that

$$C(y) = \int (-4y)\, dy = -2y^2 + C_1,$$

where C_1 is an arbitrary constant. Therefore

$$u = 2x^2 + x - 2y^2 + C_1 \quad \text{and}$$
$$f(z) = u + iv = (2x^2 + x - 2y^2) + i(4xy + y) + C_1.$$

We can rewrite $f(z)$ in a different form:

$$\begin{aligned} f(z) &= 2(x^2 + 2ixy - y^2) + x + iy + C_1 \\ &= 2(x + iy)^2 + (x + iy)C_1 = 2z^2 + z + C_1. \end{aligned}$$

If the simply connected domain contains a segment of the real axis, then the last step—writing the obtained expression for $f(z)$ in the variables x, y as an expression with the variable z—can be done in the following much simpler way.[4] Namely, it's sufficient to replace x by z and y by 0. For example,

$$\begin{aligned} f(z) &= (2x^2 + x - 2y^2) + i(4xy + y) + C_1 \\ &= (2z^2 + z - 0) + i(4z \cdot 0 + 0) + C_1 = 2z^2 + z + C_1. \end{aligned}$$

The justification of this rule is based on the uniqueness theorem—Theorem 6.28—which will be proved later.

Problems

1. Construct an analytic function $f(z)$ for which the real part is

$$u(x, y) = e^{-y} \sin x + y,$$

and $f(0) = i$.

2. Reconstruct the analytic function $f(z)$ from its real part $u(x, y)$ and its value at the point $z_0 = 0$.

[4] I am grateful to Professor M. P. Ovchintsev for this useful remark.

a) $u(x,y) = x^2 - y^2 + x, \quad f(0) = 0;$

b) $u(x,y) = x^3 - 3xy^2 + 1, \quad f(0) = 1.$

3. Reconstruct the analytic function $f(z)$ from its imaginary part $v(x,y)$ and its value at the point $z_0 = 0$.

a) $v(x,y) = e^x \cos y, \quad f(0) = 1 + i;$

b) $v(x,y) = 2xy + 2x, \quad f(0) = 0;$

c) $v(x,y) = 3x^2y - y^3, \quad f(0) = 1.$

4. Give a detailed proof of Theorem 3.9.

3.3 The Geometric Meaning of the Derivative. Conformal Mappings

3.3.1 The geometric meaning of the argument of the derivative

Let $z(t) = x(t) + iy(t)$, $\alpha < t < \beta$, be a smooth curve (the definition of smooth curves is given in Section 2.1.1). We take two values, t_0 and $t_0 + \Delta t$, from the interval (α, β); they correspond to the points $z(t_0)$ and $z(t_0 + \Delta t)$ on the curve. The vector

$$\Delta z = z(t_0 + \Delta t) - z(t_0) = \Delta x + i\Delta y$$

is the direction of the secant line through these two points (Fig. 23). If we multiply Δz by the real number $1/\Delta t$, we get the vector $\Delta z/\Delta t$, parallel to the vector Δz. Now we start to reduce Δt; then the point $z(t_0 + \Delta t)$ approaches $z(t_0)$ along the curve, and the vector $\Delta z/\Delta t$ will turn, approaching the vector

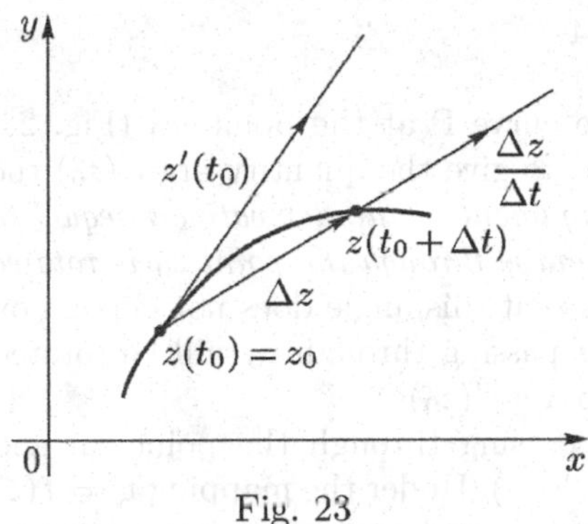

Fig. 23

$$\lim_{\Delta t \to 0} \frac{\Delta z}{\Delta t} = \lim_{\Delta t \to 0} \left(\frac{\Delta x}{\Delta t} + i\frac{\Delta y}{\Delta t} \right) = x'(t_0) + iy'(t_0) = z'(t_0).$$

The limit of the secant line through the point $z(t_0)$ is called the *tangent line* to the curve at that point. So the vector $z'(t_0)$ is in the direction of the tangent line at $z(t_0)$.

Now suppose that the function $f(z)$ is analytic at the point z_0, and that $f'(z_0) \neq 0$. Also assume that a curve γ passes through the point z_0, and that the parametric equation for γ is $z(t) = x(t) + iy(t)$, with $z(t_0) = z_0$. The curve γ is mapped by the function $w = f(z)$ to the Γ lying in the plane of the variable w; the equation of the curve Γ can be written $w(t) = f(z(t))$; then z_0 is mapped to the point $w_0 = f(z_0)$. By the chain rule

$$w'(t_0) = f'(z_0) \cdot z'(t_0). \tag{3.14}$$

From this it follows (see (1.11) that

$$\arg w'(t_0) = \arg f'(z_0) + \arg z'(t_0). \tag{3.15}$$

But $z'(t_0)$ is a vector tangent to the curve γ at the point z_0 (Fig. 24,

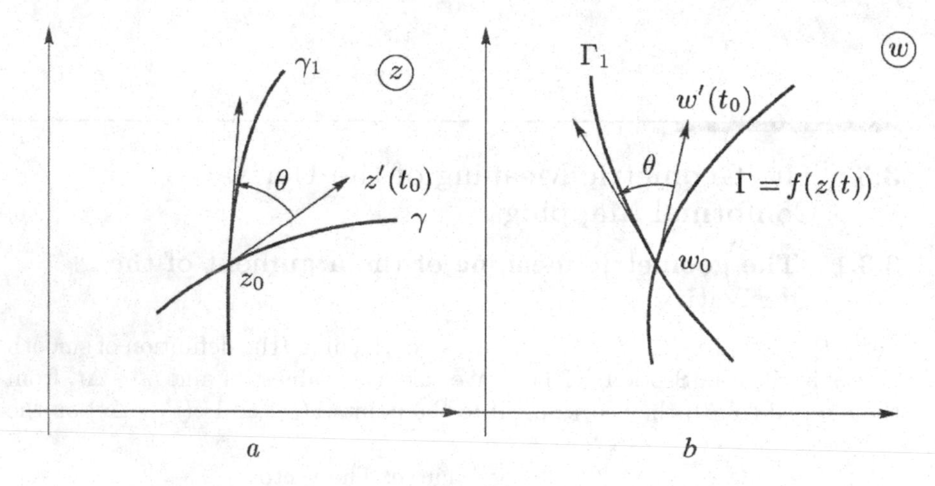

Fig. 24

a), and $w'(t_0)$ is a vector tangent to the curve Γ at the point w_0 (Fig. 24, *b*). Therefore the equality (3.15) allows us to give the quantity $\arg f'(z_0)$ the following geometric interpretation: *the argument of the derivative is equal to the angle through which the tangent to a curve through the point z_0 is rotated under the mapping $w = f(z)$*. This means that this angle does not depend on the curve γ, i.e. the tangents of all curves passing through z_0 will be rotated through one and the same angle, equal to $\arg f'(z_0)$.

Now take any two curves γ and γ_1 passing through the point z_0, and introduce the tangents to the curves (Fig. 24, *a*). Under the mapping $w = f(z)$ the curves γ and γ_1 go to the curves Γ and Γ_1 and each of the tangents to γ and γ_1 are rotated through the same angle. Therefore the angle θ between the tangents to γ and γ_1 will be equal (not only in size, but also in orientation, i.e. clockwise or counterclockwise) to the angle between the tangents to Γ and Γ_1. Recall that the angle between two curves at a point of intersection is defined to be the angle between their tangents at that point. Therefore, *if $f'(z_0) \neq 0$, the mapping $w = f(z)$ preserves the angles between curves.*

3.3.2 The geometric meaning of the modulus of the derivative

Fix a point z_0 and take an increment Δz; obviously the modulus $|\Delta z|$ is equal to the distance between the points z_0 and $z = z_0 + \Delta z$ (Fig. 25, *a*). Let $w = f(z)$, $w_0 = f(z_0)$, and $\Delta w = w - w_0$. Then the quantity $|\Delta w|/|\Delta z|$ represents the ratio by which the distance between the points z_0 and z changes as a result of the mapping $w = f(z)$. The limit of $|\Delta w|/|\Delta z|$ as $z_0 \to 0$ is

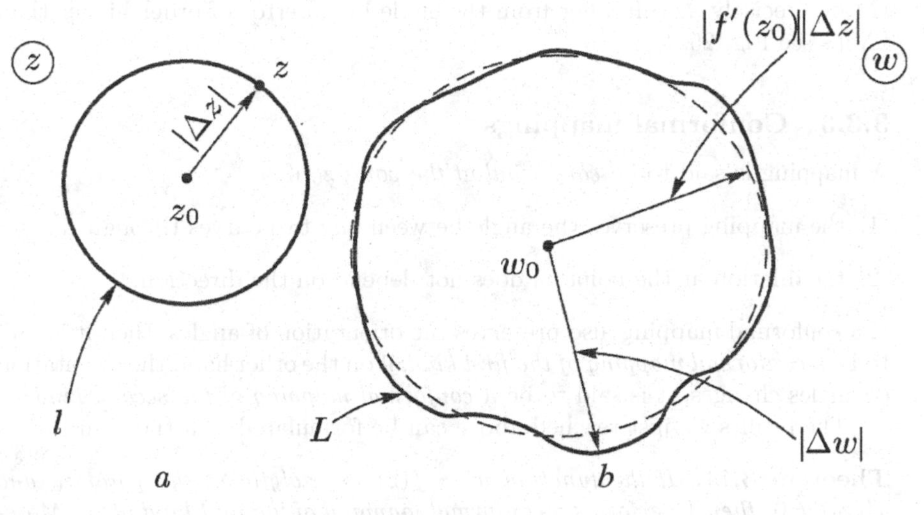

Fig. 25

called the *dilation coefficient* at the point z_0 under the mapping $w = f(z)$. Since

$$\lim_{z_0 \to 0} \frac{|\Delta w|}{|\Delta z|} = |f'(z_0)|,$$

the modulus $|f'(z_0)|$ *is equal to the dilation coefficient at* z_0 *under the mapping* $w = f(z)$. If $|f'(z_0)| > 1$, then in a sufficiently small neighborhood of the point z_0 the distance between points is increased under the mapping, and the result is a dilation or expansion (in the geometric sense); but if $|f'(z_0)| < 1$, the mapping leads to a contraction (although $|f'(z_0)|$ is still referred to as the dilation coefficient).

Because the derivative $|f'(z_0)|$ does not depend on the path by which the point $z_0 + \Delta z$ approaches z_0, the dilation coefficient is the same for all directions. This property can be illustrated in the following way. Take a circle l with center z_0 and radius $|\Delta z|$ (here we assume that Δz has a fixed modulus, but may vary in direction—Fig. 25). Under the mapping $w = f(z)$ the circle goes to the curve L (Fig. 25, *b*). The distance from the point $w = f(z_0 + \Delta z)$ to the point $w_0 = f(z_0)$ equals

$$|\Delta w| = |w - w_0| = |f(z_0 + \Delta z) - f(z_0)|.$$

Since

$$\Delta w = f'(z_0)\Delta z + \alpha(\Delta z)\Delta z,$$

where $\alpha(\Delta z) \to 0$ as $\Delta z \to 0$, then

$$|w - w_0| = |f'(z_0)\Delta z + \alpha(\Delta z)\Delta z|.$$

This equation means that the points of the curve L will deviate very little from the circle $|w - w_0| = |f'(z_0)||\Delta z|$ with center w_0 and radius $|f'(z_0)||\Delta z|$. (More precisely, L will differ from the circle by an error of order higher than $|\Delta z|$—see Fig. 25, *b*.)

3.3.3 Conformal mappings

A mapping is said to be *conformal at the point* z_0 if

1. the mapping preserves the angle between any two curves through z_0;

2. the dilation at the point z_0 does not depend on the direction.

If a conformal mapping also preserves the orientation of angles, then it is said to be a *conformal mapping of the first kind;* if on the other hand the orientation of angles changes, it is said to be a *conformal mapping of the second kind.*

The results we just reached above can be formulated as a theorem.

Theorem 3.11. *If the function* $w = f(z)$ *is analytic at the point* z_0 *and* $f'(z_0) \neq 0$, *then* f *performs a conformal mapping of the first kind at* z_0. *Moreover* $\arg f'(z_0)$ *is the angle of rotation, and* $|f'(z_0)|$ *is the dilation coefficient, of the mapping.*

An example of a conformal mapping of the second kind is given by the function (which is not analytic!) $w = \overline{z}$, mapping every domain D to its reflection E about the x-axis.

If $f'(z_0) = 0$, then in general the mapping is not conformal at the point z_0. For example, the mapping $w = z^2$ doubles every angle whose vertex is at the origin—see Example 2.3 and equations (2.7).

4

Conformal Mappings

In this chapter we will look more deeply into the geometric aspect of functions of a complex variable.

A function $w = f(z)$ is said to be *one-to-one (1-1)* on the domain D if distinct points in D are assigned distinct values by f; i.e. if $z_1 \neq z_2$, then $f(z_1) \neq f(z_2)$. For example, the function $w = z^2$ is not 1-1 on the entire complex plane, because $(-1)^2 = 1^2$. However it is 1-1 on the half-plane $D = \{ z : \operatorname{Re} z > 0 \}$ (prove it!). If a function is analytic and 1-1 on D, it is said to be *univalent* on D.

Suppose that $f(z)$ maps the domain D onto the domain E. Then if f is also 1-1, it means that each point $w \in E$ has only one preimage in D. Therefore in this case the mapping from D onto E, carried out by the function $w = f(z)$, is a *bijection:* each point $z \in D$ is associated with a point $w \in E$, and conversely each point $w \in E$ is associated with a unique point $z \in D$.

A mapping from the domain D onto the domain E is said to be *conformal on the domain* D if it is 1-1 on D and conformal at each point of D.

From Theorem 3.11 it follows that if a function $f(z)$ is univalent on the domain D and $f'(z)$ is never zero on points of D, then this function performs a conformal mapping on the domain D.[1] Notice that if a mapping is conformal at each point of D, it's not necessarily conformal on D since it might be not 1-1 on D. For example, as we show in Section 4.3, the mapping $w = e^z$ is conformal at each $z \in \mathbb{C}$ but is not 1-1 in any domain containing points z_1, z_2 with $z_1 - z_2 = 2\pi n i$, where n is a positive integer.

We will look now at some important elementary functions and the properties of their mappings.

4.1 Linear and Möbius Transformations

4.1.1 Linear functions

The function

$$w = az + b \tag{4.1}$$

[1]In Section 7.4 we will prove that if a function $f(z)$ is univalent, that is 1-1 and analytic on the domain D, then $f'(z) \neq 0$ on points of D.

DOI: 10.1201/9780367810283-4

where a and b are given complex numbers, is called a *linear function.* As long as $w' = a \neq 0$, the mapping (4.1) is conformal on the whole complex plane $\mathbb{C}$, since it is also 1-1 on $\mathbb{C}$: for if

$$w_1 = az_1 + b \quad \text{and} \quad w_2 = az_2 + b$$

then

$$w_1 - w_2 = a(z_1 - z_2),$$

so that if $z_1 \neq z_2$ then $w_1 \neq w_2$.

If we additionally define $w(\infty) = \infty$, we obtain a 1-1 mapping of the extended complex plane $\overline{\mathbb{C}}$ onto $\overline{\mathbb{C}}$.

In studying the geometric properties of the mapping (4.1), we first consider the case when $b = 0$, i.e. $w = az$. Let

$$a = |a|e^{i\alpha} \quad \text{and} \quad z = |z|e^{i\phi};$$

then

$$w = |a||z|e^{i(\alpha+\phi)}.$$

Therefore to obtain the vector $w = az$ we must perform the following two actions.

1. Multiply the given vector z by $|a|$. In doing this the direction of the vector z remains unchanged, but its length is multiplied by $|a|$. In other words, multiplying by $|a|$ performs a geometric dilation about the origin, with dilation coefficient $|a|$.

2. Rotate the vector obtained from the previous step through the angle α.

Turning to the general case of (4.1), note that adding the vector b to the vector az simply performs a parallel translation of the endpoints of the vector az along the vector b.

Thus the mapping (4.1) is effected by means of the following operations: 1) a geometric dilation about the origin with coefficient $|a|$; 2) a rotation about the origin through angle α; and 3) a parallel translation along vector b.

Example 4.1 Find the linear function which maps the square D with side 2 onto the square E with side 4 (Fig. 26) and maps point A to point B.

Solution We will determine each of the three steps that make up the linear mapping.

1. A geometric dilation which takes D to a square D_1 with side 4. Evidently the dilation coefficient is 2, so the transformation must be $w_1 = 2z$. Then the point $A = (\sqrt{2}, \sqrt{2})$ goes to $A_1 = (4\sqrt{2}, 2\sqrt{2})$, and A' goes to A_1'.

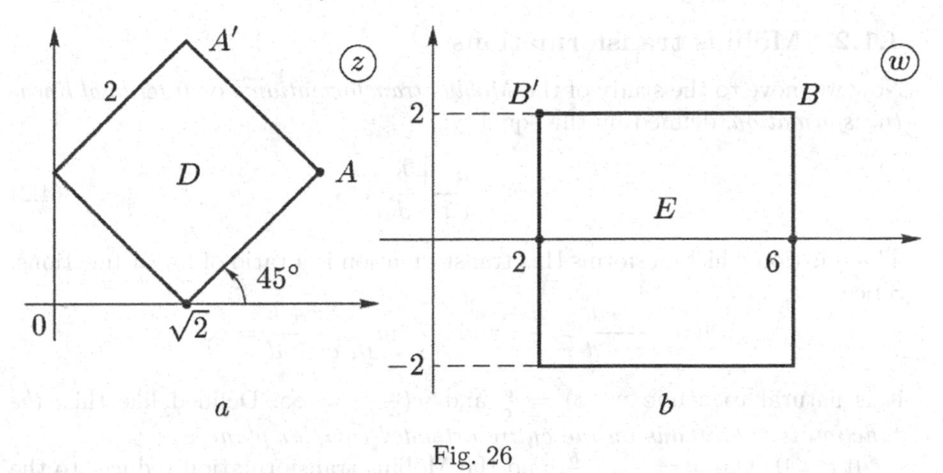

Fig. 26

2. A rotation of D_1 about the origin to a square D_2, which makes side $A_1'A_1$ parallel to side $B'B$. Obviously the angle of rotation must be 45° counter-clockwise; this transformation can be written in the form $w_2 = w_1 e^{i\pi/4}$. Point A_1 then goes to

$$A_2 = (4\sqrt{2} + i2\sqrt{2})\left(\cos\frac{\pi}{4} + i\sin\frac{\pi}{4}\right) = (4+2i)(1+i) = 2+6i.$$

3. There remains a parallel translation of the square D_2 along the vector $\mathbf{A_2B}$. Considering that $B = 6 + 2i$, we have $\mathbf{A_2B} = 4 - 4i$ and $w = w_2 + 4 - 4i$.

   Thus

$$w = w_2 + 4 - 4i = w_1 e^{\pi/4} + 4 - 4i = 2ze^{\pi/4} + 4 - 4i,$$

and the linear mapping we sought is $w = az + b$ where

$$a = 2e^{\pi/4} = \sqrt{2} + i\sqrt{2}, \quad \text{and} \quad b = 4 - 4i.$$

Let us note that each one of the three steps of a linear mapping (dilation, rotation, and translation) preserves lines and circles, i.e. lines are mapped to lines, and circles to circles. Consequently linear mappings (4.1) also possess these properties.

4.1.2 Möbius transformations

Now we move to the study of the *Möbius transformation*[2], or *fractional linear transformation*, defined by the equality:

$$w = \frac{az+b}{cz+d}. \tag{4.2}$$

The function which performs this transformation is a ratio of linear functions. Since

$$\lim_{z\to\infty} \frac{az+b}{cz+d} = \frac{a}{c} \quad \text{and} \quad \lim_{z\to -d/c} \frac{az+b}{cz+d} = \infty,$$

it is natural to define $w(\infty) = \frac{a}{c}$ and $w(-\frac{d}{c}) = \infty$. Defined like this, *the function is continuous on the entire extended complex plane* $\overline{\mathbb{C}}$.

If $c = 0$ then $w = \frac{a}{d}z + \frac{b}{d}$, and the Möbius transformation reduces to the linear case which we have already explored. So in the rest of this subsection we assume that $c \neq 0$.

Multiplying the numerator and denominator of (4.2) by c, and adding $ad - ad$ to the numerator, allows us write the function in the form

$$w = \frac{az+b}{cz+d} = \frac{a(cz+d)+(bc-ad)}{c(cz+d)} = \frac{a}{c} + \frac{bc-ad}{c(cz+d)}. \tag{4.3}$$

If $bc - ad = 0$, then $w = \frac{a}{c}$ and the function (4.2) reduces to a constant. So we will now assume that the conditions

$$c \neq 0 \quad \text{and} \quad bc - ad \neq 0 \tag{4.4}$$

are both satisfied.

We will show that *the Möbius transfomation (4.2) is a bijective mapping of* $\overline{\mathbb{C}}$ *onto* $\overline{\mathbb{C}}$. With this in mind we solve the equation (4.2) for z in terms of w (which is possible when $z \neq -\frac{d}{c}$, $z \neq \infty$, $w \neq \frac{a}{c}$, and $w \neq \infty$):

$$z = \frac{-dw+b}{cw-a} = -\frac{d}{c} + \frac{bc-ad}{c(cw-a)}. \tag{4.5}$$

Therefore each value $w \neq \frac{a}{c}$ and $w \neq \infty$ has exactly one preimage $z \neq -\frac{d}{c}$ and $z \neq \infty$. But by definition the value $w = \frac{a}{c}$ is associated to $z = \infty$, and the value $w = \infty$ is associated to the value $z = -\frac{d}{c}$. Thus every point $w \in \overline{\mathbb{C}}$ has only one preimage in $z \in \overline{\mathbb{C}}$, as we needed to show.

Now we establish the conformality of the mapping (4.2). Since

$$w' = \frac{ad-bc}{(cz+d)^2},$$

[2] August Möbius (1790–1868) was a German mathematician and theoretical astronomer best known for his discovery of the Möbius strip.

then when $z \neq \frac{-d}{c}$ and $z \neq \infty$ the derivative w' exists and is nonzero. By Theorem 3.11, therefore, the Möbius transformation is a conformal mapping everywhere except those for two points.

To clarify the situation at $z = \frac{-d}{c}$ and $z = \infty$ we need the following definition.

The angle value between two curves meeting at the point at infinity is the angle value between the images of those two curves under the mapping $w = \frac{1}{z}$, at the origin.

Example 4.2 Find the angle value between the arm of the parabola $y = x^2$, where $x \geq 0$, and the ray $y = x/\sqrt{3}$ at the point $z = \infty$ (Fig. 27, *a*).

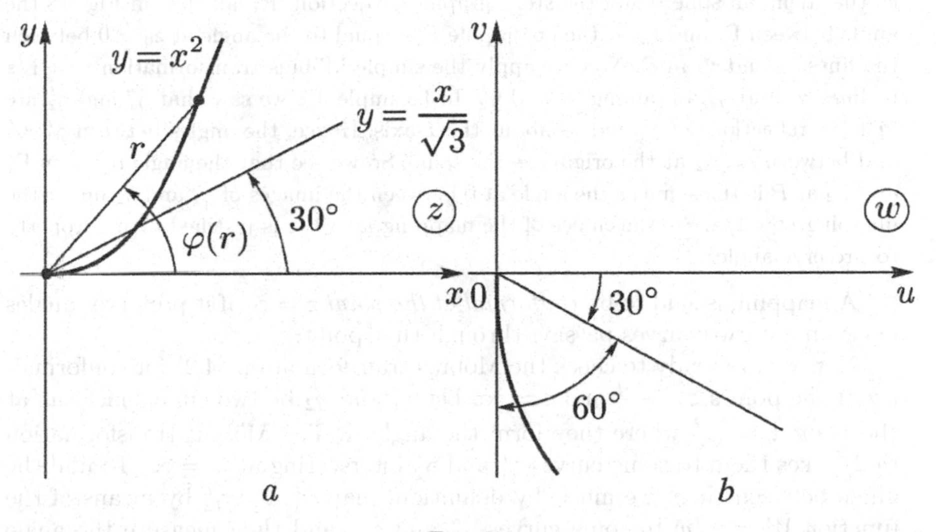

Fig. 27

Solution Any point of the parabola may be written in the form

$$z = re^{i\phi(r)},$$

where $\phi(r) \to \pi/2$ as $r \to \infty$. Under the mapping $w = \frac{1}{z}$ this point of the parabola goes to

$$w = \frac{1}{z} = \frac{1}{r}e^{-i\phi(r)}.$$

If we let $r \to \infty$ then $|w| = \frac{1}{r} \to 0$, and $\arg w = -\phi(r) \to -\frac{\pi}{2}$. This means that the arm of the parabola is mapped to a curve which is tangent to the negative y-axis (Fig. 27, *b*). For points of the ray, $z = re^{i\pi/6}$. Hence,

$$w = \frac{1}{z} = \frac{1}{r}e^{-i\pi/6}.$$

This means that ray is mapped to new ray which is the reflection of the original about the x-axis. The angle between the images of the curves in the plane of the variable w is, evidently, $\pi/3$; so this is, by definition, the angle value between the original curves at the points $z = \infty$.

Why was the angle value between curves meeting at $z = \infty$ defined as it was? For example, why do we use the mapping $w = 1/z$ rather than, say, $w = 1/z^2$? Consider for simplicity two straight lines γ_1 and γ_2 passing through the origin. Referring back to Fig. 18 with $z_0 = Z_0 = 0$, the natural definition of the angle at $z = \infty$ between lines γ_1 and γ_2 would be the angle at P between their images Γ_1 and Γ_2 under the stereographic projection; but it is preferable to give a definition that does not refer to the Riemann sphere and the stereographic projection. Recall that in Fig. 18 the angle between Γ_1 and Γ_2 at the north pole P is equal to the angle at $z_0 = 0$ between the lines γ_1 and γ_2 in $\mathbb{C}$. Now we apply the simple Möbius transformation $w = 1/z$ to lines γ_1 and γ_2, obtaining γ_1' and γ_2'. In Example 4.2 we saw that γ_1' and γ_2' are just the reflections of γ_1 and γ_2 about the x-axis. Hence, the angles between γ_1', γ_2' and between γ_1, γ_2 at the origin are the same. So we see that the angle between Γ_1 and Γ_2 at P is the same as the angle at 0 between the images of γ_1 and γ_2 under the mapping $w = 1/z$. So, the choice of the mapping $w = 1/z$ is justified by its property to preserve angles.

A mapping is said to be *conformal at the point* $z = \infty$ if it preserves angles between any two curves passing through that point.

Now we are ready to check the Möbius transformation (4.2) for conformality at the points $z = -\frac{d}{c}$ and $z = \infty$. Let γ_1 and γ_2 be two curves meeting at the point $z = -\frac{d}{c}$ where they form the angle α. The Möbius transformation (4.2) takes them to some curves γ_1' and γ_2' intersecting at $w = \infty$. To find the angle between them we must, by definition, map γ_1' and γ_2' by means of the function $W = 1/w$ to some curves γ_1'' and γ_2'', and then measure the angle between them at the point $W = 0$. The mapping from γ_1 and γ_2 to γ_1'' and γ_2'' has the form

$$W = \frac{1}{w} = \frac{cz + d}{az + b}. \tag{4.6}$$

The derivative

$$\frac{dW}{dz} = \frac{bc - ad}{(az + b)^2}$$

at the point $z = -\frac{d}{c}$ exists and is nonzero. Consequently the angle between γ_1'' and γ_2'' at the point $W = 0$ is equal to α; this means that the angle between γ_1' and γ_2' is also equal to α. The dilation coefficient under the Möbius transformation at this point is ∞ in all directions, and therefore independent of direction. So the Möbius transformation (4.2) is conformal at the point $z = -\frac{d}{c}$.

Now we check conformality at the point $z = \infty$. Let γ_1 and γ_2 be two curves meeting at the point $z = \infty$. We will let γ_1' and γ_2' denote the curves to which γ_1 and γ_2 are mapped under the Möbius transformation (4.2), denoted here by

f; in particular the intersection at $z = \infty$ is mapped to $w = f(\infty) = \frac{a}{c}$. Now consider the inverse function $f^{-1}(w) = z$; we have already shown that f^{-1} is itself a Möbius transformation—see (4.5); moreover it maps $\frac{a}{c}$ to $z = \infty$. By the result from the previous paragraph, f^{-1} is conformal at $\frac{a}{c}$. That means the angle at $w = \frac{a}{c}$ between γ_1' and γ_2' is equal to the angle at $z = \infty$ between γ_1 and γ_2. And that is the condition for the conformality of f at $z = \infty$.

We can formulate this result as a theorem.

Theorem 4.3. *Any Möbius transformation*

$$w = \frac{az+b}{cz+d}, \quad \text{where } ad - bc \neq 0, \quad w(\infty) = \frac{a}{c}, \quad \text{and} \quad w\Big(-\frac{d}{c}\Big) = \infty, \tag{4.7}$$

performs a bijective conformal mapping from the extended complex plane $\overline{\mathbb{C}}$ *onto itself* $\overline{\mathbb{C}}$.

Note that this theorem covers also the case when $c = 0$, because then the function becomes linear, and so possesses the same properties claimed in the theorem.

Now we will see that Möbius transformations have an additional property. To state the following theorem as simply and generally as possible, it's convenient to think of lines as being circles of infinitely large radius. We will use the term $\overline{\text{circle}}$ for this extended notion of circle (that is $\overline{\text{circle}}$ means circle or line).

Theorem 4.4. *Under Möbius transformations* $\overline{\textit{circles}}$ *are preserved, i.e. they are always mapped onto other* $\overline{\textit{circles}}$.

(Of course, it is possible for a $\overline{\text{circle}}$ of infinite radius, that is a line, to be mapped onto a circle of finite radius, and vice versa.)

Proof. Consider the equation

$$A(x^2 + y^2) + Bx + Cy + D = 0, \tag{4.8}$$

where A, B, C, and D are real coefficients. When $A = 0$ we get $Bx+Cy+D = 0$, which is the equation of a line. If $A \neq 0$, then dividing by A, and then completing the squares, produces an equation of the form

$$(x - x_0)^2 + (y - y_0)^2 = \pm R^2,$$

which defines some circle, if the right side is $+R^2$, some point, if $R = 0$, and the empty set, if the right side is $-R^2$. On the other hand, any $\overline{\text{circle}}$ can be expressed in the form of equation (4.8).

First we show that the mapping $w = 1/z$ preserves $\overline{\text{circles}}$. Take an arbitrary $\overline{\text{circle}}$ in the complex plane, and express it in the form of equation (4.8). Writing $z = x + iy$ and $w = u + iv$, and using $z = 1/w$, we get

$$x + iy = \frac{1}{u+iv} = \frac{u - iv}{u^2 + v^2}.$$

From this it follows that

$$x = \frac{u}{u^2+v^2} \quad \text{and} \quad y = -\frac{v}{u^2+v^2}.$$

To get the equation of the curve to which the circle is mapped under $w = 1/z$, we plug these expressions for x and y into (4.8):

$$\frac{A}{u^2+v^2} + \frac{Bu}{u^2+v^2} - \frac{Cv}{u^2+v^2} + D = 0,$$

or

$$A + Bu - Cv + D(u^2+v^2) = 0.$$

We have arrived at an equation of the same form as (4.8), except in the plane of variable $w = u + iv$. As we saw earlier, such an equation defines either a circle, a point, or the empty set. But any Möbius transformation is a bijection; so the image of a circle cannot be either a single point or the empty set. That means the image must be a circle, so the mapping $w = 1/z$ preserves circles.

Now consider the general case of a Möbius transformation of the form (4.2). If $c = 0$, then we get a linear mapping of the form $w = a_1 z + b_1$, which breaks down to a dilation, a rotation, and a translation. Clearly each of these component mappings preserve circles; and therefore so does their composition, $w = a_1 z + b_1$.

Now suppose that $c \neq 0$. Making use of the equality (4.3), let us put the Möbius transformation in the form

$$w = \frac{az+b}{cz+d} = \frac{a}{c} + \frac{bc-ad}{c^2(z+\frac{d}{c})} = E + \frac{F}{z+G}, \tag{4.9}$$

where

$$E = \frac{a}{c}, \quad F = \frac{bc-ad}{c^2}, \quad \text{and} \quad G = \frac{d}{c}.$$

From the equality (4.9) it follows that the Möbius transformation can be broken down into the following three components:

$$\begin{aligned} 1)\ w_1 &= z + G \\ 2)\ w_2 &= \tfrac{1}{w_1} \\ 3)\ \ w &= E + Fw_2. \end{aligned}$$

We have already established that each of these mappings preserve circles. Therefore the composition of the three, i.e. the original Möbius transformation, also preserves circles, which is what we were to prove. □

In order to state the next theorem, we need a new definition. The points A and A' are said to be *symmetric about a circle* of radius $R < \infty$, if they both lie on the same line through the center O of the circle, and

$$OA \cdot OA' = R^2. \tag{4.10}$$

Here are a few simple facts to note: if a point A approaches the circle, i.e. if $OA \to R$, then OA' also approaches R; every point on the circle is symmetric to itself; and if $OA \to 0$ then $OA' \to \infty$, and therefore the point O is symmetric with the point at infinity. Points are *symmetric about a circle of radius* $R = \infty$ (that is about a line) if they are symmetric in the usual sense, i.e. one is the image of the other under reflection over the line.

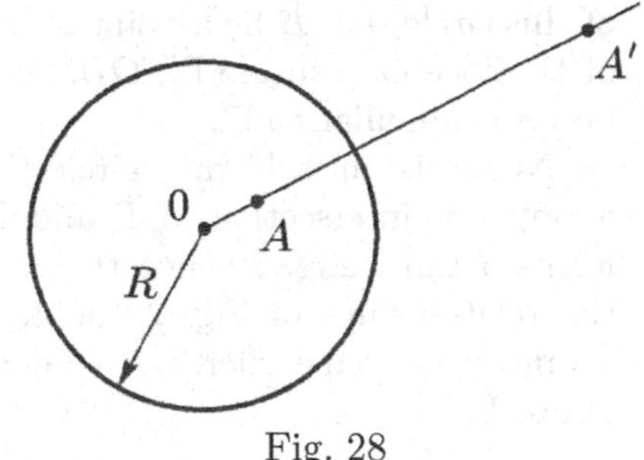

Fig. 28

Lemma 4.5. *Two points A and A' are symmetric about the circle Γ if and only if any circle passing through A and A' is perpendicular to Γ (*Fig. 29*).*

Proof. The proof is simple for symmetric points about lines, and that is left to the reader. In what follows we deal only with the case of symmetry about a proper circle.

Only if (necessity): Let points A and A' be symmetric about the circle Γ. Take an arbitrary circle Γ' through points A and A'. Then Γ' might be a line, but by definition, the line through A and A' must pass through the center O, and therefore is perpendicular to Γ. So we may assume that Γ' is a proper circle.

Let B be a point of Γ' at which the segment OB is tangent to Γ'. A well known theorem of elementary geometry says that

$$OA \cdot OA' = OB^2.$$

By assumption, the left side of this equation is equal to R^2; therefore we must have $OB = R$. So evidently the point B lies on the circle Γ, and hence B is a point of intersection of the two circles. Because OB is a radius of Γ, it is perpendicular to Γ; and because OB is tangent to Γ', the angle it makes with Γ is the same as the angle that Γ and Γ' make with each other. Hence the circles intersect at right angles.

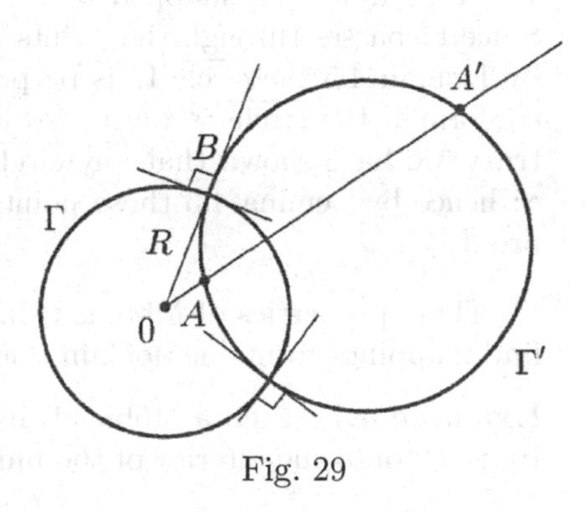

Fig. 29

If (Sufficiency): Now let A and A' be such that any circle passing through both, also meets Γ in a right angle; we will show that A and A' are symmetric points about Γ. By assumption, the line through A and A' must be perpendicular to Γ, so it must pass through the point O. So A, A', and O are collinear. In fact A and A' must lie on the same side of O. For if not, consider the circle Γ' with diameter AA'; the point O would be in the interior of this circle. Let B be a point of intersection of Γ and Γ', then OB is a radius of Γ. Since O is inside Γ', OB cannot be tangent to Γ'; and hence Γ cannot be perpendicular to Γ'.

Now take an arbitrary circle Γ' passing through A and A', and let B be a point of intersection of Γ and Γ'. By assumption, the intersection at B forms a right angle, hence the radius OB of Γ is tangent to Γ'. Therefore the relationships of Fig. 29 hold, and by that same theorem of elementary geometry quoted earlier, $OA \cdot OA' = R^2$, which means A and A' are symmetric about Γ. □

Now we are ready for the following property of Möbius transformations which we will call *preservation of symmetry.*

Theorem 4.6. *Any pair of points symmetric about a circle are mapped by any Möbius transformation to a pair of points symmetric about the image of the circle.*

Proof. Let z_1 and z_2 be symmetric about the circle Γ. Under a Möbius transformation f of the form (4.2) Γ will be mapped to a curve γ, which, by Theorem 4.4, is also a circle. We write the images of z_1 and z_2 as w_1 and w_2; we must show that w_1 and w_2 are symmetric about γ. Take any circle γ' passing through w_1 and w_2, and consider its preimage Γ' under f, i.e. the set of points in the plane of the variable z which are mapped to γ'. Recall that the inverse function $f^{-1}(w)$ is also a Möbius transformation, and under f^{-1} the circle γ' is mapped to Γ'; and by Theorem 4.4, Γ' is also a circle. Since Γ' passes through the points z_1 and z_2, which are symmetric about Γ, by Lemma 4.5 the circle Γ' is perpendicular to Γ. Because the mapping f is conformal, the circle γ' must also be perpendicular to γ. Since γ' was arbitrary, we have shown that any circle through w_1 and w_2 is perpendicular to γ; hence by Lemma 4.5 those points are symmetric about γ, completing the proof. □

These properties of Möbius transformations just established enable us to find mappings from one domain bounded by a circle to another.

Example 4.7 Find a Möbius transformation mapping the upper half-plane $\operatorname{Im} z > 0$ onto the interior of the unit disk $|w| < 1$.

Solution Let z_0 be the point in the upper half-plane which is mapped to the center of the unit disk, i.e. $w(z_0) = 0$. By Theorem 4.6, the point $\overline{z_0}$, which is symmetric to z_0 about the real axis, must go to $w = \infty$, the point symmetric to $w = 0$ about the unit circle. Any Möbius transformation which satisfies $w(z_0) = 0$ and $w(\overline{z_0}) = \infty$ must look like

$$w = A\frac{z - z_0}{z - \overline{z_0}},$$

where A is a complex constant. But this constant is not completely arbitrary, since for any real value $z = x$ its image w must lie on the unit circle, hence $|w| = 1$. Since

$$|x - z_0| = |x - \overline{z_0}|,$$

we get

$$1 = |w| = \left| A\frac{x - z_0}{x - \overline{z_0}} \right| = |A| \left| \frac{x - z_0}{x - \overline{z_0}} \right| = |A|.$$

Therefore $A = e^{i\phi}$, and the desired function must have the form

$$w = e^{i\phi}\frac{z - z_0}{z - \overline{z_0}}, \quad \text{where} \quad \operatorname{Im} z_0 > 0. \tag{4.11}$$

Note that indeed $|w(z)| = 1$ for real z, and any function of the form (4.11) maps the upper half-plane onto the unit disk.

We see that there is an infinite family of such Möbius transformations, performing the required mapping. Each of these functions is determined by a real number ϕ and a complex number z_0.

Example 4.8 Find a Möbius transformation which maps the unit disk $|z| < 1$ onto itself, i.e. $|w| < 1$.

Solution Let z_0 be the point in the unit disk $|z| < 1$ which is mapped to the point $w = 0$. Then the point z_0', which is symmetric to z_0 about the unit circle $|z| = 1$, must go to the point $w = \infty$, the point symmetric to $w = 0$ about the unit circle $|w| = 1$. Let us express z_0' in terms of z_0. Since z_0 and z_0' lie on the same ray starting at $z = 0$, we can write $z_0' = kz_0$, where k is a positive real number. From the definition of symmetric points we know that $|z_0||z_0'| = 1$, or $k|z_0|^2 = 1$, which means that $k = 1/|z_0|^2$ and

$$z_0' = kz_0 = \frac{z_0}{|z_0|^2} = \frac{z_0}{z_0\overline{z_0}} = \frac{1}{\overline{z_0}}.$$

Any Möbius transformation satisfying the conditions $w(z_0) = 0$ and $w(1/\overline{z_0}) = \infty$, must have the form

$$w = A\frac{z - z_0}{z - 1/\overline{z_0}} = -A\overline{z_0}\frac{z - z_0}{1 - z\overline{z_0}},$$

where A is complex constant. As in the previous example, the value of A is constrained by the condition that points on the unit circle $|z| = 1$ must be mapped to the unit circle $|w| = 1$. In particular, if $z = 1$ then

$$1 = |w| = \left| -A\overline{z_0}\frac{1 - z_0}{1 - \overline{z_0}} \right| = |-A\overline{z_0}|,$$

(using $|1 - z_0| = |\overline{z_0} - 1|$). Therefore $-A\overline{z_0} = e^{i\phi}$, and the desired mapping must have the form

$$w = e^{i\phi}\frac{z - z_0}{1 - z\overline{z_0}}. \tag{4.12}$$

On the other hand, any function of the form (4.12) maps the unit disk $|z| < 1$ onto itself. The reader can prove this fact directly, substituting $z = e^{it}$ into (4.12), or use Theorems 4.4 and 4.6 (problem 7).

Here again the mapping is not unique. In Fig. 30 you can see the preimage of the polar grid in the plane of w. The preimages of radial lines (Fig. 30,

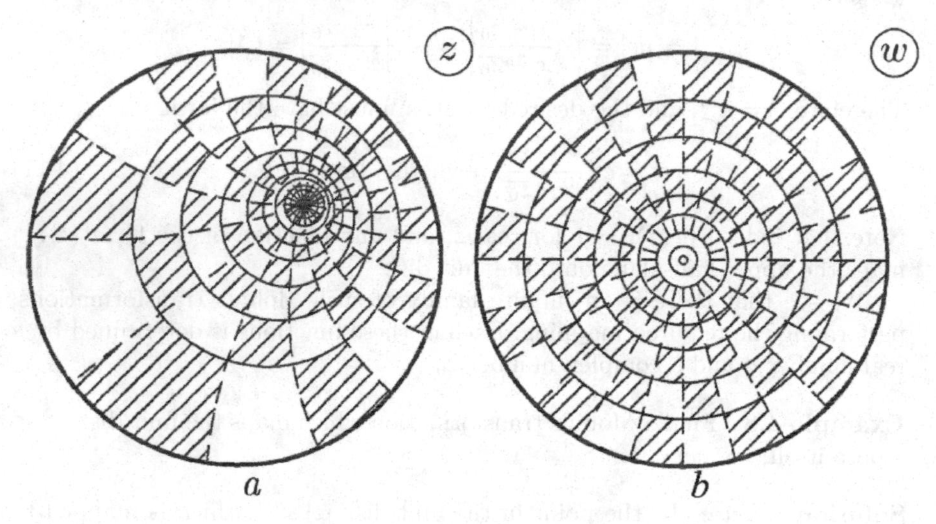

Fig. 30

b) are arcs perpendicular to the circle $|z| = 1$ (Fig. 30, a); the preimages of circles $|w| = r < 1$, are similar circles in the z-plane, without however being concentric.

Problems

1. Find an analytic function which maps the triangle with vertices at

$$A = (-1, 0), \quad B = (-4, -1), \quad C = (-4, 1)$$

onto the triangle with vertices at

$$A' = (3, 0), \quad B' = (5, -6), \quad C' = (1, -6)$$

(see Fig. 31).

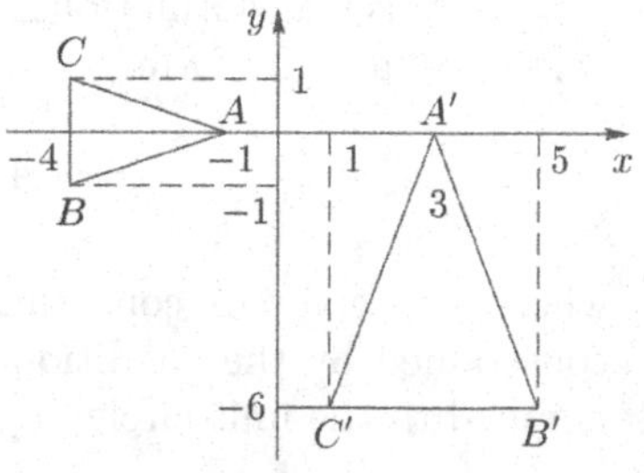

Fig. 31

2. Find an analytic function which maps the triangle with vertices at $(0, 0)$, $(-1, 2)$, $(1, 2)$ onto the triangle with vertices at $(2, 0)$, $(6, 2)$, $(6, -2)$.

3. How many distinct linear functions mapping the square D onto the square E are there? In Example 4.1 we have found one so far.

4. To what domain is the disk $|z+2i|<2$ mapped under the transformation $w=\frac{1}{z}$?

5. To what domain is the disk $|z+i|<3$ mapped under the transformation $w=\frac{1}{z}$?

6. To what domains do the following disks go under the mapping $w=\frac{1}{z}$?

a) $|z-i|<1$; b) $|z-i|<2$.

7. Prove that any function of the form (4.12) maps the unit disk $|z|<1$ onto itself.

8. Find a conformal mapping $w=f(z)$ of the disk $|z|<1$ onto the disk $|w|<1$ which satisfies the conditions

$$w\left(-\frac{i}{2}\right)=0 \quad \text{and} \quad \operatorname{Arg} w'\left(-\frac{i}{2}\right)=\frac{\pi}{4}.$$

Also find the dilation coefficient at the point $z_0=-\frac{i}{2}$.

9. Find a conformal mapping $w=f(z)$ of the unit disk onto itself which satisfies the conditions

$$w\left(\frac{1}{2}\right)=0, \quad \operatorname{Arg} w'\left(\frac{1}{2}\right)=\frac{\pi}{2}.$$

Determine the dilation coefficient at the point $z_0=\frac{1}{2}$.

10. Find a mapping of the upper half-plane $\operatorname{Im} z>0$ onto the disk $|w|<1$, satisfying the conditions

$$w(2i)=0 \quad \text{and} \quad \operatorname{Arg} w'(2i)=\pi.$$

Also find the dilation coefficient at the point $z_0=2i$.

11. Find a conformal mapping $w=f(z)$ of the upper half-plane $\operatorname{Im} z>0$ onto the unit disk $|w|<1$ which satisfies the condtions

$$w(i)=0, \quad \operatorname{Arg} w'(i)=-\frac{\pi}{2}.$$

Determine the dilation coefficient at the point $z_0=i$.

12. Find a Möbius transformation $f(z)$ mapping the interior of the unit disk $|z|<1$ onto the upper half-plane $\operatorname{Im} w>0$.

13. Find a Möbius transformation $f(z)$ mapping the interior of the unit disk $|z|<1$ onto the upper half-plane $\operatorname{Im} w>0$ in such a way that the upper half of the unit circle is mapped onto the positive real axis.

4.2 The Power Function. The Concept of Riemann Surface

Let us look carefully at the power function

$$w = z^n, \tag{4.13}$$

where n is a positive integer. The derivative $w' = nz^{n-1}$ exists and is nonzero at all points except 0 and ∞. Therefore the mapping performed by this function is conformal everywhere except $z = 0$ and $z = \infty$. Writing z and w in exponential form, $z = re^{i\phi}$ and $w = \rho e^{i\theta}$, leads to the equations

$$\rho = r^n \quad \text{and} \quad \theta = n\phi$$

(we have already considered the power function in the case $n = 2$ in Example 2.3). From this it is clear that the circle $|z| = r$ goes to the circle $|w| = r^n$. Also, an angle of measure α, with $0 < \alpha < \frac{2\pi}{n}$, with vertex at the origin in the z-plane, goes to an angle of measure $n\alpha$ in the w-plane; so we see that conformality is violated at $z = 0$, since angles at that point are not preserved, but rather multiplied by n. It is easy to show also that the conformality fails at $z = \infty$ (the reader should try it).

Suppose points z_1 and z_2 satisfy $z_2 = z_1 e^{i2\pi/n}$, assuming $n > 1$; in other words, z_2 is the result of rotating z_1 about the origin through an angle of $\frac{2\pi}{n}$. Then obviously $z_1 \neq z_2$, and

$$z_2^n = z_1^n e^{i2\pi} = z_1^n,$$

so the function $f(z) = z^n$ is not 1-1 in $\mathbb{C}$. However, it is 1-1 on any sector of the complex plane with angle $\alpha < \frac{2\pi}{n}$ and the vertex at the origin. We are almost ready to introduce the inverse to the power function, but we need one more definition.

A multivalued function of a complex variable is a rule that associates to a complex number z, a *set* of complex numbers $\{w_k\}$; this set may be of any size, including infinite.

All functions considered so far have been single-valued, with the exception of the $\arg z$ function; that function is multivalued:

$$\arg z = \operatorname{Arg} z + 2\pi k,$$

where $\operatorname{Arg} z$ is the principal value of the argument, and k is any integer. From this point in the text, when the term *function* is used without any qualification, it is understood to refer to a single-valued function; if a function is to be multivalued, that will always be stipulated explicitly.

Now let the function $w = f(z)$ maps the domain D onto the domain E. The *inverse of the function* $w = f(z)$ is the function (which is generally speaking multivalued) $z = g(w)$, defined on the domain E, which associates to every $w \in E$ the set of $z \in D$ for which $w = f(z)$:

$$g(w) = \{\, z : f(z) = w \,\}.$$

In other words, the inverse to the function $w = f(z)$ associates to each $w \in E$ all the preimages of w in D.

If the function $w = f(z)$ is 1-1 on D, then the inverse function will be single-valued (and also 1-1) on E; if $w = f(z)$ is not 1-1, then the inverse function will be multivalued. For example, the inverse to the function $w = z^n$ is the multivalued function $z = \sqrt[n]{w}$: each value of w, besides 0 and ∞, is associated with n different nth roots, defined in the formula (1.17). The numbers 0 and ∞ each have only one root: $\sqrt[n]{0} = 0$ and $\sqrt[n]{\infty} = \infty$.

Theorem 4.9. *Let the function $w = f(z)$ be univalent on the domain D, mapping D onto a domain E, and suppose $f'(z) \neq 0$. Then the inverse function $z = g(w)$ is also univalent on the domain E, and*

$$g'(w) = \frac{1}{f'(z)}. \tag{4.14}$$

Proof. Let us fix an arbitrary point $z \in D$, and consider an increment $\Delta z \neq 0$. Then, because the function $w = f(z)$ is 1-1, the corresponding increment $\Delta w = f(z + \Delta z) - f(z)$ is also nonzero. Therefore

$$\frac{g(w + \Delta w) - g(w)}{\Delta w} = \frac{\Delta z}{\Delta w} = \frac{1}{\frac{\Delta w}{\Delta z}}.$$

Since $f'(z) = \lim_{\Delta z \to 0} \frac{\Delta w}{\Delta z} \neq 0$ and the function $f(z)$ is a bijection, it is true for the inverse function that $\Delta g(w) = \Delta z \to 0$ as $\Delta w \to 0$. Hence

$$g'(w) = \lim_{\Delta w \to 0} \frac{\Delta z}{\Delta w} = \frac{1}{\lim\limits_{\Delta z \to 0} \frac{\Delta w}{\Delta z}} = \frac{1}{f'(z)},$$

as required. □

Now we have an awkward notational issue: the input to the inverse function $z = g(w)$ is the variable w, although by convention we like to denote the input to a function by z. So for consistency with the convention we will write $w = g(z)$; for example, the inverse to the function $w = z^n$ is written as $w = \sqrt[n]{z}$.

Let us investigate the function $\sqrt[n]{z}$ further. As mentioned above, it is multivalued. Nevertheless it is possible to define this function on a set different from, and more complicated than, $\mathbb{C}$, but which allows the function $\sqrt[n]{z}$ to be a continuous bijection.

Let us construct this set. Take n copies of the complex plane, denoted D_0, D_1, ..., D_{n-1}, called *sheets*. We cut each of these sheets along the positive real axis, that is we *remove* the positive real axis, producing two edges in each sheet: an upper edge (leading to $y > 0$) and a lower edge (leading to $y < 0$). Then we lay all the sheets on top of each other (Fig. 32 illustrates the case $n = 4$). Next we take the lower edge of D_0 and glue it to the upper edge of D_1; the "glue" consists of copy of the positive real axis which binds to both

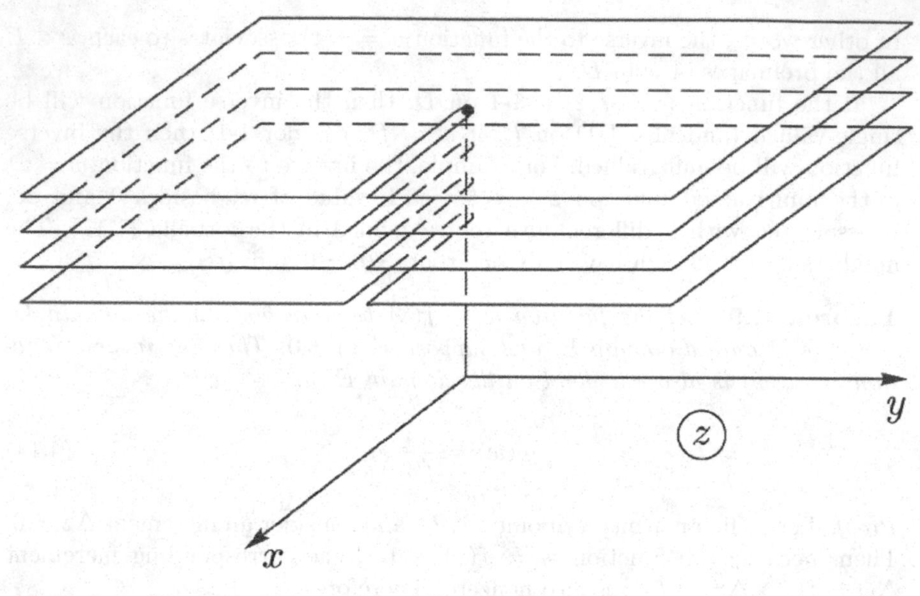

Fig. 32

of these sheets. Then glue the lower edge of D_1 to the upper edge of D_2; and continue in this way until the lower edge of D_{n-2} is glued to the upper edge of D_{n-1}. Finally we glue the one remaining free lower edge, of D_{n-1}, to the one remaining free upper edge, of D_0.

If you are following these instructions at home with paper and glue, you will find that last step is physically impossible (at least in three dimensions), because after the gluings in the previous steps, D_0 is now separated from D_{n-1} by the intervening sheets. But we will agree to consider the lower edge of D_{n-1} as glued to the upper edge of D_0, and to consider that there are no intersections between different sheets—in particular, the points of the glue, i.e. copies of the positive real axis, are distinct at each level.

The finished construction is shown in Fig. 33; this is called the *Riemann surface* for the function $w = \sqrt[n]{z}$. Over each point in the complex plane, except 0 and ∞, lie exactly n points in the Riemann surface. This includes each point on the positive real axis $x > 0$, since all the gluings which lie over it are considered nonintersecting. Only two points do not have this property: $z = 0$ and $z = \infty$. All sheets of the Riemann surface are considered glued together at the points lying above $z = 0$ and $z = \infty$.

Now we define the function $w = \sqrt[n]{z}$ on the new Riemann surface. Recall that if $z = re^{i\phi}$ then the nth roots of z are given by (formula (1.17))

$$w = \sqrt[n]{z} = \sqrt[n]{r}\left(\cos\frac{\phi + 2\pi k}{n} + i\sin\frac{\phi + 2\pi k}{n}\right), \tag{4.15}$$

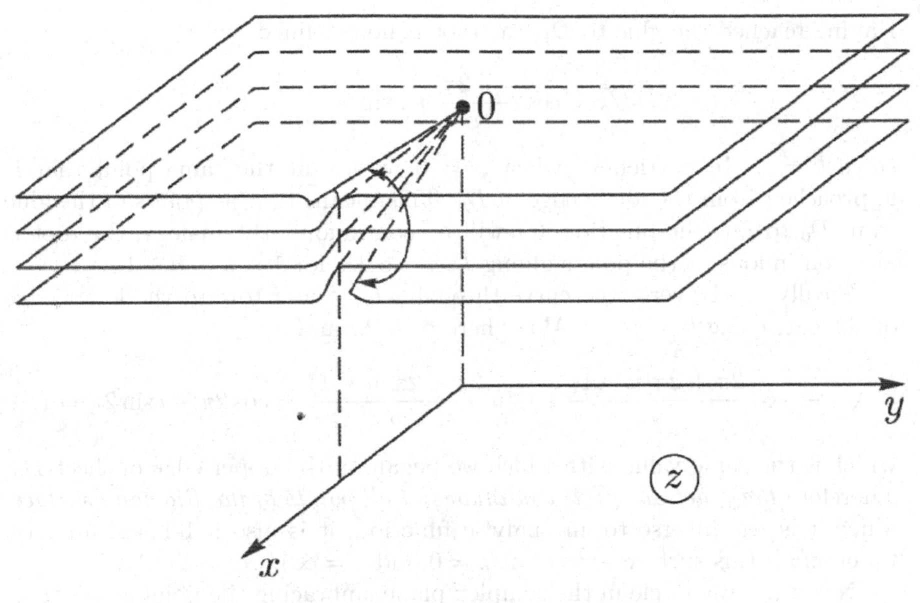

Fig. 33

where $k = 0, \pm 1, \pm 2, \ldots$. The angle ϕ can be chosen from any interval of width 2π; we find it convenient to use $0 \le \phi < 2\pi$.

The points $z = re^{i\phi}$ lying in the sheet D_0 along with the glue to D_{n-1}, we associate with the value of the root obtained using $k = 0$; the points lying in the sheet D_1 along with the glue to D_0, we associate with the value of the root using $k = 1$. And so on: points lying in D_k, $1 \le k \le n-1$, along with the glue to D_{k-1}, are associated with the root using the given k. The function constructed this way is a single-valued function on the Riemann surface.

It is not hard to show that this function is a bijective mapping from the Riemann surface onto the complex plane. In fact, the sheet D_k will be mapped to the sector $k\frac{2\pi}{n} < \phi < (k+1)\frac{2\pi}{n}$, and the glue will be mapped to the rays between adjacent sectors. So the entire complex plane is filled by images of points in the Riemann surface.

Now we will show that this mapping is continuous. Let z be a point in D_k (*not* in the glue); then continuity at this point follows immediately from formula (4.15) with fixed k. To demonstrate continuity at points in the glue, consider a curve on the Riemann surface consisting of points lying above the circle $|z| = 1$ in the complex plane. We begin traversing this curve at the point z which lies on the glue bound to the upper edge of sheet D_0; here $r = 1$, $\phi = 0$, and $k = 0$, so $\sqrt[n]{r} = 1$. From this point we traverse the curve along its branch through sheet D_0, so that $\phi \to 2\pi$ and

$$\sqrt[n]{z} = \cos\frac{\phi}{n} + i\sin\frac{\phi}{n} \to \cos\frac{2\pi}{n} + i\sin\frac{2\pi}{n}.$$

Having reached the glue to D_1, the root is now defined by

$$\sqrt[n]{z} = \cos\frac{\phi + 2\pi}{n} + i\sin\frac{\phi + 2\pi}{n},$$

since $k = 1$. In particular, when $\phi = 0$ we are at the same point that is approached from the lower edge of D_0. This means that at points of the glue from D_0 to D_1, the function $\sqrt[n]{z}$ will be continuous. By analogy, the root is also continuous at the points gluing D_{k-1} to D_k for $1 \le k \le n-1$.

Finally, we traverse the curve through D_{n-1}, and toward the lower edge of the cut, using $k = n-1$. Along here $\phi \to 2\pi$ and

$$\sqrt[n]{z} \to \cos\frac{2\pi + 2\pi(n-1)}{n} + i\sin\frac{2\pi + 2\pi(n-1)}{n} = \cos 2\pi + i\sin 2\pi = 1,$$

which is the same value with which we began on the upper edge of sheet D_0. Therefore *the function* $\sqrt[n]{z}$ *is continuous at all points of the Riemann surface.* Since it is the inverse to an analytic function, it is also a 1-1 and analytic function on this surface (except at $z = 0$ and $z = \infty$).

Now take any circle in the complex plane embracing the point $z = 0$. This circle can also be regarded as embracing $z = \infty$. As we traverse the curve in the Riemann surface consisting of points above this circle, we move from one sheet of the Riemann surface to another. For this reason the points $z = 0$ and $z = \infty$ are called *branch points* of the Riemann surface. No other points in $\mathbb{C}$ have this property: any point other than $z = 0$ and $z = \infty$ will have circles centered around it and not containing 0. When lifted to the Riemann surface, they become n disconnected circles, each never exiting a single sheet, and not a connected path through all the sheets.

Let $F(z)$ be a multivalued function on a domain D. A function f which is single-valued and analytic on the same domain D is said to be a *regular branch* (or sometimes just branch) of F if the value of $f(z)$, at each point z in D, matches one of the values of $F(z)$.

A multivalued function $F(z)$ is single-valued and analytic on its Riemann surface, with the exception of branch points. So if a regular branch exists on a domain D, it must also be possible to place this domain somewhere in the Riemann surface, without cutting it or touching any branch points. The domain D must either fit entirely within one sheet, or straddle several sheets around their glue binding (like a carpet over adjacent stair steps).

For example, the ring $1 < |z| < 2$ could not be placed in the Riemann surface of the function $F(z) = \sqrt[n]{z}$ $(n \ge 2)$, because the points over the positive real axis would have to be in different sheets at the same time, which is impossible. But if we cut the ring along some radius, then it would work; moreover the placement could then be done in n distinct ways, allowing n distinct branches of the function $\sqrt[n]{z}$ to be constructed.

To construct a particular regular branch, it suffices to set the value of the function at any one point in the domain D. That would determine which sheet of the Riemann surface that point would be placed in, and thus determine the placement of the entire domain D.

Example 4.10 Work out a regular branch $f(z)$ of the function $w = \sqrt[4]{z}$ on the domain

$$D = \left\{ z = re^{i\phi} : -\frac{3\pi}{2} < \phi < \frac{\pi}{2} \right\},$$

and satisfying the condition that $f(1) = i$. Also find $f(-1)$.

Solution The domain D is the complex plane with a cut removing the positive imaginary axis ($y \geq 0$); so defining a regular branch on D is possible. By formula (4.15)

$$F(z) = \sqrt[4]{z} = \sqrt[4]{r}\left(\cos\frac{\phi + 2\pi k}{4} + i\sin\frac{\phi + 2\pi k}{4}\right),$$

$$\text{where} \quad k = 0, 1, 2, 3 \quad \text{and} \quad -\frac{3\pi}{2} < \phi < \frac{\pi}{2}.$$

In order to define the branch $f(z)$ we must find the right value of k. Since $f(1) = i$, then $\phi = 0$ and $r = 1$; and we get

$$i = \cos\frac{2\pi k}{4} + i\sin\frac{2\pi k}{4},$$

from which it follows that $k = 1$. So the required branch is

$$f(z) = \sqrt[4]{r}\left(\cos\frac{\phi + 2\pi}{4} + i\sin\frac{\phi + 2\pi}{4}\right) \quad \text{for} \quad -\frac{3\pi}{2} < \phi < \frac{\pi}{2}.$$

In particular,

$$f(-1) = \sqrt[4]{1}\left(\cos\frac{-\pi + 2\pi}{4} + i\sin\frac{-\pi + 2\pi}{4}\right) = \frac{\sqrt{2}}{2} + i\frac{\sqrt{2}}{2}.$$

We carried out our construction of the Riemann surface for the function $w = \sqrt[n]{z}$ by cutting the complex plane along the positive real axis. It is worth pointing out that the choice of which curve to cut along is not critical: an analogous construction can be carried out by cutting along, for example, any ray starting at the origin.

Problems

1. Work out a regular branch $f(z)$ of the function $w = \sqrt[3]{z}$ satisfying the condition $f(-1) = \frac{1}{2} - i\frac{\sqrt{3}}{2}$, and find $f(1)$ if

(a) $D = \left\{ z = re^{i\phi} : -\frac{\pi}{2} < \phi < \frac{3\pi}{2} \right\}$; (b) $D = \left\{ z = re^{i\phi} : -\frac{3\pi}{2} < \phi < \frac{\pi}{2} \right\}$.

2. Let D be the complex plane $\mathbb{C}$ with a cut removing the closed interval $[-1, 1]$ (that is D is the exterior of this interval). Let $f(z)$ be the branch of the function $w = \sqrt{z^2 - 1}$ defined by the formula

$$f(z) = |z^2 - 1|^{1/2} e^{i\phi}, \quad \text{where } \phi = \tfrac{1}{2}\big(\operatorname{Arg}(z-1) + \operatorname{Arg}(z+1)\big).$$

Prove that: (a) $f(z)$ is analytic and single-valued in D; (b) $f(z)$ is not continuous (and therefore is not analytic) in $\mathbb{C}$.

3. What is wrong in the following chain of equalities?

$$1 = \sqrt{1} = \sqrt{(-1)^2} = -1.$$

4.3 Exponential and Logarithmic Functions

4.3.1 Exponential function

The *exponential function* $f(z) = e^z$ can be defined for any $z = x + iy$ by

$$e^z = e^{x+iy} = e^x(\cos y + i \sin y). \tag{4.16}$$

The second equality in (4.16) holds if take $e^{x+iy} = e^x e^{iy}$ to be true by definition, and then apply the Euler formula (1.19). From (4.16) it follows that

$$|e^z| = |e^{x+iy}| = e^x \quad \text{and} \quad \arg e^z = y + 2\pi n.$$

Using definition (4.16) and the properties of the function $e^{i\phi}$ it is easy to show that the exponential function retains its usual properties:

$$e^{z_1+z_2} = e^{z_1} e^{z_2}; \quad e^{z_1-z_2} = \frac{e^{z_1}}{e^{z_2}}; \quad (e^z)^n = e^{nz}. \tag{4.17}$$

We will prove the first of these. Let $z_1 = x_1 + iy_1$ and $z_2 = x_2 + iy_2$. Applying (4.16) and (1.20) we get

$$e^{z_1} e^{z_2} = e^{x_1} e^{iy_1} e^{x_2} e^{iy_2} = e^{x_1} e^{x_2} e^{iy_1} e^{iy_2} = e^{x_1+x_2} e^{i(y_1+y_2)} = e^{z_1+z_2}.$$

The reader can easily prove the other two (problem 6).

Now we will show that the function e^z is analytic on the entire complex plane $\mathbb{C}$. For this we must verify that the Cauchy-Riemann conditions (3.4) are satisfied. So let $w = u + iv$, and then from (4.16) we have

$$u + iv = e^x \cos y + ie^x \sin y,$$

and

$$u = e^x \cos y, \qquad v = e^x \sin y;$$

$$\frac{\partial u}{\partial x} = \frac{\partial v}{\partial y} = e^x \cos y, \qquad \frac{\partial u}{\partial y} = -\frac{\partial v}{\partial x} = -e^x \sin y.$$

So the Cauchy-Riemann conditions are satisfied, and therefore the function e^z is analytic. To compute the derivative $(e^z)'$, we will exploit the independence of the derivative from the direction in which it is taken, and simply compute the derivative in the direction of the x-axis:

$$(e^z)' = \frac{\partial}{\partial x}(e^x(\cos y + i \sin y)) = e^x(\cos y + i \sin y) = e^z.$$

Therefore we get the usual formula for the derivative:

$$(e^z)' = e^z.$$

However, as a function of a complex variable, e^z has a property not shared by it real-valued cousin e^x: *the function e^z is periodic with purely imaginary period $2\pi i$.* Indeed, for an arbitrary integer n,

$$e^{z+2\pi ni} = e^x(\cos(y + 2\pi n) + i \sin(y + 2\pi n)) = e^x(\cos y + i \sin y) = e^z.$$

One consequence of the periodicity of this function $w = e^z$ is that it is not

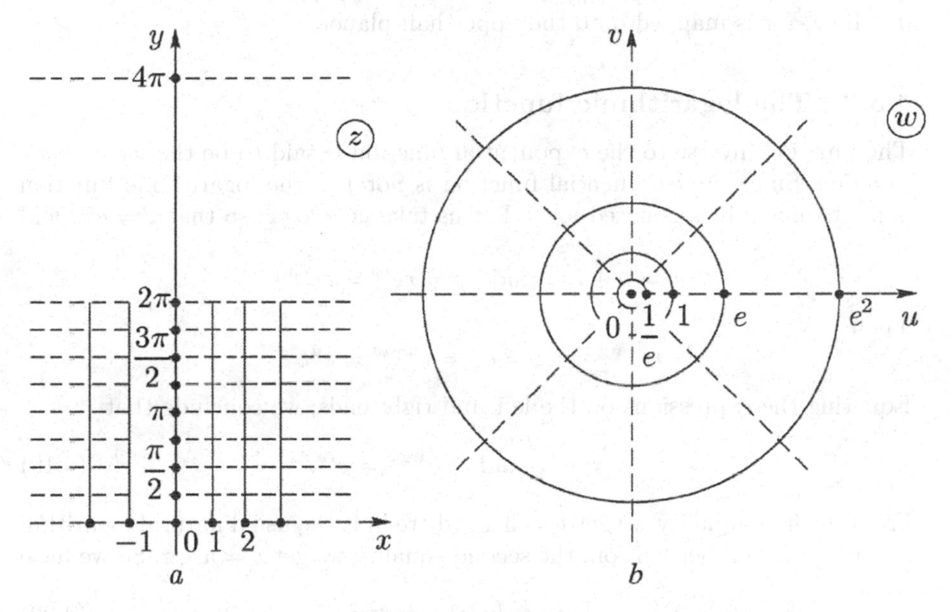

Fig. 34

1-1 on the entire complex plane. On what kinds of domains *is* this function 1-1? To explore this, let us again set $z_1 = x_1 + iy_1$ and $z_2 = x_2 + iy_2$. By (4.16), the equality $e^{z_1} = e^{z_2}$ is equivalent to the following conditions:

$$e^{x_1} = e^{x_2}, \quad \cos y_1 = \cos y_2, \quad \text{and} \quad \sin y_1 = \sin y_2,$$

which implies that $x_1 = x_2$ and $y_1 = y_2 + 2\pi n$, where n is an arbitrary integer; or, in other words,

$$z_1 - z_2 = 2\pi ni. \tag{4.18}$$

So the mapping $w = e^z$ will be bijective on the domain D if and only if D contains no pair of points for which (4.18) holds. An example of such a domain is any horizontal strip of width 2π, for example

$$\{ z : -\infty < x < \infty,\ 2\pi k < y < 2\pi(k+1) \},$$

where k is any integer. Each such strip is mapped to a set of values $w = e^z = e^x e^{iy} = \rho e^{i\theta}$ for which

$$0 < \rho < \infty \quad \text{and} \quad 2\pi k < \theta < 2\pi(k+1).$$

(Here, $\rho = e^x$ and $\theta = y$.) This set is the entire complex plane, but with the positive real axis cut out. The horizontal lines $y = y_0$ (shown as dotted in Fig. 34, *a*) go to the rays $\theta = y_0$ (Fig. 34, *b*), and the vertical segments $x = x_0$, $2\pi k < y < 2\pi(k+1)$ (shown for the the case $k = 0$) go to the circles $\rho = e^{x_0}$ (which however omit the points on the positive real axis); the strip $0 < \operatorname{Im} z < h < 2\pi$ is mapped to sector $0 < \theta < h$. Note that the strip $0 < \operatorname{Im} z < \pi$ is mapped onto the upper half-plane.

4.3.2 The logarithmic function

The function inverse to the exponential function is said to be the *logarithmic function.* Since the exponential function is not 1-1, the logarithmic function is multivalued; it is denoted $\log z$. Let us take $w = \log z$, so that $z = e^w$, and set

$$w = u + iv \quad \text{and} \quad z = re^{i\phi} = re^{i \arg z}.$$

Then

$$re^{i \arg z} = z = e^w = e^{u+iv} = e^u e^{iv}.$$

Equating the expressions on the left and right ends, we conclude that

$$r = e^u \quad \text{and} \quad e^{i \arg z} = e^{iv}. \tag{4.19}$$

From the first equality we get $u = \ln r$, where ln is the usual natural logarithm of a positive number r. From the second equality we get $v = \arg z$. So we have

$$\log z = \ln |z| + i \arg z. \tag{4.20}$$

This formula associates every complex number different from 0 and ∞ with an infinite set of values $\log z$, each of which differs from the others by a mulitple of $2\pi i$. Recall that $\arg z$ can be written in terms of its principal value $-\pi < \operatorname{Arg} z \leq \pi$,

$$\arg z = \operatorname{Arg} z + 2\pi k, \quad k = 0 \pm 1, \pm 2, \ldots.$$

This allows us to write $\log z$ in a similar way:

$$\log z = \ln |z| + i(\operatorname{Arg} z + 2\pi k), \quad k = 0 \pm 1, \pm 2, \ldots. \tag{4.21}$$

For any fixed value of k, the function $\log z$ is single-valued and continuous on the complex plane minus the negative real axis; it also analytic since it is the inverse of the analytic function e^z. So for each value of k, the formula (4.21) defines a regular branch of the multivalued function $\log z$. This branch maps bijectively the plane cut along the negative real axis onto the strip

$$-\pi + 2\pi k < \operatorname{Im} w < \pi + 2\pi k.$$

The branch defined when $k = 0$ is denoted by $\operatorname{Log} z$ and is called *the principal value of the logarithmic function*:

$$\operatorname{Log} z = \ln|z| + i \operatorname{Arg} z. \tag{4.22}$$

For example,

$$\operatorname{Log} i = \ln 1 + i\tfrac{\pi}{2} = i\tfrac{\pi}{2},$$
$$\operatorname{Log}(-i) = \ln 1 - i\tfrac{\pi}{2} = -i\tfrac{\pi}{2}.$$

If we approach the point $z = -1$ from the upper half-plane $y > 0$, then

$$\lim_{y\to 0^+} \operatorname{Log}(-1 + iy) = \ln 1 + i\pi = i\pi;$$

But if we approach from below, then

$$\lim_{y\to 0^-} \operatorname{Log}(-1 + iy) = \ln 1 - i\pi = -i\pi.$$

To form the Riemann surface of the logarithmic function $\log z$, we need infinitely many copies of the complex plane to serve as sheets. We cut each one along the negative real axis, and then glue them together (with copies of the negative real axis) as shown in Fig. 35.

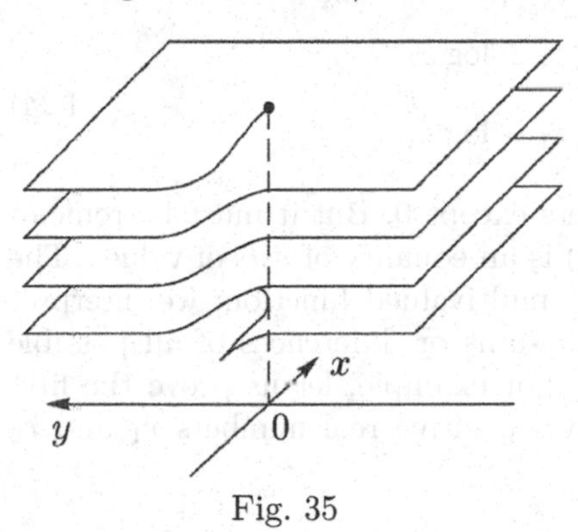

Fig. 35

Lying over (and under!) every point of the complex plane, except 0 and ∞, are infinitely many points of the Riemann surface. At the points $z = 0$ and $z = \infty$, the function $\operatorname{Log} z$ is undefined, and there are no points corresponding to them in the Riemann surface. Those two points are called *branch points of infinite order*.

Fig. 35 illustrates the fact that

$$\lim_{y\to 0^+} \operatorname{Log}(-1+iy) \neq \lim_{y\to 0^-} \operatorname{Log}(-1+iy).$$

For if we assume that the points $-1 + ih$ and $-1 - ih$, $h > 0$, move over a single sheet of the Riemann surface as $h \to 0$, then the limit points turn out to be on different sheets.

As with the power function, for the logarithmic function there are different ways that $\mathbb{C}$ can be cut in order to permit a regular branch to be defined; a cut along any ray starting at the origin would work. Suppose the cut is made along a ray starting at the origin and making the angle θ with the positive x-axis. Then a regular branch can be defined by, for $z = re^{i\phi}$,

$$\operatorname{Log} z = \ln r + i(\phi + 2\pi k), \tag{4.23}$$

where $\theta < \phi < \theta + 2\pi$ and k is any fixed integer. The formula in (4.22) is just the case when $\theta = -\pi$ and $k = 0$.

Example 4.11 Work out a formula for a regular branch $f(z)$ of the function $\log z$, defined on $\mathbb{C}$ with a cut along the ray $\operatorname{Arg} z = \frac{\pi}{2}$; it should also satisfy the condition that $f(1) = 0$. Then find $f(-1)$.

Solution The formula for $f(z)$ will follow that of equation (4.23) with $\frac{\pi}{2} < \phi < \frac{5\pi}{2}$. Then $1 = 1e^{2\pi i}$, and we get

$$0 = f(1) = \ln 1 + i(2\pi + 2\pi k) = i(2\pi + 2\pi k),$$

so evidently $2\pi + 2\pi k = 0$ and k must be -1. So this branch is given by the formula

$$f(re^{i\phi}) = \ln r + i(\phi - 2\pi), \quad \frac{\pi}{2} < \phi < \frac{5\pi}{2}.$$

When $z = -1$, then $r = 1$ and $\phi = \pi$; therefore

$$f(-1) = \ln 1 + i(\pi - 2\pi) = -i\pi.$$

Note that the choice $\frac{\pi}{2} < \phi < \frac{5\pi}{2}$ of the interval for ϕ is not unique—see problem 7.

The multivalued function $\log z$ retains the usual properties of the logarithmic function:

$$\begin{aligned} \log(z_1 z_2) &= \log z_1 + \log z_2, \\ \log\left(\frac{z_1}{z_2}\right) &= \log z_1 - \log z_2, \end{aligned} \tag{4.24}$$

where z_1 and z_2 are any complex numbers except 0. But it must be remembered that each of the equalities in (4.24) is an equality of *sets* of values. The left side of each is the set of values of a multivalued function; we interpret the right side of each as being the set of sums or differences of all possible values of the functions $\log z_1$ and $\log z_2$. For example, let us prove the first equality. Of course we know that any two positive real numbers r_1 and r_2 satisfy $\ln(r_1 r_2) = \ln r_1 + \ln r_2$; therefore

$$\begin{aligned} \log(z_1 z_2) &= \ln |z_1 z_2| + i \arg(z_1 z_2) \\ &= \ln |z_1| + \ln |z_2| + i \arg z_1 + i \arg z_2 = \log z_1 + \log z_2, \end{aligned}$$

(using equality (1.11)), as required.

But at the same time, a regular branch of $\log z$ will violate, for some values of z_1 and z_2, the corresponding two properties for $\log z$.

Example 4.12 Check the validity of $\operatorname{Log}(z \cdot z) = \operatorname{Log} z + \operatorname{Log} z$, or $\operatorname{Log}(z^2) = 2\operatorname{Log} z$, for (a) $z = \frac{e}{\sqrt{2}}(1-i)$, (b) $z = \frac{e}{\sqrt{2}}(-1-i)$.

Solution (a) We see that $|z| = e$ and $\operatorname{Arg} z = -\frac{\pi}{4}$, so

$$\operatorname{Log} z = \ln e + i\left(-\frac{\pi}{4}\right) = 1 - i\frac{\pi}{4}, \quad 2\operatorname{Log} z = 2 - i\frac{\pi}{2}.$$

Now we find $\operatorname{Log} z^2$: since $z^2 = -ie^2$, we have $|z|^2 = e^2$ and $\operatorname{Arg}(z^2) = -\frac{\pi}{2}$, so

$$\operatorname{Log}(z^2) = \ln e^2 + i\left(-\frac{\pi}{2}\right) = 2 - i\frac{\pi}{2}.$$

Thus the equality $\operatorname{Log}(z^2) = 2\operatorname{Log} z$ holds.

(b) Here $|z| = e$ and $\operatorname{Arg} z = -\frac{3\pi}{4}$, so

$$\operatorname{Log} z = 1 - i\frac{3\pi}{4}, \quad 2\operatorname{Log} z = 2 - i\frac{3\pi}{2}.$$

But $z^2 = ie^2$, so we have $|z|^2 = e^2$ and $\operatorname{Arg}(z^2) = \frac{\pi}{2}$, making $\operatorname{Log}(z^2) = 2 + i\frac{\pi}{2}$. Thus,

$$\operatorname{Log}(z^2) = 2 + i\frac{\pi}{2} \neq 2 - i\frac{3\pi}{2} = 2\operatorname{Log} z.$$

Note that $2 + i\frac{\pi}{2}$ and $2 - i\frac{3\pi}{2}$ are both values of the multivalued function $\log(z^2)$ when $z^2 = ie^2$; so this example does not contradict equations (4.24).

To conclude this section, we will show that derivative of any regular branch $f(z)$ of the logarithm is given by the formula

$$f'(z) = \frac{1}{z},$$

analogous to the formula in the case of the real logarithm ln. This comes from the equation $(e^z)' = e^z$ and the formula (4.14) for the derivative of an inverse function. Here the inverse to $w = f(z)$ is the function $z = e^w$, and so from (4.14) we get

$$f'(z) = \frac{1}{g'(w)} = \frac{1}{(e^w)'} = \frac{1}{e^w} = \frac{1}{z}. \tag{4.25}$$

Problems

1. Find the principal value and all values of the logarithmic function at the given point: (*a*) $z = 1 + i$; (*b*) $z = -1 + i$; (*c*) $z = -1 - i\sqrt{3}$.

2. Find all roots of the following equations:

$$(a)\ e^z = -1; \quad (b)\ e^z = e; \quad (c)\ e^z = -ie.$$

3. Find a conformal mapping of the strip between the lines $y = x + 1$ and $y = x + 3$ (Fig. 107) onto the unit disk.

4. Find conformal mappings of the given strips onto the unit disk $|w| < 1$.

a) between the lines $y = -x$ and $y = -x + \pi$;

b) between the lines $y = \sqrt{3}x$ and $y = \sqrt{3}x + \pi$.

5. Find conformal mappings from the given strips to the unit disk.

a) $-1 < \operatorname{Im} z < 3$; b) $1 < \operatorname{Re} z < 3$; c) $-\pi < \operatorname{Re} z < 3\pi$.

6. Prove the last two identities in (4.17).

7. Use reasoning similar to that used in the Example 4.11 to find the formula for the regular branch of the function $\log z$ on $-\frac{3\pi}{2} < \phi < \frac{\pi}{2}$, still satisfying the condition $f(1) = 0$. Verify that value of $f(-1)$ will be the same as in Example 4.11. (In fact the solution can be derived from the previous one by making the substitution $\phi_1 = \phi - 2\pi$.)

8. Show that $\operatorname{Log}(z_1 z_2) = \operatorname{Log} z_1 + \operatorname{Log} z_2$ if and only if $\operatorname{Arg}(z_1 z_2) = \operatorname{Arg} z_1 + \operatorname{Arg} z_2$. Formulate and prove similar criteria for $\operatorname{Log}(z_1/z_2)$ and $\operatorname{Log}(z^n) = n \operatorname{Log} z$, where n is a positive integer.

9. What is wrong in the following chain of equalities?

$$0 = \operatorname{Log} 1 = \operatorname{Log}(-1)^2 = 2 \operatorname{Log}(-1) = 2\pi i.$$

10. For which z the equality $\operatorname{Log} z = \operatorname{Log}(-1) + \operatorname{Log}(-z)$ is valid?

4.4 Power, Trigonometric, and Other Functions

4.4.1 The general power function

This is the function of the form $w = z^a$, where a is a fixed complex number $a = \alpha + i\beta$; it is defined by

$$z^a = e^{a \log z}. \tag{4.26}$$

If $z = re^{i\phi}$, we have $\log z = \ln r + i(\phi + 2\pi k)$ for all $k \in \mathbb{Z}$. Plugging this in to (4.26),

$$z^a = e^{(\alpha+i\beta)(\ln r + i(\phi+2\pi k))} = e^{\alpha \ln r - \beta(\phi+2\pi k)} e^{i(\alpha(\phi+2\pi k)+\beta \ln r)},$$

where we have separated the real and imaginary parts of the exponent. From this we see that, as long as $\beta \neq 0$, the modulus $|z^a| = e^{\alpha \ln r - \beta(\phi + 2\pi k)}$ assumes an infinite number of different values. So, like the logarithm, the function z^a is infinitely multivalued.

Example 4.13 Find all values of the function $w = z^i$ at the point $z = i$.

Solution Since $\log i = \ln 1 + i(\frac{\pi}{2} + 2\pi k)$, we get

$$i^i = e^{i \log i} = e^{i^2(\frac{\pi}{2} + 2\pi k)} = e^{-\frac{\pi}{2} - 2\pi k}, \quad k = 0, \pm 1, \pm 2, \ldots.$$

So all the values of i^i turn out, rather ironically, to be real.

When $\beta = 0$ we get

$$z^a = e^{\alpha \ln r} e^{i\alpha(\phi + 2\pi k)}. \tag{4.27}$$

Here we see the values vary only in the argument $\theta_k = \alpha(\phi + 2\pi k)$. If α is a rational number, i.e. expressible as an irreducible fraction $\alpha = \frac{m}{n}$ (m and n are integers without common factors), then the θ_k produce only n distinct values of $z^a = e^{\alpha \ln r} e^{i\theta_k}$; in fact we can restrict the values of θ_k to just

$$\theta_k = \frac{m}{n}\phi + \frac{2\pi k m}{n}, \quad \text{where} \quad k = 0, 1, 2, \ldots, n-1,$$

since when $k = n, n+1, \ldots$ the new values produced differ from the first n by multiples of 2π, and therefore lead to no new values of $e^{i\theta_k}$ or z^a. So when $\alpha = \frac{m}{n}$ the formula (4.27) gives us

$$z^{\frac{m}{n}} = r^{\frac{m}{n}} \left(\cos \frac{m(\phi + 2\pi k)}{n} + i \sin \frac{m(\phi + 2\pi k)}{n} \right), \quad k = 0, 1, 2, \ldots, n-1$$

(we have used the equality $e^{\alpha \ln r} = r^\alpha$ for real numbers r and α). Comparing this formula with the root formula (1.17), we see that

$$z^{\frac{m}{n}} = \sqrt[n]{z^m}.$$

So for rational powers a, the function z^a is *finitely* multivalued.

When the exponent $a = \alpha$ is real and irrational, then no two distinct values of $\theta_k = \alpha(\phi + 2\pi k)$ differ from one another by a multiple of 2π. For if θ_{k_1} and θ_{k_2} were such values, then $\theta_{k_1} - \theta_{k_2} = 2\pi k_1 \alpha - 2\pi k_2 \alpha = 2\pi l$, where l is an integer and $k_1 \neq k_2$; which would mean

$$\alpha = \frac{l}{k_1 - k_2}$$

is rational, contrary to our assumption. Therefore when the power a is real and irrational, the power function z^a is infinitely multivalued; and its Riemann surface will be similar to that of the logarithm.

In fact, because of its definition in terms of the log function, the general power function $w = r^a$ admits the family of regular branches in the same domains as the logarithm: for example, in the plane cut along a ray. The branch

$$e^{a\,\mathrm{Log}\,z} = e^{a\ln|z|+i\,\mathrm{Arg}\,z},$$

defined on the plane cut along the negative real axis, is called the *principal branch of the power function.* If z^{a+b}, z^a, and z^b are regular branches of the power function, all defined using the same regular branch of the logarithm, then the equalities

$$z^{a+b} = z^a z^b \quad \text{and} \quad z^{a-b} = \frac{z^a}{z^b}$$

are valid. In fact, let $f(z)$ be a regular branch of the logarithm $\log z$, and we will derive the first equality (the second can be derived analogously):

$$z^{a+b} = e^{(a+b)f(z)} = e^{af(z)+bf(z)} = e^{af(z)}e^{bf(z)} = z^a z^b.$$

Applying the chain rule and using equalities $f'(z) = 1/z$, $z = e^{f(z)}$, and properties of the exponential function, we get

$$(z^a)' = \left(e^{af(z)}\right)' = e^{af(z)}\cdot af'(z) = ae^{af(z)}\frac{1}{z} = \frac{ae^{af(z)}}{e^{f(z)}} = ae^{(a-1)f(z)} = az^{a-1}$$

for every regular branch of the power function. We have arrived at the usual formula for the derivative of the power function:

$$(z^a)' = az^{a-1}.$$

4.4.2 The trigonometric functions

For real values of x, Euler's formula (1.19) gives us

$$e^{ix} = \cos x + i\sin x,$$
$$e^{-ix} = \cos x - i\sin x.$$

From these equalities we easily get

$$\cos x = \frac{e^{ix}+e^{-ix}}{2}, \quad \sin x = \frac{e^{ix}-e^{-ix}}{2i}.$$

These formulas will serve as the basis for the following definitions:

$$\cos z = \frac{e^{iz}+e^{-iz}}{2}, \quad \sin z = \frac{e^{iz}-e^{-iz}}{2i},$$
$$\tan z = \frac{\sin z}{\cos z}, \quad \cot z = \frac{\cos z}{\sin z}. \tag{4.28}$$

Defined in this way, these functions preserve many of the properties of the trigonometric functions of real variables. From the periodicity of the function

e^z it follows that $\sin z$ and $\cos z$ are periodic with period 2π, and that $\tan z$ and $\cot z$ are periodic with period π. It is also clear from the definitions that $\sin z$ is an odd function, i.e. $\sin(-z) = -\sin z$:

$$\sin(-z) = \frac{e^{i(-z)} - e^{-i(-z)}}{2i} = -\frac{e^{iz} - e^{-iz}}{2i} = -\sin z;$$

while $\cos z$ is an even function, i.e. $\cos(-z) = \cos z$ (problem 5). Also, the usual trig identities hold for the functions defined in (4.28); in particular

$$\begin{aligned} \sin^2 z + \cos^2 z &= 1, \\ \sin(z_1 + z_2) &= \sin z_1 \cos z_2 + \cos z_1 \sin z_2, \\ \sin(\tfrac{\pi}{2} - z) &= \cos z, \end{aligned} \tag{4.29}$$

etc. These identities all flow algebraically from (4.28). For example, let us prove the second of these formulas:

$$\begin{aligned} &\sin z_1 \cos z_2 + \cos z_1 \sin z_2 \\ &= \frac{e^{iz_1} - e^{-iz_1}}{2i} \cdot \frac{e^{iz_2} + e^{-iz_2}}{2} + \frac{e^{iz_1} + e^{-iz_1}}{2} \cdot \frac{e^{iz_2} - e^{-iz_2}}{2i} \\ &= \frac{e^{i(z_1+z_2)} - e^{-i(z_1+z_2)}}{2i} = \sin(z_1 + z_2). \end{aligned}$$

The functions $\sin z$ and $\cos z$ are analytic on the entire plane $\mathbb{C}$, and they have the familiar differentiation formulas:

$$(\sin z)' = \cos z, \quad (\cos z)' = -\sin z.$$

Let us prove, as an example, the formula for the derivative of $\sin z$:

$$\begin{aligned} (\sin z)' &= \left(\frac{e^{iz} - e^{-iz}}{2i}\right)' = \frac{1}{2i}((e^{iz})' - (e^{-iz})') \\ &= \frac{1}{2i}(ie^{iz} + ie^{-iz}) = \frac{1}{2}(e^{iz} + e^{-iz}) = \cos z. \end{aligned}$$

Using the quotient rule we get

$$(\tan z)' = \frac{1}{\cos^2 z} = \sec^2 z, \quad (\cot z)' = -\frac{1}{\sin^2 z} = -\csc^2 z.$$

However, not every familiar property of the real trig functions are preserved in their extension to the complex plane. In particular, $\sin z$ and $\cos z$ can take values whose modulus is greater than 1. For example,

$$\begin{aligned} \cos i &= \frac{e^{i^2} + e^{-i^2}}{2} = \frac{e^{-1} + e^{1}}{2} \approx 1.54, \\ \sin i &= \frac{e^{-1} - e^{1}}{2i} \approx -1.17i. \end{aligned}$$

Note, however that $1.54^2 + (-1.17i)^2 \approx 2.37 - 1.37 = 1$!

4.4.3 Inverse trig functions

These, of course, are the inverse functions to the trig functions just defined. Since those trig functions were periodic, their inverse functions will be infinitely multivalued. Given that the definitions in (4.28) used simple exponential expressions, their inverses will use logarithms. Let us start with $w = \cos^{-1} z$. From the definition of cos we have

$$z = \cos w = \frac{e^{iw} + e^{-iw}}{2},$$

and after multiplying through by $2e^{iw}$ we get

$$e^{2iw} - 2ze^{iw} + 1 = 0,$$

which is of quadratic type in e^{iw}. The solutions are

$$e^{iw} = z + \sqrt{z^2 - 1}$$

(we omit the $\pm$ before the root, because the square root of a complex number is a two-valued function, not single-valued). Now applying log to both sides we get[3]

$$w = \cos^{-1} z = -i \log(z + \sqrt{z^2 - 1}). \tag{4.30}$$

Note here that

$$(z + \sqrt{z^2 - 1})(z - \sqrt{z^2 - 1}) = 1,$$

i.e. the two different solutions of the quadratic equation are multiplicative inverses of one another, so their logarithms differ only in sign. Because the square root is a two-valued function, the expression $i \log(z + \sqrt{z^2 - 1})$ is equivalent to $\pm i \log(z + \sqrt{z^2 - 1})$, so without a minus sign in front we get the same set of values:

$$w = \cos^{-1} z = i \log(z + \sqrt{z^2 - 1}).$$

Similar formulas can be given for the inverses to the other trig functions:

$$\begin{aligned} \sin^{-1} z &= \frac{\pi}{2} - \cos^{-1} z = \frac{\pi}{2} + i \log(z + \sqrt{z^2 - 1}) \\ &= -i \log(iz + \sqrt{1 - z^2}), \\ \tan^{-1} z &= \frac{\pi}{2} - \cot^{-1} z = \frac{1}{2i} \log \frac{i - z}{i + z}. \end{aligned} \tag{4.31}$$

Example 4.14 Find $\sin^{-1} 2$.

Solution By formula (4.31)

$$\sin^{-1} 2 = \frac{\pi}{2} + i \log(2 \pm \sqrt{3})$$

[3]It is rather unexpected that even the elementary real-valued function $\cos^{-1} x$ of a real variable x, $-1 \leq x \leq 1$, can be written in terms of the complex logarithmic function: $\cos^{-1} x = -i \operatorname{Log}(x + i\sqrt{1 - x^2})$.

(here we have used the $\pm$ sign because $\sqrt{3}$ is usually understood as the positive value). To calculate the logarithm we will use formula (4.20), plugging the values $|z| = 2 \pm \sqrt{3}$ and $\operatorname{Arg} z = 0$, which results in

$$\sin^{-1} 2 = \frac{\pi}{2} + i \ln(2 \pm \sqrt{3}) + 2\pi k = \frac{\pi}{2} \pm i \ln(2 + \sqrt{3}) + 2\pi k,$$

where $\sqrt{3}$ is the positive value, and k is any integer.

Differentiation formulas for regular branches of inverse trig functions are similar to the corresponding functions of real variable. For example, let us derive the formula for the derivative of $w = \sin^{-1} z$. Then $z = \sin w$, and the formula (4.14) implies

$$(\sin^{-1} z)' = \frac{1}{(\sin w)'} = \frac{1}{\cos w} = \frac{1}{\sqrt{1 - \sin^2 w}} = \frac{1}{\sqrt{1 - z^2}}. \tag{4.32}$$

Therefore, the derivative depends on the choice of the branch of the square root. The same result follows from the formulas (4.31) (see problem 11).

Other elementary functions worth mentioning are the *hyperbolic functions* $\sinh z$, $\cosh z$, $\tanh z$, and $\coth z$, defined by

$$\begin{gathered} \sinh z = \frac{e^z - e^{-z}}{2}, \quad \cosh z = \frac{e^z + e^{-z}}{2}, \\ \tanh z = \frac{\sinh z}{\cosh z} = \frac{e^z - e^{-z}}{e^z + e^{-z}}, \quad \coth z = \frac{\cosh z}{\sinh z} = \frac{e^z + e^{-z}}{e^z - e^{-z}}. \end{gathered} \tag{4.33}$$

These all can be easily expressed through trig functions:

$$\begin{aligned} \sinh z &= -i \sin(iz), \quad & \cosh z &= \cos(iz), \\ \tanh z &= -i \tan(iz), \quad & \coth z &= i \cot(iz), \end{aligned}$$

and therefore are not essentially different from the earlier ones.

4.4.4 The Zhukovsky function

The function

$$w = \frac{1}{2}\left(z + \frac{1}{z}\right) \tag{4.34}$$

is called the *Zhukovsky*[4] *function*. This function has many applications in the theory of airfoils (see Sections 8.3.4, 8.3.5), and is also very useful in the construction of conformal mappings. It is analytic on $\overline{\mathbb{C}}$ except at the points $z = 0$ and $z = \infty$. The derivative

$$w' = \frac{1}{2}\left(1 - \frac{1}{z^2}\right)$$

[4]Nikolay Yegorovich Zhukovsky (Joukowsky) (1847–1921) was a Russian mathematician and engineer, a founder of modern aero- and hydrodynamics.

exists everywhere in $\overline{\mathbb{C}}$ except at the points $z = 0$ and $z = \infty$, and it becomes zero at $z = \pm 1$. Therefore the mapping (4.34) is conformal everywhere in $\overline{\mathbb{C}}$ except at the points 0, ± 1, and ∞.[5]

Let us see when it is possible for two distinct points to be mapped to the same image. Let $z_1 \neq z_2$, and supppose that

$$\frac{1}{2}\left(z_1 + \frac{1}{z_1}\right) = \frac{1}{2}\left(z_2 + \frac{1}{z_2}\right).$$

Then

$$0 = \left(z_1 + \frac{1}{z_1}\right) - \left(z_2 + \frac{1}{z_2}\right) = (z_1 - z_2)\left(1 - \frac{1}{z_1 z_2}\right).$$

Since we assume $z_1 \neq z_2$, this must mean that

$$z_1 z_2 = 1. \tag{4.35}$$

Therefore the Zhukovsky transform is 1-1 on a domain D, if and only if D contains no pair of distinct points satisfying (4.35). Examples of such domains are the exterior $|z| > 1$ of the unit disk (where $|z_1 z_2| > 1$) and the interior $|z| < 1$ of the same disk (where $|z_1 z_2| < 1$).

To get a visual grasp of the Zhukovsky transform, let us see what circles and rays are mapped to under it (see Fig. 36 a; circles are solid, rays are dotted). Setting $z = re^{i\phi}$, we can rewrite (4.34) as

$$w = u + iv = \frac{1}{2}\left(re^{i\phi} + \frac{1}{r}e^{-i\phi}\right) = \frac{1}{2}\left(r + \frac{1}{r}\right)\cos\phi + i\frac{1}{2}\left(r - \frac{1}{r}\right)\sin\phi,$$

from which we get

$$\begin{cases} u = \dfrac{1}{2}\left(r + \dfrac{1}{r}\right)\cos\phi, \\ v = \dfrac{1}{2}\left(r - \dfrac{1}{r}\right)\sin\phi. \end{cases} \tag{4.36}$$

Now consider the circle $r = r_0$. From (4.36) we get

$$\cos\phi = \frac{u}{\frac{1}{2}\left(r_0 + \frac{1}{r_0}\right)}, \quad \sin\phi = \frac{v}{\frac{1}{2}\left(r_0 - \frac{1}{r_0}\right)}. \tag{4.37}$$

Squaring both equations and then adding them together, we get

$$\frac{u^2}{\frac{1}{4}\left(r_0 + \frac{1}{r_0}\right)^2} + \frac{v^2}{\frac{1}{4}\left(r_0 - \frac{1}{r_0}\right)^2} = 1. \tag{4.38}$$

Here we recognize the equation of an ellipse, with the semiaxes

$$a_{r_0} = \frac{1}{2}\left(r_0 + \frac{1}{r_0}\right), \quad b_{r_0} = \frac{1}{2}\left(r_0 - \frac{1}{r_0}\right).$$

[5] Using the definition of an angle at the point at infinity it is possible to show that the Zhukovsky transform is also conformal at 0 and ∞; conformality fails only at ± 1.

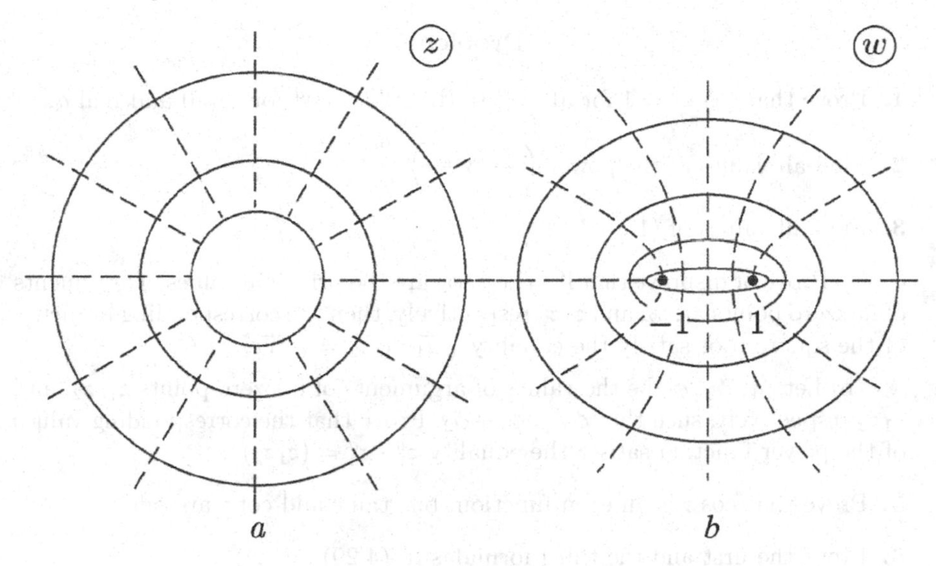

Fig. 36

So: circles $|z| = r_0$ in the z-plane become ellipses in the w-plane (Fig. 36 b). If $r_0 \to 1$, then $a_{r_0} \to 1$ and $b_{r_0} \to 0$: in the limit the ellipse is pinched to the line segment $[-1, 1]$. On the other hand, when r_0 is large, the difference $a_{r_0} - b_{r_0} = \frac{1}{r_0}$ is small, so large ellipses will be nearly circular.

To find the images of the rays $\phi = \phi_0$, we first rework equations (4.36) into the form

$$\frac{u}{\cos\phi_0} = \frac{1}{2}\left(r + \frac{1}{r}\right), \quad \frac{v}{\sin\phi_0} = \frac{1}{2}\left(r - \frac{1}{r}\right).$$

Again squaring both equations, but this time subtracting the second from the first, we get

$$\frac{u^2}{\cos^2\phi_0} - \frac{v^2}{\sin^2\phi_0} = 1. \tag{4.39}$$

This time we recognize the equation of a hyperbola with semiaxes $a_{\phi_0} = |\cos\phi_0|$ and $b_{\phi_0} = |\sin\phi_0|$. Therefore radial lines are mapped to arms of hyperbolas (Fig. 36, b).

So we see that the Zhukovsky transform is a conformal mapping of the exterior of the unit disk onto the exterior of the interval $[-1, 1]$, i.e. to $\mathbb{C}$ with the segment $-1 \le x \le 1$ cut out.

From the original definition (4.34) of the Zhukovsky transform w, it is easy to see that $w(z) = w(1/z)$. The function $z \mapsto 1/z$ is a conformal mapping of the interior of the unit disk onto the exterior of the same disk; therefore the Zhukovsky transform is also a conformal mapping of the interior of the unit disk onto the exterior of $[-1, 1]$.

Problems

1. Prove that (a) $z^0 = 1$ for all $z \neq 0$; (b) $|z^a| = |z|^a$ for $z \neq 0$ and real a.

2. Find all values of the power $\left(-\sqrt{3}+i\right)^{-6i}$.

3. Find all values of $(1+i)^i$.

4. (a) Prove or disprove that if ϕ_1, ϕ_2, ϕ_3 are the principal values of arguments of nonzero points z_1, z_2, and $z_1 z_2$ respectively, then the corresponding branches of the square root satisfy the equality $\sqrt{z_1} \cdot \sqrt{z_2} = \sqrt{z_1 z_2}$.

(b) Let ϕ_1, ϕ_2, ϕ_3 be the values of arguments of nonzero points z_1, z_2, and $z_1 z_2$ respectively, such that $\phi_3 = \phi_1 + \phi_2$. Prove that the corresponding values of the power function satisfy the equality $z_1^a \cdot z_2^a = (z_1 z_2)^a$.

5. Prove that $\cos z$ is an even function, but $\tan z$ and $\cot z$ are odd.

6. Prove the first and the third formulas in (4.29).

7. Prove the identities (4.31).

8. Find all values of the expression $\cos^{-1}(-5i)$.

9. Find all values of the expression $\tan^{-1} \dfrac{-2\sqrt{3}+3i}{3}$.

10. Find all solutions of the given equations.

$$a)\ \sin z = 2i; \quad b)\ \tan z = \frac{2-i}{5}; \quad c)\ \cos z = -5; \quad d)\ \tan z = \frac{2\sqrt{3}+3i}{7}.$$

11. Derive the formulas for derivatives of $\sin^{-1} z$, $\cos^{-1} z$, and $\tan^{-1} z$ (a) directly from (4.31) using the chain rule; (b) using the formula (4.14). Show that all branches of $\tan^{-1} z$ have the same derivative.

4.5 General Properties of Conformal Mappings

In the preceding sections we have looked at some elementary functions and the conformal mappings carried out by them. A natural question arises: is it possible to map any domain D, by means of a conformal mapping, onto any other domain D'? The answer, in general, is no: using the continuity of conformal mappings we see that, for example, a multiply connected domain can never be conformally mapped onto a simply connected domain. But if we restrict our attention to simply connected domains, then we have the following theorem.

Theorem 4.15 (Riemann Mapping Theorem). *Let D and D' be simply connected domains in the extended complex planes of the variables z and w, respectively. Also assume that their boundaries consist of more than a single point. Then there exists an analytic function that performs a conformal mapping of D onto D'.*

The proof is omitted. The interested reader may find it for example in the books by L. Ahlfors [1] and D. Marshall [6].

It follows from the Riemann's theorem that a simply connected domain D can only fail to be conformally mappable onto the unit disk in two cases: a) if D is the extended plane $\overline{\mathbb{C}}$ (for which the boundary is empty) or b) if D is the extended plane from which a single point has been removed (for which the boundary is that single point). An example of the second case would the complex plane $\mathbb{C}$, which is $\overline{\mathbb{C}}$ from which the point at infinity has been removed.

The theorem says that such a mapping of D onto D' exists, but not that it is unique. To define such a conformal mapping uniquely, it is necessary to give additional conditions, called normalization conditions, which consist of specifying three real parameters. For example, we could specify for some point z_0 in D, some point w_0 in D', and some real number β, that

$$f(z_0) = w_0 \quad \text{and} \quad \beta = \operatorname{Arg} f'(z_0). \tag{4.40}$$

Here the three real parameters consist in the two coordinates of the point w_0, and β. The conditions (4.40) specify that the image of z_0 is to be w_0, and the angle of rotation under f, of infinitessimal vectors at z_0, is to be β; the mapping satisfying those conditions is unique.

It is possible to give the normalization conditions in other forms than (4.40). For example, they could be given by specifying the images of one interior point and one boundary point of the domain D, i.e.

$$f(z_0) = w_0 \quad \text{and} \quad f(z_1) = w_1,$$

where z_0 and w_0 are as above, and z_1 and w_1 are points on the boundaries of D and D' respectively. The value w_1 counts as only one real parameter, because it may be identified by, e.g., the arc length that separates it along the boundary of D' from a certain fixed point. So here again exactly three real parameters have been specified.

One more example of normalization conditions:

$$f(z_k) = w_k \quad \text{for} \quad k = 1, 2, 3,$$

where the z_k and w_k are boundary points of D and D', respectively, and the points z_k and w_k are traversed in the same order.

As an example, recall the mapping of the unit disk $|z| < 1$ onto itself, $|w| < 1$, the Möbius transformation (4.12). This function has exactly three real parameters: the two coordinates of z_0, and ϕ. Hence no other conformal mappings of the disk to itself are possible—the only ones that exist are in this family of Möbius transformations.

In the next paragraphs we state some important properties of conformal mappings.

Property 4.16 (preservation of domains). *If the function $w = f(z)$ is analytic on the domain D and is not constant, then the set D', to which f maps D, is also a domain (i.e. a connected open set).*

The proof of this property is proposed as problem 3 after Section 7.4.

Next is a statement about the relationship between boundaries under conformal mappings.

Property 4.17 (Caratheodory's[6] theorem—preservation of boundaries). *Let D and D' be simply connected domains bounded by Jordan closed curves Γ and Γ'. Also let the function $w = f(z)$ be a conformal mapping of D onto D'. Then this function can be defined on the boundary Γ in such a way that it is continuous on the closed domain $\overline{D}$ and maps Γ bijectively and continuously onto Γ'.*

Definition of Jordan curve is given in Section 3.3.1. Property 4.17 means that when there is a conformal mapping of one domain onto another, there must also be an (induced) bijective continuous mapping between their boundaries.

Property 4.18 (preservation of orientation). *Under a conformal mapping between domains D and D', the orientation of traversal of the boundary is preserved.*

In other words, if a path along the boundary keeps the domain D on the left, then the image of that path along the boundary of D' also keeps that domain on the left.

The next property is very important in the construction of conformal mappings. It is roughly converse to Property 4.17.

Property 4.19. *Let D and D' be simply connected domains which are bounded by Jordan closed curves Γ and Γ' in $\mathbb{C}$. Also let $w = f(z)$ be a function analytic on D and continuous on $\overline{D}$, which maps Γ bijectively onto Γ' and preserves orientation in the mapping from Γ onto Γ'. Then the function $w = f(z)$ carries out a conformal mapping of domain D onto domain D'.*

We will prove this property in Section 7.4.[7]

[6] Constantin Carathéodory (1873–1950) was a Greek mathematician who spent most of his professional career in Germany.

[7] One may extend Property 4.19 to an unbounded domain which is mapped conformally onto a domain bounded by a Jordan closed curve. But the boundedness of Γ' is essential. Indeed, consider the function $w = f(z) = z^3$ in the upper half-plane $D = \{ z : \operatorname{Im} z > 0 \}$, and let $D' = \{ w : \operatorname{Im} w > 0 \}$. Obviously $f(z)$ maps bijectively the real axis $\{\operatorname{Im} z = 0\}$ which is the boundary of D, onto the real axis $\{\operatorname{Im} w = 0\}$, the boundary of D', and orientation is preserved under this mapping. But the mapping $f(z) = z^3$ is not bijective in D, and even does not map D into D' (prove it!).

One consequence of this is that when we want to determine what the image of a certain domain D is under a certain mapping $w = f(z)$, it suffices to follow the boundary of D and determine the contour to which it is mapped by $f(z)$.

Example 4.20 Find the domain to which the function

$$w = \cosh z = \frac{1}{2}(e^z + e^{-z})$$

maps the half-strip $D = \{ z : \operatorname{Re} z > 0,\ 0 < \operatorname{Im} z < \pi \}$.

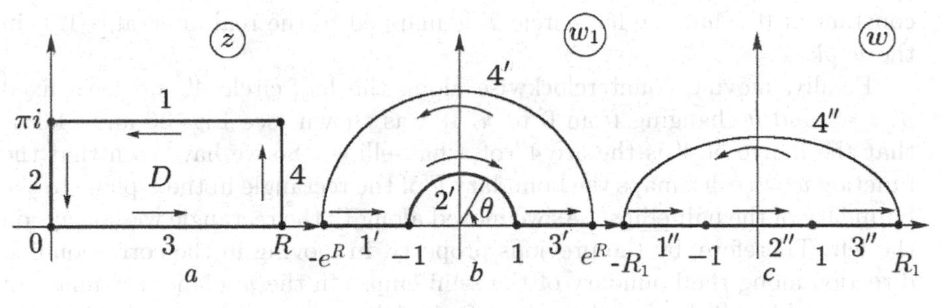

Fig. 37

Solution The given function is unbounded on the semi-axes $x > 0$. Hence, Γ' is not bounded and Theorem 4.17 is not applicable directly. So we start by considering a rectangle those sides are the intervals 1,2,3,4—see Fig. 37, *a*.

The function is a composition of the following two functions: 1) the exponential function $w_1 = e^z$ and 2) the Zhukovsky transform $w = \frac{1}{2}(w_1 + 1/w_1)$. In order to sketch the desired domain, we will trace the image of the given rectangle.

Let us consider the first mapping. Let $w_1 = e^z = e^x e^{iy} = \rho e^{i\theta}$. Then

$$\rho = e^x \quad \text{and} \quad \theta = y.$$

In moving left along the first piece of the boundary (Fig. 37, *a*), the value of θ is fixed at $\theta = y = \pi$, while x goes from R to 0. This means that the value of $\rho = e^x$ goes from e^R to 1, and therefore the first piece of the boundary is mapped to the interval $1'$ on the negative x-axis (Fig. 37, *b*).

Moving down along the second piece of the boundary, x is fixed at 0, while y goes from π to 0. Therefore ρ is fixed at $\rho = e^0 = 1$, and θ goes from π to 0. Thus the point w_1 follows the half-circle $2'$ in Fig. 37, *b*.

The third piece of the boundary is similar to the first: as z moves right along the interval 3, w_1 traces the interval $3'$, which is the mirror image of $1'$. Here $0 \le x \le R$, $\theta = 0$.

Finally, moving up along the fourth the piece of the boundary we have $x = R$, $0 \le \theta \le \pi$. Hence the image of the interval 4 is the half-circle $4'$ of radius $\rho = e^R$.

Now we find the images of the curves in the w_1-plane just determined, under the Zhukovsky transform. Moving right along $1'$, $\theta = \pi$ and ρ goes from e^R to 1; plugging these values into equations (4.36) for ϕ and r, respectively, we see that u goes from $-\frac{1}{2}(e^R + e^{-R})$ to -1, while v is constant at 0. This means that the interval $1'$ is mapped to the interval $1''$ in the w-plane with endpoints $-R_1$ and -1; for short we denote $R_1 = \frac{1}{2}(e^R + e^{-R})$. Analogously, the interval $3'$ is also mapped to the interval $3''$ in the w-plane symmetric to $1''$. In both cases, the direction of movement is preserved.

Moving clockwise along the half-circle $2'$, we have $\rho = 1$ and θ going from $-\pi$ to 0; plugging these into (4.36), we get u running from -1 to 1, and v constant at 0. Thus the half-circle $2'$ is mapped to the real interval $[-1, 1]$ in the w-plane.

Finally, moving counterclockwise along the half-circle $4'$, we have fixed $\rho = e^R$ and θ changing from 0 to π. It was shown (see Fig. 36 and (4.38)) that the image of $4'$ is the arc $4''$ of a half-ellipse. So we have seen that the function $w = \cosh z$ maps the boundary Γ of the rectangle in the z-plane to the boundary of the half-ellipse. As we moved along Γ, the rectangle was always on the left. Therefore, by the previous property, in moving in the corresponding direction along the boundary of the semi-ellipse in the w-plane, the image of the rectangle will be found to the left, i.e. the region inside the half-ellipse. Therefore the function $w = \cosh z$ conformally maps the rectangle to the half-ellipse.

Note that R could be any positive number. Passing to the limit as $R \to \infty$ we see that also $R_1 \to \infty$, and therefore the function $w = \cosh z$ conformally maps the domain D to the upper half-plane.

Problems

1. Prove that the function $w = \cos z$ conformally maps the half-strip $\{ z : 0 < \operatorname{Re} z < \pi,\ \operatorname{Im} z < 0 \}$ onto the upper half-plane.

2. Determine a half-strip in $\mathbb{C}$ which is conformally mapped by $w = \sin z$ onto the upper half-plane.

5

Integration

5.1 Definition of the Contour Integral

Consider a smooth curve Γ in the complex plane, given by the parametric equations

$$x = x(t), \quad y = y(t), \quad \alpha \le t \le \beta. \tag{5.1}$$

The definition of a smooth curve was given in Section 2.1.1; as already mentioned there, these two equations may be written together in the form

$$z(t) = x(t) + iy(t), \quad \alpha \le t \le \beta. \tag{5.2}$$

As the parameter t increases from α to β, the corresponding point $z(t)$ will traverse the curve Γ. Thus these equations define not only the points in the curve Γ, but also a direction of movement along it. The curve Γ together with its direction of movement is called an *oriented curve.*

Let $D \subset \mathbb{C}$ be a domain containing the curve Γ, and let

$$f(z) = u(x, y) + iv(x, y)$$

be a continuous function defined on D, and we define $dz = dx + i\,dy$. We may now introduce the notion of the integral of f along the curve Γ, denoted by

$$\int_\Gamma f(z)\,dz.$$

Expanding the integrand, we get

$$f(z)\,dz = (u + iv)(dx + i\,dy) = (u\,dx - v\,dy) + i(v\,dx + u\,dy).$$

In this way, the integral can be interpreted naturally by the equality

$$\int_\Gamma f(z)\,dz = \int_\Gamma u\,dx - v\,dy + i\int_\Gamma v\,dx + u\,dy. \tag{5.3}$$

Those who have taken multivariate calculus will recognize, on the right-hand side, two real path integrals in two dimensions. To evaluate these integrals,

DOI: 10.1201/9780367810283-5

we must replace x and y by the functions $x(t)$ and $y(t)$, and the differentials dx and dy by

$$dx = x'(t)\,dt, \quad dy = y'(t)\,dt.$$

Then the two integrals on the right-hand side of (5.3) reduce to integrals with respect to t, on the interval (α, β):

$$\int_{\alpha}^{\beta} (u(z(t))x'(t) - v(z(t))y'(t))\,dt + i\int_{\alpha}^{\beta} (v(z(t)) + u(z(t))y'(t))\,dt. \qquad (5.4)$$

This, in turn, is easily seen to be equivalent to

$$\int_{\alpha}^{\beta} (u(z(t)) + iv(z(t)))(x'(t) + iy'(t))\,dt = \int_{\alpha}^{\beta} f(z(t))z'(t)\,dt.$$

Now we are ready to give the following formal definition.

The contour integral of the complex function f along the curve Γ is the number which is denoted by $\int_{\Gamma} f(z)\,dz$ and is calculated by the formula

$$\int_{\Gamma} f(z)\,dz \overset{\text{def}}{=} \int_{\alpha}^{\beta} f(z(t))z'(t)\,dt. \qquad (5.5)$$

Here $z(t)$ is the parametrization of the curve Γ,

$$z(t) = x(t) + iy(t), \quad \alpha \le t \le \beta,$$

and

$$z'(t) = x'(t) + iy'(t).$$

Example 5.1 Calculate the integral of the function $f(z) = (z - a)^n$ over the circle Γ of radius r with center a, and traversing it counterclockwise.

Solution This circle is easily parametrized by

$$z(t) = a + re^{it}, \quad \text{where} \quad 0 \le t \le 2\pi,$$

which also has the correct orientation of counterclockwise. Then we have

$$f(z(t)) = (z(t) - a)^n = r^n e^{int},$$
$$z'(t) = \left(re^{it}\right)' = ire^{it}.$$

So applying formula (5.5) we get

$$\int_{\Gamma} (z - a)^n\,dz = ir^{n+1} \int_{0}^{2\pi} e^{i(n+1)t}\,dt.$$

Using the Euler formula, this integral may be split into two real integrals, representing the real and imaginary parts of the original:

$$ir^{n+1} \int_0^{2\pi} (\cos(n+1)t + i \sin(n+1)t) \, dt$$

$$= -r^{n+1} \int_0^{2\pi} \sin(n+1)t \, dt + ir^{n+1} \int_0^{2\pi} \cos(n+1)t \, dt.$$

As long as $n \neq -1$ this evaluates to zero. But in the case when $n = -1$ the integral becomes

$$i \int_0^{2\pi} 1 \, dt = 2\pi i.$$

We have obtained a result which, despite its simplicity, will be very important for the development of the theory:

$$\int_{|z-a|=r} (z-a)^n \, dz = \begin{cases} 0 & \text{for} \quad n \neq -1, \\ 2\pi i & \text{for} \quad n = -1. \end{cases} \tag{5.6}$$

Note especially that the value never depends on the radius r of the circle!

Example 5.2 Calculate the integral of the function $f(z) = 1$ over a smooth curve starting at the point a and ending at the point b.

Solution We assume the curve is defined by a smooth parametrization $z(t) = x(t) + iy(t)$ on $\alpha \leq t \leq \beta$, with $z(\alpha) = a$ and $z(\beta) = b$. Then using formula (5.5) and the fundamental theorem for real integration we get

$$\int_\Gamma 1 \, dz = \int_\alpha^\beta z'(t) \, dt = \int_\alpha^\beta x'(t) \, dt + i \int_\alpha^\beta y'(t) \, dt = x(t)\Big|_\alpha^\beta + iy(t)\Big|_\alpha^\beta$$

$$= x(\beta) - x(\alpha) + i(y(\beta) - y(\alpha)) = z(\beta) - z(\alpha) = b - a.$$

Here we see that the integral of the function $f(z) = 1$ does not depend on the path Γ, but only on its endpoints a and b.

We will now briefly lay out another way of defining the contour integral of a complex function, analogous to the definition of a real integral over an interval.

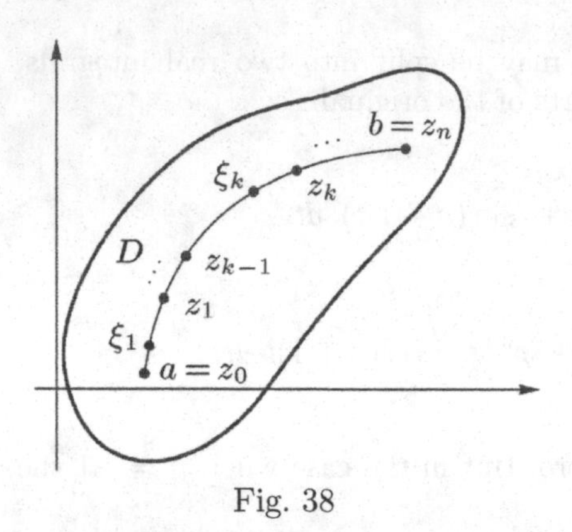

Fig. 38

Divide the curve Γ into n pieces by selecting $n+1$ points $z_0 = a$, z_1, ..., z_{n-1}, $z_n = b$, numbered in order of the direction of Γ (Fig. 38). For $1 \leq k \leq n$, denote by Δz_k the displacement $z_k - z_{k-1}$; also for each such k choose an arbitrary point ζ_k on Γ between z_{k-1} and z_k. Now we form the sum

$$\sum_{k=1}^{n} f(\zeta_k)\Delta z_k.$$

This sum is analogous to the Riemann sum in real integration. Denote by λ the maximum length of the n pieces of Γ. Consider the sequence of partitions of Γ for which $\lambda \to 0$ (and hence $n \to \infty$), and take the limit of the Riemann sums as $\lambda \to 0$. Then the value of the integral of f over the contour Γ is defined to be that limit:

$$\int_{\Gamma} f(z)\,dz \overset{\text{def}}{=} \lim_{\lambda\to 0} \sum_{k=1}^{n} f(\zeta_k)\Delta z_k. \tag{5.7}$$

It can be shown that this definition leads also to the formula (5.3), and consequently is equivalent to definition (5.5) given earlier.

5.1.1 Properties of the contour integral

Here are some important properties of the contour integral.

Property 5.3 (Linearity). *For any complex constants a and b,*

$$\int_{\Gamma} (af(z) + bg(z))\,dz = a\int_{\Gamma} f(z)\,dz + b\int_{\Gamma} g(z)\,dz.$$

Linearity is inherited through formula (5.5) from the corresponding property of real integration.

Property 5.4 (Path Additivity). *If a path Γ is divided into into parts Γ_1 and Γ_2, so that orientation is preserved, then*

$$\int_{\Gamma} f(z)\,dz = \int_{\Gamma_1} f(z)\,dz + \int_{\Gamma_2} f(z)\,dz.$$

This property is also inherited from real integration.

Proof. Let Γ be a path starting at a and ending at b, and let Γ be divided into two pieces by the point c; a path Γ_1 from a to c, and another path Γ_2 from c

to b. Let $z(t)$ for $\alpha \le t \le \beta$ be a parametrization of Γ, and let γ be the point in $[\alpha, \beta]$ that is mapped to c; then on $[\alpha, \gamma]$ the function $z(t)$ parametrizes Γ_1, and on $[\gamma, \beta]$ it parametrizes Γ_2. Now we apply the formula (5.5) and also use the additivity of real integration:

$$\int_{\Gamma} f(z)\, dz = \int_{\alpha}^{\beta} f(z(t))z'(t)\, dt = \int_{\alpha}^{\gamma} f(z(t))z'(t)\, dt + \int_{\gamma}^{\beta} f(z(t))z'(t)\, dt$$

$$= \int_{\Gamma_1} f(z)\, dz + \int_{\Gamma_2} f(z)\, dz,$$

what was required. □

One consequence of this property is that we can calculate integrals not only over smooth curves, but also over curves that are only *piecewise smooth*, meaning that they are composed of a finite number of smooth curves that have been chained together. In what follows, we will consider integrals only over piecewise smooth curves.

Property 5.5. *When the orientation of a path is reversed, i.e. the start and end points are reversed, the value of an integral over that path changes sign.*

Proof. Let Γ be a path starting at a and ending at b, and let $z(t)$ for $\alpha \le t \le \beta$ be a parametrization of Γ. We will denote by Γ^- the curve with same points as Γ but with the opposite orientation. Then Γ^- is parametrized by $z_1(t) = z(\alpha + \beta - t)$ for $\alpha \le t \le \beta$: if we define a new variable $\tau = \alpha + \beta - t$, then as t runs from α to β, τ runs from β to α, so $z_1(t)$ traverses Γ^- in the correct direction.

We see easily that $dt = -d\tau$ and

$$z_1'(t) = z'(\alpha + \beta - t)(-1) = -z'(\tau),$$

so we can make the change of variable from t to τ:

$$\int_{\Gamma^-} f(z)\, dz = \int_{\alpha}^{\beta} f(z_1(t))z_1'(t)\, dt = \int_{\beta}^{\alpha} f(z(\tau))(-z'(\tau))(-d\tau)$$

$$= -\int_{\alpha}^{\beta} f(z(\tau))z'(\tau)\, d\tau = -\int_{\Gamma} f(z)\, dz.$$

Property 5.5 is proved. □

Note that this property follows almost immediately from the definition of the contour integral (5.7) given in terms of Riemann sums: in reversing the orientation of the path, each displacement Δz_k also changes sign.

Property 5.6. *The modulus of the integral does not exceed the integral of the modulus:*

$$\left|\int_\Gamma f(z)\,dz\right| = \left|\int_\alpha^\beta f(z(t))\,z'(t)\,dt\right| \le \int_\alpha^\beta |f(z(t))|\,|z'(t)|\,dt$$
$$= \int_\alpha^\beta |f(z(t))|\sqrt{(x'(t))^2 + (y(t))^2}\,dt.$$

Proof. We will use the corresponding result for the integrals over an interval $[\alpha, \beta]$ when $\alpha \le \beta$:

$$\left|\int_\alpha^\beta g(t)\,dt\right| \le \int_\alpha^\beta |g(t)|\,dt. \tag{5.8}$$

This inequality follows directly from the definition (5.7), since

$$\left|\sum_{k=1}^n g(\zeta_k)\Delta t_k\right| \le \sum_{k=1}^n |g(\zeta_k)|\Delta t_k.$$

Passing to the limit as $\lambda \to 0$, we get (5.8). Then the desired result follows immediately from applying this to the formula (5.5). □

Note also that

$$\int_\alpha^\beta |z'(t)|\,dt = \int_\alpha^\beta \sqrt{(x'(t))^2 + (y'(t))^2}\,dt$$

is the formula for the arc length of the path; so this property provides an easy estimate for the value of the integral if $|f(z)|$ is bounded. Namely, if $|f(z)| \le M$ as $z \in \Gamma$, then by Property 5.6 we have

$$\left|\int_\Gamma f(z)\,dz\right| \le M\int_\alpha^\beta \sqrt{(x'(t))^2 + (y'(t))^2}\,dt = M\ell, \tag{5.9}$$

where ℓ is the arc length of Γ.

Problems

1. Calculate the integral of the function $f(z) = \bar{z}^2$ along the piecewise linear curve ABC, where $A = (-1, 0)$, $B = (0, 1)$, $C = (1, 0)$.

2. Compute the integral of the function $f(z) = \bar{z}^2$ along the upper half-circle $|z| = 1$, $\operatorname{Im} z \ge 0$, in the clockwise direction.

3. Compute the integral of the function $f(z) = z|z|$ along the lower half-circle $|z| = 1$, $\operatorname{Im} z \leq 0$, in the counterclockwise direction. (Cf. the path in the previous problem.)

4. Compute the integrals of the function $f(z) = z \operatorname{Re} z$ along the given paths.

a) The line segment from the origin to $(2, 4)$.

b) The arc of the parabola $y = x^2$ where $0 \leq x \leq 2$.

5. Prove that

$$a) \left| \int_{|z|=2} \frac{dz}{z^2+1} \right| \leq \frac{4\pi}{3}; \quad b) \left| \int_{|z|=2} (z^2+1)\, dz \right| \leq 20\pi; \quad c) \left| \int_{|z|=1} \frac{e^{iz}\, dz}{z^2} \right| \leq 2\pi e.$$

5.2 Cauchy-Goursat Theorem

The following theorem plays an important role in contour integrals of analytic functions.

Theorem 5.7 (Cauchy-Goursat theorem). *Let $f(z)$ be an analytic function on the simply connected domain D. Then for any piecewise smooth closed path Γ lying within D,*

$$\int_{\Gamma} f(z)\, dz = 0.$$

Proof. For simplicity we will make the additional assumption that the derivative $f'(z)$ is continuous on D. By virtue of the formula (5.3) for the contour integral, the integral above is 0 if and only if

$$\int_{\Gamma} u\, dx - v\, dy = 0 \quad \text{and} \quad \int_{\Gamma} v\, dx + u\, dy = 0. \tag{5.10}$$

Now recall that if $P(x, y)$ and $Q(x, y)$ are continuous real functions on a simply connected domain and have continuous first-order partial derivatives, then the equality

$$\int_{\Gamma} P\, dx + Q\, dy = 0$$

on all closed curves Γ, is equivalent to the equality

$$\frac{\partial P}{\partial y} = \frac{\partial Q}{\partial x} \tag{5.11}$$

(see for example [9], Section 16.3). Since the continuity of the partial derivatives of u and v follows from the continuity of $f'(z)$, we may apply this to the equalities (5.10), and find they are equivalent to

$$\frac{\partial u}{\partial y} = -\frac{\partial v}{\partial x} \quad \text{and} \quad \frac{\partial v}{\partial y} = \frac{\partial u}{\partial x}. \tag{5.12}$$

But these are just the Cauchy-Riemann conditions (3.4), which hold by the assumption that f is analytic on D. So we have proved the equalities (5.10), from which it follows that $\int\limits_{\Gamma} f\, dz$ is 0, as required. □

Theorem 5.7 has been proved by Cauchy in 1825 under the additional assumption about continuity of $f'(z)$. Goursat[1] established that this assumption can be omitted. The removal of this condition is important. For example, based on Theorem 5.7 we will show that analytic functions have derivatives of any order (Theorem 5.18); therefore, derivatives of all analytic functions are continuous "automatically". It is impossible to establish this important fact based on the original weakened version of the theorem. There are many proofs of Theorem 5.7, and all of them are much more complicated than the above arguments. One of the proofs is given in the Appendix.

Notice that even if Γ does intersect itself, then it can still be expressed as a sum of closed Jordan curves, so Theorem 5.7 holds even in that case.

In Section 5.4 we will see that the converse of the theorem is also true (Morera's theorem).

In the case that a function $f(z)$ is analytic not just on D but on the closed simply connected domain $\overline{D}$, then the boundary of $\overline{D}$ can play the role of the closed path. We will use this in the next theorem, which generalizes Cauchy's theorem to multiply connected domains.

Theorem 5.8 (Cauchy's theorem for multiply connected domains). *Assume that the function $f(z)$ is analytic on a closed multiply connected domain $\overline{D}$, and a total boundary Γ of $\overline{D}$ consists of a single exterior boundary Γ_1 and several interior boundaries $\Gamma_2, \Gamma_3, \ldots, \Gamma_n$. Also assume that all boundary curves of $\overline{D}$ are oriented so that D remains on the left during traversal. Then the integral of f over the entire boundary of $\overline{D}$ is zero.*

Proof. In this proof we will assume that $n = 3$ (Fig. 39), but the idea can easily be extended to any n. We cut the domain D along the arcs AB and CE, i.e. we remove the points on those arcs from D; the resulting domain we denote by D^*, and its boundary by Γ^*.

[1] Édouard Jean-Baptiste Goursat (1858–1936) was a French mathematician.

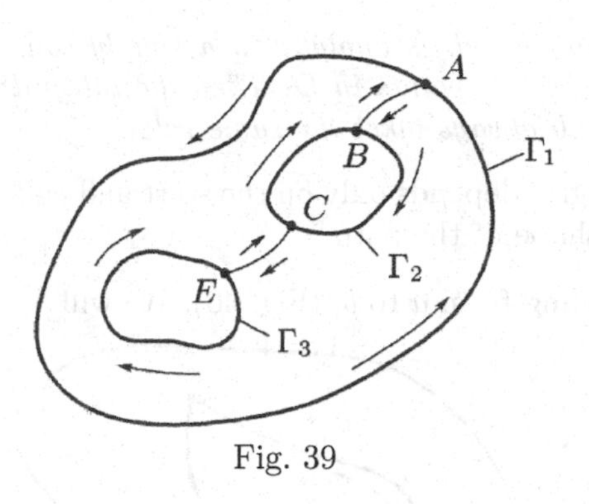

Fig. 39

We see that Γ^* is composed of the contours Γ_1, Γ_2, and Γ_3, as well as the arcs AB and CE; and in traversing Γ^*, the contours of Γ_1, Γ_2, and Γ_3 are traversed exactly once each; but the arcs AB and CE are traversed twice each, once in each direction. Therefore Γ^* is a closed curve, and D^* is a simply connected domain. So by the previous theorem the integral of f over Γ^* is zero. Applying Property 5.4, we can express the integral of f over Γ^* as:

$$0 = \int\limits_{\Gamma^*} f(z)\,dz = \int\limits_{\Gamma_1} f(z)\,dz + \int\limits_{\Gamma_2} f(z)\,dz + \int\limits_{\Gamma_3} f(z)\,dz$$

$$+ \int\limits_{AB} f(z)\,dz + \int\limits_{BA} f(z)\,dz + \int\limits_{CE} f(z)\,dz + \int\limits_{EC} f(z)\,dz.$$

But by Property 5.5, the integrals over AB and BA differ only in sign; so they cancel. Similarly for the integrals over CE and EC. So we are left with

$$0 = \int\limits_{\Gamma_1} f(z)\,dz + \int\limits_{\Gamma_2} f(z)\,dz + \int\limits_{\Gamma_3} f(z)\,dz, \tag{5.13}$$

as desired. □

This theorem can be reformulated as an identity between the integrals over the exterior and interior boundaries.

Theorem 5.9. *If $f(z)$ is an analytic function on an closed n-connected domain, and the boundary contours are all oriented the same way, either clockwise or counterclockwise, then*

$$\int\limits_{\Gamma_1} f(z)\,dz = \int\limits_{\Gamma_2} f(z)\,dz + \cdots + \int\limits_{\Gamma_n} f(z)\,dz, \tag{5.14}$$

where Γ_1 is the exterior boundary, and the others are the interior boundaries.

Proof. This follows immediately from the previous theorem: in equation (5.13) we reverse the orientation of the contours $\Gamma_2, \Gamma_3, \ldots, \Gamma_n$ (or that of Γ_1), and move the corresponding terms to the other side of the equation. □

Theorem 5.7 is equivalent to the following important property, *independence of the integral $\int\limits_{\Gamma} f(z)\,dz$ from the path of integration.*

Corollary 5.10. *Let $f(z)$ be a function which is analytic in a simply connected domain D, and let a and b be any two points in D. Then the integral along any path Γ lying in D, from a to b always takes the same value.*

In other words, the value of the integral depends only on the start and end points of the contour, and *not* on the shape of the path.

Proof. Let Γ_1 and Γ_2 be two paths leading from a to b, (Fig. 40). We will denote by Γ_1^- the path otherwise identical to Γ_1 but with its orientation reversed, i.e. Γ_1^- goes from b to a. If we set $\Gamma = \Gamma_2 \cup \Gamma_1^-$, then Γ is a closed curve, and f is analytic on it and its interior. Therefore by Theorem 5.7,

$$\int_{\Gamma} f(z)\,dz = 0.$$

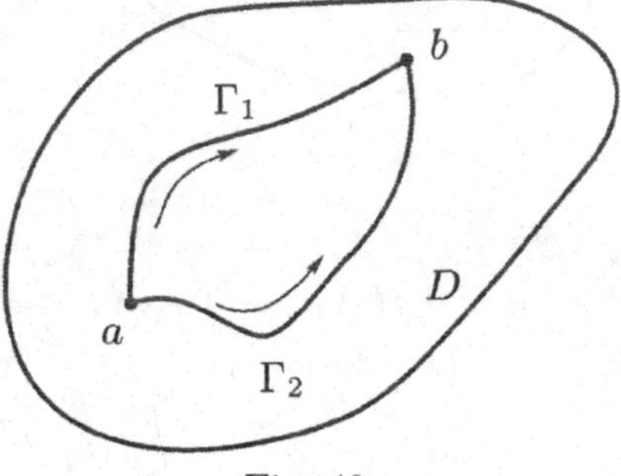

Fig. 40

Then

$$0 = \int_{\Gamma} f(z)\,dz = \int_{\Gamma_2} f(z)\,dz + \int_{\Gamma_1^-} f(z)\,dz = \int_{\Gamma_2} f(z)\,dz - \int_{\Gamma_1} f(z)\,dz,$$

so that

$$\int_{\Gamma_1} f(z)\,dz = \int_{\Gamma_2} f(z)\,dz.$$

Since Γ_1 and Γ_2 were arbitrary, the value of the integral is the same for any path. □

A point z at which $f(z)$ is analytic, is called a *regular point.* Points at which $f(z)$ is not analytic, including the points at which $f(z)$ is undefined, are called *singular points* or *singularities.* A singularity z_0 is said to be *isolated* if there is some neighborhood of z_0 in which all points except z_0 are regular.

Consider a continuous deformation of a curve Γ which may or may not be closed. If the curve is not closed, we assume that its initial and terminal points stay still under the deformation.[2]

[2] The formal definition of continuous deformation is the following. We say that a curve Γ_2 parametrized by $z_2(t)$, $0 \le t \le 1$, is obtained by a *continuous deformation* of a curve Γ_1 parametrized by $z_1(t)$, $0 \le t \le 1$, if there is a continuous function $z(s,t)$ of two real variables on the square $0 \le s \le 1$, $0 \le t \le 1$, such that

$$z(0,t) = z_1(t), \quad z(1,t) = z_2(t), \quad 0 \le t \le 1,$$

and

$$z(s,0) = a, \quad z(s,1) = b, \quad 0 \le s \le 1, \text{ if } \Gamma_1, \Gamma_2 \text{ have endpoints } a, b;$$
$$z(s,0) = z(s,1), \quad 0 \le s \le 1, \text{ if } \Gamma_1, \Gamma_2 \text{ are closed.}$$

Corollary 5.11 (Invariance of the integral under path deformation). *Let $f(z)$ be a function which is analytic on a domain including Γ. The integral of $f(z)$ over Γ does not change its value under continuous deformations of Γ which never passes through a singularity of f.*

Proof. Suppose that Γ_1 is obtained by a continuous deformation of Γ, and Γ is not closed. Since the deformation passes through no singularities, the region between Γ and Γ_1 contains no singularities. Therefore, f is analytic on a simply-connected domain containing Γ and Γ_1—see Fig. 40 with Γ instead of Γ_2. So this result follows from the previous corollary.

If Γ is closed, the equality $\int_\Gamma f(z)\,dz = \int_{\Gamma_1} f(z)\,dz$ follows immediately from the Theorem 5.9 with $n = 2$, since $f(z)$ is analytic on the 2-connected domain between Γ and Γ_1. □

Example 5.12 The function $f(z) = 1/z$ is analytic on the entire complex plane $\mathbb{C}$ with the exception of $z = 0$. Let Γ be the path corresponding to the unit circle $|z| = 1$, taken counterclockwise. By virtue of formula (5.6),

$$\int\limits_{|z|=1} \frac{1}{z}\,dz = 2\pi i \neq 0. \tag{5.15}$$

This shows that the requirement in Theorem 5.7 that the domain be simply connected is essential. From formula (5.15) it follows that the two different paths from -1 to 1, the first following the half-circle over the origin, the other following the half-circle below it, would produce different values for the integral of $f(z) = 1/z$—in fact, the reader should do those two calculations, guided by Example 5.1 (problem 1 below).

At the same time, in deformations of the unit circle which do not pass through $z = 0$ (for example, changing the radius) the integral over the resulting path will have the same value, $2\pi i$—recall that in Example 5.1 the result turned out to be independent of the radius r of the circle. And if we integrate the function $1/z$ over a closed curve which does not contain $z = 0$ in its interior, then by Cauchy's theorem the result will be zero.

Problems

1. Evaluate the integral $\int_\Gamma \frac{1}{z}\,dz$ from $A = (-1, 0)$ to $B = (1, 0)$ along the curve Γ, where (a) Γ is the upper half-circle $|z| = 1$, $\operatorname{Im} z \geq 0$; (b) Γ is the lower half-circle $|z| = 1$, $\operatorname{Im} z \leq 0$.

2. Compute the integral of the function $f(z) = \bar{z}^2$ along the given curves Γ. a) The piecewise linear path $AOBC$, where $A = (-1, 0)$, $O = (0, 0)$, $B = (0, 1)$, $C = (1, 0)$. b) The segment AC.

If the results are different, please explain why. Note that all three curves, $AOBC$, AC, and ABC, the latter having been considered in problem 1 to

Section 5.1, as well as the path in problem 2 to Section 5.1, are paths from A to C.

3. What is the easiest way to calculate the integrals of the function $f(z) = z^2$ along the given curves Γ (A, B, C, O are as in the previous problem)? Must all three integrals agree? Compute the integrals and explain your answer.

a) The piecewise linear path ABC;

b) The piecewise linear path $AOBC$;

c) The interval AC.

4. Prove that

$$a) \int_{|z|=10} \frac{dz}{z^2 - 3z + 1} = 0; \quad b) \int_{|z|=2} \frac{z\,dz}{z^4 + z - 2} = 0.$$

5.3 Indefinite Integral

Suppose that $f(z)$ is an analytic function on some domain D. Then another analytic function $F(z)$ on D is called an *antiderivative* of $f(z)$, if $F'(z) = f(z)$ at all points z in D. Clearly if we add any constant C to $F(z)$, then the resulting function is another antiderivative of $f(z)$. We will show that these are the only antiderivatives, that is, *all antiderivatives of $f(z)$ can be obtained through the addition of constants to $F(z)$.*

We will show that if $F_1(z)$ and $F_2(z)$ are both antiderivatives of $f(z)$, then they differ by only an additive constant, i.e. $F_2(z) = F_1(z) + C$, for some constant C. Indeed, if we define a new function as the difference of the two

$$\phi(z) = F_1(z) - F_2(z),$$

then ϕ is also analytic on D, and

$$\phi'(z) = F_1'(z) - F_2'(z) = f(z) - f(z) = 0$$

everywhere on D. Therefore by Theorem 3.7, $\phi(z)$ is constant. Since ϕ was defined as the difference of $F_1(z)$ and $F_2(z)$, this shows that $F_1(z) = F_2(z) + C$.

So if F is an antiderivative of f, then all other antiderivatives of f can be written in the form $F(z) + C$, where C is any constant. The entire set of antiderivatives of f can be referred to as the *indefinite integral* of f, and is denoted by $\int f(z)\,dz$. Therefore,

$$\int f(z)\,dz = F(z) + C. \tag{5.16}$$

Now let $f(z)$ be a analytic function on a simply connected domain D, and let z_0 and z be two points in D. Consider the integral

$$\Phi(z) = \int_{z_0}^{z} f(\zeta)\, d\zeta, \tag{5.17}$$

where the integral is calculated using any path from z_0 to z which lies in D; since D is simply connected, by Corollary 5.10 all such paths produce the same value for the integral. If z_0 is fixed, the value of the integral depends only on z, and therefore $\Phi(z)$ is a single-valued function on D.

Theorem 5.13. *Suppose that $f(z)$ is an analytic function on the simply connected domain D, and z_0 is a fixed point in D. Then the function $\Phi(z)$ defined in (5.17) is also analytic on D, and is an antiderivative of f, that is*

$$\frac{d}{dz}\int_{z_0}^{z} f(\zeta)\, d\zeta = f(z).$$

Proof. Let us take an arbitrary point z in D (Fig. 41). The function f is analytic, and consequently continuous, at z. So for any $\epsilon > 0$ there is some $\delta > 0$ for which

$$|f(\zeta) - f(z)| < \epsilon \quad \text{when} \quad |\zeta - z| < \delta. \tag{5.18}$$

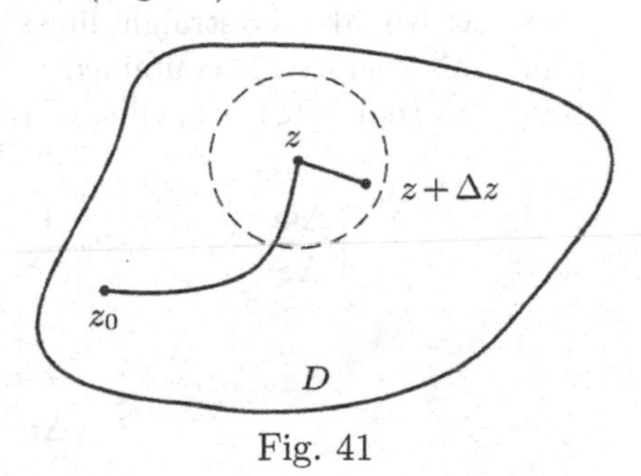

Fig. 41

Now give to z some increment Δz; then the function $\Phi(z)$ takes the corresponding increment

$$\Delta\Phi = \Phi(z+\Delta z) - \Phi(z) = \int_{z_0}^{z+\Delta z} f(\zeta)\, d\zeta - \int_{z_0}^{z} f(\zeta)\, d\zeta$$

$$= \int_{z_0}^{z} f(\zeta)\, d\zeta + \int_{z}^{z+\Delta z} f(\zeta)\, d\zeta - \int_{z_0}^{z} f(\zeta)\, d\zeta = \int_{z}^{z+\Delta z} f(\zeta)\, d\zeta.$$

From Example 5.2 we know that

$$\int_{z}^{z+\Delta z} 1\, d\zeta = (z+\Delta z) - z = \Delta z.$$

Using this identity, and simultaneously adding and subtracting $f(z)\Delta z$, we can write $\Delta\Phi$ as

$$\begin{aligned}\Delta\Phi &= \int_z^{z+\Delta z} f(\zeta)\,d\zeta - f(z)\Delta z + f(z)\Delta z \\ &= \int_z^{z+\Delta z} f(\zeta)\,d\zeta - f(z)\int_z^{z+\Delta z} d\zeta + f(z)\Delta z \\ &= \int_z^{z+\Delta z} (f(\zeta) - f(z))\,d\zeta + f(z)\Delta z.\end{aligned}$$

Next, we move the term $f(z)\Delta z$ to the left side, and divide by Δz:

$$\frac{\Delta\Phi}{\Delta z} - f(z) = \frac{1}{\Delta z}\int_z^{z+\Delta z} (f(\zeta) - f(z))\,d\zeta.$$

Now we need an upper bound for the size of the integral. For the path from z to $z+\Delta z$ we take the straight line segment connecting the two points (Fig. 41). Since this segment is contained within a disk of radius δ around z, we know by (5.18) that $|f(\zeta) - f(z)| < \epsilon$. Then applying Property 5.6, we get

$$\begin{aligned}\left|\frac{\Delta\Phi}{\Delta z} - f(z)\right| &\le \frac{1}{|\Delta z|}\int_\alpha^\beta |f(\zeta) - f(z)||\zeta'(t)|\,dt \\ &\le \frac{\epsilon}{|\Delta z|}\int_\alpha^\beta |\zeta'(t)|\,dt = \frac{\epsilon}{|\Delta z|}|\Delta z| = \epsilon,\end{aligned}$$

where we have used the fact that $\int_\alpha^\beta |\zeta'(t)|\,dt$ gives the length of the line segment, that is $|\Delta z|$.

To summarize what we have shown: for any $\epsilon > 0$ there is some $\delta > 0$ for which, whenever $|\Delta z| < \delta$,

$$\left|\frac{\Delta\Phi}{\Delta z} - f(z)\right| \le \epsilon.$$

This means that the limit of $\frac{\Delta\Phi}{\Delta z}$ exists and is $f(z)$:

$$\lim_{\Delta z\to 0}\frac{\Delta\Phi}{\Delta z} = f(z),$$

i.e. $\Phi'(z) = f(z)$, which is what we wanted. □

Corollary 5.14. *Suppose that $f(z)$ is analytic on a simply connected domain D, and that $F(z)$ is any antiderivative of f. Then*

$$\int_{z_0}^{z_1} f(\zeta)\, d\zeta = F(z_1) - F(z_0), \tag{5.19}$$

where z_0 and z_1 are any two points in D, and the path of integration is any path from z_0 to z_1 lying in D.

Proof. Because the function $\Phi(z)$ defined by (5.17) is an antiderivative of $f(z)$, then any antiderivative $F(z)$ can be written in the form $F(z) = \Phi(z) + C$, that is

$$F(z) = \int_{z_0}^{z} f(\zeta)\, d\zeta + C,$$

where C is a constant. Plugging z_0 in for z in this equation, we find that $F(z_0) = C$. Then plugging in $z = z_1$, we get

$$F(z_1) = \int_{z_0}^{z_1} f(\zeta)\, d\zeta + F(z_0),$$

from which (5.19) follows. □

Remark 5.15 So we see that the notions of antiderivative and the fundamental theorem of calculus for functions of a real variable are fully preserved for complex variables. Thanks to this, integrals of elementary functions can be calculated using the same formulas and methods as in real analysis; for example, the same tables of antiderivatives can be used, the standard integration by parts formula

$$\int_a^b u\, dv = uv\Big|_a^b - \int_a^b v\, du$$

is valid for analytic functions u, v as well; it follows from the product rule for differentiation and from the fundamental theorem of calculus.

Example 5.16 Evaluate the integral $\int_0^{3i} z^2\, dz$.

Solution $\displaystyle\int_0^{3i} z^2\, dz = \frac{z^3}{3}\Big|_0^{3i} = \frac{(3i)^3}{3} - 0 = -9i.$

The fundamental theorem provides a convenient way of evaluating integrals, but it must be remembered that it only applies when *the domain of the* *integrand $f(z)$ is simply connected.* If that is not the case, the value of the integral may depend on the path of integration, and then the function $\Phi(z)$ will be multivalued.

To illustrate this, consider again the function $f(z) = 1/z$, which is analytic everywhere except at the point $z = 0$. Let D be the complex plane with the negative real axis removed. This is a simply-connected domain, and the principal value of the logarithm

$$\operatorname{Log} z = \ln|z| + i \operatorname{Arg} z \quad (-\pi < \operatorname{Arg} z < \pi)$$

on this domain is an antiderivative of the function $1/z$ (recall formula (4.25)). Applying the fundamental theorem on D, we get

$$\int_1^z \frac{1}{\zeta}\, dz = \operatorname{Log} \zeta|_1^z = \operatorname{Log} z = \ln|z| + i \operatorname{Arg} z,$$

where the integral is taken along any curve lying within D.

Now consider again the integral

$$\Phi(z) = \int_1^z \frac{1}{\zeta}\, dz, \tag{5.20}$$

but this time along a path Γ from the point 1 to the point z, which avoids 0 but is otherwise arbitrary (Fig. 42). In particular, Γ may encircle the point $z = 0$, even multiple times (something which is not possible on D). For a $\zeta \in \Gamma$ we will denote by ϕ the angle between the position vector $\overrightarrow{0\zeta}$ and the positive x-axis; so if $\zeta = 1$, then $\phi = 0$. In moving along the path Γ, the angle ϕ will vary continuously until finally reaching the value $\operatorname{Arg} z + 2\pi n$ at the endpoint z, where n will be the number of revolutions of ζ about the origin. We assume for simplicity that $\operatorname{Im} z > 0$, and introduce now the linear path Γ_1 from $\zeta = 1$ to $\zeta = z$; this path lies within D (in Fig. 42 Γ_1 is dotted). Let Γ_1^- be the same path as Γ_1, but with orientation reversed. In moving along Γ_1^- from z to 1, the angle ϕ changes value by $-\operatorname{Arg} z$; while in moving along Γ from 1 to z, the angle changes by $\operatorname{Arg} z + 2\pi n$. Therefore in traversing the combined path $\Gamma_1^- \cup \Gamma$ from z to z, the angle will change by $2\pi n$. This means that $\Gamma_1^- \cup \Gamma$ is a path which encircles the origin n times, and therefore it will be continuously

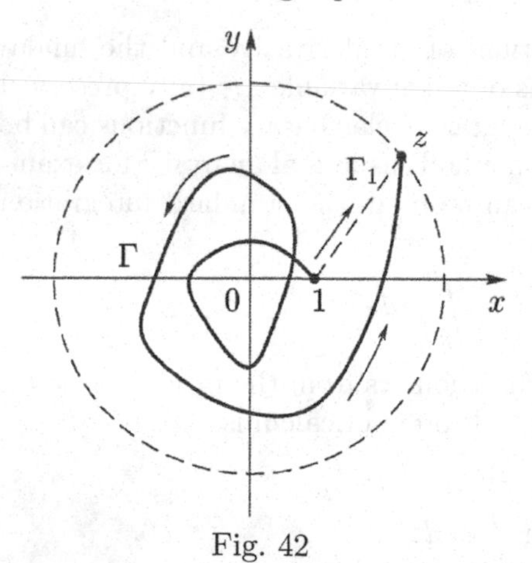

Fig. 42

deformable, without passing through the origin, to n repetitions of a circle centered at 0—in Fig. 42 the circle is dotted, and $n = 2$. According to formula (5.15), when integrating the function $1/z$ each revolution around the origin contributes $\pm 2\pi i$ to the value, where the sign depends on the orientation of the path (clockwise or counterclockwise). Therefore

$$\int\limits_{\Gamma_1^- \cup \Gamma} \frac{1}{\zeta}\, d\zeta = 2\pi n i.$$

From this we get

$$2\pi n i = \int\limits_{\Gamma_1^-} \frac{d\zeta}{\zeta} + \int\limits_{\Gamma} \frac{d\zeta}{\zeta} = -\int\limits_{\Gamma_1} \frac{d\zeta}{\zeta} + \int\limits_{\Gamma} \frac{d\zeta}{\zeta} = -\operatorname{Log} z + \int\limits_1^z \frac{d\zeta}{\zeta},$$

which finally yields

$$\int\limits_1^z \frac{d\zeta}{\zeta} = \operatorname{Log} z + 2\pi n i = \log z.$$

So the integral (5.20), with no path of integration specified, is equivalent to the multivalued function $\log z$.

Problems

1. Check whether the function $f(z) = z \sin 5z$ is analytic in $\mathbb{C}$, and if so, find an antiderivative F in $\mathbb{C}$ of f, satisfying the condition $F(\frac{\pi}{2}) = 0$.

2. Check whether the function $f(z)$ is analytic in $\mathbb{C}$, and if so, find an antiderivative $F(z)$ satisfying the given condition.

a) $f(z) = ze^z$, $F(0) = 0$;
b) $f(z) = (z+2)(z^2+4z+3)^3$, $F(0) = 0$;
c) $f(z) = ze^{z^2+2}$, $F(i\sqrt{2}) = 0$;
d) $f(z) = z\cos 2z$, $F(0) = 0$.

5.4 The Cauchy Integral Formula

The formula we are about to give expresses a fundamental property of analytic functions. It turns out that an analytic function on a closed domain $\overline{D}$ is completely determined by its values on the boundary of $\overline{D}$: the values the function takes on the entire interior can be calculated from its values on the boundary.

Theorem 5.17. *Let $f(z)$ be an analytic function on a closed domain $\overline{D}$ (either simply or multiply connected). Then the value $f(z)$ the function takes at an interior point $z \in D$ can be expressed in terms of the values $f(\zeta)$ at points ζ in the boundary Γ of $\mathbb{C}$, by*

$$f(z) = \frac{1}{2\pi i} \int_{\Gamma} \frac{f(\zeta)}{\zeta - z}\, d\zeta, \tag{5.21}$$

the Cauchy integral formula.

Proof. Let us fix an arbitrary interior point $z \in D$ (Fig. 43). Let $\Gamma_1, \Gamma_2, \ldots, \Gamma_n$ be the contours that bound D, all with the same standard orientation, i.e. D remains on the left during traversal. Let

$$\Gamma = \Gamma_1 \cup \Gamma_2 \cdots \cup \Gamma_n$$

be the boundary as a whole. Consider an arbitrary $\epsilon > 0$; since f is continuous at z, there is some disk γ centered at z, and lying in D, such that

$$|f(\zeta) - f(z)| < \epsilon \quad \text{whenever} \quad \zeta \in \gamma. \tag{5.22}$$

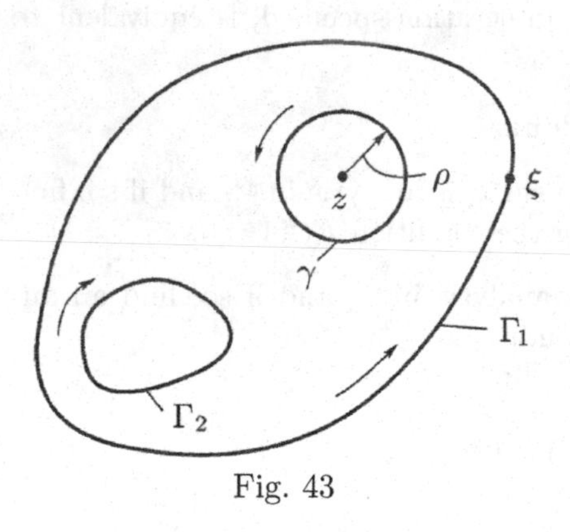

Fig. 43

Let ρ denote the radius of γ. We will regard γ as a path with orientation counterclockwise; and we will denote by γ^- the path with clockwise orientation but otherwise the same as γ.
Now consider the function

$$\phi(\zeta) = \frac{f(\zeta)}{\zeta - z}$$

of the variable ζ. It is analytic on all of $\overline{D}$, with the exception of the point $\zeta = z$. In particular, it is analytic on the domain $\overline{D}^*$ formed from $\overline{D}$ by removing $|\zeta - z| < \rho$. This domain $\overline{D}^*$ is $(n+1)$-connected, its boundary being composed of the paths $\Gamma_1, \Gamma_2, \ldots, \Gamma_n$, and γ^-, for all of which $\overline{D}^*$ is on the left during traversal. To the function $\phi(\zeta)$ on the domain $\overline{D}^*$ we apply Cauchy's theorem, formula (5.13):

$$\int_{\Gamma_1} \frac{f(\zeta)}{\zeta - z}\, d\zeta + \cdots + \int_{\Gamma_n} \frac{f(\zeta)}{\zeta - z}\, d\zeta + \int_{\gamma^-} \frac{f(\zeta)}{\zeta - z}\, d\zeta = 0,$$

which can be rearranged into

$$\int_{\Gamma} \frac{f(\zeta)}{\zeta - z}\, d\zeta = \int_{\Gamma_1} \frac{f(\zeta)}{\zeta - z}\, d\zeta + \cdots + \int_{\Gamma_n} \frac{f(\zeta)}{\zeta - z}\, d\zeta = \int_{\gamma} \frac{f(\zeta)}{\zeta - z}\, d\zeta. \tag{5.23}$$

Now we look at the integral on the right. By formula (5.6) with $n = -1$, we get

$$1 = \frac{1}{2\pi i} \int_{\gamma} \frac{1}{\zeta - z} \, d\zeta.$$

Since $f(z)$ is constant with respect to ζ, we may multiply through by it to get

$$f(z) = \frac{1}{2\pi i} \int_{\gamma} \frac{f(z)}{\zeta - z} \, d\zeta. \tag{5.24}$$

After dividing equation (5.23) through by $2\pi i$, we can take the term on its left side and subtract the left side of (5.24); then take the term on the right side of (5.23) and subtract the right side of (5.24); and then set the results equal:

$$\begin{aligned} \frac{1}{2\pi i} \int_{\Gamma} \frac{f(\zeta)}{\zeta - z} \, d\zeta - f(z) &= \frac{1}{2\pi i} \int_{\gamma} \frac{f(\zeta)}{\zeta - z} \, d\zeta - \frac{1}{2\pi i} \int_{\gamma} \frac{f(z)}{\zeta - z} \, d\zeta \\ &= \frac{1}{2\pi i} \int_{\gamma} \frac{f(\zeta) - f(z)}{\zeta - z} \, d\zeta. \end{aligned}$$

Using Property 5.6 and the inequality (5.22), and assuming γ is parametrized by $\zeta(t) = \rho e^{it}$, we get the estimate

$$\left| \frac{1}{2\pi i} \int_{\Gamma} \frac{f(\zeta)}{\zeta - z} \, d\zeta - f(z) \right| = \left| \frac{1}{2\pi i} \int_{\gamma} \frac{f(\zeta) - f(z)}{\zeta - z} \, d\zeta \right|$$

$$\leq \frac{1}{2\pi} \int_{0}^{2\pi} \frac{|f(\zeta(t)) - f(z)|}{\rho} |\zeta'(t)| \, dt \leq \frac{1}{2\pi} \frac{\epsilon}{\rho} \int_{0}^{2\pi} |\zeta'(t)| \, dt \leq \frac{1}{2\pi} \frac{\epsilon}{\rho} 2\pi\rho = \epsilon.$$

Since ϵ may be chosen as small as desired, and the left side of the previous relation is independent of ϵ, then it must in fact be zero; so the Cauchy integral formula is proved. □

In fact Cauchy's integral formula is also valid for functions which are analytic only in the interior D of the domain, and only continuous on $\overline{D}$. This is to be expected, since the boundary of $\overline{D}$ can be thought of as a limit of contours in D; but we omit a formal proof.

The Cauchy integral formula has numerous important applications.

Theorem 5.18. *A function $f(z)$ which is analytic on a closed domain $\overline{D}$, has at every interior point $z \in D$ derivatives of all orders. These derivatives can be expressed by*

$$f^{(n)}(z) = \frac{n!}{2\pi i} \int_{\Gamma} \frac{f(\zeta)}{(\zeta - z)^{n+1}} \, d\zeta \quad \text{where} \quad n = 1, 2, 3, \ldots, \tag{5.25}$$

where Γ is boundary of D.

This means that, from the existence of the first derivative of a function on a domain, we can infer the existence of all derivatives! As one consequence, *the derivative of an analytic function is itself analytic*, because it is itself differentiable. This property is a point of essential difference between complex functions and real ones.

We can derive formula (5.25) (informally) as follows. Differentiate both sides of equation (5.21) with respect to z; on the right side, we must differentiate the expression under the integral sign, in which ζ is considered a constant:

$$\frac{d}{dz}\frac{f(\zeta)}{\zeta - z} = \frac{f(\zeta)}{(\zeta - z)^2},$$

$$\left(\frac{d}{dz}\right)^2 \frac{f(\zeta)}{\zeta - z} = \frac{2f(\zeta)}{(\zeta - z)^3},$$

$$\vdots$$

$$\left(\frac{d}{dz}\right)^n \frac{f(\zeta)}{\zeta - z} = \frac{n!f(\zeta)}{(\zeta - z)^{n+1}}.$$

Plugging these derivatives into (5.21) yields (5.25). Unfortunately differentiation under the integral sign has not been justified; so for the reader who is (rightly!) skeptical of this argument we give a full proof.

Proof. First we prove formula (5.25) for $n = 1$. Fix an interior point $z \in D$, and let $\delta > 0$ be the distance of z from the boundary Γ. We give to z an increment Δz such that $|\Delta z| < \delta$, so that $z + \Delta z$ is still in D, and the Cauchy formula (5.21) for this point is valid:

$$f(z + \Delta z) = \frac{1}{2\pi i}\int_\Gamma \frac{f(\zeta)}{\zeta - (z + \Delta z)}\, d\zeta\,.$$

From this and (5.21) we get

$$\frac{\Delta f}{\Delta z} = \frac{f(z+\Delta z) - f(z)}{\Delta z} = \frac{1}{2\pi i \Delta z}\int_\Gamma \left(\frac{f(\zeta)}{\zeta - (z+\Delta z)} - \frac{f(\zeta)}{\zeta - z}\right) d\zeta$$
$$= \frac{1}{2\pi i}\int_\Gamma \frac{f(\zeta)}{(\zeta - z - \Delta z)(\zeta - z)}\, d\zeta.$$

We want to show that this is close to formula (5.25), so let us examine their difference:

$$\frac{\Delta f}{\Delta z} - \frac{1}{2\pi i}\int_\Gamma \frac{f(\zeta)\, d\zeta}{(\zeta - z)^2} = \frac{1}{2\pi i}\int_\Gamma \left(\frac{1}{(\zeta - z - \Delta z)(\zeta - z)} - \frac{1}{(\zeta - z)^2}\right) f(\zeta)\, d\zeta$$
$$= \frac{\Delta z}{2\pi i}\int_\Gamma \frac{f(\zeta)}{(\zeta - z - \Delta z)(\zeta - z)^2}\, d\zeta.$$

Since $f(z)$ is analytic, it is continuous and bounded on Γ. So there is M such that $|f(\zeta)| < M$ as $\zeta \in \Gamma$; that provides an upper estimate on the numerator of the integrand. We also need a lower estimate on the denominator; by a previous assumption $|\zeta - z| \geq \delta$, and

$$|\zeta - z - \Delta z| \geq |\zeta - z| - |\Delta z| \geq \delta - |\Delta z|.$$

Therefore, if we denote by L the arc length of the boundary Γ, we get

$$\left| \frac{\Delta f}{\Delta z} - \frac{1}{2\pi i} \int\limits_{\Gamma} \frac{f(\zeta)}{(\zeta - z)^2}\, d\zeta \right| \leq \frac{|\Delta z|}{2\pi} \cdot \frac{M}{\delta^2(\delta - |\Delta z|)} \cdot L.$$

This approaches zero as $\Delta z \to 0$, so the limit of $\Delta f / \Delta z$ exists and

$$f'(z) = \lim_{\Delta z \to 0} \frac{\Delta f}{\Delta z} = \frac{1}{2\pi i} \int\limits_{\Gamma} \frac{f(\zeta)}{(\zeta - z)^2}\, d\zeta. \tag{5.26}$$

This establishes the theorem for the case $n = 1$.

Now we use a similar strategy to estimate the difference

$$\frac{f'(z + \Delta z) - f'(z)}{\Delta z} - \frac{2!}{2\pi i} \int\limits_{\Gamma} \frac{f(\zeta)}{(\zeta - z)^3}\, d\zeta$$

and prove that it tends to 0 as $\Delta z \to 0$; instead of the Cauchy formula (5.21) we apply (5.26). It gives (5.25) for $n = 2$. Continuing in this way, we obtain (5.25) for $n = 3, 4$, and so on. This last part of the proof can be done more accurately by the method of mathematical induction—see problem 4. □

The Cauchy integral formulas (5.21) and (5.25) can sometimes be used for computing integrals over closed curves which enclose singularities.

Example 5.19 Evaluate the integral

$$\int\limits_{\Gamma} \frac{\sin z}{z^2 + 4}\, dz,$$

where Γ is the circle centered at i with radius 2.

Solution Since

$$z^2 + 4 = (z - 2i)(z + 2i),$$

the integrand has two singularities, at $\pm 2i$, of which $2i$ lies in the interior of Γ (we trust the reader to confirm this with a sketch). Let us write the integral in the form

$$\int\limits_{\Gamma} \frac{\sin z}{(z - 2i)(z + 2i)}\, dz = \int\limits_{\Gamma} \frac{\frac{\sin z}{(z+2i)}}{(z - 2i)}\, dz.$$

We denote the numerator $\frac{\sin z}{z+2i}$ of the this integrand by $f(z)$; this function is analytic on the closed disk bounded by Γ. Therefore by formula (5.21), when we replace z by $2i$ and the variable ζ by z, we get

$$\int_\Gamma \frac{\sin z}{z^2+4}\,dz = \int_\Gamma \frac{f(z)}{z-2i}\,dz = 2\pi i f(2i)$$

$$= \frac{\pi}{2}\sin(2i) = \frac{\pi}{2}\cdot\frac{e^{-2}-e^2}{2i} = i\frac{\pi}{4}(e^2-e^{-2}).$$

Example 5.20 Evaluate the integral

$$\int_\Gamma \frac{e^z}{(z-2)^4}\,dz,$$

where Γ is the circle $|z| = 3$.

Solution The point $z = 2$ lies inside Γ, while the function $f(z) = e^z$ is analytic on the entire closed disk $|z| \leq 3$ (in fact, on all of $\mathbb{C}$). Using formula (5.25) with $n = 3$ we get

$$\int_\Gamma \frac{e^z}{(z-2)^4}\,dz = \int_\Gamma \frac{f(z)}{(z-2)^4}\,dz = \frac{2\pi i}{3!}f'''(2) = \frac{\pi i}{3}e^2.$$

We have seen in Section 3.2 that any harmonic function in a simply connected domain can be regarded as the real part of an analytic function on the same domain. According to Theorem 5.18 therefore, any harmonic function has partial derivatives of all orders, and these derivatives in their turn are harmonic functions.

Theorem 5.21 (Average Value Theorem). *Suppose the function $f(z)$ is analytic on a closed disk $|z - z_0| \leq R$ of radius R and centered at z_0. Then the value $f(z_0)$ the function takes at the center of the disk is equal to average of the values it takes on the circle $|z - z_0| = R$, that is*

$$f(z_0) = \frac{1}{2\pi}\int_0^{2\pi} f(z_0 + Re^{i\phi})\,d\phi. \tag{5.27}$$

Proof. Applying the Cauchy integral formula produces

$$f(z_0) = \frac{1}{2\pi i}\int_\Gamma \frac{f(\zeta)}{\zeta - z_0}\,d\zeta,$$

where Γ is the circle centered at z_0 of radius R. Then using the parametrization $\zeta(\phi) = z_0 + Re^{i\phi}$, $0 \le \phi \le 2\pi$, for which $\zeta'(\phi) = iRe^{i\phi}$, the integral becomes

$$f(z_0) = \frac{1}{2\pi i}\int_0^{2\pi} \frac{f(z_0 + Re^{i\phi})}{Re^{i\phi}} iRe^{i\phi}\, d\phi = \frac{1}{2\pi}\int_0^{2\pi} f(z_0 + Re^{i\phi})\, d\phi$$

as desired. □

Corollary 5.22. *A property analogous to* (5.27) *holds also for harmonic functions: if $u(z)$ is harmonic on a closed disk $|z - z_0| \le R$, then*

$$u(z_0) = \frac{1}{2\pi}\int_0^{2\pi} u(z_0 + Re^{i\phi})\, d\phi.$$

Proof. Let u be a harmonic function in a disk. There is a conjugate harmonic function v in the disk. Now we apply (5.27) to the analytic function $f = u+iv$, and separate real and imaginary parts in both sides of the equality. □

Cauchy's formula for derivatives also provides easy bounds on the size of the modulus of any derivative.

Theorem 5.23 (Cauchy inequality for derivatives). *Suppose $f(z)$ is an analytic function on a closed domain $|z - z_0| \le R$, and let M be the maximum modulus $|f(z)|$ that f assumes on the circle $|z - z_0| = R$. Then the derivatives of f at z_0 obey the inequality*

$$|f^{(n)}(z_0)| \le \frac{Mn!}{R^n}, \quad \text{for} \quad n = 1, 2, 3, \ldots \tag{5.28}$$

Proof. Applying formula (5.25), where we take Γ to be the circle $|z - z_0| = R$, we get

$$|f^{(n)}(z_0)| = \left| \frac{n!}{2\pi i}\int_\Gamma \frac{f(\zeta)}{(\zeta - z)^{n+1}}\, d\zeta \right| \le \frac{n!}{2\pi} \cdot \frac{M}{R^{n+1}} \cdot 2\pi R = \frac{Mn!}{R^n},$$

as desired. □

Theorem 5.24 (Liouville's[3] theorem). *If the function $f(z)$ is analytic and bounded in modulus on the entire complex plane $\mathbb{C}$, then in fact $f(z)$ is constant, i.e. $f(z) = C$.*

[3]Joseph Liouville (1809–1882) was a French mathematician and engineer.

Proof. Let M be an upper bound for $|f(z)|$ on $\mathbb{C}$. Take any point $z_0 \in \mathbb{C}$, and let Γ be the circle $|z - z_0| = R$ of radius R centered at z_0. Applying the previous theorem with $n = 1$ in formula (5.28) we get

$$|f'(z_0)| \leq \frac{M}{R}.$$

Since R could be any positive number, this inequality is only possible if $|f'(z_0)| = 0$. And since z_0 is arbitrary, we must have $f'(z) = 0$ everywhere in $\mathbb{C}$. Therefore $f(z) = C$ for some constant C. □

As an important application of Liouville's theorem we prove the following version of the fundamental theorem of algebra.

Theorem 5.25. *Every polynomial*

$$P(z) = a_n z^n + a_{n-1} z^{n-1} + \cdots + a_0, \quad a_n \neq 0,$$

of degree $n \geq 1$ has at least one root in the complex plane.

Proof. Assume that the statement is incorrect, that is there exists a polynomial $P(z)$ of degree $n \geq 1$ without zeros. We are going to show that this assumption will lead us to contradiction. Let

$$f(z) = \frac{1}{P(z)}.$$

Since $P(z) \neq 0$ for all z in $\mathbb{C}$, the function $f(z)$ is analytic on $\mathbb{C}$. Obviously, $\lim_{z\to\infty} P(z) = \infty$. Hence, $\lim_{z\to\infty} f(z) = 0$, and $f(z)$ is not constant. Moreover, $f(z)$ is bounded on $\mathbb{C}$. Indeed, since $f(z) \to 0$ as $z \to \infty$, there is a positive number R such that $|f(z)| < 1$ as $|z| > R$. But $f(z)$ is bounded on the closed disk $|z| \leq R$ as well, because $f(z)$ is analytic, and therefore continuous in this disk. Thus, $f(z)$ is analytic and bounded on $\mathbb{C}$, and by Liouville's theorem $f(z)$ is constant. We came to contradiction. Hence, our assumption that $P(z) \neq 0$ for all z in $\mathbb{C}$, is wrong, and the theorem is proved. □

An alternative proof of this celebrated result will be given later—see Theorem 7.40. Using this theorem, it's not difficult to prove that every polynomial of degree n has *precisely* n roots—see problem 5.

By Cauchy's Theorem 5.7, if f is an analytic function on a simply connected domain D, then the integral of f over any closed curve lying in D is zero. The following theorem is the converse.

Theorem 5.26 (Morera's[4] theorem). *Suppose the $f(z)$ is a continuous function on a simply connected domain D, and that every integral of f over any closed curve lying in D is zero. Then f is analytic on D.*

---

[4]Giacinto Morera (1856–1909), was an Italian engineer and mathematician.

Proof. We fix some point $z_0 \in D$. Then for any point $z \in D$ we define

$$\Phi(z) = \int_{z_0}^{z} f(\zeta)\, d\zeta. \tag{5.29}$$

Since $\int_\Gamma f(\zeta)\, d\zeta = 0$ for every closed curve Γ lying in D, for the same reason as in Corollary 5.10, the integral is path independent. So the integral in (5.29) defines a single-valued function of the variable $z \in D$. We now want to show that $\Phi(z)$ is analytic, and here the argument given in Theorem 5.13 works unchanged, and also shows that $\Phi'(z) = f(z)$. Since the derivative of an analytic function is itself analytic (Theorem 5.18), then f is analytic. □

As an application of Morera's theorem we prove the symmetry principle, which is useful for constructing conformal mappings from one symmetric domain onto another. This principle is a special case of analytic continuation.[5] We need the following lemma.

Lemma 5.27. *Let D_1 and D_2 be two disjoint simply connected domains whose boundaries intersect in a common smooth curve γ. Suppose $f_1(z)$ is analytic in D_1 and continuous in $D_1 \cup \gamma$, while $f_2(z)$ is analytic in D_2 and continuous in $D_2 \cup \gamma$. Moreover, suppose that $f_1(z)$ and $f_2(z)$ coincide on γ. Then the function*

$$g(z) = \begin{cases} f_1(z), & z \in D_1, \\ f_1(z) = f_2(z), & z \in \gamma, \\ f_2(z), & z \in D_2 \end{cases} \tag{5.30}$$

is analytic in $D = D_1 \cup \gamma \cup D_2$.

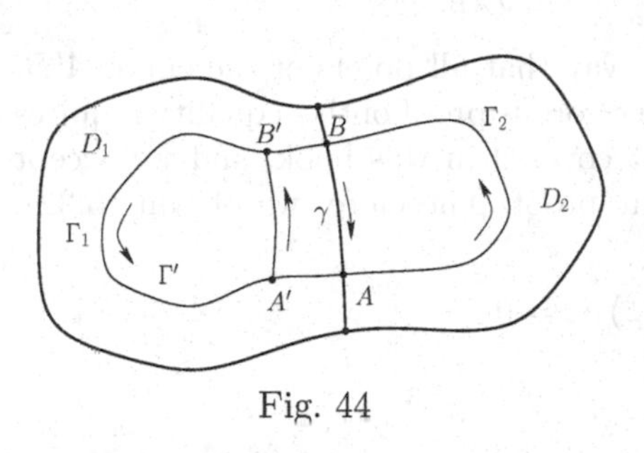

Fig. 44

In other words, under the conditions of Lemma 5.27, the function f_2 is an analytic continuation of f_1 through γ. This lemma shows us one more specific property of analytic functions which has no analogue for real-valued functions of one or two variables. Indeed, the function of two variables $f(x, y) = |y|$ is differentiable in the upper and in the lower half planes and is continuous on $\mathbb{R}^2$, but is not differentiable for $y = 0$.

Proof. By Morera's theorem, it is sufficient to show that every integral of g over any closed curve lying in D is zero. Let Γ be a closed curve contained in

[5] We will discuss a notion of analytic continuation in a more detailed way in Section 6.6.

D. If Γ does not intersect γ, then either Γ is contained in D_1, or Γ is contained in D_2. In both cases

$$\int_{\Gamma} g(z)\,dz = 0 \tag{5.31}$$

by Cauchy's Theorem 5.7. Suppose now that Γ intersects γ at points A and B. We divide Γ into two curves Γ_1 and Γ_2 with endpoints A and B (Fig. 44), and denote by AB and BA the part of γ between points A and B traversed from A to B and from B to A, respectively. First we prove that

$$\int_{\Gamma_1 \cup AB} g(z)\,dz = \int_{\Gamma_1 \cup AB} f_1(z)\,dz = 0. \tag{5.32}$$

Choose points A' and B' on Γ_1 which are close to A and B, respectively, and consider an auxiliary curve $A'B'$ in D_1. Let Γ' be a part of Γ_1 between B' and A'. The closed curve $\Gamma' \cup A'B'$ is contained in D_1, and by Cauchy's Theorem 5.7,

$$\int_{\Gamma' \cup A'B'} g(z)\,dz = \int_{\Gamma'} g(z)\,dz + \int_{A'B'} g(z)\,dz = 0.$$

Hence,

$$\lim_{A' \to A,\ B' \to B} \left(\int_{\Gamma'} g(z)\,dz + \int_{A'B'} g(z)\,dz \right) = 0.$$

Obviously,

$$\lim_{A' \to A,\ B' \to B} \int_{\Gamma'} g(z)\,dz = \int_{\Gamma_1} g(z)\,dz.$$

Since g is continuous in $D_1 \cup \gamma$,

$$\lim_{A' \to A,\ B' \to B} \int_{A'B'} g(z)\,dz = \int_{AB} g(z)\,dz,$$

if we choose the curves $A'B'$ in such a way that all points of the curve $A'B'$ approach AB as $A' \to A,\ B' \to B$. The rigorous proof of this equality requires notions in Real Analysis which are not covered in this book, and we accept this (intuitively clear) equality without proof. Therefore, we obtain (5.32). Analogously,

$$\int_{\Gamma_2 \cup BA} g(z)\,dz = 0.$$

Therefore,

$$\begin{aligned}
\int_{\Gamma} g(z)\,dz &= \int_{\Gamma_1} g(z)\,dz + \int_{\Gamma_2} g(z)\,dz \\
&= \int_{\Gamma_1} g(z)\,dz + \int_{AB} g(z)\,dz + \int_{BA} g(z)\,dz + \int_{\Gamma_2} g(z)\,dz \\
&= \int_{\Gamma_1 \cup AB} g(z)\,dz + \int_{\Gamma_2 \cup BA} g(z)\,dz = 0,
\end{aligned}$$

since the integrals over AB and BA cancel each other. Thus, we have (5.31), and lemma is proved. The reader may consider for himself the case when Γ intersects γ in more than two points. □

Theorem 5.28 (Symmetry principle). *Suppose that a simply-connected domain D_1 is bounded by a Jordan curve, a function $f_1(z)$ maps conformally D_1 onto a domain G_1, and the following conditions are satisfied:*

(a) the boundary of D_1 contains a line segment or a circular arc γ;

(b) the image Γ of γ under the mapping $f_1(z)$ is also a line segment or a circular arc (which is a part of the boundary of G_1).

Then there exists a function $f_2(z)$ such that:

(1) $f_2(z)$ coincides with $f_1(z)$ on γ and is analytic on $D_2 \cup \gamma$, where D_2 is a domain symmetric to D_1 with respect to γ (in other words, $f_2(z)$ is an analytic continuation of $f_1(z)$ into D_2 through γ);

(2) $f_2(z)$ maps D_2 conformally onto the domain G_2 symmetric to G_1 with respect to Γ;

(3) if in addition, domains G_1, G_2 are disjoint and also domains D_1, D_2 are disjoint, then the function $g(z)$ defined in (5.30) performs a conformal mapping of $D_1 \cup \gamma \cup D_2$ onto $G_1 \cup \Gamma \cup G_2$.

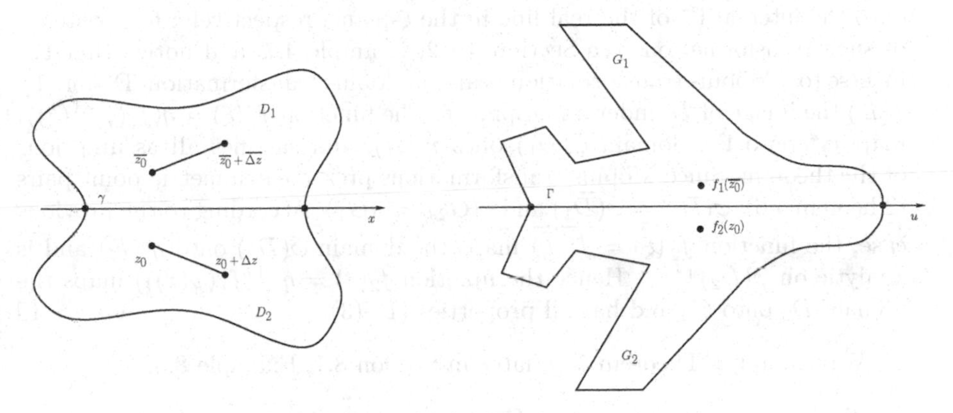

Fig. 45

Proof. We start with the special case when γ and Γ are segments of the real axis (Fig. 45). Let

$$f_2(z) = \overline{f_1(\overline{z})}.$$

Clearly $f_2(z)$ is 1-1 in D_2 and maps D_2 onto the domain G_2 which is symmetric to G_1. Prove that $f_2(z)$ is analytic in D_2. Let z_0 be a point in D_2, and $z_0 + \Delta z \in D_2$. Then

$$\frac{f_2(z_0 + \Delta z) - f_2(z_0)}{\Delta z} = \frac{\overline{f_1(\overline{z_0} + \overline{\Delta z}) - f_1(\overline{z_0})}}{\Delta z} = \overline{\left(\frac{f_1(\overline{z_0} + \overline{\Delta z}) - f_1(\overline{z_0})}{\overline{\Delta z}}\right)}. \tag{5.33}$$

Since $f_1(z)$ is differentiable at $\overline{z_0}$, there exists the limit

$$\lim_{\overline{\Delta z}\to 0} \frac{f_1(\overline{z_0}+\overline{\Delta z}) - f_1(\overline{z_0})}{\overline{\Delta z}} = f_1'(\overline{z}_0).$$

Hence there exists the limit of the left hand side of (5.33) which is equal to $f_2'(z_0) = \overline{f_1'(\overline{z}_0)}$. Therefore, $f_2(z)$ is differentiable in D_2.[6]

If $z = x \in \gamma$, then $f_1(z) \in \Gamma$ and therefore, is real as well. Hence, $f_2(x) = \overline{f_1(x)} = f_1(x)$. By the Caratheodory's theorem (Property 4.17), $f_1(z)$ is continuous in $D_1 \cup \gamma$. Hence, the function $f_2(z)$ is continuous in $D_2 \cup \gamma$. Therefore we may apply Lemma 5.27 to restrictions of f_1 and f_2 on disjoint subdomains of D_1 and D_2 near γ whose boundaries contain γ, and see that f_1 and f_2 are analytic on γ.

If domains G_1 and G_2 are disjoint, then by Lemma 5.27 the function $g(z)$ defined in (5.30) is single-valued and analytic in $D_1 \cup \gamma \cup D_2$. By construction, $g(z)$ maps $D_1 \cup \gamma \cup D_2$ conformally onto $G_1 \cup \Gamma \cup G_2$. So, in the case when γ and Γ are segments of the real axis, the theorem is proved.

The general case is reduced to the previously considered by the Möbius transformation. Let $\xi = \varphi(z)$ and $\zeta = \eta(w)$ be Möbius transformations which map the arc γ onto the interval γ^* of the real line in the ξ-plane, and Γ onto the interval Γ^* of the real line in the ζ-plane, respectively; for existence of such transformations see Section 4.1.2, Example 4.7, and notice that the inverse to a Möbius transformation is also a Möbius transformation. Denote by $f(D)$ the image of D under a mapping f. The function $f_1^*(\xi) = \eta(f_1(\varphi^{-1}(\xi)))$ maps γ^* onto Γ^*, domain $\varphi(D_1)$ onto $\eta(G_1)$, and satisfies all assumptions of the theorem. Since Möbius transformations preserve symmetric point pairs (Theorem 4.6), $\varphi(D_2) = \overline{\varphi(D_1)}$ and $\eta(G_2) = \overline{\eta(G_1)}$. According to the previous case, the function $f_2^*(\xi) = \overline{f_1^*(\overline{\xi})}$ maps the domain $\varphi(D_2)$ onto $\eta(G_2)$ and is analytic on $\varphi(D_2) \cup \gamma^*$. Hence the function $f_2(z) = \eta^{-1}(f_2^*(\varphi(z)))$ maps the domain D_2 onto G_2 and has all properties (1)–(3). □

We will apply Theorem 5.28 later in Section 8.1, Example 8.5.

Problems

1. Using either Cauchy's theorem or Cauchy's formula for derivatives, find the value of the integral

$$\int_\Gamma \frac{\cos 2z}{z^2+\pi^2}\,dz,$$

where Γ is the closed curve $|z+i\pi| = 2$, taken in the counterclockwise direction.

---

[6] Another proof of differentiability of $f_2(z)$ is based on the Cauchy-Riemann equations.

2. Using either Cauchy's theorem or Cauchy's formula for derivatives, find the value of the integral,

$$\int_{\Gamma} \frac{z^5 - 2z^4 + 3}{(z-i)^4}\, dz$$

where Γ is the closed curve $|z| = 5$, taken in the counterclockwise direction.

3. Using either Cauchy's Theorem or Cauchy's formula for derivatives, find the values of the integrals along the closed curve Γ; the orientation of the path is counterclockwise in each case.

a) $\displaystyle\int_{\Gamma} \frac{\cos z}{z^2 - \pi^2}\, dz, \ \Gamma : |z - \pi| = 1;$ b) $\displaystyle\int_{\Gamma} \frac{e^z}{z^2 + 1}\, dz, \ \Gamma : |z - i| = 1;$

c) $\displaystyle\int_{\Gamma} \frac{\sin z}{(z-1)^3}\, dz, \ \Gamma : |z| = 2;$ d) $\displaystyle\int_{\Gamma} \frac{z^3 + 1}{(z+1)^4}\, dz, \ \Gamma : |z| = 3;$

e) $\displaystyle\int_{\Gamma} \frac{z \sin z}{(z + \frac{\pi}{2})^2}\, dz, \ \Gamma : |z| = 4.$

4. Prove formula (5.25) using mathematical induction.

5. Using Theorem 5.25, prove that every polynomial

$$P(z) = a_n z^n + a_{n-1} z^{n-1} + \cdots + a_0, \quad a_n \neq 0,$$

of degree $n \geq 0$ has precisely n roots in the complex plane; some of these roots may coincide.

6

Series

6.1 Definitions

If we have an infinite sequence z_1, z_2, ... of complex numbers, we can use it to form an expression

$$z_1 + z_2 + \cdots + z_n + \cdots = \sum_{n=1}^{\infty} z_n, \tag{6.1}$$

which is called *numerical series.* The numbers z_1, z_2, ... are called the terms of the series. For the moment the expression (6.1) is purely formal, in that we cannot add up infinitely many terms. But if we restrict ourselves to a finite number of terms, for example choose the first n terms, then we can really add all of them whatever the value of n. We define the *partial sums* S_n of the series to be the sums of only the first n terms:

$$S_n = z_1 + z_2 + \cdots + z_n = \sum_{k=1}^{n} z_k.$$

If the limit of this sequence $\{S_n\}$ exists and is finite, then the series is said to be *convergent.* The number $S = \lim_{n\to\infty} S_n$ is called *the sum of the series.* If the series converges and its sum equals S, then we write

$$\sum_{n=1}^{\infty} z_n = S.$$

It does not mean that we added all terms of the series (that is impossible). But if we add up sufficiently many terms of the series, the partial sum becomes as close we like to S.

If the limit does not exist or is infinite, the series is *divergent.*

As should be expected by now, many of the definitions and properties of complex series are inherited from those of real series. The following theorem formalizes the connection.

DOI: 10.1201/9780367810283-6

Theorem 6.1. *A series (6.1), with $z_n = x_n + iy_n$ for $n = 1, 2, \ldots$, is convergent if and only if the two real series $\sum_{n=1}^{\infty} x_n$ and $\sum_{n=1}^{\infty} y_n$ are both convergent. If so, then*

$$\sum_{n=1}^{\infty} z_n = \sigma + i\tau$$

if and only if

$$\sum_{n=1}^{\infty} x_n = \sigma \quad \text{and} \quad \sum_{n=1}^{\infty} y_n = \tau.$$

Proof. Let us use the following notation for the partial sums:

$$\begin{aligned} S_n &= z_1 + z_2 + \cdots + z_n, \\ \sigma_n &= x_1 + x_2 + \cdots + x_n, \\ \tau_n &= y_1 + y_2 + \cdots + y_n, \end{aligned}$$

so that $S_n = \sigma_n + i\tau_n$. The theorem now follows immediately by applying Theorem 2.2 to the two sequences $\{\sigma_n\}$ and $\{\tau_n\}$: $\{S_n\}$ converges if and only if these do; and their limits are the real and imaginary parts, respectively, of the limit of $\{S_n\}$. But convergence of the sequences $\{S_n\}$, $\{\sigma_n\}$, and $\{\tau_n\}$ is equivalent to the convergence of the series $\sum_{n=1}^{\infty} z_n$, $\sum_{n=1}^{\infty} x_n$, and $\sum_{n=1}^{\infty} y_n$, respectively. □

With the help of this theorem, many properties and formulas which are valid for real series can be transferred to complex series. Here are a few:

1. (Necessary condition for convergence) If $\sum_{n=1}^{\infty} z_n$ converges, then

$$\lim_{n\to\infty} z_n = 0.$$

The converse, of course, is false: e.g. $\frac{1}{n} \to 0$, but $\sum_{n=1}^{\infty} \frac{1}{n} = \infty$.

2. If

$$\sum_{n=1}^{\infty} z_n = S \quad \text{and} \quad \sum_{n=1}^{\infty} w_n = T,$$

are convergent complex series, then

$$\sum_{n=1}^{\infty} (z_n + w_n) = S + T.$$

3. Suppose that $\sum_{n=1}^{\infty} z_n$ converges to the sum S; then for any complex constant λ,

$$\sum_{n=1}^{\infty} \lambda z_n = \lambda S.$$

4. If any finite number of terms are either added to or removed from a convergent series, then the resulting series is still convergent.

5. (Cauchy's criterion for convergence) A series $\sum_{n=1}^{\infty} z_n$ converges if and only if, for any $\epsilon > 0$ there is some N such that

$$\text{whenever } n_2 > n_1 > N \quad \text{then} \quad \left| \sum_{k=n_1}^{n_2} z_k \right| < \epsilon.$$

As with real series, the concept of absolute convergence will be important for us. A series $\sum_{n=1}^{\infty} z_n$ is said to be *absolutely convergent* if the series of the moduli of its terms,

$$\sum_{n=1}^{\infty} |z_n| = |z_1| + |z_2| + \cdots + |z_n| + \ldots,$$

is convergent. This is a strictly stronger condition than mere convergence.

Theorem 6.2. *If the series $\sum_{n=1}^{\infty} |z_n|$ converges, the so does the series $\sum_{n=1}^{\infty} z_n$. In other words, absolute convergence implies convergence.*

Proof. The Cauchy convergence criterion applies to both real and complex series, so first we apply it to the convergent series $\sum_{n=1}^{\infty} |z_n|$: let $\epsilon > 0$ be arbitrary, then we can find an N such that whenever $n_2 > n_1 > N$,

$$\sum_{k=n_1}^{n_2} |z_k| < \epsilon.$$

Recall from the first chapter that for any complex numbers z and w, $|z+w| \leq |z| + |w|$; this inequality is easy to extend to any number of summands. So comparing the above with the series $\sum_{n=1}^{\infty} z_n$, we see that

$$\left| \sum_{k=n_1}^{n_2} z_k \right| \leq \sum_{k=n_1}^{n_2} |z_k| < \epsilon.$$

Since ϵ was arbitrary, applying the Cauchy convergence criterion again, now to the series $\sum_{n=1}^{\infty} z_n$, we see that this series converges. □

The converse of this theorem fails: convergence does not imply absolute convergence. For example, the alternating real series $1 - \frac{1}{2} + \frac{1}{3} - \frac{1}{4} + \ldots$ converges, while $1 + \frac{1}{2} + \frac{1}{3} + \frac{1}{4} + \ldots$ does not.

We say that a series $\sum_{n=1}^{\infty} z_n$ *converges conditionally,* if this series converges, but the series $\sum_{n=1}^{\infty} |z_n|$ diverges (i.e. a series converges conditionally if it converges, but *not* absolutely).

Since the series $\sum_{n=1}^{\infty} |z_n|$ is composed of nonegative real terms, all the tools we have acquired for real series can be used to help determine absolute convergence. Here we recall a few (proofs are omitted).

Comparison test. Suppose that $\{z_n\}$ and $\{w_n\}$ are sequences for which $|z_n| \le |w_n|$ for all sufficiently big n; in other words, there is some N such that $|z_n| \le |w_n|$ holds for all $n \ge N$. Then

1. if $\sum_{n=1}^{\infty} |w_n|$ converges then so does $\sum_{n=1}^{\infty} |z_n|$;
2. if $\sum_{n=1}^{\infty} |z_n|$ diverges then so does $\sum_{n=1}^{\infty} |w_n|$.

D'Alembert's test (the ratio test). Suppose that the ratios between successive terms approach a limit:

$$\lim_{n\to\infty} \frac{|z_{n+1}|}{|z_n|} = l.$$

Then,

1. if $l < 1$ then $\sum_{n=1}^{\infty} z_n$ converges absolutely;
2. if $l > 1$ then $\sum_{n=1}^{\infty} z_n$ diverges;
3. if $l = 1$ then the test is inconclusive: $\sum_{n=1}^{\infty} z_n$ may either converge or diverge.

Cauchy's root test. Suppose that

$$\lim_{n\to\infty} \sqrt[n]{|z_n|} = l.$$

Then we have the same conclusions as for D'Alembert's test:

1. if $l < 1$ then $\sum_{n=1}^{\infty} z_n$ converges absolutely;
2. if $l > 1$ then $\sum_{n=1}^{\infty} z_n$ diverges;
3. if $l = 1$ then the test is inconclusive: $\sum_{n=1}^{\infty} z_n$ may either converge or diverge.

Example 6.3 Determine whether the following series are convergent.

$$(a) \sum_{n=1}^{\infty} \frac{\cos(in)}{2^n}; \qquad (b) \sum_{n=1}^{\infty} \frac{\cos(i+n)}{2^n}.$$

Solution (a) We will try to compare the given series to simpler one. From the definition of $\cos z$ in (4.28),

$$\cos(in) = \frac{e^{-n} + e^n}{2} > \frac{e^n}{2}.$$

Therefore

$$\frac{\cos(in)}{2^n} > \frac{e^n}{2 \cdot 2^n} = \frac{1}{2}\left(\frac{e}{2}\right)^n,$$

so we can use the series

$$\sum_{n=1}^{\infty} \frac{1}{2}\left(\frac{e}{2}\right)^n$$

in a comparison test. This being a geometric series with growth ratio $\frac{e}{2} > 1$, it diverges.

We could also have applied D'Alembert's test:

$$l = \lim_{n\to\infty} \frac{1}{2}\left(\frac{e}{2}\right)^{n+1} \Big/ \frac{1}{2}\left(\frac{e}{2}\right)^n = \lim_{n\to\infty} \frac{e}{2} = \frac{e}{2} > 1.$$

Or, more simply, the failure of the series' terms to approach zero, the necessary condition mentioned above, implies divergence. Then by the comparison test, the original series (a) also diverges.

(b) First we show that the modulus of $\cos(i+n)$ is bounded, and the bound does not depend on n. For

$$\begin{aligned} |\cos(i+n)| &= |\cos i \cos n - \sin i \sin n| \\ &\leq |\cos i|\,|\cos n| + |\sin i|\,|\sin n| \leq |\cos i| + |\sin i| = M, \end{aligned}$$

where M is some positive constant. Then

$$\left|\frac{\cos(i+n)}{2^n}\right| \leq \frac{M}{2^n},$$

so we can again compare the original series (b) to a geometric series, $\sum_{n=1}^{\infty} \frac{M}{2^n}$. This time the geometric series converges; and then by the comparison test, the original series (b) converges absolutely.

The tail end of a series, in other words what is left over after removing the first n terms, is called the *nth remainder*, or just the remainder, of the original series. If we started with a convergent series, then any remainder will also be convergent, and its sum is denoted by r_n:

$$r_n = \sum_{k=n+1}^{\infty} z_k.$$

It is easy to see that for any n, $S = S_n + r_n$, where S is the sum and S_n is the partial sum of the series $\sum_{n=1}^{\infty} z_n$. From this we see that *for a convergent series, the remainders* $r_n \to 0$ *as* $n \to \infty$. For if S is the sum of the series, then $\lim_{n\to\infty} S_n = S$, hence

$$\lim_{n\to\infty} r_n = \lim_{n\to\infty} (S - S_n) = S - S = 0.$$

Problems

1. Prove properties 1–4 of series listed after Theorem 6.1.

2. Determine whether the following series are convergent. If a series converges, is the convergence absolute or conditional?

$$a)\ \sum_{n=0}^{\infty} \frac{1}{(1+i)^n};\quad b)\ \sum_{n=0}^{\infty} \frac{e^{in}}{(1+i)^n};\quad c)\ \sum_{n=0}^{\infty} \frac{e^{n}}{(1+i)^n};$$

$$d)\ \sum_{n=0}^{\infty} e^{in\pi/4};\quad e)\ \sum_{n=1}^{\infty} \frac{i^n}{n};\quad f)\ \sum_{n=0}^{\infty} \frac{\sin(i+n)}{(1+i)^n}.$$

6.2 Function Series

A *function series* is an expression of the form

$$f_1(z) + f_2(z) + \cdots + f_n(z) + \cdots = \sum_{n=1}^{\infty} f_n(z), \tag{6.2}$$

where $f_1(z),\ f_2(z), \ldots$ are all functions defined on one and the same domain D.

When we fix a point $z \in D$, the function series becomes a numerical series, which may converge for some values of z and diverge for others. Such points are called, respectively, *points of convergence* and *points of divergence* of the series. The set of all points of convergence is called the *set of convergence* of the function series. For any z in the set of convergence, we denote by $S(z)$ the sum of the corresponding numerical series. In this way, $S(z)$ defines a function on the set of convergence.

Now suppose that the function series (6.2) happens to converge on some domain D, and let $S(z)$ be as above; also let $S_n(z) = \sum_{k=1}^{n} f_k(z)$ be the partial sum of the numerical series at z. This function series is said to be *uniformly convergent* on D, if for any $\epsilon > 0$ there is some number N, dependent on ϵ, for which

$$\text{whenever} \quad n > N, \quad \text{then} \quad |S(z) - S_n(z)| < \epsilon \quad \text{for any } z \in D.$$

The essential feature of uniform convergence is that N may depend on ϵ but *not* on z; in ordinary convergence N may depend on both.

Let us explore this. Suppose that $S(z)$ is the sum of a uniformly convergent function series, and let $\epsilon > 0$ be arbitrarily small. Then for a sufficiently large n, the partial sum $S_n(z)$ will differ from $S(z)$ by less then ϵ *at all points z in D*. In other words, the partial sums $S_n(z)$ approach their limits $S(z)$ uniformly in the entire domain D.

Of course, that function series converges on a domain D does not necessarily mean that it converges uniformly, as this example shows.

Example 6.4 Consider the series

$$1 + z + z^2 + \cdots = \sum_{k=0}^{\infty} z^k, \tag{6.3}$$

the terms of which form a geometric sequence with ratio z. Recall the formula for the sum of the first n terms of such a series:

$$S_n(z) = \sum_{k=0}^{n-1} z^k = \frac{1 - z^n}{1 - z}, \quad z \neq 1.$$

When $|z| < 1$, the limit exists:

$$S(z) = \lim_{n\to\infty} S_n(z) = \lim_{n\to\infty} \frac{1 - z^n}{1 - z} = \frac{1}{1 - z}.$$

So this series (6.3) converges in the disk $|z| < 1$, and the sum is $S(z) = 1/(1 - z)$.

Now let us see whether or not the convergence is uniform. Fix some $\epsilon > 0$. For uniform convergence we must find some N for which $|S(z) - S_n(z)| < \epsilon$ whenever $n > N$ and $|z| < 1$. So let us calculate:

$$|S(z) - S_n(z)| = \left| \frac{1}{1 - z} - \frac{1 - z^n}{1 - z} \right| = \left| \frac{z^n}{1 - z} \right|$$

which must be less that ϵ for all z, $|z| < 1$. But regardless of the value of n, we can choose a z sufficiently close to 1 so that the right hand side will be as large as we want; certainly we can always make it bigger than ϵ. So there is no n large enough to satisfy $|S(z) - S_n(z)| < \epsilon$ at all points in the unit disk, and therefore the function series does not converge uniformly there.

At the same time the function series (6.3) will converge uniformly on every disk $|z| < r$ of radius $r < 1$. For in this case

$$|S(z) - S_n(z)| = \left| \frac{z^n}{1 - z} \right| \leq \frac{|z|^n}{1 - |z|} < \frac{r^n}{1 - r} \quad \text{whenever} \quad |z| < r.$$

We leave it as an exercise for the reader to fill in the missing details.

Theorem 6.5 (Weierstrass[1] uniform convergence test). *Let $\sum_{n=1}^{\infty} a_n$, where all $a_n > 0$, be an (absolutely) convergent real series; let $\sum_{n=1}^{\infty} f_n(z)$ be a function series, defined on some domain D. If there is N such that $|f_n(z)| \leq a_n$ for all $n \geq N$ and all $z \in D$, then the function series converges absolutely and uniformly on D.*

[1] Karl Theodor Wilhelm Weierstrass (1815–1897) was a German mathematician who made an enormous contribution to modern analysis.

When the conditions of this theorem are satisfied, the series $\sum_{n=1}^{\infty} a_n$ is said to be a *majorizing* or *dominating* series for $\sum_{n=1}^{\infty} f_n(z)$.

Proof. Absolute convergence follows from our earlier comparison test (1). To prove uniformity, fix $\epsilon > 0$. Since the series $\sum_{n=1}^{\infty} a_n$ converges, the remainder sum

$$r_n = \sum_{k=n+1}^{\infty} a_k$$

tends to 0 as $n \to \infty$. Hence there is $K > N$ such that $r_n < \epsilon$ whenever $n > K$. Then the remainder for the function series,

$$\begin{aligned}|S(z) - S_n(z)| = |f_{n+1}(z) + f_{n+2}(z) + \ldots| &\leq |f_{n+1}(z)| + |f_{n+2}(z)| + \ldots \\ &\leq |a_{n+1}| + |a_{n+2}| + \cdots = r_n < \epsilon,\end{aligned}$$

for any $z \in D$. So for any $\epsilon > 0$ there is some number K, dependent on ϵ, for which $|S(z) - S_n(z)| < \epsilon$ for any $z \in D$ whenever $n > K$. Therefore the function series converges uniformly on D. □

Here we list some of the fundamental properties of uniformly convergent function series.

Theorem 6.6. *If the functions in the series (6.2) are continuous on the domain D, and if the series converges uniformly on D, then the sum $S(z)$ is also continuous on D.*

Proof. Take any number $\epsilon > 0$, and a point z_0 in D. Then by uniform convergence we can find an N such that $|S(z) - S_n(z)| < \epsilon/3$ whenever $n > N$ and $z \in D$. Let us fix one such $n > N$; then the partial sum $S_n(z)$ is a finite sum of continuous functions, and therefore is itself continuous on D. Therefore we can find a real number $\delta > 0$ such that $|S_n(z) - S_n(z_0)| < \epsilon/3$ whenever $|z - z_0| < \delta$. Then

$$\begin{aligned}|S(z) - S(z_0)| &= |S(z) - S_n(z) \;+\; S_n(z) - S_n(z_0) \;+\; S_n(z_0) - S(z_0)| \\ &\leq |S(z) - S_n(z)| + |S_n(z) - S_n(z_0)| + |S_n(z_0) - S(z_0)| \\ &\leq \frac{\epsilon}{3} + \frac{\epsilon}{3} + \frac{\epsilon}{3} = \epsilon.\end{aligned}$$

Thus, for any $\epsilon > 0$ we can find a $\delta > 0$ such that $|S(z) - S(z_0)| < \epsilon$ whenever $|z - z_0| < \delta$; which is the definition of continuity for S at z_0. Since z_0 was arbitrary, this establishes that S is continuous everywhere in D. □

This theorem extends the fact known from calculus that a finite sum of continuous functions is continuous, to an infinite sum—as long as the convergence is uniform. In general, if the convergence is not uniform this property does not hold—see problem 4.

Now we will extend another such property of finite sums to uniformly convergent series: the integrsal of a sum is the sum of the integrals.

Theorem 6.7. *Suppose that the functions in the series (6.2) are continuous on the domain D, and that the series converges uniformly on D to the function $S(z)$. Then the integral of S, along any curve Γ lying within D, can be evaluated termwise:*

$$\int_\Gamma S(z)\,dz = \int_\Gamma f_1(z)\,dz + \int_\Gamma f_2(z)\,dz + \cdots = \sum_{n=1}^{\infty} \int_\Gamma f_n(z)\,dz. \tag{6.4}$$

Proof. By the previous theorem we know that $S(z)$ is continuous, and therefore the integral $\int_\Gamma S(z)\,dz$ exists. We must prove the equality (6.4). So let

$$\sigma_n = \sum_{k=1}^{n} \int_\Gamma f_k(z)\,dz$$

be the partial sums of the series of integrals. We must show that

$$\lim_{n\to\infty} \sigma_n = \int_\Gamma S(z)\,dz.$$

Since termwise integration is valid for finite sums, we have

$$\sigma_n = \sum_{k=1}^{n} \int_\Gamma f_k(z)\,dz = \int_\Gamma \left(\sum_{k=1}^{n} f_k(z) \right) dz = \int_\Gamma S_n(z)\,dz,$$

where $S_n(z)$ is the partial sum of the function series (6.2). Let l be the arc length of the path Γ; and let $\epsilon > 0$ be arbitrary. By the uniform convergence of the function series, we can find an N such that $|S(z) - S_n(z)| < \epsilon/l$ whenever $n > N$ and $z \in D$. For such n and z we have

$$\left| \int_\Gamma S(z)\,dz - \sigma_n \right| = \left| \int_\Gamma S(z)\,dz - \int_\Gamma S_n(z)\,dz \right|$$

$$= \left| \int_\Gamma (S(z) - S_n(z))\,dz \right| \le \int_\Gamma |S(z) - S_n(z)|\,|z'(t)|\,dt \le \frac{\epsilon}{l} \cdot l = \epsilon,$$

where we have again used Property 5.6 for integrals. Thus, we have shown that for any $\epsilon > 0$ there is some N such that

$$\left| \int_\Gamma S(z)\,dz - \sigma_n \right| < \epsilon \quad \text{whenever } n > N.$$

According to the definition of limit,

$$\lim_{n\to\infty} \sigma_n = \int_\Gamma S(z)\,dz,$$

which means the series of integrals converges to the integral of the sum. □

We need one further simple property.

Property 6.8. *If the function series (6.2) converges uniformly on a domain D to the function $S(z)$, and the function $g(z)$ is bounded on D, then the series $\sum_{n=1}^{\infty} g(z)f_n(z)$ converges uniformly on D and its sum is $g(z)S(z)$.*

Proof. Since $g(z)$ is bounded, there is some real M for which $|g(z)| < M$ on all of D. Let

$$S_n(z) = \sum_{k=1}^{n} f_k(z), \quad \sigma_n(z) = \sum_{k=1}^{n} g(z)f_k(z)$$

be the partial sums of the two series. Since the partial sums are of only finitely many terms,

$$\sigma_n(z) = g(z)\sum_{k=1}^{n} f_k(z) = g(z)S_n(z).$$

Now we take an arbitrary $\epsilon > 0$; by uniform convergence, we can find an N such that $|S(z) - S_n(z)| < \epsilon/M$ whenever $n > N$ and $z \in D$. Then

$$\begin{aligned} |g(z)S(z) - \sigma_n(z)| &= |g(z)S(z) - g(z)S_n(z)| \\ &= |g(z)|\,|S(z) - S_n(z)| \leq M \cdot \frac{\epsilon}{M} = \epsilon. \end{aligned}$$

This shows that $\sum_{n=1}^{\infty} g(z)f_n(z)$ converges uniformly to $g(z)S(z)$. □

Theorems 6.5–6.7 and Property 6.8 are valid not only for series defined on domains, but also for series defined on curves; and the proofs would be the same.

So far in our study of function series we have never had to require the functions $f_n(z)$ to be analytic. But if, in addition to converging uniformly, the series (6.2) is composed of functions which are analytic, then the sum of the series will possess additional important properties; properties which would not hold for the sum of a uniformly convergent series of differentiable *real* functions.

Theorem 6.9 (Weierstrass' theorem for the sum of uniformly convergent series). *Suppose that the functions $f_n(z)$, for $n = 1, 2, 3 \ldots$, are all analytic on the same domain D; and that the series*

$$\sum_{n=1}^{\infty} f_n(z) \tag{6.5}$$

is uniformly convergent on D. Then the sum $S(z)$ is also analytic on D. In addition, for each $k = 1, 2, 3 \ldots$ the series

$$\sum_{n=1}^{\infty} f_n^{(k)}(z),$$

obtained by termwise k-fold differentiation of the original series, will converge on D to the function $S^{(k)}(z)$. In other words, the series (6.5) can be differentiated termwise any number of times.

Proof. Let z_0 be any point in D. Since the domain D is open, we can find a neighborhood U of z_0 lying wholly within D; that is we can find a disk centered at z_0 lying within D. We denote by γ the circle forming the boundary of this disk.

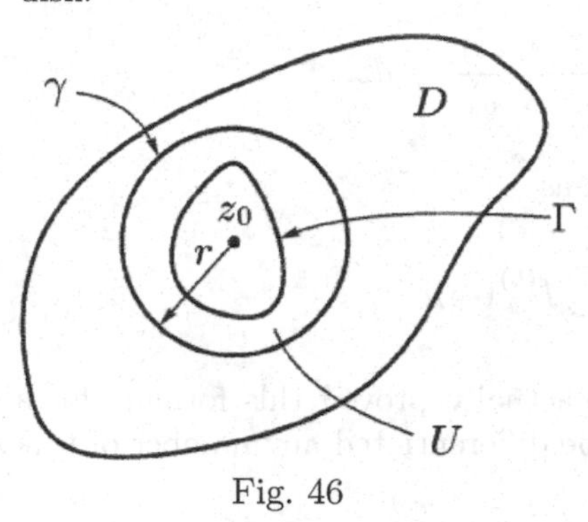

Fig. 46

Now let Γ be any closed contour lying in D (Fig. 46). By assumption the series (6.5) converges uniformly on D, therefore the same is true on U. The functions $f_n(z)$, being analytic, are certainly continuous on D and U as well. Therefore we can be apply Theorem 6.7, according to which

$$\int_\Gamma S(z)\,dz = \sum_{n=1}^{\infty} \int_\Gamma f_n(z)\,dz.$$

The functions $f_n(z)$ being analytic all the integrals on the right side are zero, by Cauchy's Theorem 5.7; therefore the integral on the left is also zero. By Theorem 6.6, the function $S(z)$ is continuous on a simply connected domain U. And since Γ was arbitrary, we can now apply Morera's Theorem 5.26 to conclude that $S(z)$ is analytic on U, and in particular at the point z_0. Since z_0 was arbitrary, we have actually proved that $S(z)$ is analytic on the entire domain D.

Now we want to show that the series (6.5) can be differentiated termwise. Denote by r the radius of the disk U, so that the circle γ is $|z - z_0| = r$. Consider the function

$$g(z) = \frac{k!}{2\pi i} \frac{1}{(z - z_0)^{k+1}},$$

where k is any counting number. When $z \in \gamma$ the modulus of $g(z)$ is

$$|g(z)| = \frac{k!}{2\pi} \frac{1}{|z - z_0|^{k+1}} = \frac{k!}{2\pi r^{k+1}},$$

so that $g(z)$ is bounded on γ. Since the series (6.5) converges uniformly on γ to $S(z)$, we can apply Property 6.8 to conclude that the series

$$\sum_{n=1}^{\infty} g(z) f_n(z)$$

converges uniformly on γ to $g(z)S(z)$, i.e.

$$\frac{k!}{2\pi i} \frac{S(z)}{(z - z_0)^{k+1}} = \sum_{n=1}^{\infty} \frac{k!}{2\pi i} \frac{f_n(z)}{(z - z_0)^{k+1}}.$$

Next we integrate along γ, applying Theorem 6.7 on the right side to integrate termwise:

$$\frac{k!}{2\pi i}\int_{\gamma}\frac{S(z)}{(z-z_0)^{k+1}}\,dz=\sum_{n=1}^{\infty}\frac{k!}{2\pi i}\int_{\gamma}\frac{f_n(z)}{(z-z_0)^{k+1}}\,dz.$$

Here we recognize the terms on the left and right from Cauchy's formula (5.25) for derivatives:

$$f^{(k)}(z_0)=\frac{k!}{2\pi i}\int_{\gamma}\frac{f_n(z)}{(z-z_0)^{k+1}}\,dz.$$

So we can rewrite the previous equation as

$$S^{(k)}(z_0)=\sum_{n=1}^{\infty}f^{(k)}(z_0).$$

Again, since z_0 was arbitrary, we have actually proved this formula for all $z\in D$. So the function series (6.5) can be differentiated any number of times termwise throughout the domain D. □

Problems

1. Prove that the series

$$\sum_{n=1}^{\infty}\frac{z^n}{n^2}$$

converges uniformly on the closed disk $|z|\leq 1$ and diverges when $|z|>1$.

2. a) Find the set of convergence of the series

$$\sum_{n=0}^{\infty}e^{nz}.$$

b) Prove that this series converges uniformly on any half-plane $\operatorname{Re} z\leq -\delta$ with $\delta>0$, and does not converge uniformly on the set $\operatorname{Re} z<0$.

3. a) Find the set of convergence of the series

$$\sum_{n=1}^{\infty}\frac{\sin(nz)}{n^2}.$$

b) Does the series converge uniformly on the set of convergence?

4. Consider the series $\sum\limits_{n=1}^{\infty}f_n(z)$, where

$$f_1(z)=e^{-|z|^2}=e^{-(x^2+y^2)},\quad f_n(z)=e^{-n|z|^2}-e^{-(n-1)|z|^2},\quad n=2,3,\ldots$$

Note that the functions $f_n(z)$ are continuous and even differentiable on $\mathbb{C}$ as real-valued functions of two variables.

a) Show that the partial sums $S_n(z)$ of the series are equal to

$$S_n(z) = e^{-n|z|^2}.$$

b) Prove that the series $\sum\limits_{n=1}^{\infty} f_n(z)$ converges on $\mathbb{C}$, but its sum $S(z)$ is not continuous on $\mathbb{C}$. Why this fact does not contradict Theorem 6.6 ?

6.3 Power Series

Power series is a series of the form

$$\sum_{n=0}^{\infty} c_n(z-z_0)^n = c_0 + c_1(z-z_0) + c_2(z-z_0)^2 + \cdots + c_n(z-z_0)^n + \ldots, \quad (6.6)$$

where z_0, c_0, c_1, ... are given complex constants, and z is a complex variable. The numbers c_0, c_1, ... are called the *coefficients* of the power series, and z_0 is called the *center* of the series.

Obviously power series are a special case of function series, and they turn out to play a very important role in function theory. The following theorem addresses some aspects of the set of convergence for power series.

Theorem 6.10 (Abel's theorem).
1. If the power series (6.6) converges at some point z_1 other than z_0, then it converges absolutely on the open disk of radius $|z_1 - z_0|$ centered at z_0, i.e. on the set

$$|z - z_0| < |z_1 - z_0|.$$

2. If the power series (6.6) diverges at some point z_2, then it diverges at all z outside the circle of radius $|z_2 - z_0|$ centered at z_0, i.e. on the set

$$|z - z_0| > |z_2 - z_0|, \text{ see Fig. 47}$$

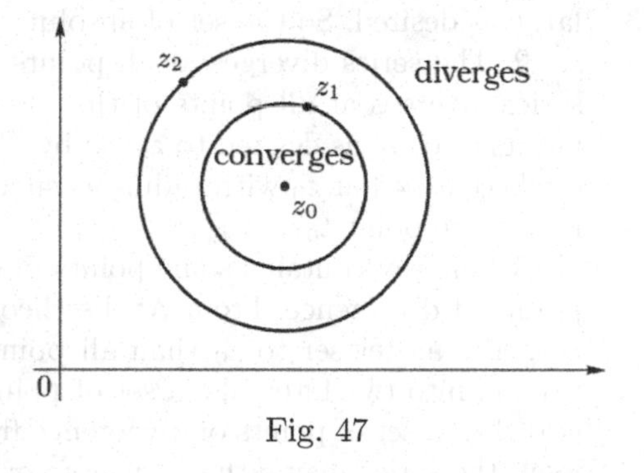

Fig. 47

Proof. 1. If (6.6) converges at $z = z_1$, then the necessary condition

$$\lim_{n\to\infty} c_n(z_1 - z_0)^n = 0$$

must hold. Any sequence having a limit must be bounded, so we can find a real M such that $|c_n(z_1 - z_0)^n| < M$ for all n. Now take any z for which $|z - z_0| < |z_1 - z_0|$, and set $q = |z - z_0|/|z_1 - z_0|$; of course $0 \leq q < 1$. Then

$$\begin{aligned} |c_n(z - z_0)^n| &= |c_n(z - z_0)^n| \cdot \left|\frac{z_1 - z_0}{z_1 - z_0}\right|^n \\ &= |c_n(z_1 - z_0)^n| \cdot \left|\frac{z - z_0}{z_1 - z_0}\right|^n \leq Mq^n. \end{aligned}$$

Since $\{Mq^n\}$ is a geometric sequence with ratio $0 \leq q < 1$, the corresponding series $\sum_{n=0}^{\infty} Mq^n$ is convergent; and then, by the comparison test, so is $\sum_{n=0}^{\infty} |c_n(z - z_0)^n|$. This proves part 1.

2. We could follow a similar strategy for this part. But there is an easier trick: we simply use part 1 to prove part 2 by contradiction. So assume, contrary to the theorem, that there is some z' for which $|z' - z_0| > |z_2 - z_0|$, but that the power series (6.6) converges at $z = z'$. Then by part 1, since z_2 is closer to z_0 than z' is, the series would have to converge also at z_2; but we assumed it diverged there. So no z' satisfying $|z' - z_0| > |z_2 - z_0|$ could be a point of convergence. □

With the help of Abel's theorem, we can make out a more precise geometric characterization of the set of convergence of a power series. First note that the series (6.6) always converges at $z = z_0$, since then all terms but the first are 0. Now consider any ray starting at z_0. When we ask about convergence at points along the ray aside from z_0, there are three possibilities.

1. The series converges at all points on the ray. Then it must in fact converge absolutely on the entire complex plane. Indeed, the ray contains points as distant as desired from z_0; and then by Abel's theorem the series converges absolutely at all points inside a circle centered at z_0 of radius as large as desired. So the set of absolute convergence is all of $\mathbb{C}$.

2. The series diverges at all points of the ray except z_0. In this case, the series diverges at all points of the plane except z_0. Indeed, the ray contains points as close as desired to z_0; so by Abels theorem it must diverge outside a circle centered at z_0 with radius as small as desired. So the set of convergence is just the point z_0.

3. The ray contains some points of convergence besides z_0, and also some points of divergence. From Abel's theorem we see that all the points of convergence are closer to z_0 than all points of divergence, so the points on the ray fall into two intervals: a set of points of convergence from z_0 to some z^*, and then a set of points of divergence from z^* to ∞. Note that at the point z^* itself the series may either converge or diverge, but at all other points along the ray the question is settled. Then by Abel's theorem, the series converges absolutely inside the disk $|z - z_0| < |z^* - z_0|$, which is bounded by the circle through z^* centered at z_0, and the series diverges at all points in the exterior of that circle. In this case, the open disk $|z - z_0| < |z^* - z_0|$ is called the *disk of convergence* of the power series, and the radius $|z^* - z_0|$ of the disk is called

the *radius of convergence*. At points on the circle $|z - z_0| = |z^* - z_0|$ itself, the series may either converge or diverge.

The notions of disk and radius of convergence can be extended to cases 1 and 2 above: If the series converges on the entire plane, we can take the radius of the disk of convergence to be ∞; if the series converges only at z_0, then we can take the radius to be 0.

So to summarize: *The set of convergence of the series (6.6) always consists of the interior of some disk centered at z_0, with the possible addition of some or all points on the circle that bounds it. This disk may be the whole plane or just the point z_0.*

In many cases the disk and radius of convergence can be found with the help of either D'Alembert's test or the Cauchy root test. Let us try applying D'Alembert's test to the general power series (6.6), under the assumption that the corresponding limit exists:

$$\lim_{n\to\infty}\left|\frac{c_{n+1}(z-z_0)^{n+1}}{c_n(z-z_0)^n}\right| = \lim_{n\to\infty}\left|\frac{c_{n+1}(z-z_0)}{c_n}\right| = |z-z_0|\lim_{n\to\infty}\left|\frac{c_{n+1}}{c_n}\right| = l.$$

Now we define

$$R = \frac{1}{\lim\limits_{n\to\infty}\left|\frac{c_{n+1}}{c_n}\right|}, \tag{6.7}$$

noting that $|z-z_0|/R = l$. According to D'Alembert's test, the series converges when $l < 1$ and diverges when $l > 1$. This is equivalent to: the series converges when $|z - z_0| < R$, and it diverges when $|z - z_0| > R$. So evidently this R is the radius of convergence for the series.

Using similar reasoning with the Cauchy root test, we can get the formula

$$R = \frac{1}{\lim\limits_{n\to\infty}\sqrt[n]{|c_n|}}. \tag{6.8}$$

Thus either formula (6.7) or (6.8) may be used to calculate the radius of convergence of a power series, under the condition that the limit appearing in the formula exists. But it is not necessary to memorize the formulas, since the indicated test can be applied directly.

Example 6.11 Find the disk of convergence of the series

$$1 + z + \frac{z^2}{2!} + \cdots + \frac{z^n}{n!} + \cdots = \sum_{n=1}^{\infty}\frac{z^n}{n!}. \tag{6.9}$$

Solution Applying D'Alembert's test,

$$\lim_{n\to\infty}\left|\frac{z^{n+1}}{(n+1)!} \div \frac{z^n}{n!}\right| = \lim_{n\to\infty}\left|\frac{z^{n+1}n!}{z^n(n+1)!}\right| = \lim_{n\to\infty}\left|\frac{z}{n+1}\right| = 0 = l$$

for any z. So $l = 0 < 1$ for all $z \in \mathbb{C}$, and therefore this series converges absolutely in the entire complex plane, and the radius of convergence is $R = \infty$.

Example 6.12 Find the disk of convergence of the series

$$1 + z + 2!\, z^2 + \cdots + n!\, z^n + \cdots = \sum_{n=1}^{\infty} n!\, z^n. \tag{6.10}$$

Solution Again applying D'Alembert's test,

$$\begin{aligned} \lim_{n\to\infty} \left| \frac{(n+1)!\, z^{n+1}}{n!\, z^n} \right| &= \lim_{n\to\infty} (n+1)|z| \\ &= |z| \lim_{n\to\infty} (n+1) = \begin{cases} 0, & \text{if } z = 0, \\ \infty, & \text{if } z \neq 0. \end{cases} \end{aligned}$$

So $l = 0 < 1$ for $z = 0$, and this series converges; while $l = \infty > 1$ for $z \neq 0$, and the series diverges. So the disk of convergence reduces to the single point $z = 0$, and the radius of convergence is $R = 0$.

Example 6.13 Find the set of convergence for the series

$$\sum_{n=1}^{\infty} \frac{(z-2)^n}{n^2} = (z-2) + \frac{(z-2)^2}{4} + \ldots. \tag{6.11}$$

Solution D'Alembert's test again:

$$\begin{aligned} l = \lim_{n\to\infty} \left| \frac{(z-2)^{n+1}}{(n+1)^2} \div \frac{(z-2)^n}{n^2} \right| &= \lim_{n\to\infty} \left| \frac{(z-2)n^2}{(n+1)^2} \right| \\ = |z-2| \lim_{n\to\infty} \left(\frac{n}{n+1} \right)^2 &= |z-2|. \end{aligned}$$

So when $l = |z-2| < 1$ the series converges absolutely; and when $l = |z-2| > 1$, the series diverges. Therefore the disk of convergence is $|z-2| < 1$, which is centered at $z_0 = 2$ and of radius $R = 1$.

To completely determine the set of convergence, we must examine what happens on the boundary, i.e. when $|z-2| = 1$. At points on that circle we have

$$\left| \frac{(z-2)^n}{n^2} \right| = \frac{|z-2|^n}{n^2} = \frac{1}{n^2}.$$

You will probably recall from calculus that the series $\sum_{n=1}^{\infty} \frac{1}{n^2}$ is convergent (see for example [8], Section 11.3); therefore at all points on the circle the series (6.11) converges, in fact absolutely. So the complete set of convergence is the closed disk $|z-2| \leq 1$.

On the other hand, the disk of convergence of the series $\sum_{n=1}^{\infty}(z-2)^n$ is the same disk $|z-2| < 1$ as in the previous example, but in this case the series diverges at all points on the circle $|z-2| = 1$—see problem 1.

Examples 6.11–6.13 show that all three cases considered above—$R = \infty$, $R = 0$, and $0 < R < \infty$—can in fact occur.

Here are some basic properties of power series.

Property 6.14. *A power series (6.6), with radius of convergence R, converges absolutely and uniformly on any disk $|z-z_0| < r < R$ lying within the disk of convergence.*

Proof. Let z_1 be a point for which

$$r < |z_1 - z_0| < R$$

(Fig. 48). Then, if z is any point in the disk $|z-z_0| < r$,

$$|c_n(z-z_0)^n| < |c_n(z_1-z_0)^n|$$

for each n. Since z_1 lies within the disk of convergence, the power series must converge absolutely at $z = z_1$, that is the series $\sum_{n=0}^{\infty}|c_n(z_1-z_0)^n|$ converges. Therefore by the Weierstrass test (Theorem 6.5), the series $\sum_{n=0}^{\infty} c_n(z-z_0)^n$ must converge absolutely and uniformly on the disk $|z-z_0| < r$. □

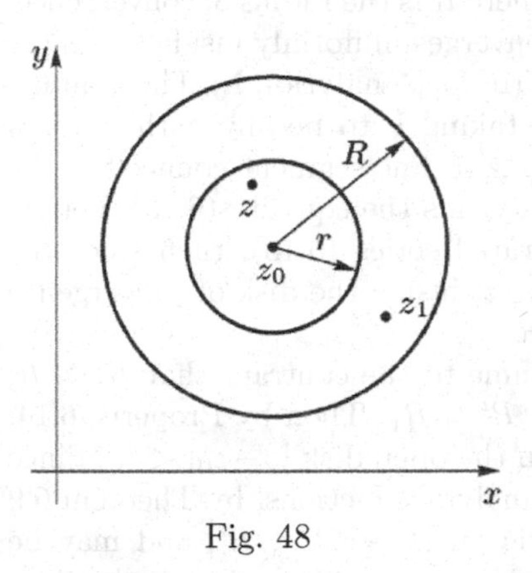

Fig. 48

Property 6.15. *The sum $S(z)$ of a power series (6.6) is a function which is analytic within the disk of convergence.*

Proof. Take any point z' inside the disk of convergence, and choose a real number r such that $|z'-z_0| < r < R$, where R is the radius of convergence. By Property 6.14 the power series converges uniformly inside $|z-z_0| < r$. Then since each term $c_n(z-z_0)^n$ is analytic (on the entire plane), by Weierstrass' Theorem 6.9, the sum $S(z)$ is an analytic function on the open disk $|z-z_0| < r$, so in particular it is analytic at z'. Since z' was arbitrary, the sum $S(z)$ is in fact analytic on the entire disk of convergence of the power series. □

Property 6.16. *A power series*

$$\begin{aligned} S(z) &= c_0 + c_1(z-z_0) + c_2(z-z_0)^2 + \dots + c_n(z-z_0)^n + \dots \\ &= \sum_{n=0}^{\infty} c_n(z-z_0)^n \end{aligned} \tag{6.12}$$

can, within the disk of convergence, be integrated termwise, that is

$$\int_{z_0}^{z'} S(z)\,dz = c_0(z'-z_0) + \frac{c_1}{2}(z'-z_0)^2 + \frac{c_2}{3}(z'-z_0)^3 + \cdots + \frac{c_n}{n+1}(z'-z_0)^{n+1} + \cdots = \sum_{n=0}^{\infty} \frac{c_n}{n+1}(z'-z_0)^{n+1}, \tag{6.13}$$

whenever z' lies within the disk of convergence of the original power series (6.12). The power series obtained by integration has the same disk of convergence as the original one.

Proof. As in the proof of the preceding property, we start by taking any point z' inside the disk of convergence of the original series (6.12), and choosing a real number r such that $|z'-z_0| < r < R$, where R is the radius of convergence. By Property 6.14 the power series (6.12) converges uniformly inside $|z-z_0| < r$. Since its terms are continuous functions (in fact analytic), by Theorem 6.7 the series can be integrated termwise. By taking Γ to be any path from z_0 to z' lying inside the disk of convergence (e.g. a line segment connecting the two points), an application of that theorem yields the equality (6.13). Let R_1 be the radius of convergence of the integrated series (6.13); then since the integrated series still converges at all points z' inside the disk of convergence of the original series, we know that $R_1 \geq R$.

We will show that in fact $R_1 = R$. Assume to the contrary that $R_1 > R$; then there is some R' in between, i.e. $R < R' < R_1$. Then by Property 6.14, the integrated series converges uniformly on the open disk $|z-z_0| < R'$. Since the terms of the integrated series are still analytic functions, by Theorem 6.9 the sum of the integrated series is analytic on $|z-z_0| < R'$, and may be differentiated termwise on this same disk. However, differentiating the integrated series yields the original series (6.12); and we assumed its radius of convergence was $R < R'$. This contradiction means that $R_1 \not> R$, so in fact $R_1 = R$. □

Property 6.17. *A power series (6.12) can be, within the disk of convergence, differentiated termwise any number of times:*

$$S'(z) = c_1 + 2c_2(z-z_0) + 3c_3(z-z_0)^2 + \cdots + nc_n(z-z_0)^{n-1} + \ldots, \tag{6.14}$$

$$S''(z) = 2c_2 + 3\cdot 2c_3(z-z_0) + \cdots + n(n-1)c_n(z-z_0)^{n-2} + \ldots, \tag{6.15}$$

$$\vdots$$

$$S^{(n)}(z) = n(n-1)(n-2)\cdots 1c_n + (n+1)n(n-1)\cdots 2c_{n+1}(z-z_0) + \ldots. \tag{6.16}$$

All series obtained through termwise differentiation in this way have the same disk of convergence as the original power series.

Proof. We again take any point z' inside the disk of convergence $|z-z_0| < R$ of the series (6.12), and choose a positive number r such that $|z' - z_0| < r < R$. Since the power series (6.12) converges uniformly inside $|z - z_0| < r$, by Weierstrass Theorem 6.9, it may be differentiated termwise on this disk (and in particular at the point z') any number of times. Hence, the series (6.14)–(6.16) converge at all points inside the disk of convergence of the original series to the corresponding derivatives of $S(z)$. Therefore, the radii of convergence of the series (6.14)–(6.16) are greater than or equal to R.

First we will show that the radius of convergence R_1 of the series (6.14) equals R. Assume to the contrary that $R_1 > R$. Integrating this series inside the disk $|z - z_0| < R_1$ yields the original series (6.12) except the first term c_0; but the presence of this term does not affect convergence. By Property 6.16, the integrated series has the same disk of convergence $|z - z_0| < R_1$. Therefore, the radius of convergence of the original series equals $R_1 > R$. This contradiction means that $R_1 = R$, that is single differentiation of a power series preserves its radius of convergence. But the series (6.15) is obtained by termwise differentiation of the series (6.14). Hence, again its radius of convergence equals R, and so on. □

Although the disk and radius of convergence are unaltered under termwise integration or differentiation, nevertheless the convergence/divergence at points on the boundary of the disk may change.

Properties 6.16 and 6.17 can be helpful in finding the sum of certain series.

Example 6.18 Find the sum $S(z)$ of the series

$$\sum_{n=1}^{\infty} nz^n.$$

Solution We know the sum of a similar series, the geometric series:

$$\sum_{n=0}^{\infty} z^n = \frac{1}{1-z} \quad \text{when} \quad |z| < 1,$$

see Example 6.4. Differentiating this series termwise we get

$$\sum_{n=1}^{\infty} nz^{n-1} = \frac{d}{dz}\frac{1}{1-z} = \frac{1}{(1-z)^2},$$

and then multiplying on both sides by z we arrive at

$$S(z) = \sum_{n=1}^{\infty} nz^n = \frac{z}{(1-z)^2} \quad \text{when} \quad |z| < 1.$$

Problems

1. Show that the disk of convergence of the series $\sum_{n=1}^{\infty}(z-2)^n$ is the same disk $|z-2|<1$ as in the Example 6.13, but that in this case the series diverges at all points on the circle $|z-2|=1$.

2. Find the radius of convergence of the following power series:

$$a)\ \sum_{n=1}^{\infty}\frac{n(z-2)^n}{3^n};\quad b)\ \sum_{n=1}^{\infty}n^n z^n;\quad c)\ \sum_{n=1}^{\infty}\frac{n^n}{n!}z^n;$$

$$d)\ \sum_{n=0}^{\infty}\frac{(z+1)^n}{n!};\quad e)\ \sum_{n=0}^{\infty}\frac{3^n(z-1)^n}{2n+1};\quad f)\ \sum_{n=0}^{\infty}\frac{4^n(z-1)^{2n}}{n+1}.$$

3. Use termwise integration to find the sum of the series

$$\sum_{n=1}^{\infty}\frac{z^n}{n},$$

and determine its disk of convergence.

4. Find the sums of the following series and determine their radii of convergence:

$$a)\ \sum_{n=2}^{\infty}n(n-1)z^n,\quad b)\ \sum_{n=1}^{\infty}n^2z^n.$$

5. Let $R<\infty$ be the radii of convergence of the series $\sum\limits_{n=0}^{\infty}c_nz^n$. Prove that the sequence $\{c_nz^n\}$ is unbounded for every z with $|z|>R$.

6. Let R_1, R_2 be the radii of convergence of the series $\sum\limits_{n=0}^{\infty}a_nz^n$ and $\sum\limits_{n=0}^{\infty}b_nz^n$, respectively. Prove that the radius of convergence R of each of the following series satisfies the inequality:

a) for $\sum\limits_{n=0}^{\infty}(a_n\pm b_n)z^n,\quad R\geq\min(R_1,R_2);$

b) for $\sum\limits_{n=0}^{\infty}(a_nb_n)z^n,\quad R\geq R_1R_2;$

c) for $\sum\limits_{n=0}^{\infty}\frac{a_n}{b_n}z^n,\quad R\leq\frac{R_1}{R_2}.$

6.4 Power Series Expansion

We say that a function $f(z)$ has a power series expansion on a domain U, if the function can be written as the sum of a power series which converges at each point of U. Or, in other words, if there exist numbers z_0, c_0, c_1, c_2, ..., such that

$$f(z) = c_0 + c_1(z - z_0) + c_2(z - z_0)^2 + \cdots + c_n(z - z_0)^n + \dots \\ = \sum_{n=0}^{\infty} c_n(z - z_0)^n. \tag{6.17}$$

The following theorem allows the values of these coefficients to be found, under the assumption that such a power series expansion exists.

Theorem 6.19 (Uniqueness of the power series expansion of a function). *Suppose that a function $f(z)$ has a power series expansion on some open disk $U = \{|z - z_0| < R\}$. Then f is analytic on U, and the coefficients c_0, c_1, c_2, ... of the power series are determined uniquely by the formulas*

$$c_0 = f(z_0), \quad c_1 = f'(z_0), \quad c_2 = \frac{f''(z_0)}{2!}, \quad \dots \quad c_n = \frac{f^{(n)}(z_0)}{n!}. \tag{6.18}$$

Proof. By assumption some power series (6.17) converges on U to the sum $S(z) = f(z)$. That f is analytic therefore follows from Property 6.15.

To find the value of c_0 we simply substitute $z = z_0$ into the equation (6.17), making all terms of the series zero except the first. To find the remaining coefficients, we apply Property 6.17 and formulas (6.14)–(6.16) with $S(z) = f(z)$. Then we again substitute $z = z_0$ into the resulting series, to get

$$f'(z_0) = c_1, \quad f''(z_0) = 2c_2, \; \dots \; f^{(n)}(z_0) = n(n-1)(n-2)\cdots 1 c_n = n!\, c_n,$$

after removing the zero terms. The equalities (6.18) then follow immediately. □

With the conventions that $f^{(0)}(z_0) = f(z_0)$ and $0! = 1$, we may write all equalities (6.18) as the general formula for the coefficients:

$$c_n = \frac{f^{(n)}(z_0)}{n!} \quad \text{for} \quad n = 0, 1, 2, \dots. \tag{6.19}$$

The formulas (6.18) and (6.19) are called *Taylor's formula.*[2] A power series whose coefficients are determined by Taylor's formula is called a *Taylor series.* Combining these results we can write Taylor series as

$$f(z) = f(z_0) + f'(z_0)(z - z_0) + \frac{f''(z_0)}{2!}(z - z_0)^2 + \dots \\ = \sum_{n=0}^{\infty} \frac{f^{(n)}(z_0)}{n!}(z - z_0)^n. \tag{6.20}$$

[2]Brook Taylor (1685–1731) was a famous English mathematician.

Theorem 6.19 shows that, if $f(z)$ has a power series expansion on some neighborhood U of the point z_0, then it will be the Taylor series (6.20) and no other; and $f(z)$ will be analytic on U. The following fundamental theorem states that all analytic functions have such a power series expansion.

Theorem 6.20 (Existence of the power series expansion of an analytic function). *Let the function $f(z)$ be analytic on some domain D, and let z_0 be any point of D. If the disk $U = \{ z : |z - z_0| < R \}$ lies within D, then $f(z)$ can be expressed as a sum of Taylor series which converges on U.*

Proof. We take an arbitrary point z in the disk U, and choose a real r such that $|z - z_0| < r < R$ (Fig. 49). We let Γ denote the circle $|\zeta - z_0| = r$.

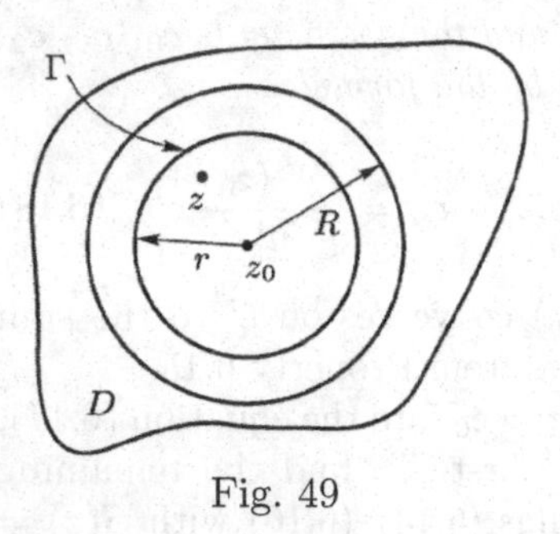

Fig. 49

By the Cauchy integral formula (5.21),

$$f(z) = \frac{1}{2\pi i} \int_{\Gamma} \frac{f(\zeta)}{\zeta - z}\, d\zeta. \tag{6.21}$$

As a first step toward the power series expansion of $f(z)$, we will expand the function $1/(\zeta - z)$. Recall that when $|q| < 1$,

$$\sum_{n=0}^{\infty} q^n = \frac{1}{1-q} \tag{6.22}$$

(geometric series)—see Example 6.4. In our case,

$$\frac{1}{\zeta - z} = \frac{1}{(\zeta - z_0) - (z - z_0)} = \frac{1}{(\zeta - z_0)} \cdot \frac{1}{1 - \frac{z-z_0}{\zeta - z_0}},$$

so we apply the formula above with $q = \frac{z-z_0}{\zeta - z_0}$; notice that $|q| < 1$ since $|z - z_0| < |\zeta - z_0|$:

$$\frac{1}{\zeta - z} = \frac{1}{(\zeta - z_0)} \sum_{n=0}^{\infty} \left(\frac{z - z_0}{\zeta - z_0} \right)^n = \sum_{n=0}^{\infty} \frac{(z - z_0)^n}{(\zeta - z_0)^{n+1}}. \tag{6.23}$$

For fixed z and z_0, and any ζ on Γ, we denote by a_n the modulus of the nth term of this series:

$$a_n = \left| \frac{(z - z_0)^n}{(\zeta - z_0)^{n+1}} \right| = \frac{|z - z_0|^n}{r^{n+1}} = \frac{1}{r} \left(\frac{|z - z_0|}{r} \right)^n. \tag{6.24}$$

Note that a_n does not depend on ζ. Since the growth ratio

$$\frac{|z - z_0|}{r} = |q| < 1,$$

the geometric series $\sum_{n=0}^{\infty} a_n$ converges. So by Weierstrass' uniform convergence test (Theorem 6.5), the series (6.23) converges absolutely and uniformly in ζ on Γ; we have already noticed that Theorems 6.5–6.7 and Property 6.8 are valid not only for series defined on domains, but also for series defined on curves.

Multiplying both sides of (6.23) by $\frac{1}{2\pi i}f(\zeta)$ we get

$$\frac{1}{2\pi i}\frac{f(\zeta)}{\zeta - z} = \sum_{n=0}^{\infty} \frac{1}{2\pi i}\frac{f(\zeta)(z-z_0)^n}{(\zeta - z_0)^{n+1}}.$$

Since the function $\frac{1}{2\pi i}f(\zeta)$ is bounded on Γ, by Property 6.8 the last series converges uniformly in ζ. We now integrate on both sides along Γ, using Theorem 6.7 to integrate termwise on the right:

$$\begin{aligned} f(z) &= \frac{1}{2\pi i}\int_{\Gamma} \frac{f(\zeta)}{\zeta - z}\, d\zeta = \sum_{n=0}^{\infty} \frac{1}{2\pi i}\int_{\Gamma} \frac{f(\zeta)(z-z_0)^n}{(\zeta - z_0)^{n+1}}\, d\zeta \\ &= \sum_{n=0}^{\infty} (z-z_0)^n \frac{1}{2\pi i}\int_{\Gamma} \frac{f(\zeta)}{(\zeta - z_0)^{n+1}}\, d\zeta = \sum_{n=0}^{\infty} c_n (z-z_0)^n, \end{aligned} \tag{6.25}$$

where

$$c_n = \frac{1}{2\pi i}\int_{\Gamma} \frac{f(\zeta)}{(\zeta - z_0)^{n+1}}\, d\zeta, \quad n = 0, 1, 2, \ldots. \tag{6.26}$$

(Note here that $(z-z_0)^n$ is independent of ζ, and therefore it can be removed from under the integral sign.)

Thus we have established the existence of a power series expansion of the function $f(z)$. By Theorem 6.19 this series is unique, and must be the Taylor series expansion. □

The fact that the power series expansion (6.25) of $f(z)$ we obtained here is the same as the Taylor series can be verified without referring to Theorem 6.19. Indeed, the reader may already have noticed that the integral in our formula (6.26) for c_n bears a resemblance to the one used in the Cauchy integral formula (5.25) for derivatives:

$$f^{(n)}(z_0) = \frac{n!}{2\pi i}\int_{\Gamma} \frac{f(\zeta)}{(\zeta - z_0)^{n+1}}\, d\zeta \quad \text{where} \quad n = 0, 1, 2, \ldots,$$

(which is valid also for $n = 0$ with the conventions mentioned earlier). From this equality and (6.26) we see that the coefficients

$$c_n = \frac{f^{(n)}(z_0)}{n!},$$

as in the Taylor series.

According to Theorems 6.19 and 6.20, a function $f(z)$ is analytic in a domain D if and only if it has a power series expansion in a neighborhood of any point in D. Therefore, we proved that our definition of analytic function given in Section 3.1.3 (namely a function $f(z)$ is said to be analytic in an open domain D if it is differentiable at each point in D) is equivalent to the following classical definition: *A function $f(z)$ is said to be analytic in an open domain D if it has a power series expansion in a neighborhood of any point in D.*

Also, it has a nice corollary.

Corollary 6.21. *The radius of convergence of the Taylor series of a function $f(z)$ is equal to the shortest distance from the center z_0 to a singularity.*

Recall that a singularity is a point where the function is not analytic.

Proof. Let z^* be the closest singularity of f to z_0, and let $R = |z^* - z_0|$. Then $f(z)$ is analytic on $U = |z - z_0| < R$, so by the Theorem 6.20 the Taylor series converges on U. Therefore the radius of convergence is at least R. On the other hand, it cannot be larger than R, for then the point z^* would lie within the disk of convergence of the series, and therefore f would be analytic at z^*, contrary to our assumption. □

Taylor's formula (6.20), together with the formulas for the derivatives

$$(e^z)' = e^z, \quad (\sin z)' = \cos z, \quad (\cos z)' = -\sin z,$$

allow us to easily obtain the Taylor series expansions of these functions, which are analogous to the expansions of the corresponding real functions familiar from calculus:

$$e^z = 1 + z + \frac{z^2}{2!} + \cdots + \frac{z^n}{n!} + \cdots = \sum_{n=0}^{\infty} \frac{z^n}{n!}, \tag{6.27}$$

$$\sin z = z - \frac{z^3}{3!} + \frac{z^5}{5!} - \cdots + (-1)^n \frac{z^{2n+1}}{(2n+1)!} + \cdots = \sum_{n=0}^{\infty} (-1)^n \frac{z^{2n+1}}{(2n+1)!}, \tag{6.28}$$

$$\cos z = 1 - \frac{z^2}{2!} + \frac{z^4}{4!} - \cdots + (-1)^n \frac{z^{2n}}{(2n)!} + \cdots = \sum_{n=0}^{\infty} (-1)^n \frac{z^{2n}}{(2n)!}, \tag{6.29}$$

(here $z_0 = 0$). These series converge in the entire complex plane $\mathbb{C}$, which can be verified directly using D'Alembert's test (compare with Example 6.11, where we considered the series (6.27)), or it can be inferred from the Corollary 6.21—none of the three functions have any singularities in $\mathbb{C}$ (see Sections 4.3.1 and 4.4.2).

Remark 6.22 The power series above may even be taken as the *definitions* of these three functions. In that case Euler's formula $e^{iy} = \cos y + i \sin y$ and definitions of $\sin z$ and $\cos z$,

$$\sin z = \frac{e^{iz} - e^{-iz}}{2i}, \quad \cos z = \frac{e^{iz} + e^{-iz}}{2},$$

which we introduced earlier, will be consequences of the formulas (6.27)–(6.29). We will show, as an example, how to obtain Euler's formula: multiply the series (6.28) for $\sin z$ by i, and then add to it the series (6.29) for $\cos z$, placing the terms in ascending order of degree:

$$\begin{aligned}
&\cos z + i \sin z \\
&\quad = 1 + iz - \frac{z^2}{2!} - \frac{iz^3}{3!} + \dots + (-1)^n \frac{z^{2n}}{(2n)!} + (-1)^n \frac{iz^{2n+1}}{(2n+1)!} + \dots \\
&\quad = 1 + (iz) + \frac{(iz)^2}{2!} + \frac{(iz)^3}{3!} + \dots + \frac{(iz)^{2n}}{(2n)!} + \frac{(iz)^{2n+1}}{(2n+1)!} + \dots \\
&\quad = \sum_{n=0}^{\infty} \frac{(iz)^n}{n!} = e^{iz},
\end{aligned}$$

where we have used the equality $i^{2n} = (i^2)^n = (-1)^n$.

Consider the function $f(z) = \frac{1}{1+z}$. What is its Taylor series expansion around $z_0 = 0$? To get it, we can use either the formula (6.22) for the sum of a geometric series with $q = -z$, or we can use the Taylor formula (6.20); in either case, we get

$$\frac{1}{1+z} = 1 - z + z^2 - \dots + (-1)^n z^n + \dots = \sum_{n=0}^{\infty} (-1)^n z^n. \tag{6.30}$$

This function has a unique singularity at the point $z^* = -1$, so the distance from the point $z_0 = 0$ to $z^* = -1$ equals 1, and by the Corollary 6.21 the Taylor series expansion around $z_0 = 0$ has a radius of convergence of 1; this could also have been determined by D'Alembert's test. So the disk of convergence is $|z| < 1$.

The formulas (6.27)–(6.29) and (6.30) can be used to obtain the Taylor series expansions of many other functions. Two techniques are often helpful: substitution, and termwise integration or differentiation.

Example 6.23 Find the Taylor series expansion of the function

$$f(z) = \frac{1}{1+z^2}$$

centered at the point $z_0 = 0$, and determine the radius of convergence.

Solution The Taylor series expansion (6.30) is valid for any z in the open unit disk $|z| < 1$. So we can substitute z^2 for z under condition $|z^2| < 1$:

$$\frac{1}{1+z^2} = 1 - z^2 + z^4 - \cdots + (-1)^n z^{2n} + \cdots = \sum_{n=0}^{\infty} (-1)^n z^{2n}. \tag{6.31}$$

The condition $|z^2| < 1$ is equivalent to $|z| < 1$, so the radius of convergence is $R = 1$.

The series in this example is also interesting for demonstrating how the complex series can shed light on the behavior of the corresponding real one. For the latter,

$$\frac{1}{1+x^2} = \sum_{n=0}^{\infty} (-1)^n x^{2n},$$

converges on the interval $(-1, 1)$, and diverges when $|x| > 1$. But the reason for this might remain mysterious if one looked only at the graph of the function $f(x) = 1/(1 + x^2)$ on the real line: it is defined and infinitely differentiable everywhere, even at $x = \pm 1$, and its value never exceeds 1. But when looking at it in the context of the complex plane, one sees immediately that $f(z) = 1/(1 + z^2)$ has singularities at $z^* = \pm i$, the distance from which to the origin equals 1, and thus the radius of convergence around $z_0 = 0$ must be 1.

Example 6.24 Find the power series expansion of the function $f(z) = \operatorname{Log}(1 + z)$, and its disk of convergence.

Solution We have seen in Section 4.3.2 that the principal branch of the logarithmic function is defined by

$$\operatorname{Log} w = \ln |w| + i \operatorname{Arg} w,$$

and that

$$(\operatorname{Log} w)' = \frac{1}{w}.$$

Therefore our $f(z) = \operatorname{Log}(1 + z)$ is an antiderivative of the function $f'(z) = 1/(1 + z)$ for which $f(0) = \ln 1 = 0$. We have seen that $f'(z)$ has the power series expansion (6.30) which converges inside of the disk $|z| < 1$. Using the fundamental formula (5.19),

$$\int_0^z \frac{1}{1+\zeta}\, d\zeta = \operatorname{Log}(1 + z)\Big|_0^z = \operatorname{Log}(1 + z) - \operatorname{Log} 1 = \operatorname{Log}(1 + z) \quad \text{when} \quad |z| < 1.$$

Since the power series expansion (6.30) for $f'(z)$ converges within the disk $|z| < 1$, we may integrate termwise there:

$$\begin{aligned}\operatorname{Log}(1+z) &= \int_0^z \frac{1}{1+\zeta}\,d\zeta = \int_0^z \sum_{n=0}^{\infty}(-1)^n\zeta^n\,d\zeta = \sum_{n=0}^{\infty}\int_0^z (-1)^n\zeta^n\,d\zeta \\ &= \sum_{n=0}^{\infty}(-1)^n\frac{z^{n+1}}{n+1} = z - \frac{z^2}{2} + \frac{z^3}{3} - \frac{z^4}{4} + \dots.\end{aligned}$$

If we change the index from n to $k = n+1$, this can be written also as

$$\operatorname{Log}(1+z) = \sum_{k=1}^{\infty}(-1)^{k-1}\frac{z^k}{k}. \tag{6.32}$$

The disk of convergence of the integrated series is the same as that of the original, $|z| < 1$.

Example 6.25 Find the Taylor series expansion of the function

$$f(z) = \frac{1}{z-3}$$

in the neighborhood of the point $z_0 = 1$, and find the disk of convergence.

Solution Our strategy is to once again to make use of the formula for the geometric series. First let us make a change of variables:

$$w = z - z_0 = z - 1, \quad z = w + 1.$$

Then

$$\begin{aligned}\frac{1}{z-3} &= \frac{1}{w+1-3} = \frac{1}{-2+w} = \frac{1}{-2(1-\frac{w}{2})} \\ &= -\frac{1}{2}\cdot\frac{1}{1-\frac{w}{2}} = -\frac{1}{2}\sum_{n=0}^{\infty}\left(\frac{w}{2}\right)^n = -\sum_{n=0}^{\infty}\frac{w^n}{2^{n+1}};\end{aligned}$$

we used the formula (6.22) with $q = w/2$. This series converges when $|w/2| < 1$, or $|w| < 2$. Substituting $w = z - 1$, we can express the series in terms of the original variable z:

$$\frac{1}{z-3} = -\sum_{n=0}^{\infty}\frac{(z-1)^n}{2^{n+1}},$$

which converges when $|z-1| < 2$.

Note that we could also have determined the disk of convergence by noting that the only singularity of $f(z)$ is at $z^* = 3$, and the distance from $z_0 = 1$ to 3 is, in fact, 2. So the radius of convergence must be $R = 2$, and the disk of convergence is $|z-1| < 2$.

Problems

1. Expand the function

$$f(z) = \left(\frac{e^{z^2} - 1}{z}\right)'$$

in a Taylor series in the neighborhood of the point $z_0 = 0$, and find the disk of convergence of the resulting series.

2. Expand the function

$$f(z) = \frac{z-6}{(z+2)(z-5)}$$

in a Taylor series in the neighborhood of the point $z_0 = 3$, and find the disk of convergence of the resulting series.

3. Expand the function $f(z)$ in a Taylor series in the neighborhood of the point z_0, and determine the radius of convergence of the series obtained.

a) $f(z) = \left(\frac{\sin z}{z}\right)'$, $z_0 = 0$; b) $f(z) = \frac{z+6}{z(z+3)}$, $z_0 = -1$;

c) $f(z) = \frac{z-1}{(z+1)(z-3)}$, $z_0 = 2$; d) $f(z) = \frac{z}{z^2-4}$, $z_0 = 1$;

e) $f(z) = \int_0^z e^{-t^2}\,dt$, $z_0 = 0$; f) $f(z) = \frac{1}{(1-z)^2}$, $z_0 = 0$.

4. Prove the following generalization of Liouville's theorem: If the function $f(z)$ is analytic on the entire complex plane $\mathbb{C}$, and

$$|f(z)| \le M|z|^k,$$

where M is a positive constant and k is a fixed positive integer, then $f(z)$ is a polynomial of degree no higher than k.

5. Prove that

$$(\tan^{-1} z)' = \frac{1}{1+z^2}$$

(see (4.31)), and find the Taylor series expansion of the branch $f(z)$ of the function $\tan^{-1} z$ with $f(0) = 0$, centered at the point $z_0 = 0$; determine the radius of convergence.

6.5 Uniqueness Property

The point z_0 is called a *zero of the function* f if $f(z_0) = 0$. Suppose that $f(z)$ is analytic on some neighborhood U of z_0, that $f(z_0) = 0$, but that $f(z)$ is not identically zero on U. Then $f(z)$ will have a Taylor expansion on U centered at z_0, and the coefficient c_0 in formula (6.20) must be 0, since $f(z_0) = 0$. But since $f(z)$ is not identically zero, among the remaining coefficients there must be at least one which is nonzero. The number n of the first nonzero coefficient c_n is called the *multiplicity* or the *order of the zero* z_0. This series expansion of $f(z)$ at the point z_0 will look like

$$f(z) = c_n(z - z_0)^n + c_{n+1}(z - z_0)^{n+1} + \cdots = \sum_{k=n}^{\infty} c_k(z - z_0)^k, \qquad (6.33)$$

where $c_n \neq 0$ and $n \geq 1$. Also from the Taylor formula (6.20) we see that n is the order of the first nonzero derivative of f at z_0, i.e. the smallest n for which $f^{(n)}(z_0) \neq 0$. When $n = 1$ we say the zero at z_0 is *simple.*

The previous expansion (6.33) may be rewritten as

$$f(z) = (z - z_0)^n(c_n + c_{n+1}(z - z_0) + \dots) = (z - z_0)^n \phi(z), \qquad (6.34)$$

where

$$\phi(z) = c_n + c_{n+1}(z - z_0) + \dots,$$

and $\phi(z_0) = c_n \neq 0$.

Example 6.26 Locate the zeros of the function $f(z) = (z^2 - 4)^3 e^z$, and find their orders.

Solution The equation $f(z) = 0$ is equivalent to $z^2 - 4 = 0$, or $z = \pm 2$. We will put $f(z)$ in the form

$$f(z) = (z + 2)^3(z - 2)^3 e^z.$$

First we look at the zero $z_0 = 2$, and we set $\phi(z) = (z + 2)^3 e^z$, so that formula (6.34) at $z_0 = 2$ would be

$$f(z) = (z - 2)^3 \phi(z),$$

where $\phi(2) = 4^3 e^2 \neq 0$. Therefore the point $z_0 = 2$ is a zero of order 3.

When we look at the other zero, $z_0 = -2$, a similar procedure gives us $\phi(z) = (z - 2)^3 e^z$ and

$$f(z) = (z + 2)^3 \phi(z),$$

where $\phi(-2) = -4^3 e^{-2} \neq 0$, so that $z_0 = -2$ is also a zero of order 3.

Theorem 6.27. *Suppose that $f(z)$ is analytic on some neighborhood U of z_0, that $f(z_0) = 0$, but that $f(z)$ is not identically zero on U. Then there is some neighborhood of the point z_0 in which $f(z)$ has no other zeros, besides z_0.*

Proof. Since $f(z)$ is not identically zero on U, z_0 must be a zero of finite order; let n be that order. Then we can express $f(z)$ in U in the form of (6.34). The function ϕ will be analytic and so of course continuous, and $\phi(z_0) \neq 0$. Therefore we can find some neighborhood U_1 of z_0 possibly smaller than U, in which $\phi(z) \neq 0$. Then since the factor $(z - z_0)^n$ becomes zero only at the point z_0, the function $f(z) = (z - z_0)^n \phi(z)$ is also never zero in U_1 except at z_0. □

We say that a point z_0 is an *isolated zero* of the function $f(z)$, if $f(z_0) = 0$, and there is some neighborhood of z_0 in which $f(z)$ has no other zeros, besides z_0. Theorem 6.27 shows that if an analytic function is not identically zero on some neighborhood of its zero z_0, then the zero z_0 is isolated. From this we can derive the following important theorem.

Theorem 6.28. (Uniqueness theorem) *Let D be a (connected) domain, and let $\{z_n\}$ be an infinite sequence of pairwise distinct points which converge to a point $a \in D$. If two analytic functions f_1 and f_2 coincide at each point of $\{z_n\}$, i.e. $f_1(z_n) = f_2(z_n)$ for all $n = 1, 2, \ldots$, then $f_1(z) = f_2(z)$ everywhere in D.*

Proof. Consider the function $f(z) = f_1(z) - f_2(z)$. It is analytic and hence continuous on D. Because it has zeros at all the points of the sequence $\{z_n\}$, and $z_n \to a$ as $n \to \infty$, by continuity $f(a)$ must also be zero. However, the point a would not be an isolated zero of f because any neighborhood of a contains infinitely many other zeros z_n.

Now take any neighborhood U, lying in D, of the point a; then $f(z)$ is analytic on U. Suppose that $f(z)$ is not identically zero on U. By the previous theorem, the zero a of $f(z)$ must be isolated. This contradiction means that $f(z)$ must be identically zero on U. We now wish to show that $f(z)$ is in fact identically zero throughout the domain D. Let b be an arbitrary point of the domain D. Since D is connected, there must be some continuous path Γ, lying in D, from a to b.

Let ρ be the shortest distance from the points of Γ to the boundary of D, and let δ be some positive number less than ρ, for example $\delta = \frac{1}{2}\rho$. Then we can select a finite sequence of points $w_0, w_1, \ldots, w_n$ with $w_0 = a$ and $w_n = b$, and such that the distance between successive points in the sequence is less than δ.

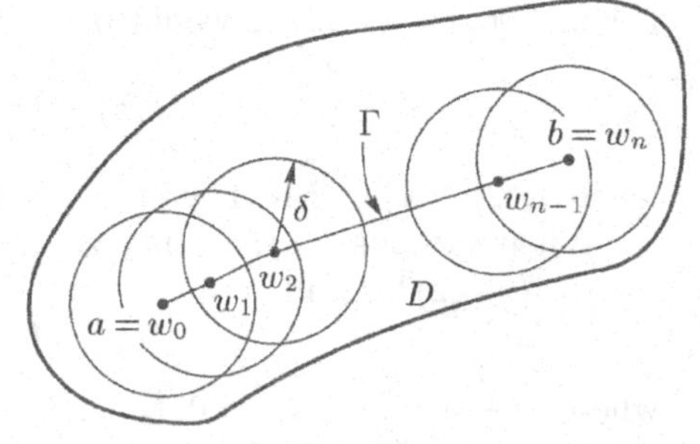

Fig. 50

For each point w_k we construct a disk U_k centered at w_k with radius δ (Fig. 50). Because $\delta < \rho$ all these disks lie entirely within the domain D. Since U_0 is a neighborhood of the point $w_0 = a$, then $f(z)$ must be identically zero on U_0. By construction, the distance between w_0 and w_1 must be less than δ. Therefore w_1 lies in the interior of U_0; that means than some (small) neighborhood of w_1 lies within U_0. On this neighborhood, $f(z) \equiv 0$, and therefore w_1 is a nonisolated zero of f. Since f is analytic in the neighborhood U_1 of the point w_1, then, by the same reasoning used to show $f(z) \equiv 0$ in all neighborhoods of a, f must in fact be identically zero on U_1 also.

Using the same reasoning as we pass from U_1 to U_2, we get that f is identically zero on U_2. Continuing through the sequence of disks, we eventually arrive at U_n and see that on it too, $f(z) \equiv 0$. This means, in particular, that $f(b) = 0$. Since b was arbitrary, we have shown that f is identically zero on all of the domain D.

Of course, wherever $f(z) = 0$, we have $f_1(z) = f_2(z)$, i.e. the functions f_1 and f_2 take the same value. So f_1 and f_2 are equal everywhere in D. □

This theorem shows another essential difference between differentiable complex functions and differentiable real functions: two infinitely differentiable real functions may agree on a "big" set within a domain, and yet not be identically equal on that domain. For example, in Fig. 51 we see the functions

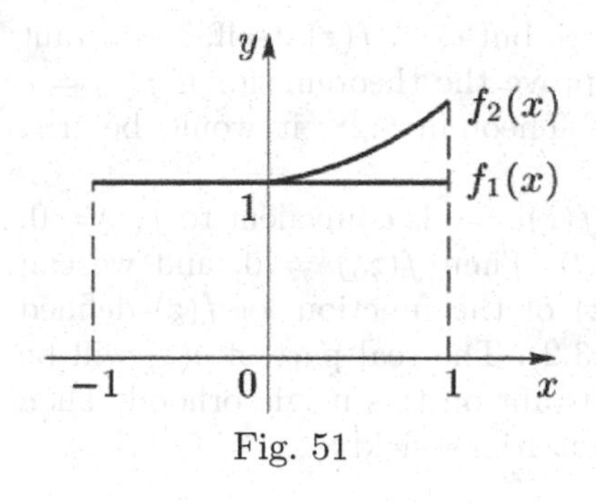

Fig. 51

$$f_1(x) = 0 \quad \text{when} \quad x \in (-1, 1),$$

$$f_2(x) = \begin{cases} 0 & \text{when} \quad x \in (-1, 0] \\ e^{-1/x} & \text{when} \quad x \in (0, 1) \end{cases}$$

which are both infinitely differentiable, and agree on the set $(-1, 0]$, but nevertheless $f_1(x) \not\equiv f_2(x)$ on $(-1, 1)$. Whereas any two functions analytic on a domain D, which coincide on a subset of D containing a convergent sequence of points with the limit in D, by Theorem 6.28 would have to be identically equal on the whole domain D. This subset can be, for example, a small disk or an interval lying in D.

From this we can derive yet another important property of analytic functions.

Theorem 6.29 (Maximum Modulus Principle). *Let $f(z)$ be an analytic function on a bounded domain D, continuous on the closed domain $\overline{D}$. Then either $f(z)$ is constant on $\overline{D}$, or the maximum modulus $|f(z)|$ of the function f on $\overline{D}$ is attained only on the boundary of D.*

Proof. The real-valued function $|f(z)|$ is continuous on the closed and bounded set $\overline{D}$, and therefore it must take a maximum value M at some point z_0 in $\overline{D}$. We will show that if z_0 is in the interior of $\overline{D}$, i.e. in D, then $f(z)$ is constant.

So assume that $z_0 \in D$, and $|f(z_0)| = M$. We will show that in any neighborhood of z_0 lying within D, the value of $|f(z)|$ is always M. For that consider the circle $|z-z_0| = R$ of radius R centered at z_0, where R is sufficiently small for the closed disk $|z - z_0| \leq R$ to lie within D. By the Average Value Theorem 5.21,

$$f(z_0) = \frac{1}{2\pi} \int_0^{2\pi} f(z_0 + Re^{i\phi})\, d\phi,$$

from which we get the inequality

$$M = |f(z_0)| \leq \int_0^{2\pi} |f(z_0 + Re^{i\phi})|\, d\phi. \tag{6.35}$$

By assumption, the integrand $|f(z_0 + Re^{i\phi})|$ never exceeds the maximum M. On the other hand, it can never be strictly less than M either: for then, by continuity, there would have to be an interval $\phi_1 \leq \phi \leq \phi_2$ in which $|f(z_0 + Re^{i\phi})| \leq M - \epsilon$ for some $\epsilon > 0$; and that would mean the right side of (6.35) would be strictly less than M, which contradicts (6.35). So in fact $|f(z_0+Re^{i\phi})| = M$ everywhere on the circle. Since R was arbitrary sufficiently small number, this means that $|f(z)| = M$ whenever z lies in a neighborhood of z_0 contained in D.

Now we want to show that not just $|f(z)|$, but even $f(z)$ itself, is constant on some neighborhood of z_0. This would prove the theorem: for if $f(z) = c$ on such a neighborhood, by the previous Theorem 6.28, it would be true everywhere in D.

Of course, if $M = 0$ then the equality $|f(z)| = 0$ is equivalent to $f(z) = 0$, and we are done. So assume that $M > 0$. Then $f(z_0) \neq 0$, and we can find a regular branch $g(z) = u(z) + iv(z)$ of the function $\log f(z)$ defined in a neighborhood U of z_0 (see Section 4.3.2). The real part of $g(z)$ will be $u(z) = \ln |f(z)| = \ln M$, that is $u(z)$ is constant on this neighborhood. Then since g is analytic, the Cauchy-Riemann conditions yield

$$\frac{\partial v}{\partial y} = \frac{\partial u}{\partial x} = 0 = \frac{\partial u}{\partial y} = -\frac{\partial v}{\partial x},$$

and so $v(z)$ is constant on U as well. Therefore, $g(z) = c_1$. Hence $f(z) = e^{g(z)} = e^{c_1} = c$, that is $f(z)$ is constant on U as desired. □

Yet another way to prove this theorem would be to use the Property 4.16 that analytic functions preserve domains, that is, a nonconstant analytic function, which is defined on an open and connected set, takes as its image another open and connected set. The reader should try to prove Theorem 6.29 using that method. We preferred to give a different (more complicated) proof since the property of preservation of domains was stated without proof.

From this theorem we see that *if an analytic function does take its maximum modulus, even locally, at an interior point of its domain, then it must be constant on the entire domain.* In fact this statement is equivalent to the theorem.

Corollary 6.30. *If a harmonic function $u(z)$ in a domain D takes its maximum or minimum value at an interior point of D, then u must be constant on the entire domain D.*

Proof. Suppose that u attains its maximum value at a point $z_0 \in D$. Then there is a number $r > 0$ such that the disk $B = \{|z - z_0| < r\}$ is contained in D. Let v be the harmonic conjugate to u in B. The function

$$f(z) = e^{u(z)+iv(z)}$$

is analytic in B, and $|f(z)| = e^{u(z)}$. If $u(z_0)$ is the maximum value of u, then $e^{u(z_0)}$ is the maximum value of $|f(z)|$ in B. By the Theorem 6.29, $f(z)$ is constant in B and hence in D.

If u has a minimum value at z_0, we consider the function $u_1 = -u$ and reduce the problem to the case of maximum. □

Another proof is based on the Mean Value property of harmonic functions (Corollary 5.22) and just repeats the proof of Theorem 6.29—see the arguments after the inequality (6.35). We leave details for the reader.

Problems

1. Find the order of the zero at $z_0 = 0$ of the functions:

$$a)\ \sin(z^3); \quad b)\ 1 - \cos(z^3); \quad c)\ \mathrm{Log}(1 + z^3); \quad d)\ e^{z^2} - z^2 - 1.$$

2. Suppose that a function $f(z)$ is analytic at a point z_0 and has zero at z_0 of order n. Find the order of the zero at z_0 of the following functions:

$$a)\ \sin(f(z)); \quad b)\ 1 - \cos(f(z)); \quad c)\ e^{f(z)} - 1.$$

3. Suppose that functions $f(z)$ and $g(z)$ are analytic at a point z_0 and have zeros at z_0 of order m and n, respectively, and $m < n$. Find the order of the zero at z_0 of the following functions: a) $f(z) + g(z)$; b) $f(z)g(z)$.

4. Let

$$f(z) = \begin{cases} \sin \frac{\pi}{z}, & z \neq 0, \\ 0, & z = 0. \end{cases}$$

a) Prove that $f(\frac{1}{n}) = 0$, $n = 1, 2, \ldots$

b) According to the part a), $f(z_n) = 0$ on a sequence of points which converges to the point $z_0 = 0$; but $f(z)$ is not equal to zero identically in any neighborhood of z_0. Why this fact does not contradict to Theorem 6.28?

5. a) Prove that the "Minimum Modulus Principle" does not take place in general: we cannot replace "maximum" by "minimum" in Theorem 6.29.

b) Prove the following "Minimum Modulus Principle". *Let $f(z)$ be an analytic function on a bounded domain D, continuous on the closed domain $\overline{D}$. Suppose that $f(z)$ does not have zeros in D. Then either $f(z)$ is constant on $\overline{D}$, or the minimum modulus $|f(z)|$ of the function f on $\overline{D}$ is attained only on the boundary of D.*

6.6 Analytic Continuations

The notion of analytic continuation arises from various motivations. One of them is the situation where a function which was originally defined with a specific formula and on a specific domain D, in fact can be defined as an analytic function on a wider domain. For instance, consider the series

$$\sum_{n=0}^{\infty} z^n.$$

The set of convergence of this power series is the disk $D = \{|z| < 1\}$, and its sum $S(z)$ is analytic in D (Property 6.15). Therefore, by this formula we have defined an analytic function whose domain is D. In this specific case we know the sum of the series, namely $S(z) = (1 - z)^{-1}$ (see Example 6.32 below). The function $f(z) = (1 - z)^{-1}$ is analytic in $\mathbb{C} \setminus \{1\}$. Thus, there exists a function $f(z)$ that is analytic in a wider domain than D and coincides with $S(z)$ in D. Therefore, the function $S(z)$, defined initially in D, can be extended to an analytic function $f(z)$ on a wider domain. We call $f(z)$ an analytic continuation of $S(z)$. We will return to this example later—see Example 6.32.

Another incentive for introducing the concept of analytic continuation, and the associated concept of a complete analytic function, is the need to work with multivalued functions. For example, how to interpret the expression $\sqrt{z} + \sqrt{z}$, if each term allows two values? It is all the more unclear how to understand a derivative of the function $\sqrt{z}$. However, we already have some experience working with multivalued functions—see Sections 4.2 and 4.3. From that we see we should consider regular branches of multivalued functions. Here we will study that concept in a more general setting.

In Fig. 52 we see different ways that two domains D_1 and D_2 could share a third domain (shaded). In case a, the intersection $D_1 \cap D_2 = G$ is itself a domain; in case b, $D_1 \cap D_2$ has two components, and either G or G' can be regarded as the shared domain.

Suppose that the analytic functions f_1 and f_2 are defined on D_1 and D_2, respectively; and suppose that these two functions agree on the shared domain G. Then f_2 is called *the direct analytic continuation* of f_1 to the domain D_2

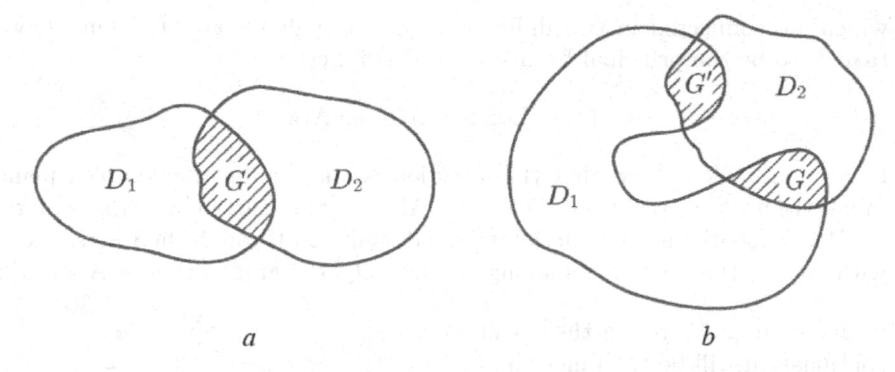

Fig. 52

through G. It is easy to show, using the Uniqueness Theorem 6.28, that f_2 is unique for given f_1, D_1, D_2, and G. (Indeed, if f_2^* is another direct analytic continuation of f_1 to D_2 through G, then f_2^* coincides with f_2 in G which is a part of the domain D_2. Hence, by the Uniqueness Theorem f_2^* coincides with f_2 everywhere in D_2). However, if as in Fig. 52 *b* the intersection $D_1 \cap D_2$ has two components G and G', then the continuation through G may not agree with the continuation through G'; we will see this in Example 6.31.

In Fig. 53 we see a chain of domains D_1, D_2, ..., D_n, in which each consecutive pair D_k and D_{k+1} share the domain G_k. Suppose that on each D_k is defined an analytic function f_k, in such a way that f_k and f_{k+1} agree on G_k. In other words, each f_{k+1} is the direct analytic continuation of f_k through G_k. Then f_n is called *the analytic continuation of f_1 to the domain D_n through the domains $G_1, \dots, G_{n-1}$*; and once again, f_n is unique (why?) for the given f_1 and chains D_2, ..., D_{n-1}, G_1, ..., G_{n-1}.

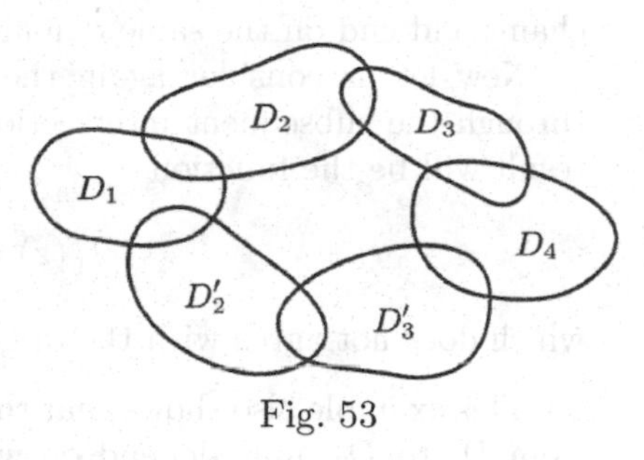

Fig. 53

However, if we change any of the intermediate domains D_2, ..., D_{n-1}, or any of the shared domains G_k, then the continuation of f_1 to D_n through the domains G_1, ..., G_{n-1} may also change; and this is possible even when the first and last domains D_1 and D_n remain the same.

It is possible that the last link in the chain of domains could be the same as the first; for example, consider the chain D_1, D_2, D_3, D_4, D_3', D_2', D_1, in Fig. 53. In this case it is quite possible that the continuation of f_1 through the chain will not agree with the original f_1 on D_1. This is illustrated in the following example.

Example 6.31 In Fig. 54 we see the domains

$$D_1 = \{\, z : |z-1| < \tfrac{1}{2} \,\}, \quad D_3 = \{\, z : |z+1| < \tfrac{1}{2} \,\}$$

which are connected in two different ways: through D_2 and D_2'. On D_1 we take f_1 to be the principal branch of the log function

$$f_1(z) = \operatorname{Log} z = \ln|z| + i \operatorname{Arg} z.$$

In Section 4.3.2 we saw that this function is analytic on the complex plane minus the negative real axis, with $-\pi < \operatorname{Arg} z < \pi$; in particular, f_1 is analytic on D_1. We continue this function to the domain D_3 through the intersections with D_2. As the point z is moving through D_2, the argument $\phi = \operatorname{Arg} z$ will be increasing. Therefore the resulting continuation will be the function

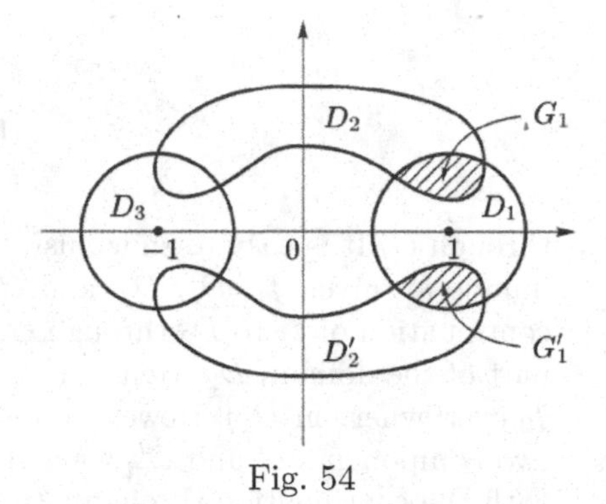

Fig. 54

$$f_3(z) = \ln|z| + i\phi, \quad \text{where} \quad \phi > 0,$$

which is analytic on D_3.
However, if we continue f_1 to D_3 through the intersections with D_2', then we get the function

$$f_3^*(z) = \ln|z| + i\phi^*, \quad \text{where} \quad \phi^* < 0,$$

since ϕ^* is decreasing on D_2'. In fact, ϕ and ϕ^* differ by 2π, so that $f_3(-1) = i\pi$, while $f_3^*(-1) = -i\pi$. So the two continuations, which go through different chains and end on the same domain, nevertheless differ from each other.

Now let us consider again the continuation of f_1 on D_1, but this time through the subsequent intersections of D_1, D_2, D_3, D_2', and D_1. Then the result will be the function

$$f_1^*(z) = f_1(z) + 2\pi i,$$

which does not agree with the original function f_1.

This example also shows that the direct analytic continuation of a function from D_1 to D_2, may depend on which part of the intersection $D_1 \cap D_2$ the continuation passes through. Indeed, looking again at Fig. 54, denote by D_2^* the union $D_2 \cup D_3 \cup D_2'$, and let us consider the continuation of f_1 on D_1 to D_2^*. Here the intersection $D_1 \cap D_2^*$ has two components, G_1 and G_1', and the direct analytic continuations $f(z)$ and $f^*(z)$ will be different depending on whether it is through G_1 or G_1': in the first case we would get $f(-1) = i\pi$, and in the second, $f^*(-1) = -i\pi$.

Suppose that f_1 is a (single-valued) function on the domain D_1. Consider the set D of points z' for which some analytic continuation (direct or through a chain of domains) can be found of f_1 to some domain containing z'; in other words, D is the set of points to which it is possible to extend the function f_1. Then D must contain D_1, by virtue of the continuation using the trivial chain consisting of D_1 alone. But there may be points $z' \in D$ outside of D_1.

Note that there may be several different analytic continuations from D_1 to the same point z'; thus we may have multiple distinct values $f(z')$ associated with the point z', even possibly infinitely many.

If we associate to each z' in D all values $f(z')$ obtained by continuations via all possible chains and their shared domains, then generally speaking we define a multivalued function. As we remarked at the beginning of this section, it's quite complicated to work with such functions. To overcome this difficulty, we introduce the notion of a function element.

A function element $\mathcal{F}$ is the pair (D, f), where D is a domain in $\overline{\mathbb{C}}$ and f is a single-valued analytic function on D. A function element $\mathcal{F}_n = (D_n, f_n)$ is called *the analytic continuation of the element* $\mathcal{F}_1 = (D_1, f_1)$ through the domains $G_1, \dots, G_{n-1}$, if the function f_n is the analytic continuation of f_1 to D_n through the domains $G_1, \dots, G_{n-1}$. (One can easily see that, conversely, $\mathcal{F}_1$ is the analytic continuation of $\mathcal{F}_n$ through the domains $G_{n-1}, \dots, G_1$).

Returning to Example 6.31 we see that $\mathcal{F}_1 = (D_1, f_1)$, $\mathcal{F}_2 = (D_1, f_1 + 2\pi)$, $\dots$, $\mathcal{F}_n = (D_1, f_1 + 2\pi(n-1)i)$ are different elements, despite the fact that the domains D_1 of these elements coincide. Each of these elements is an analytic continuation of any other of them.

A *complete analytic function* is the family of all function elements obtained by analytic continuation of some function element via all possible chains. Two complete analytic functions F and H are considered equal if there exist an element (D, f) of a function F and an element (G, h) of a function H such that h is the direct analytic continuation of f from D to the domain G through some component of the intersection $D \cap G$. By the Uniqueness Theorem, in this case every element of F is also an element of H, and vice versa. In other words, the families of function elements of F and H coincide.

If there is a point z_0 and two elements (D_1, f_1) and (D_2, f_2) of F such that $z_0 \in D_1 \cap D_2$ and $f_1(z_0) \neq f_2(z_0)$, we say that F is *multivalued*; otherwise F is *single-valued*. For example, the function $F(z) = \frac{1}{z-1}$ is single-valued, and $F(z) = \log z$ is multivalued.

Single-valued analytic functions f in function elements (D, f) of a complete analytic function F are called *regular branches* of F. (Notice that we have already met the notion of regular branch in Section 4.2. For example, a single-valued function $\operatorname{Log} z$ is a regular branch of $\log z$.) Now we may define operations with complete analytic functions as the corresponding operations with their branches. For instance, derivative F' of a complete analytic function F with elements (D, f) is the family of function elements (D, f'). Therefore, the notions of function element and of regular branch of a complete analytic function allow us to apply all methods developed for single-valued analytic functions, to multivalued functions.

For the rest of this section we will use F to denote a complete analytic function, generally multivalued, and f to denote one of its regular branches.

Note on terminology. As we have said, in this book we use the phrase "analytic function" to mean one that is single-valued, and we will add "multivalued" explicitly when that is intended. But in some other books "analytic function" includes complete analytic functions, and the terms "regular" or "holomorphic" are used when single-valued functions are intended.

The possibility of forming an analytic continuation leads to the following construction, due to Weierstrass. First we illustrate the method with an example.

Example 6.32 Consider the function

$$f_1(z) = \sum_{n=0}^{\infty} z^n,$$

which is defined on $D_1 = \{ z : |z| < 1 \}$, since that is the disk of convergence of the series. Continue this function to the largest domain possible.

Solution We recognize this as the geometric series, so on D_1 the function is equal to $1/(1-z)$; but for $|z| \geq 1$ the series diverges and therefore does not define a function. So, we will pretend we do not know the values of function outside D_1, and work out the continuation "blind", so as to demonstrate the technique of continuing functions defined only by a power series.

First we will find the Taylor series expansion of this function at $a_2 = 0.9i$. Since a_2 lies in D_1, f_1 is analytic there, and therefore has a Taylor series expansion in a neighborhood of that point lying in D_1. Since in D_1 the function $f_1(z)$ is defined, we may use the Taylor formula (6.20) for $f_1(z) = 1/(1-z)$: the derivatives are

$$f_1(z)' = \frac{1}{(1-z)^2}, \quad f_1(z)'' = \frac{2}{(1-z)^3}, \quad \ldots \quad f_1^{(n)}(z) = \frac{n!}{(1-z)^{n+1}},$$

and we get the coefficients

$$c_n = \frac{f_1^{(n)}(a_2)}{n!} = \frac{1}{(1-0.9i)^{n+1}}$$

and the power series

$$f_2(z) = \sum_{n=0}^{\infty} \frac{(z-0.9i)^n}{(1-0.9i)^{n+1}}. \tag{6.36}$$

This series obviously converges to $f_1(z)$ in small neighborhoods of a_2 lying inside D_1, for example $|z - a_2| < 0.1$; so $f_1(z) = f_2(z) = 1/(1-z)$ in these neighborhoods. Let D_2 be the disk for convergence for the series (6.36); then by virtue of the Uniqueness Theorem 6.28, $f_2(z) = 1/(1-z)$ on D_2. To determine the radius of convergence R_2 for this second series, we could use the root or D'Alembert's test from Section 6.1 as well as the formulas (6.7), (6.8). But since in this case we know what the series sums to, it is easier to use Corollary 6.21: R_2 is equal to the distance from the center $a_2 = 0.9i$ to the singularity $z = 1$ of the function $f_2(z)$, so

$$R_2 = |1 - 0.9i| = \sqrt{1 + 0.9^2} \approx 1.35.$$

We see here that R_2 is larger than the distance from a_2 to the boundary of D_1, which would be 0.1. Therefore the second series (6.36) continues the function f_1 to D_2, part of which lies outside of D_1.

We can continue this continuation: we could choose a point a_3 in D_2 which is even further from $z = 1$, and then expand f_2 in a power series at a_3 in a disk D_3 of radius $R_3 = |a_3 - 1|$. And so on, choosing a sequence of points a_2, $a_3, \ldots$, and a corresponding sequence of disks; and this could be done in such a way that the disks would cover the entire complex plane with the exception of $z = 1$. So the function f_1, originally defined only on D_1, could be extended to the entire complex plane minus the point $z = 1$, i.e. the entire domain of definition of the function $1/(1 - z)$. Because of the Uniqueness Theorem, this continuation would not depend on the choice of the points a_2, $a_3, \ldots$, and it would turn out to be single-valued.

Example 6.33 Continue the function

$$f_1(z) = \sum_{n=0}^{\infty}(-1)^{n-1}\frac{z^n}{n}$$

to a disk containing the point $z = -2$.

Solution This series we recognize from (6.32) as the expansion of the function $\mathrm{Log}(1 + z)$, and whose disk of convergence is $D_1 = \{|z| < 1\}$—see (6.32).

The reader should make their own drawing, and verify the following steps of the construction.

There are infinitely many ways to continue this function to a disk containing $z = -2$; we will show two of them. First, take the point $a_2 = 0.9i$ and expand f_1 into a power series there. The radius of convergence R_2 for the new series will be

$$R_2 = |-1 - a_2| \approx 1.35,$$

and the disk of convergence will be $D_2 = \{\, z \,:\, |z - a_2| < R_2 \,\}$; we denote this continuation by f_2. For the next continuation we take $a_3 = -1.34+0.9i$, which lies within D_2; the series expansion f_3 at a_3 will then have $R_3 = |1+a_3| \approx 0.96$, and $D_3 = \{\, z \,:\, |z - a_3| < R_3 \,\}$. Since D_3 still does not yet contain $z = -2$, we extend the chain once more, with the disk D_4 centered at $a_4 = -2.29 + 0.9i \in D_3$ with radius $R_4 = |1 + a_4|$; this domain D_4, of the continuation f_4, at last contains our target $z = -2$.

The power series expansion of f_4 will be, by the Uniqueness Theorem, the Taylor series of some branch of $\log(1 + z)$ at a_4. As the point z moves from 0 to -2 along the upper arc of the circle $|z + 1| = 1$, through the domains D_1, D_2, D_3, and D_4, the argument $\mathrm{Arg}(1 + z)$ will increase continuously from 0 to π; therefore $f_4(-2) = \ln 1 + i\pi = i\pi$.

But we can also continue the function f_1 through a chain reflecting, over the real axis, the one just used. Let D_2^*, D_3^*, and D_4^*, be the sets of conjugates of the points in D_2, D_3, and D_4, and let f_4^* be the continuation to D_4^* obtained that way. Then $f_4^*(-2) = -i\pi$. So in this case the continuation yields a multivalued function. If this continuation were to be extended in all directions, we would obtain the complete multivalued analytic function $\log z$, defined on entire complex plane except $z = -1$.

The construction here was analogous to that in the Example 6.31, but more concrete, and at the same time, as we will see, more general.

Now let us move to the general case. Suppose we are given a power series

$$\sum_{n=0}^{\infty} c_n (z - a_1)^n,$$

converging on some disk $D_1 = \{ z : |z - a_1| < r_1 \}$, where $r_1 > 0$. Then by Property 6.15 the sum $f_1(z)$ will be an analytic function on D_1. We then choose another point $a_2 \in D_1$, distinct from a_1. The function f_1 will be analytic on a disk D_1', centered at a_2, and tangent to the circle $|z - a_1| = r_1$, so that D_1' lies within D_1. Therefore f_1 can be expanded to a Taylor series at a_2,

$$\sum_{n=0}^{\infty} \frac{f_1^{(n)}(a_2)}{n!}(z - a_2)^n,$$

which obviously converges on D_1' to $f_1(z)$. We will denote by D_2 the actual disk of convergence of this second series, which may extend outside of D_1 (Fig. 55). In that case, the second series provides a direct analytic continuation $f_2(z)$ of the function $f_1(z)$ to the domain D_2.

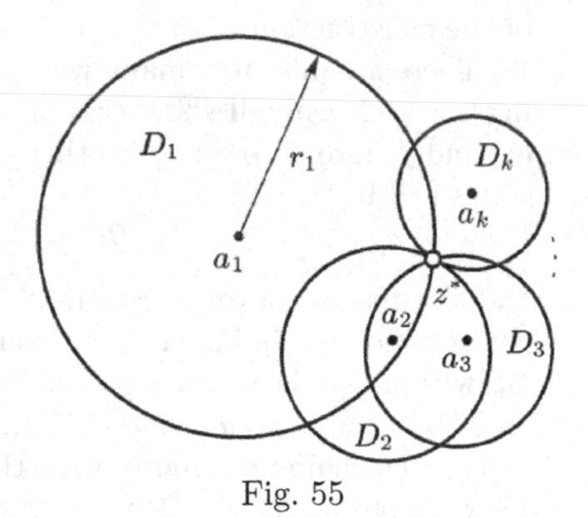

Fig. 55

The procedure just described can be iterated an infinite number of times, each time choosing some new point a_3, a_4, ... as the center of the next Taylor expansion. The result will be a complete analytic function $F(z)$; and any particular expansion $f_k(z)$, on a disk D_k, will be a regular branch of F.

The radius r_k of the disk D_k can be increased until the boundary of the disk meets the closest singularity z^* (see Fig. 55). So if D is the domain of the complete analytic function F, the boundary Γ of D will consist entirely of singularities—for if any part of Γ were free of singularities, it would be possible to continue F there to a larger domain. The boundary Γ may consist of lines, segments, or other combinations of points. In the simplest case, Γ consists of isolated singular points, at each of which there is some sufficiently small punctured disk on which the complete function F can be defined.

If $F(z)$ is single-valued in some punctured neighborhood of the singularity z^*, then z^* is said to be a *a singular point of single-valued character*. If, on the other hand, $F(z)$ is multivalued on punctured neighborhoods, no matter how small, of z^*, then z^* is said to be a *singular point of multivalued character*, or a *branch point.*

For example, the function $F(z) = 1/z$ has singularity at $z^* = 0$ of single-valued character. We examine these points in detail later.

As an example of a singularity of multivalued character, we can take the point $z^* = 0$ for the function $F(z) = \sqrt[n]{z}$; this is a singularity because the derivative does not exist at that point. The multivalued character is clear because the continuation of $\sqrt[n]{z} = \sqrt[n]{|z|}e^{i\frac{\phi}{n}}$ from any domain, through a chain that goes counterclockwise around the origin and back to the same domain, will produce the function $\sqrt[n]{|z|}e^{i\frac{\phi+2\pi}{n}}$, differing from the original.

Another example of a singularity of multivalued character, or a branch point, is the function $\log z$ at $z^* = 0$.

Of course, it may happen that it is impossible to extend an analytic function in a domain D to any domain G not contained in D. Here is an example.

Example 6.34 Consider the function

$$f_1(z) = \sum_{n=0}^{\infty} z^{2^n}. \tag{6.37}$$

We prove that $f_1(z)$ is analytic in the unit disk $D = \{|z| < 1\}$, but that there no direct analytic continuation of f_1 to any domain having points outside D.

Solution If $|z| < r < 1$ then $|z^{2^n}| < r^{2^n}$. The series $\sum_{n=0}^{\infty} r^{2^n}$ converges since

$$\sum_{n=0}^{\infty} r^{2^n} < \sum_{k=0}^{\infty} r^k = \frac{1}{1-r}.$$

By the Weierstrass uniform convergence test, Theorem 6.5, the series (6.37) converges uniformly in any disk $|z| < r$ with $r < 1$, and by Theorem 6.9 (Weierstrass' theorem) its sum is analytic in $|z| < r$. But r might be arbitrarily close to 1. Hence, the series (6.37) converges in the whole disk D (not necessarily uniformly), and its sum $f_1(z)$ is analytic in D.

Let G be a domain which has a non-empty intersection with D and is not contained in D. We show that the direct analytic continuation of $f_1(z)$ to G does not exist.

Define the points $z_{n,k}$ by the equality

$$z_{n,k} = e^{2\pi ki/2^n}, \quad \text{where } n = 1, 2, \ldots, \quad k = 0, 1, \ldots, 2^n - 1.$$

Notice that the points $z_{n,k}$ are situated on the unit circle $|z| = 1$, and form the regular polygon with 2^n vertices centered at the origin. Hence the distances $|z_{n,k} - z_{n,k+1}|$ between neighboring points $z_{n,k}$ and $z_{n,k+1}$ tend to zero as $n \to \infty$. Thus, points $z_{n,k}$ are dense on the unit circle. Namely, every (arbitrarily small) disk centered at a point z on the unit circle, contains a point $z_{n,k}$.

We claim that for every point $z_{n,k}$, as $z \to z_{n,k}$ along the radius from 0 to $z_{n,k}$, that is when $z = re^{2\pi ki/2^n}$ and $r \to 1^-$, we have $f_1(z) \to \infty$. In fact

$$f_1(re^{2\pi ki/2^n}) = \sum_{j=0}^{\infty}(re^{2\pi ki/2^n})^{2^j} = \sum_{j=0}^{\infty} r^{2^j} e^{2\pi ki\cdot 2^{j-n}}$$
$$= \sum_{j=0}^{n-1} r^{2^j} e^{2\pi ki\cdot 2^{j-n}} + \sum_{j=n}^{\infty} r^{2^j} e^{2\pi ki\cdot 2^{j-n}}.$$

The first sum in the last line has modulus less then or equal to n as $0 < r < 1$. Thus this sum is bounded. In the second sum, the factor

$$e^{2\pi ki\cdot 2^{j-n}} = 1$$

since $j - n \geq 0$; therefore this sum equals $\sum_{j=n}^{\infty} r^{2^j}$. Since each term tends to 1 as $r \to 1^-$, this sum tends to infinity as $r \to 1^-$. Hence,

$$\lim_{r \to 1^-} f_1(re^{2\pi ki/2^n}) = \infty.$$

Therefore, the function $f_1(z)$ is unbounded in the intersection of D with any arbitrarily small disk centered at any point $z_{n,k}$.

Let $f(z)$ be a direct analytic continuation of f_1 to G. Since G contains both points in D and points outside D, and G is connected, there is a point $z \in G$ with $|z| = 1$. Since G is open, there is a disk Δ centered at z whose closure $\overline{\Delta}$ (that is Δ together with its boundary) is contained in G. Insofar as f is analytic in G, f is bounded in Δ. But Δ contains a point $z_{n,k}$ and Δ is open. Hence, Δ contains a disk Δ_1 centered at $z_{n,k}$. We have proved above that f_1 is unbounded in the intersection of Δ_1 with D. Because $f = f_1$ in this intersection, f is unbounded in Δ_1 and hence in Δ. We arrive to a contradiction. Thus, a direct analytic continuation of f_1 to a domain G with points outside D does not exist.

Informally, the singular points $z_{n,k}$ are so dense on the circle $|z| = 1$, that they do not allow $f_1(z)$ to be continued through the boundary of D.

Finally, we note that in the process of analytic continuation, the disks of convergence D_k may form a Riemann surface of the complete analytic function F, on which $F(z)$ will be single-valued. For this it must be understood that if a newly constructed disk happens to intersect with an older one, like D_k and D_1 in Fig. 55, and the new continuation does not agree with the older one, then the new disk must be added as sheet above or below the old one, *not* overlapping it; recall the method of constructing Riemann surfaces in Section 4.2.

Like a tiny sprout which contains within itself all the information necessary for constructing a huge tree, any regular branch of an analytic function, no matter how small its domain, includes within itself enough information about the complete analytic function to completely reconstruct it by means of analytic continuation.

Problems

1. Let $f_0(z)$ be the regular branch of the multivalued function $F(z) = \sqrt{z^2 - 1}$ in the disk $D_0 = \{|z - 2| < 1\}$ such that $f_0(2) = \sqrt{3}$.

a) Let $f_1(z)$, $f_2(z)$ be the analytic continuations of $f_0(z)$ to the disk $D_1 = \{|z + 2| < 1\}$ through chains of small disks centered at points on the upper half and on the lower half of the circle $|z| = 2$, respectively. Prove that $f_1(z) = f_2(z)$.

b) Let $f_3(z)$, $f_4(z)$ be the analytic continuations of $f_0(z)$ to the disk $D_2 = \{|z| < 1\}$ through chains of small disks centered at points on the upper half and on the lower half of the circle $|z - 1| = 1$, respectively. Prove that $f_3(z) = -f_4(z)$.

c) Find all branch points of the Riemann surface of the function $F(z)$. How many copies of the complex plane are needed, and how to cut and glue them together to construct the Riemann surface?

2. Prove that the following functions are analytic in the unit disk $D = \{|z| < 1\}$, but that there no direct analytic continuation of them to any domain having points outside D:

$$a)\ f(z) = \sum_{n=0}^{\infty} z^{3^n}; \quad b)\ f(z) = \sum_{n=0}^{\infty} z^{n!}.$$

6.7 Laurent Series

In this section we will see how to create series expansions of a wider class of functions than considered previously: we will investigate functions which are not analytic on an entire disk, but only on a ring or *annulus* $r < |z - z_0| < R$. A very important special case is when $r = 0$, that is, the expansion of a function defined on a punctured neighborhood of z_0. This will allow us to study functions in the neighborhood of points where analyticity fails, i.e. singularities.

For this, power series will no longer be sufficient, since they can represent only functions which are analytic in an entire disk $|z - z_0| < R$ (by Theorem 6.19). But to the terms $c_n(z - z_0)^n$ with nonnegative n, we will add the terms when $n = -1, -2, \ldots$, and consider the sum of the two series

$$\sum_{n=0}^{\infty} c_n(z - z_0)^n \quad \text{and} \quad \sum_{n=-1}^{-\infty} c_n(z - z_0)^n.$$

We will look for the expansion of a function $f(z)$ on an annulus in the form

$$f(z) = \sum_{n=-\infty}^{\infty} c_n(z - z_0)^n = \sum_{n=0}^{\infty} c_n(z - z_0)^n + \sum_{n=-1}^{-\infty} c_n(z - z_0)^n, \qquad (6.38)$$

in which the convergence of the series on the left signifies the convergence of both series on the right. As in Section 4.2, we will prove the existence and uniqueness of such expansions. We start out with an existence theorem.

Theorem 6.35. *Suppose that $f(z)$ is analytic on an annulus*

$$V = \{ z : r < |z - z_0| < R \}.$$

Then f may be represented on V as the sum of the series

$$f(z) = \sum_{n=-\infty}^{\infty} c_n (z - z_0)^n, \tag{6.39}$$

where the coefficients are defined by

$$c_n = \frac{1}{2\pi i} \int\limits_{|\zeta - z_0| = \rho} \frac{f(\zeta)}{(\zeta - z_0)^{n+1}}\, d\zeta, \tag{6.40}$$

where $n = 0, \pm 1, \pm 2, \ldots$, and ρ is any real value between r and R.

Proof. In Fig. 56 we see an arbitrary point z in V. Choose any r' and R' so that $r < r' < |z - z_0| < R' < R$; we define

$$V' = \{ \zeta : r' < |\zeta - z_0| < R' \}.$$

Obviously, V' lies in V and contains the point z. Denote by Γ_1 and Γ_2 the circles $|z - z_0| = R'$ and $|z - z_0| = r'$, respectively, oriented counterclockwise; and let Γ_2^- be the same circle as Γ_2 except oriented clockwise. Then we can take $\Gamma' = \Gamma_1 \cup \Gamma_2^-$ to be the boundary of V', so that the domain V' remains on the left while traversing the boundary Γ'.

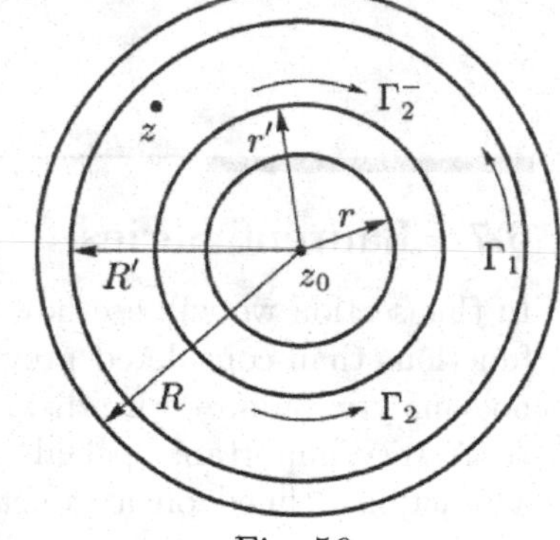

Fig. 56

The function $f(z)$ is analytic on the closed domain $\overline{V'}$, and therefore we can apply the Cauchy integral Theorem 5.17 to get

$$\begin{aligned} f(z) &= \frac{1}{2\pi i} \int\limits_{\Gamma_1} \frac{f(\zeta)}{\zeta - z}\, d\zeta + \frac{1}{2\pi i} \int\limits_{\Gamma_2^-} \frac{f(\zeta)}{\zeta - z}\, d\zeta \\ &= \frac{1}{2\pi i} \int\limits_{\Gamma_1} \frac{f(\zeta)}{\zeta - z}\, d\zeta - \frac{1}{2\pi i} \int\limits_{\Gamma_2} \frac{f(\zeta)}{\zeta - z}\, d\zeta. \end{aligned} \tag{6.41}$$

Now the series expansion of the first integral can obtained just as in the proof of Theorem 6.20. The function $\frac{1}{\zeta - z}$ can be put in the form

$$\frac{1}{\zeta - z} = \frac{1}{(\zeta - z_0) - (z - z_0)} = \frac{1}{(\zeta - z_0)} \cdot \frac{1}{1 - \frac{z - z_0}{\zeta - z_0}} = \frac{1}{(\zeta - z_0)} \cdot \frac{1}{1 - q},$$

$$\text{where } q = \frac{z - z_0}{\zeta - z_0}.$$

For all $\zeta \in \Gamma_1$, we have $|z - z_0| < |\zeta - z_0|$, hence $|q| < 1$. Therefore the formula (6.22) for the sum of a geometric series may be applied, according to which

$$\frac{1}{\zeta - z} = \frac{1}{\zeta - z_0} \sum_{n=0}^{\infty} \left(\frac{z - z_0}{\zeta - z_0} \right)^n = \sum_{n=0}^{\infty} \frac{(z - z_0)^n}{(\zeta - z_0)^{n+1}}. \tag{6.42}$$

Exactly as in the proof of Theorem 6.20 (see (6.24)) we show that the series (6.42) converges absolutely and uniformly in the variable ζ on Γ_1. We multiply this equation by the function $\frac{1}{2\pi i} f(\zeta)$; because this function is bounded on Γ_1, according to Property 6.8 the resulting series remains uniformly convergent. Then we can integrate termwise along Γ_1, obtaining

$$\frac{1}{2\pi i} \int_{\Gamma_1} \frac{f(\zeta)}{\zeta - z} \, d\zeta = \sum_{n=0}^{\infty} \frac{1}{2\pi i} \int_{\Gamma_1} \frac{f(\zeta)(z - z_0)^n}{(\zeta - z)^{n+1}} \, d\zeta = \sum_{n=0}^{\infty} c_n (z - z_0)^n, \tag{6.43}$$

where

$$c_n = \frac{1}{2\pi i} \int_{\Gamma_1} \frac{f(\zeta)}{(\zeta - z_0)^{n+1}} \, d\zeta, \quad n = 0, 1, 2, \dots \tag{6.44}$$

Thus we have expanded the first integral of (6.41) into a convergent series of powers of $(z - z_0)$. The second integral turns out differently, because $|z - z_0| > |\zeta - z_0|$ for $\zeta \in \Gamma_2$, and consequently the series (6.42) would diverge. So making a slight change, we instead write

$$-\frac{1}{\zeta - z} = \frac{1}{(z - z_0) - (\zeta - z_0)} = \frac{1}{(z - z_0)} \cdot \frac{1}{1 - \frac{\zeta - z_0}{z - z_0}} = \frac{1}{(z - z_0)} \cdot \frac{1}{1 - q_1},$$

$$\text{where } q_1 = \frac{\zeta - z_0}{z - z_0}.$$

Then $r' = |\zeta - z_0| < |z - z_0|$ for all $\zeta \in \Gamma_2$, and we have $|q_1| = \frac{r'}{|z - z_0|} < 1$. Once again we can apply formula (6.22) to get

$$-\frac{1}{\zeta - z} = \frac{1}{(z - z_0)} \sum_{n=0}^{\infty} \left(\frac{\zeta - z_0}{z - z_0} \right)^n = \sum_{n=0}^{\infty} \frac{(\zeta - z_0)^n}{(z - z_0)^{n+1}}. \tag{6.45}$$

For all $\zeta \in \Gamma_2$ we have

$$\left| \frac{(\zeta - z_0)^n}{(z - z_0)^{n+1}} \right| = \frac{1}{|z - z_0|} \left| \frac{\zeta - z_0}{z - z_0} \right|^n = \frac{1}{|z - z_0|} |q_1|^n.$$

Since the geometric series $\sum_{n=0}^{\infty} |q_1|^n$ converges, by Weierstrass' uniform convergence test (Theorem 6.5), the series in (6.45) converges absolutely and uniformly in ζ on Γ_2. We can rewrite this series in a more convenient form by introducing a new index of summation $k = -n - 1$, or $n = -k - 1$. When n takes the values $0, 1, 2, \ldots$, the index k runs through $-1, -2, -3, \ldots$. Then the equation (6.45) becomes

$$-\frac{1}{\zeta - z} = \frac{1}{(z - z_0)} \sum_{k=-1}^{-\infty} \frac{(\zeta - z_0)^{-k-1}}{(z - z_0)^{-k}} = \sum_{k=-1}^{-\infty} \frac{(z - z_0)^k}{(\zeta - z_0)^{k+1}}. \tag{6.46}$$

Multiplying this form of the equation by $\frac{1}{2\pi i} f(\zeta)$, which again preserves uniform convergence, and integrating termwise along Γ_2 we get

$$-\frac{1}{2\pi i} \int_{\Gamma_1} \frac{f(\zeta)}{\zeta - z}\, d\zeta = \sum_{k=-1}^{-\infty} \frac{1}{2\pi i} \int_{\Gamma_2} \frac{f(\zeta)(z - z_0)^k}{(\zeta - z_0)^{k+1}}\, d\zeta = \sum_{k=-1}^{-\infty} c_k (z - z_0)^k, \tag{6.47}$$

where

$$c_k = \frac{1}{2\pi i} \int_{\Gamma_2} \frac{f(\zeta)}{(\zeta - z_0)^{k+1}}\, d\zeta, \quad k = -1, -2, -3, \ldots. \tag{6.48}$$

Of course we may replace the k in these formulas by any letter; we may even use n, where here $n = -1, -2, -3, \ldots$ Substituting the resulting series (6.47), along with the earlier (6.43), into (6.41) we arrive at equation (6.39).

Turning to the formula (6.40) for the coefficients, we have already derived formulas (6.44) and (6.48). These can be combined into a single formula by again replacing the k by n. Then the only difference with (6.40) is the path of integration. But the integrand

$$\frac{f(\zeta)}{(\zeta - z_0)^{n+1}}$$

is analytic on the annulus $r < |\zeta - z_0| < R$; therefore by the invariance of the integral under path deformation, Corollary 5.11, the integrals (6.44) and (6.48) will be equal to the formula stated in the theorem for any ρ such that $r < \rho < R$. This completes the proof. □

A series in all integral powers of $(z - z_0)$, negative as well as positive, whose coefficients are determined by the formulas (6.44), is called a *Laurent* [3] *series of a function* $f(z)$. It is frequently convenient to break up the entire Laurent series into its *regular part*,

$$\sum_{n=0}^{\infty} c_n (z - z_0)^n,$$

[3] Discovered by the French mathematician Pierre Alphonse Laurent (1813–1854), who was a military engineer before he began his mathematical research.

consisting of the terms of nonnegative powers, and the *principal part*,

$$\sum_{n=-1}^{-\infty} c_n(z-z_0)^n \quad \text{or} \quad \sum_{n=-\infty}^{-1} c_n(z-z_0)^n,$$

consisting of the terms of negative powers; we will see below the reason for these names.

As with Taylor series, the Laurent series expansion is unique.

Theorem 6.36 (Uniqueness theorem for the Laurent series expansion). *Let V be an annulus centered at z_0 with inner and outer radii r and R, respectively; and let f be a function with a series expansion of the form (6.39) on V. Then f is analytic on V, and the coefficients c_n for $n \in \mathbb{Z}$ are uniquely determined by the formula (6.40).*

Proof. Since by assumption the Laurent series (6.39) converges on V, both the regular and principal parts do so. The first of these, with nonnegative n, is a standard power series, converging on some disk centered at z_0 and diverging outside it. Since it converges on V, the entire annulus V must lie inside the disk of convergence. Because the sum of a power series is analytic on its disk of convergence (Property 6.15), the sum

$$S_1(z) = \sum_{n=0}^{\infty} c_n(z-z_0)^n$$

is analytic on V. In addition, by Property 6.14, this series converges uniformly in any disk $|z-z_0| < R'$, where $R' < R$.

Now consider the principal part of the series, with negative values of n. We make the change of variables

$$Z = \frac{1}{z-z_0} \quad \text{and} \quad k = -n,$$

so that the series takes the form

$$\sum_{n=-1}^{-\infty} c_n(z-z_0)^n = \sum_{k=1}^{\infty} c_{-k} Z^k. \tag{6.49}$$

This is now a power series in the variable Z centered at $Z_0 = 0$, which converges in some disk $|Z| < R_0$, and diverges outside it, and the sum of which is an analytic function on this disk; and as before, on any disk $|Z| < R_0'$, where $R_0' < R_0$, this series converges uniformly. In terms of the original variable z, the disk of convergence in the Z-plane becomes

$$\left|\frac{1}{z-z_0}\right| < R_0, \quad \text{or} \quad |z-z_0| > \frac{1}{R_0},$$

i.e. the exterior of a disk centered at $z = z_0$ with radius $\frac{1}{R_0}$. Thus the series (6.49) converges when $|z - z_0| > \frac{1}{R_0}$ to an analytic function $S_2(z)$, and diverges when $|z - z_0| < \frac{1}{R_0}$. Since this series converges on V, the entire annulus V must lie within the domain of convergence $|z - z_0| > \frac{1}{R_0}$. Furthermore, the convergence is uniform on any domain $|z - z_0| > \frac{1}{R_0'}$ where $R_0' < R_0$. In particular, the series converges uniformly on $|z - z_0| > r'$, whenever $r' > r$.

So we see that both series on the right side of (6.38) converge and their sums S_1 and S_2 are analytic on V. Therefore the function $f(z) = S_1(z)+S_2(z)$ is also analytic on V.

Now we show that the coefficients c_n of the Laurent series expansion are determined uniquely by the formula (6.40). Take any ρ such that $r < \rho < R$, and let Γ be the circle $|z - z_0| = \rho$. Choose two numbers r' and R' such that $r < r' < \rho < R' < R$, and let V' be the annulus $r' < |z - z_0| < R'$; then both the regular and principal parts converge uniformly on V', and the series

$$\sum_{k=-\infty}^{\infty} c_k(z - z_0)^k = f(z)$$

also converges uniformly there. This property is preserved even after multiplying by any factor $(z - z_0)^{-n-1}$ with $n \in \mathbb{Z}$, because such a factor is bounded on V' (see Property 6.8):

$$\sum_{k=-\infty}^{\infty} c_k(z - z_0)^{k-n-1} = \frac{f(z)}{(z - z_0)^{n+1}}.$$

The resulting series we may now, by Theorem 6.7 integrate termwise along Γ:

$$\sum_{k=-\infty}^{\infty} c_k \int_{\Gamma} (z - z_0)^{k-n-1}\, dz = \int_{\Gamma} \frac{f(z)}{(z - z_0)^{n+1}}\, dz. \tag{6.50}$$

We can now apply the equation (5.6)

$$\int_{|z-a|=r} (z - a)^n\, dz = \begin{cases} 0 & \text{for} \quad n \neq -1, \\ 2\pi i & \text{for} \quad n = -1. \end{cases}, \tag{6.51}$$

according to which all integrals on the left side of (6.50) are zero, with the single exception of that for which $k - n - 1 = -1$, or $k = n$, when the value is $2\pi i$. Therefore only that term remains from the sum of the left side of (6.50), and we get

$$2\pi i c_n = \int_{\Gamma} \frac{f(z)}{(z - z_0)^{n+1}}\, dz, \quad n \in \mathbb{Z},$$

which is equivalent to equation (6.40). This completes the proof. □

We saw in the proof that the regular part of the Laurent series converges on the interior of some disk, while the principal part converges on the exterior of another disk, with the same center but smaller radius—if the radius of the second disk were greater than the first, then the set of convergence for the whole series would be empty. We will generally denote by R and r the radii of these disks, respectively, without further explaining that convergence is within radius R but outside of radius r.

The properties of power series (see Section 6.3) imply the following properties of Laurent series.

Property 6.37. *The set of convergence of a Laurent series (6.39) is the annulus* $V = \{z : r < |z - z_0| < R\}$, *with the possible addition of some or all points on the boundary. Also, it is possible that* $r = 0$ *or that* $R = \infty$.

Property 6.38. *The sum* $S(z)$ *of the series (6.39) is an analytic function on the interior of* V.

Property 6.39. *A Laurent series (6.39) may be termwise integrated and termwise differentiated in the interior of* V, *any number of times. The series obtained in this way has the same annulus of convergence* V *as the original series; but the convergence of points on the boundary may not be the same.*

Property 6.40. *Suppose that* $V = \{\, z : r < |z - z_0| < R\,\}$ *is the annulus of convergence of a Laurent series representing the function* $f(z)$, *and that* $0 < r < R < \infty$. *Then on both the inner and outer boundaries of* V *lie singularities of* f.

Proof. Let us split up the Laurent series into its regular and principal parts,

$$f_1(z) = \sum_{n=0}^{\infty} c_n(z - z_0)^n \quad \text{and} \quad f_2(z) = \sum_{k=-1}^{-\infty} c_k(z - z_0)^k,$$

respectively; by the change of variable $Z = 1/(z - z_0)$ the second series can be written as

$$f_3(Z) = \sum_{k=1}^{\infty} c_{-k} Z^k.$$

The disks of convergence for f_1 and f_3 are $|z - z_0| < R$ and $|Z| < \frac{1}{r}$, respectively. By Corollary 6.21, on the boundary of each disk of convergence lies a singularity of the sum. Hence, on the circles $|z - z_0| = R$ and $|Z| = \frac{1}{r}$ lie singularities of f_1 and f_3, respectively. This means that f_2 must have a singularity on $|z - z_0| = r$. Thus, the function $f = f_1 + f_2$ has singularities lying on the outer and inner boundaries of V. □

To determine the Laurent series expansion of a particular function, we usually use the same methods as used previously for finding a Taylor series expansion: substitution and termwise integration or differentiation, starting with an existing series.

Example 6.41 Find the Laurent series expansion of the function

$$f(z) = \frac{z+1}{z(z-1)}$$

in powers of $(z-1)$.

Solution Let us first make a change of variables: $w = z - 1, \quad z = w + 1$, under which the function $f(z)$ becomes

$$g(z) = \frac{w+2}{w(w+1)}.$$

Now we rewrite this using partial fractions:

$$\frac{w+2}{w(w+1)} = \frac{A}{w} + \frac{B}{w+1} = \frac{A(w+1)+Bw}{w(w+1)} = \frac{(A+B)w+A}{w(w+1)},$$

from which we get $A = 2$ and $B = -1$. Thus

$$g(w) = \frac{w+2}{w(w+1)} = \frac{2}{w} - \frac{1}{w+1}.$$

This function has singularities at $w = 0$ and $w = -1$, so it is analytic on the annuli

$$V_1 = \{\, w : 0 < |w| < 1 \,\} \quad \text{and}$$
$$V_2 = \{\, w : 1 < |w| < \infty \,\}$$

We will find the Laurent expansion on each of these annuli.

When $|w| < 1$ we can use the geometric series:

$$g(w) = \frac{2}{w} - \frac{1}{1-(-w)} = \frac{2}{w} - \sum_{n=0}^{\infty}(-1)^n w^n = \frac{2}{w} + \sum_{n=0}^{\infty}(-1)^{n+1}w^n.$$

But when $|w| > 1$ this series fails to converge; so to expand g on V_2 we rewrite the fraction

$$-\frac{1}{w+1} = -\frac{1}{w}\cdot\frac{1}{1+\frac{1}{w}}.$$

Then since $|1/w| < 1$ on V_2, we can again use the geometric series:

$$\begin{aligned} g(w) &= \frac{2}{w} - \frac{1}{w}\cdot\frac{1}{1+\frac{1}{w}} = \frac{2}{w} - \frac{1}{w}\sum_{n=0}^{\infty}(-1)^n\frac{1}{w^n} = \frac{2}{w} + \sum_{n=0}^{\infty}(-1)^{n+1}\frac{1}{w^{n+1}} \\ &= \frac{2}{w} - \frac{1}{w} + \sum_{n=1}^{\infty}(-1)^{n+1}\frac{1}{w^{n+1}} = \frac{1}{w} + \sum_{k=-2}^{-\infty}(-1)^k w^k, \end{aligned}$$

after changing the index in the sum to $k = -(n+1)$, and noting that $(-1)^{-k} = (-1)^k$.

Returning to the original variable z, we get

$$f(z) = \frac{2}{(z-1)} + \sum_{n=0}^{\infty} (-1)^{n+1}(z-1)^n \quad \text{when} \quad 0 < |z-1| < 1;$$
$$f(z) = \frac{1}{(z-1)} + \sum_{k=-2}^{-\infty} (-1)^k (z-1)^k \quad \text{when} \quad 1 < |z-1| < \infty. \tag{6.52}$$

We see that on the smaller annulus V_1, the principal part of the expansion consists of only a single term, $2/(z-1)$; all the other coefficients of the principal part are zero. But on the larger annulus V_2, the expansion consists only of the principal part; all coefficients of the regular part are zero.

By Theorem 6.36, the series we obtained in the example are unique, although we might have used other methods to get them. Also note that the boundaries of the annuli, that is the circles $|z-1| = 0$ and $|z-1| = 1$ with radii 0 and 1, respectively, each contain one singularity of the original function $f(z)$.

Problems

1. Find all Laurent expansions of the function

$$f(z) = \frac{z-2}{(z+1)(z-3)}$$

in powers of $(z-3)$.

2. Find all Laurent series expansions of $f(z)$ in powers of $(z-z_0)$.

a) $f(z) = \dfrac{z-2}{(z-1)(z+3)}$, $z_0 = 1$; b) $f(z) = \dfrac{z-1}{z(z+1)}$, $z_0 = -1$;

c) $f(z) = \dfrac{z+2}{(z-1)(z+3)}$, $z_0 = -3$; d) $f(z) = \dfrac{z+1}{z^2-4z+3}$, $z_0 = 3$.

3. Prove the following property which is an analog of Property 6.14: *If a series of the form (6.38) converges at points z_1 and z_2, where $|z_1-z_0| < |z_2-z_0|$, then it converges absolutely and uniformly on any annulus $\{z : r < |z-z_0| < R\}$, where $|z_1-z_0| < r < R < |z_2-z_0|$.*

[illegible]

Returning to the original variables, we have

[illegible]

[illegible]

Problems

1. Find all Laurent expansions of the function

[illegible]

and all Laurent expansions in powers of [illegible]

[illegible]

7

Residue Theory

7.1 Isolated Singularities

Suppose z_0 is a singular point of the function $f(z)$, i.e. f is not analytic at that point (in particular because $f(z_0)$ is undefined). Recall that z_0 is called an *isolated singularity* if f is analytic in a punctured neighborhood $0 < |z - z_0| < R$ of z_0. This definition can be applied even in the case $z_0 = \infty$, if we take a punctured neighborhood of ∞ to be a set of the form $|z| > R$, i.e. the exterior of a disk centered at the origin. In other words, a singularity is called isolated if there is some neighborhood of it in which there are no other singularities. We have seen that a function may have multiple branches in the neighborhood of a singularity, but in this chapter we will consider only singularities of single-valued character, and the function $f(z)$ will be assumed to be single-valued.

Isolated singularities can be classified according to the behavior of $f(z)$ as $z \to z_0$. There are three types:

1. *removable singularities* are those at which $f(z)$ has a finite limit:
$$\lim_{z \to z_0} f(z) = A;$$
2. *poles* are those at which the limit is ∞:
$$\lim_{z \to z_0} f(z) = \infty;$$
3. *essential singularities* are those at which no limit exists, whether finite or infinite.

Example 7.1 To show that all three types of singular points are realized, we classify the following isolated singularities.

1. The point $z_0 = 0$ for the function
$$f_1(z) = \frac{\sin z}{z}.$$

DOI: 10.1201/9780367810283-7

2. The point $z_0 = 1$ for the function
$$f_2(z) = \frac{1}{z-1}.$$
3. The point $z_0 = 0$ for the function
$$f_3(z) = e^{1/z}.$$

Solution

1. Using the formula (6.28) for the sin function, we get
$$f_1(z) = \frac{\sin z}{z} = 1 - \frac{z^2}{3!} + \frac{z^4}{5!} - \cdots + (-1)^n \frac{z^{2n}}{(2n+1)!} + \ldots, \qquad z \neq 0,$$
from which we see that $\lim_{z\to 0} f_1(z) = 1$; so $z_0 = 0$ is a removable singularity.
2. Since
$$\lim_{z\to 1} f_2(z) = \lim_{z\to 1} \frac{1}{z-1} = \infty,$$
the point $z_0 = 1$ is a pole.
3. As $z \to 0$ along the real axis, the limits from the left and right are different:
$$\lim_{z\to 0-} e^{1/z} = 0 \quad \text{but} \quad \lim_{z\to 0+} e^{1/z} = \infty.$$
Therefore no limit exists, neither finite nor infinite, and the point $z_0 = 0$ is an essential singularity. We could also have looked at the limit along the imaginary axis; there $z = iy$ and
$$e^{1/z} = e^{-i/y} = \cos\frac{1}{y} - i\sin\frac{1}{y},$$
which also has no limit as $y \to 0$.

Singularities which are *not* isolated can exist also. For example, the function $f(z) = 1/\sin\frac{\pi}{z}$ has poles at the points $z_n = \frac{1}{n}$ for $n = \pm 1, \pm 2, \ldots$. Therefore every neighborhood, however small, of the singular point $z_0 = 0$ contains other singularities; thus the singularity $z_0 = 0$ is not isolated.

Suppose that z_0 is an isolated singularity of the function $f(z)$; then f is analytic in some punctured neighborhood $0 < |z - z_0| < R$ of z_0. Of course a punctured neighborhood can be regarded as an annulus in which the inner radius is $r = 0$, so we can apply Theorem 6.35 and conclude that $f(z)$ must have an expansion to a Laurent series there. The expansion on a punctured neighborhood of a point z_0 we call the expansion about z_0. We will show that the type of the singularity, i.e. the behavior of $f(z)$ as $z \to z_0$, is determined by the form of the principal part of the Laurent series; this explains the use of the adjective "principal".

Lemma 7.2. *Suppose that $f(z)$ is analytic on a punctured neighborhood of z_0 and also suppose that $|f(z)| < M < \infty$ on that neighborhood. Then the Laurent series coefficients c_n, for $n < 0$, are all zero.*

Proof. Take any $\rho > 0$ small enough that the circle $|z - z_0| = \rho$ lies within the punctured neighborhood, and apply the formula (6.40) for the coefficients of the Laurent series:

$$|c_n| = \left| \frac{1}{2\pi i} \int\limits_{|\zeta - z_0| = \rho} \frac{f(\zeta)}{(\zeta - z_0)^{n+1}} \, d\zeta \right| \leq \frac{1}{2\pi} \int_0^{2\pi} \frac{|f(\zeta(t))|}{|\zeta(t) - z_0|^{n+1}} |\zeta'(t)| \, dt$$
$$\leq \frac{M}{2\pi\rho^{n+1}} \int_0^{2\pi} |\zeta'(t)| \, dt = \frac{M}{2\pi\rho^{n+1}} \cdot 2\pi\rho = M\rho^{-n}.$$

Since ρ can be chosen arbitrarily small, and $\rho^{-n} \to 0$ as $\rho \to 0$ and $n < 0$, this means that all $c_n = 0$ when $n = -1, -2, \ldots$; hence the coefficients of the principal part of the Laurent series are all 0. □

Theorem 7.3. *Let z_0 be an isolated singularity of the function $f(z)$. Then z_0 is removable if and only if a Laurent series expansion of $f(z)$ aboutf z_0 has the form*

$$f(z) = \sum_{n=0}^{\infty} c_n (z - z_0)^n, \tag{7.1}$$

that is the Laurent series is an ordinary power series consisting only of the regular part, and all the coefficients of the principal part are 0.

Proof. **Only if (necessity):** Suppose that z_0 is a removable singularity of f. Then there exists a finite limit $\lim_{z \to z_0} f(z) = A$. This implies that $f(z)$ is bounded on some punctured neighborhood $0 < |z - z_0| < R$ of z_0, i.e. there is a positive number M such that $|f(z)| < M$ at all z in this neighborhood. Now the statement of the theorem follows directly from Lemma 7.2.

If (sufficiency): Now suppose that the Laurent expansion has no principal part; then it is just a power series (7.1). That means it converges not only on the punctured neighborhood, but on an entire disk $|z - z_0| < R$ including the point z_0. The sum $S(z)$ of the series is analytic on that disk, and therefore continuous. Moreover, $S(z) = f(z)$ as $0 < |z - z_0| < R$. Hence, there exist the finite limit

$$\lim_{z \to z_0} f(z) = \lim_{z \to z_0} S(z) = S(z_0);$$

therefore z_0 is a removable singularity. □

In this proof we see that in a punctured neighborhood $0 < |z - z_0| < R$ of a removable singularity z_0, the function coincides with the function $S(z)$ which is analytic in the complete disk $|z - z_0| < R$. Hence, we can extended the function $f(z)$ analytically from the punctured disk to the complete disk, simply by setting $f(z_0) = S(z_0)$, without changing the values of $f(z)$ at other points. In this sense, the singularity at z_0 is removable. It is natural to consider such points to be regular, and not singular points of the function.

This situation may arise if the definition given for $f(z)$ does not immediately work at z_0, even though f can be extended to that point. For example, recall from Example 7.1 the function

$$f_1(z) = \frac{\sin z}{z},$$

which approaches 1 as $z \to 0$. This means that by setting $f_1(0) = 1$ we remove the apparent singularity at 0, obtaining a function which is analytic at 0, and even on the entire complex plane.

Lemma 7.2 and Theorem 7.3 imply that if a function $f(z)$ is analytic and bounded on a punctured neighborhood of a point z_0, then the singularity at z_0 is removable. This is another amazing property of analytic functions which is very different from the properties of real functions. For example, the function

$$f(x) = \begin{cases} 1, & x > 0, \\ -1, & x < 0, \end{cases}$$

is differentiable and bounded as $x \neq 0$, but it even does not have limit as $x \to 0$.

Next we give a characterization of poles in terms of the Laurent series expansion.

Theorem 7.4. *An isolated singularity z_0 of the function f is a pole if and only if the Laurent series expansion of f about z_0 has at least one but no more than finitely many nonzero coefficients in the principal part. In other words, z_0 is a pole if and only if the Laurent series expansion is of the form*

$$f(z) = \sum_{n=-N}^{\infty} c_n (z - z_0)^n, \tag{7.2}$$

where $N > 0$.

Proof. **Only if (necessity):** Let z_0 be a pole of f, i.e. $\lim_{z \to z_0} f(z) = \infty$. That means there is some punctured neighborhood of z_0 in which $f(z)$ is never 0. Consider the function $g(z) = 1/f(z)$; then g is analytic on this punctured neighborhood. Moreover, $\lim_{z \to z_0} g(z) = 0$, so for g the point z_0 is a removable singularity; and we can extend the definition of g to z_0 by setting $g(z_0) = 0$, and then the extended function g is analytic on the whole (nonpunctured) neighborhood. Note that z_0 must be an isolated zero of g, for otherwise f would not have been analytic in the punctured neighborhood; let N be the order of this zero. As we saw in Section 6.5, formula (6.34), the function $g(z)$ can be written in form

$$g(z) = (z - z_0)^N \phi(z),$$

where $\phi(z_0) \neq 0$ and ϕ is analytic in some neighborhood of z_0. So by continuity, there must be some neighborhood U of z_0 in which $\phi(z)$ is never 0. Hence, $1/\phi(z)$ is analytic on U, and therefore has a power series expansion

$$\frac{1}{\phi(z)} = b_0 + b_1(z - z_0) + b_2(z - z_0)^2 + \cdots = \sum_{n=0}^{\infty} b_n(z - z_0)^n,$$

where $b_0 \neq 0$. Now we get

$$\begin{aligned} f(z) &= \frac{1}{g(z)} = \frac{1}{(z - z_0)^N} \frac{1}{\phi(z)} \\ &= \frac{1}{(z - z_0)^N}(b_0 + b_1(z - z_0) + b_2(z - z_0)^2 + \dots) \\ &= \frac{c_{-N}}{(z - z_0)^N} + \frac{c_{-N+1}}{(z - z_0)^{N-1}} + \cdots + \frac{c_{-1}}{z - z_0} + \sum_{n=0}^{\infty} c_n(z - z_0)^n \qquad (7.3) \\ &= \sum_{n=-N}^{\infty} c_n(z - z_0)^n, \end{aligned}$$

where each $c_n = b_{n+N}$, and $c_{-N} = b_0 \neq 0$. So the principal part of the Laurent series for f has only finitely many nonzero terms.

If (sufficiency): Now suppose that the Laurent series expansion of $f(z)$ about z_0 has only finitely many terms in the principal part, as in (7.2); we may assume that $c_{-N} \neq 0$. We have to prove that z_0 is a pole. Multiplying (7.2) through by $(z - z_0)^N$, we get

$$\begin{aligned} h(z) = f(z)(z - z_0)^N &= \sum_{n=-N}^{\infty} c_n(z - z_0)^{n+N} \\ &= c_{-N} + c_{-N+1}(z - z_0) + c_{-N+2}(z - z_0)^2 + \dots . \end{aligned} \qquad (7.4)$$

This is a power series, which converges to an analytic function on some whole disk D including z_0; so $h(z)$ can be extended to be analytic on D, by defining $h(z_0) = c_{-N} \neq 0$. Then

$$\lim_{z \to z_0} f(z) = \lim_{z \to z_0} \frac{h(z)}{(z - z_0)^N} = \infty.$$

So z_0 is a pole of f. □

The order N of the zero z_0 of the function $g(z) = 1/f(z)$, is called the *order of the pole* z_0 of the function $f(z)$. If z_0 is a pole of order N, then $g(z) = (z - z_0)^N \phi(z)$, where $\phi(z_0) \neq 0$. We proved—see (7.3), that in this case the Laurent expansion of $f(z)$ has the form (7.2) with $c_{-N} \neq 0$,

Conversely, if the Laurent series expansion of f at z_0 takes the form (7.2) with $c_{-N} \neq 0$, then according to (7.4),

$$\frac{1}{f(z)} = (z - z_0)^N \cdot \frac{1}{h(z)} = (z - z_0)^N \phi(z), \qquad (7.5)$$

where $\phi(z_0) = 1/h(z_0) \neq 0$. Hence, z_0 is the zero of order N of $1/f(z)$, and by definition, z_0 is the pole of order N of $f(z)$. Therefore, *the order of a pole is equal to the greatest* N *such that* $c_{-N} \neq 0$ *in the principal part of the Laurent series.*

The following corollary sums this up in a convenient form for applications.

Corollary 7.5. *A point* z_0 *is a pole of order* N *of a function* f *if and only if* $f(z)$ *can be written in the form*

$$f(z) = \frac{h(z)}{(z - z_0)^N}, \tag{7.6}$$

where h *is analytic on some neighborhood of* z_0, *and* $h(z_0) \neq 0$.

Proof. The function $\phi(z) = 1/h(z)$ is analytic in some neighborhood of z_0. The condition (7.6) with $h(z_0) \neq 0$ is equivalent to the equalities (7.5). As we have seen, these equalities mean that z_0 is the pole of order N. □

Example 7.6 Find the singularities of the function

$$f(z) = \frac{z - 1}{(z^2 + 1)(z + 3)^2}$$

and categorize them.

Solution The singularities are the points where the denominator is zero: either $z^2 + 1 = 0$, and $z = \pm i$; or $(z + 3)^2 = 0$, and $z = 3$. So there are three singular points: $z_1 = i,\ z_2 = -i,\ z_3 = -3$.

At $z_1 = i$,

$$f(z) = \frac{z - 1}{(z - i)(z + i)(z + 3)^2} = \frac{1}{z - i} \cdot \frac{z - 1}{(z + i)(z + 3)^2} = \frac{h_1(z)}{z - i},$$

where

$$h_1(z) = \frac{z - 1}{(z + i)(z + 3)^2}.$$

Since this function is analytic in a neighborhood of z_1 and nonzero at $z = z_1$, by the Corollary 7.5, z_1 is a pole of order 1. Similarly, $z_2 = -i$ is also a pole of order 1.

At $z_3 = -3$,

$$f(z) = \frac{z - 1}{(z^2 + 1)(z + 3)^2} = \frac{h_3(z)}{(z + 3)^2},$$

where

$$h_3(z) = \frac{z - 1}{(z^2 + 1)}.$$

Again, h_3 is analytic in a neighborhood of z_3 and nonzero at z_3, so by the Corollary 7.5, z_3 is a pole of order 2.

Now we turn to essential singularities.

Theorem 7.7. *An isolated singularity z_0 of a function $f(z)$ is an essential singularity if and only if the Laurent expansion of f about z_0 has a principal part with infinitely many nonzero terms.*

Proof. This theorem follows immediately from the two preceding ones. For if z_0 ia an essential singularity, then by those theorems the principal part of the Laurent series for f is neither empty nor finite (otherwise z_0 would be removable or a pole, respectively), so it must be infinite.

The converse is similar: if the principal part is infinite, then the singularity z_0 can be neither removable nor a pole, so it must be an essential singularity. □

By definition, at an essential singularity the limit of the function does not exist, either finite or infinite. The following theorem demonstrates how irregular is the behavior of a function $f(z)$ in a neighborhood of an essential singularity.

Theorem 7.8 (Casorati[1]Sokhotski[2]Weierstrass theorem). *Let z_0 be an essential singularity of the function f. Then for any complex number A, including ∞, there is some sequence of numbers $\{z_n\}$ such that $z_n \to z_0$ and* $\lim_{n\to\infty} f(z_n) = A$.

Proof. Consider first the case when $A = \infty$. If $f(z)$ is bounded on some punctured neighborhood of z_0, then by Lemma 7.2, the Laurent series expansion of $f(z)$ about z_0 has no principal part, and therefore by Theorem 7.3, z_0 is a removable singularity contrary to hypothesis. Therefore $f(z)$ is unbounded on any punctured neighborhood of z_0. Hence for any $n = 1, 2, \dots$ there exists a point z_n such that

$$0 < |z_n - z_0| < \frac{1}{n} \quad \text{and} \quad |f(z_n)| > n$$

(otherwise $f(z)$ is bounded on some punctured neighborhood of z_0). Obviously, $z_n \to z_0$ and $\lim_{n\to\infty} f(z_n) = \infty$.

Now suppose $A \neq \infty$. Two cases are possible.

(a) In any punctured neighborhood of z_0 there is a point z' such that $f(z') = A$. Then there is a sequence $\{z_n\}$ such that $z_n \to z_0$ and $f(z_n) = A$; hence $\lim_{n\to\infty} f(z_n) = A$.

(b) There exists a punctured neighborhood U of z_0 such that $f(z) \neq A$ on U. Define the function ψ by

$$\psi(z) = \frac{1}{f(z) - A};$$

---

[1]Felice Casorati (1835–1890) was an Italian mathematician; he published the theorem in 1868.

[2]Julian Sokhotski (1842–1927) was a Russian-Polish mathematician; he published the theorem in 1868. Weierstrass published the same result in 1876.

then ψ is analytic on U. We will prove that z_0 is an essential singularity of ψ. Otherwise there exists the limit $\lim_{z\to z_0}\psi(z)$ whether finite or infinite. Since

$$f(z) = A + \frac{1}{\psi(z)},$$

there is also the limit $\lim_{z\to z_0} f(z)$, contrary to hypothesis that z_0 is an essential singularity of f. Therefore, z_0 must be an essential singularity of ψ. As we have already proved, there is a sequence $\{z_n\}$ such that $z_n \to z_0$ and $\lim_{n\to\infty}\psi(z_n) = \infty$. Hence,

$$\lim_{n\to\infty} f(z_n) = A + \lim_{n\to\infty}\frac{1}{\psi(z_n)} = A,$$

and the theorem is proved. □

To paraphrase this theorem: In any neighborhood of an essential singularity, however small, the function assumes values arbitrarily close to every number in the extended complex plane $\overline{\mathbb{C}}$.

In studying isolated singularities, the Taylor series of elementary functions, which we saw earlier, are often useful.

Example 7.9 Determine the type of the singular point $z_0 = 0$ for the function

$$f(z) = \frac{e^{3z}-1}{\sin(z) - z + z^3/6}.$$

Solution Let us expand both the numerator and denominator in Taylor series in powers of z. Substituting $3z$ for z in the Taylor series (6.27) for the exponential function gives

$$e^{3z} - 1 = 3z + \frac{(3z)^2}{2!} + \frac{(3z)^3}{3!} + \cdots = z\left(3 + \frac{9z}{2} + \frac{27z^3}{6} + \ldots\right).$$

Using the Taylor expansion (6.28) of $\sin z$ we get

$$\sin(z) - z + \frac{z^3}{6} = \frac{z^5}{5!} - \frac{z^7}{7!} + \cdots = z^5\left(\frac{1}{5!} - \frac{z^2}{7!} + \ldots\right).$$

These two series converge on the entire complex plane. Putting these together, we get

$$f(z) = \frac{z\left(3 + \frac{9z}{2} + \frac{27z^3}{6} + \ldots\right)}{z^5\left(\frac{1}{5!} - \frac{z^2}{7!} + \ldots\right)} = \frac{1}{z^4}\cdot\frac{3 + \frac{9z}{2} + \frac{27z^3}{6} + \ldots}{\frac{1}{5!} - \frac{z^2}{7!} + \ldots} = \frac{1}{z^4}h(z),$$

where

$$h(z) = \frac{3 + \frac{9z}{2} + \frac{27z^3}{6} + \cdots}{\frac{1}{5!} - \frac{z^2}{7!} + \ldots}.$$

Since both the numerator and denominator of $h(z)$ are analytic on $\mathbb{C}$ and nonzero at $z_0 = 0$, then $h(z)$ is analytic in some neighborhood of 0 and $h(0) \neq 0$. So by the Corollary 7.5, $z_0 = 0$ is a pole of order $N = 4$.

Example 7.10 Find the singular points of the function

$$f(z) = \sin\frac{1}{z-1},$$

and classify them.

Solution This function has only one finite singularity, at $z_0 = 1$; it is analytic elsewhere in $\mathbb{C}$, since $w = 1/(z-1)$ is analytic when $z \neq 1$, and $\sin w$ is analytic for all w.

Substituting $1/(z-1)$ for z in the Taylor series (6.28) for $\sin z$, we get

$$\begin{aligned} f(z) = \sin\frac{1}{z-1} &= \frac{1}{z-1} - \frac{1}{3!}\cdot\frac{1}{(z-1)^3} + \frac{1}{5!}\cdot\frac{1}{(z-1)^5} - \cdots \\ &= \sum_{n=0}^{\infty} \frac{(-1)^n}{(2n+1)!}\cdot\frac{1}{(z-1)^{2n+1}}. \end{aligned}$$

This is the Laurent series for $f(z)$ in a punctured neighborhood of $z_0 = 1$. Since the principal part has infinitely many terms—in fact the whole series consists only of its principal part, the regular part is empty—the point $z_0 = 1$ is an essential singularity.

We could also have determined the type of singularity of the given function directly from the definition, without carrying out the expansion to the Laurent series. For we can easily define sequences $\{z_n\}$ and $\{z_n'\}$ that both converge to $z_0 = 1$, but for which

$$f(z_n) = 1 \quad \text{and} \quad f(z_n') = 0$$

for all n (find such sequences!). Therefore $f(z)$ has no limit as $z \to 1$, and by definition $z_0 = 1$ is an essential singularity.

Now we turn to the notion of the Laurent series expansion about $z_0 = \infty$, as well as the connection between the expansion and the behavior of the function in the neighborhood of ∞. Note that the definitions of isolated singularity and its type transfer without change to $z_0 = \infty$; but theorems 7.3, 7.4, and 7.7, relating the series to the behavior, require adjustments. The matter is that the terms $c_n(z - z_0)^n$ for negative n in the principal part of the series, which determine the "irregularities" of the function near a finite point z_0, will behave "regularly" (namely approach 0) as z tends to infinity. On the contrary, the terms in the regular part with $n > 0$, tend to infinity as $z \to \infty$; these terms determine the type of singularity at ∞. Therefore, the principal part of the expansion in a neighborhood of ∞ will consist of the terms with $n > 0$, and the regular part with $n \leq 0$.

As usual when working in a neighborhood of ∞, it is useful to introduce the transformation $w = 1/z$, extended to ∞ by $w(\infty) = 0$. This is a mutually conformal mapping from the neighborhood $|z| > R$ of $z_0 = \infty$, to the neighborhood $|w| < 1/R$ of $w_0 = 0$. If the function $f(z)$ is analytic in the punctured neighborhood $R < |z| < \infty$ of $z_0 = \infty$, then the function $G(w) = f(1/w)$ will be analytic in the punctured neighborhood $0 < |w| < 1/R$ of $w_0 = 0$. Since $w \to 0$ as $z \to \infty$, then

$$\lim_{z\to\infty} f(z) = \lim_{w\to 0} G(w).$$

Therefore the function G has, at $w_0 = 0$, the same type of singularity that the function f has at $z_0 = \infty$. Expand $G(w)$ into a Laurent series about $w_0 = 0$:

$$G(w) = \sum_{k=-\infty}^{\infty} b_k w^k = \sum_{k=0}^{\infty} b_k w^k + \sum_{k=-1}^{-\infty} b_k w^k. \tag{7.7}$$

The sums on the right represent the regular and principal parts of the expansion, respectively. We change back to the variable z by substituting $w = 1/z$:

$$G\left(\frac{1}{z}\right) = \sum_{k=0}^{\infty} b_k z^{-k} + \sum_{k=-1}^{-\infty} b_k z^{-k}.$$

By also changing the index to $n = -k$, the coefficients to $c_n = b_{-n} = b_k$, and noting that $f(z) = G(1/z)$, we get

$$f(z) = \sum_{n=0}^{-\infty} c_n z^n + \sum_{n=1}^{\infty} c_n z^n. \tag{7.8}$$

This called the *Laurent series expansion of f about $z_0 = \infty$*. It looks like the expansion about $z_0 = 0$, but in this case the first sum, with non-positive powers, is called *the regular part*, while the second sum, with positive powers, is called *the principal part*. Because those parts correspond to same parts of the expansion of G in (7.7), theorems analagous to 7.3, 7.4, and 7.7 will hold for the expansion (7.8). Here is the analog to Theorem 7.3.

Theorem 7.11. *An isolated singularity of the function f about the point $z_0 = \infty$ is removable if and only if the Laurent series (7.8) for f about ∞ has no principal part, i.e. it consists only of its regular part:*

$$f(z) = \sum_{n=0}^{-\infty} c_n z^n = c_0 + \frac{c_{-1}}{z} + \frac{c_{-2}}{z^2} + \dots \tag{7.9}$$

If the function $f(z)$ is defined by the series (7.9) in a neighborhood $|z| > R$ of the point $z_0 = \infty$ (that is the principal part of the expansion (7.8) is absent), then f can be extended to ∞ by defining $f(\infty) = c_0$; the extended function is said to be *analytic at* $z_0 = \infty$. Note that this agrees with the extension of $G(w)$ to w_0 by $G(0) = c_0$, as is used for removable singularities at finite points.

Example 7.12 Show that the singularity $z_0 = \infty$ of the function

$$f(z) = \frac{z-1}{(z^2+1)(z+3)^2}$$

is removable, and extend the function to that point. Also find the Laurent series expansion about ∞.

Solution The limit of $f(z)$ as $z \to \infty$

$$\lim_{n\to\infty} f(z) = \lim_{n\to\infty} \frac{z-1}{(z^2+1)(z+3)^2} = 0$$

is finite. Therefore the singularity at ∞ is removable. If we set $f(\infty) = 0$ then f will be analytic at $z_0 = \infty$.

To find the Laurent series expansion about $z_0 = \infty$, we change the variable to $w = 1/z$. Substituting $z = 1/w$ for z, we get

$$G(w) = f\left(\frac{1}{w}\right) = \frac{\frac{1}{w} - 1}{\left(\frac{1}{w^2}+1\right)\left(\frac{1}{w}+3\right)^2} = \frac{(1-w)w^3}{(1+w^2)(1+3w)};$$

note that this is valid also at $w = 0$ if we set $G(0) = 0$. This function G has singularities at $w = \pm i$ and $w = -1/3$, but it is analytic at $w = 0$. Using the geometric series, we may write

$$\frac{1}{1+w^2} = \sum_{n=0}^{\infty}(-1)^n w^{2n} = 1 - w^2 + w^4 - w^6 + \ldots,$$

$$\frac{1}{1+3w} = \sum_{n=0}^{\infty}(-1)^n(3w)^n = 1 - 3w + 9w^2 - 27w^3 + \ldots.$$

To find the expansion of $G(w)$, we can either to multiply these series and the factor $(1-w)w^3$ directly, or use partial fractions as in Example 6.41:

$$\begin{aligned}
\frac{1-w}{(1+w^2)(1+3w)} &= -\frac{2}{5}\frac{w}{1+w^2} - \frac{1}{5}\frac{1}{1+w^2} + \frac{6}{5}\frac{1}{1+3w} \\
&= -\frac{2}{5}\sum_{n=0}^{\infty}(-1)^n w^{2n+1} - \frac{1}{5}\sum_{n=0}^{\infty}(-1)^n w^{2n} + \frac{6}{5}\sum_{n=0}^{\infty}(-1)^n(3w)^n \\
&= \frac{1}{5}\sum_{n=0}^{\infty}[(6\cdot 3^{2n} - (-1)^n)w^{2n} - (6\cdot 3^{2n+1} + 2(-1)^n)w^{2n+1}] \\
&= 1 - 4w + 11w^2 - 32w^3 + 97w^4 - 292w^5 + \ldots
\end{aligned}$$

So then

$$G(w) = w^3 - 4w^4 + 11w^5 - 32w^6 + 97w^7 - 292w^8 + \dots$$

Finally, substituting $1/z$ for w, we get

$$f(z) = G\left(\frac{1}{z}\right) = \frac{1}{z^3} - \frac{4}{z^4} + \frac{11}{z^5} - \frac{32}{z^6} + \frac{97}{z^7} - \frac{272}{z^8} + \dots,$$

which is the Laurent series expansion of f about $z_0 = \infty$.

Here is the analog of Theorem 7.4 for poles at $z_0 = \infty$.

Theorem 7.13. *An isolated singularity at $z_0 = \infty$ of the function f is a pole if and only if the principal part of the Laurent series expansion of f about that point is nonempty but finite; in other words if the Laurent series is of the form*

$$f(z) = (c_1 z + c_2 z^2 + \cdots + c_N z^N) + \sum_{n=0}^{-\infty} c_n z^n, \tag{7.10}$$

where $N > 0$ and $c_N \neq 0$.

Here the series is a regular part, and the polynomial in parentheses is a principal part of the Laurent series. The multiplicity of a pole of $f(z)$ at ∞ is defined as the multiplicity of the pole $w_0 = 0$ of the function $G(w) = f(1/w)$. As in the case of finite singularities, the order of the pole corresponds to the highest power that appears in the principal part, which in (7.10) is N.

And finally the analog to Theorem 7.7.

Theorem 7.14. *An isolated singularity at $z_0 = \infty$ of the function f is an essential singularity if and only if the principal part of the Laurent series expansion (7.8) of f about that point has infinitely many nonzero terms.*

Problems

1. Find the expansion of the function

$$f(z) = (z+2)^2 e^{z/(z+2)}$$

in a Laurent series in the neighborhood of the point $z_0 = -2$. Determine the annulus of convergence, the regular and principal parts of the expansion, and also the type of singularity at z_0.

2. Find the Laurent series expansion of the function $f(z)$ in the neighborhood of the point z_0. Determine the annulus of convergence, the regular and principal parts of the expansion, and the type of the singularity at z_0.

a) $f(z) = e^{\frac{z}{z-2}}$, $z_0 = 2$; b) $f(z) = (z-3)\sin\dfrac{2}{z-3}$, $z_0 = 3$;

c) $f(z) = \dfrac{1}{z}\cos\dfrac{3}{z^2}$, $z_0 = 0$; d) $f(z) = \dfrac{2z}{z^2+1}$, $z_0 = i$.

3. Determine the type of singularity at the point $z_0 = 0$, of the function

$$f(z) = \frac{\cos z - 1 + z^2}{e^{z^3} - 1 - z^3}.$$

4. Determine the type of singularity at $z_0 = 0$ for the function f.

a) $f(z) = \dfrac{e^{2z} - 1}{\cos z - 1 + \frac{z^2}{2}}$; b) $f(z) = \dfrac{\sin 3z - 3z + \frac{27z^3}{6}}{\cos(z^2) - 1 + \frac{z^4}{2}}$.

5. Show that the function

$$f(z) = \frac{(z^2+1)(z+3)^2}{z-1}$$

has a pole of order 3 at $z_0 = \infty$.

6. Find all isolated singularities of the function

$$f(z) = \frac{\sin\frac{1}{z}}{(z^2+9)^2},$$

and determine their type.

7. Find all isolated singularities of the function $f(z)$ and determine their type.

a) $f(z) = \dfrac{e^{\frac{1}{z-1}}}{(z^2-4)^3}$ b) $f(z) = \dfrac{\cos\frac{1}{z-1}}{(z^2+1)(z^2+4)}$

c) $f(z) = \dfrac{e^{\frac{1}{z}}}{(z^2+16)^3}$

8. In this problem we outline another proof of Theorem 7.8 for finite A. Suppose that z_0 is an isolated singularity of the function $f(z)$, and for some complex finite number A, there are no sequences of numbers $\{z_n\}$ such that $z_n \to z_0$ and $\lim_{n\to\infty} f(z_n) = A$. Prove that z_0 is either removable singularity or a pole of $f(z)$.

(i) Prove that there exist a positive number ε and a punctured neighborhood U of z_0 such that $|f(z) - A| > \varepsilon$ on U.

(ii) Deduce that the function $\psi(z) = \frac{1}{f(z)-A}$ is bounded on U, and therefore z_0 is a removable singularity of ψ.

(iii) Using the representation $f(z) = A + \frac{1}{\psi(z)}$, conclude that z_0 is either removable singularity or a pole of $f(z)$.

7.2 Residues

According to Cauchy's integral Theorem 5.17, the integral of an analytic function along a closed path Γ is zero, if Γ lies within a simply-connected domain on which f is analytic. But if the interior of Γ contains singularities of f, then the theorem does not apply and the integral may be nonzero. It turns out that the value of the integral in this case depends only on the behavior of the function in punctured neighborhoods, however small, of those singularities. More precisely, the integral depends only on the value of integrals over circles, however small, around each singularity individually. We need a definition to start.

Let z_0 be a finite isolated singularity of the function f. Assume that f is analytic in the punctured neighborhood

$$V = \{ z : 0 < |z - z_0| < R \}$$

of z_0. Take a closed contour γ lying within V, which encircles z_0 in the counterclockwise direction, keeping the interior of γ with z_0 on the left. The *residue of the function* f *at* z_0 is the value of the integral $\frac{1}{2\pi i} \int\limits_{\gamma} f(z)\, dz$. We denote this value $\operatorname{res}_{z_0} f$:

$$\operatorname{res}_{z_0} f = \frac{1}{2\pi i} \int\limits_{\gamma} f(z)\, dz. \tag{7.11}$$

Because of the invariance of integrals under path deformation (Corollary 5.11), this integral takes the same value on all such paths, i.e. those encircling z_0 counterclockwise and lying within V; therefore $\operatorname{res}_{z_0} f$ is well-defined. In particular, we may take γ to be the circle $|z - z_0| = \rho$ centered at z_0, and with $\rho < R$.

Theorem 7.15 (Residue theorem). *Let* f *be an analytic function on the domain* G, *and let* Γ *be a closed path in* G. *Suppose that* f *is analytic on the interior of* Γ *with the exception of finitely many isolated singularities* z_1, z_2, $\ldots$, z_n *(see Fig. 57). Then the value of the integral of* f *along* Γ *is equal to the sum of the residues at the* z_k, *multiplied by* $2\pi i$:

$$\int\limits_{\Gamma} f(z)\, dz = 2\pi i \sum_{k=1}^{n} \operatorname{res}_{z_k} f. \tag{7.12}$$

Proof. We encircle the singularities inside Γ with sufficiently small circles γ_k, of radius ρ_k so that each contains only one singularity, lies entirely within the interior of Γ, and does not intersect any other circle. Then the function f will be analytic on the closed $(n+1)$-connected domain $\overline{D}$ bounded by Γ and γ_1, γ_2, $\ldots$, γ_n.

The domain $\overline{D}$ can also be thought of the set of points on Γ and in its interior, from which the open disks $|z - z_k| < \rho_k$ have been removed. By Cauchy's integral theorem for multiply connected domains (Theorem 5.9, specifically formula (5.14)),

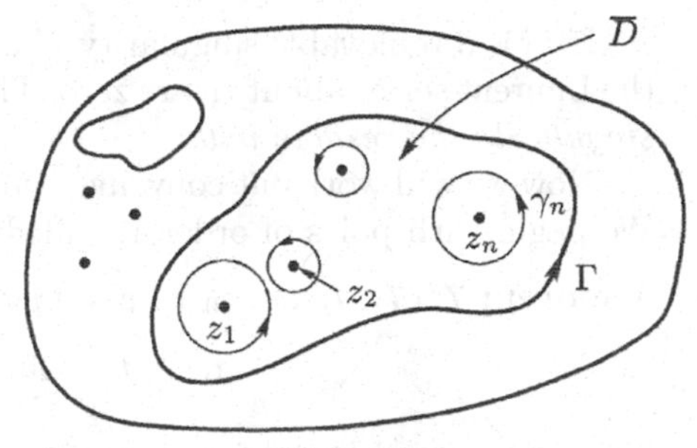

Fig. 57

$$\int_{\Gamma} f(z)\,dz = \sum_{k=1}^{n} \int_{\gamma_k} f(z)\,dz. \tag{7.13}$$

From (7.11) we easily see that $\int_{\gamma_k} f(z)\,dz = 2\pi i \operatorname{res}_{z_k} f$ for each $k = 1, \dots, n$. Substituting these equalities to (7.13) we get (7.12), as desired. □

According to this theorem, the value of the integral of a function f along the boundary of the closed domain $\overline{D}$, depends only on the local behavior of f near the singularities within $\overline{D}$. And we will see that the local behavior, in turn, is determined by the Laurent series expansion of f about those singularities; in fact, we will see in the next theorem that the residues are determined by just one coefficient of the principal part of that series. Therefore, Theorem 7.15 provides us with one of many examples when important global characteristics of analytic functions (such as the integral along the boundary of a domain) are determined by principal parts of their Laurent expansions.

It is possible to give a physical rationale for the importance of the singularities. For example, if we regard the analytic function as a complex potential of the velocity field of flow of a fluid, then singularities can be interpreted as sources, sinks, or other elements that constrain the field. The interested reader can find more on this topic in Chapter 8.

There is a simple relationship between the residue and the Laurent series expansion about a point:

Theorem 7.16. *Let f be an analytic function with an isolated singularity at z_0, and let c_{-1} be the coefficient of the term $(z - z_0)^{-1}$ the Laurent series expansion of f about z_0. Then*

$$\operatorname{res}_{z_0} f = c_{-1}. \tag{7.14}$$

Proof. This follows immediately from the integral formula (6.40) for the coefficients in the Laurent series. Setting $n = -1$ and $\gamma = \{|\zeta - z_0| = \rho\}$, we get

$$c_{-1} = \frac{1}{2\pi i} \int_{\gamma} \frac{f(z)}{(\zeta - z_0)^0}\,d\zeta = \frac{1}{2\pi i} \int_{\gamma} f(z)\,d\zeta = \operatorname{res}_{z_0} f,$$

where the last equality is by definition (7.11). □

If z_0 is a removable singularity, then all coefficients of the principal part of the Laurent series about z_0 are zero. Therefore by this theorem, *at a removable singularity the residue is 0.*

Now we will work out convenient formulas for calculating residues at poles. We begin with poles of order 1, called *simple poles.*

Theorem 7.17. *1. Let z_0 be a simple pole of a function f. Then*

$$\operatorname{res}_{z_0} f = \lim_{z\to z_0} (z - z_0)f(z). \tag{7.15}$$

2. Suppose that the function $f(z)$ can be written in the form

$$f(z) = \frac{h(z)}{\psi(z)},$$

where h and ψ are functions which are both analytic in some neighborhood of z_0, and $h(z_0) \neq 0$, while $\psi(z_0) = 0$ and $\psi'(z_0) \neq 0$ (so that z_0 is a zero of ψ of order 1, and therefore is a simple pole of f). Then

$$\operatorname{res}_{z_0} f = \frac{h(z_0)}{\psi'(z_0)}. \tag{7.16}$$

In particular, if $\psi(z) = z - z_0$, then

$$\operatorname{res}_{z_0} f = h(z_0). \tag{7.17}$$

Proof. 1. Since z_0 is a pole of order 1, the Laurent series expansion of $f(z)$ about z_0 takes the form

$$f(z) = \frac{c_{-1}}{z - z_0} + \sum_{n=0}^{\infty} c_n(z - z_0)^n$$

—see Theorem 7.4. Multiplying both sides by $(z - z_0)$ we get

$$(z - z_0)f(z) = c_{-1} + \sum_{n=0}^{\infty} c_n(z - z_0)^{n+1}.$$

Taking the limit as $z \to z_0$, and using the previous Theorem 7.16, we get

$$\lim_{z\to z_0} (z - z_0)f(z) = c_{-1} = \operatorname{res}_{z_0} f,$$

as desired.

2. Here we apply the formula (7.15) just proved and use the assumption $\psi(z_0) = 0$:

$$\begin{aligned}\operatorname{res}_{z_0} f &= \lim_{z\to z_0} \frac{(z - z_0)h(z)}{\psi(z)} = \lim_{z\to z_0} \frac{(z - z_0)h(z)}{\psi(z) - \psi(z_0)} \\ &= \lim_{z\to z_0} \frac{h(z)}{\dfrac{\psi(z) - \psi(z_0)}{z - z_0}} = \frac{\lim\limits_{z\to z_0} h(z)}{\lim\limits_{z\to z_0} \dfrac{\psi(z) - \psi(z_0)}{z - z_0}} = \frac{h(z_0)}{\psi'(z_0)},\end{aligned}$$

giving us formula (7.16). The special case (7.17) follows immediately. □

The following theorem generalizes formulas (7.15) and (7.17) to the case of a pole of any order.

Theorem 7.18. *1. Suppose that z_0 is a pole of the function f of order $n \geq 1$. Then*

$$\operatorname{res}_{z_0} f = \frac{1}{(n-1)!} \lim_{z \to z_0} \left((z - z_0)^n f(z)\right)^{(n-1)}. \tag{7.18}$$

Note that here the $^{(n-1)}$ means the $(n-1)$th derivative.

2. Suppose that on a punctured neighborhood of z_0 a function $f(z)$ takes the form

$$f(z) = \frac{h(z)}{(z - z_0)^n},$$

where h is analytic on a neighborhood of z_0, and $h(z_0) \neq 0$ (so that f has a pole of order n at z_0). Then

$$\operatorname{res}_{z_0} f = \frac{h^{(n-1)}(z_0)}{(n-1)!}. \tag{7.19}$$

Proof. 1. Since z_0 is a pole of order n, the Laurent series expansion of f about z_0 takes the form

$$f(z) = \frac{c_{-n}}{(z - z_0)^n} + \cdots + \frac{c_{-1}}{z - z_0} + \sum_{k=0}^{\infty} c_k (z - z_0)^k.$$

Then multiplying by $(z - z_0)^n$ gives

$$(z - z_0)^n f(z) = c_{-n} + \cdots + c_{-1}(z - z_0)^{n-1} + \sum_{k=0}^{\infty} c_k (z - z_0)^{k+n}.$$

Now we differentiate $n - 1$ times with respect to z. During this process, all terms with powers less than $n - 1$ will disappear:

$$\begin{aligned} &\left((z - z_0)^n f(z)\right)^{(n-1)} \\ &\quad = (n-1)!\, c_{-1} + \sum_{k=0}^{\infty} (k+n)(k+n-1)\ldots(k+2) c_k (z - z_0)^{k+1}. \end{aligned} \tag{7.20}$$

Taking the limit as $z \to z_0$ and applying the equality (7.14), we get

$$\lim_{z \to z_0} \left((z - z_0)^n f(z)\right)^{(n-1)} = (n-1)!\, c_{-1} = (n-1)! \operatorname{res}_{z_0} f,$$

from which formula (7.18) follows immediately.

2. Applying the assumption $f(z) = h(z)/(z - z_0)^n$ and formula (7.18) just proved,

$$\begin{aligned} \operatorname{res}_{z_0} f &= \frac{1}{(n-1)!} \lim_{z \to z_0} \left((z - z_0)^n \frac{h(z)}{(z - z_0)^n}\right)^{(n-1)} \\ &= \frac{1}{(n-1)!} \lim_{z \to z_0} h^{(n-1)}(z) = \frac{h^{(n-1)}(z_0)}{(n-1)!}. \end{aligned}$$

The last equality is because the analytic function h has continuous derivatives of all orders. □

Example 7.19 Find the residues at the singularities of the function

$$f(z) = \frac{z-1}{(z^2+1)(z+3)^2}.$$

Solution This is the same function as in Example 7.6, where we saw that it has poles at $z_1 = i$, $z_2 = -i$, and $z_3 = -3$; z_1 and z_2 are simple poles, while z_3 is a pole of order 2. We can use formula (7.15) to find the residues at z_1 and z_2:

$$\begin{aligned}
\text{res}_i f &= \lim_{z\to i} \frac{(z-i)(z-1)}{(z^2+1)(z+3)^2} = \lim_{z\to i} \frac{(z-i)(z-1)}{(z-i)(z+i)(z+3)^2} \\
&= \lim_{z\to i} \frac{z-1}{(z+i)(z+3)^2} = \frac{i-1}{2i(i+3)^2} = \frac{i-1}{4(-3+4i)} \\
&= \frac{(i-1)(-3-4i)}{4(-3+4i)(-3-4i)} = \frac{7+i}{100};
\end{aligned}$$

$$\begin{aligned}
\text{res}_{-i} f &= \lim_{z\to -i} \frac{(z+i)(z-1)}{(z^2+1)(z+3)^2} = \lim_{z\to -i} \frac{(z+i)(z-1)}{(z-i)(z+i)(z+3)^2} \\
&= \lim_{z\to -i} \frac{z-1}{(z-i)(z+3)^2} = \frac{-i-1}{-2i(-i+3)^2} \\
&= \frac{i+1}{4(3+4i)} = \frac{(i+1)(3-4i)}{4(3+4i)(3-4i)} = \frac{7-i}{100}.
\end{aligned}$$

We could have gotten the same result using formula (7.17), but the computation is essentially the same. For example, to evaluate $\text{res}_i f$ we write $f(z)$ in the form

$$f(z) = \frac{z-1}{(z-i)(z+i)(z+3)^2} = \frac{h(z)}{z-i}, \quad \text{where} \quad h(z) = \frac{z-1}{(z+i)(z+3)^2}.$$

By (7.17) we get

$$\text{res}_i f = h(i) = \frac{i-1}{2i(i+3)^2} = \frac{7+i}{100}.$$

Now let us find the residue at $z_3 = -3$, using formula (7.18) with $n = 2$:

$$\begin{aligned}
(z-z_0)^2 f(z) &= \frac{(z+3)^2(z-1)}{(z^2+1)(z+3)^2} = \frac{z-1}{z^2+1}, \\
\frac{d}{dz}\left((z-z_0)^2 f(z)\right) &= \frac{z^2+1-2z(z-1)}{(z^2+1)^2} = \frac{-z^2+2z+1}{(z^2+1)^2}, \\
\text{res}_{-3} f &= \frac{1}{1!} \lim_{z\to -3} \frac{-z^2+2z+1}{(z^2+1)^2} = \frac{-9-6+1}{(9+1)^2} = -\frac{14}{100}.
\end{aligned}$$

Again, we could instead have used formula (7.19). Represent $f(z)$ in the form

$$f(z) = \frac{z-1}{(z-i)(z+i)(z+3)^2} = \frac{h(z)}{(z+3)^2}, \quad \text{where } h(z) = \frac{z-1}{z^2+1}.$$

As before, we have to evaluate the derivative of the same function $h(z)$ at $z = -3$, and therefore the computation is essentially the same. This is true as a rule: neither of formulas (7.18) and (7.19) has any real advantage over the other.

Example 7.20 Find the residue of the function $f(z) = 1/\sin z$ at the point $z_0 = \pi$.

Solution Let $\psi(z) = \sin z$; then $\psi(\pi) = 0$, while $\psi'(\pi) = \cos\pi = -1 \neq 0$, so $z_0 = \pi$ is a simple pole. Therefore we can apply formula (7.16):

$$\operatorname{res}_\pi f = \frac{1}{\psi'(\pi)} = \frac{1}{-1} = -1.$$

This formula is more convenient than (7.15), which would require multiplying by $(z-\pi)$ and taking a limit.

Unfortunately these formulas do not work if we wish to find the residue at an essential singularity. In that case we must either use the integral formula (7.11), or find the principal part of the Laurent series.

Example 7.21 Find the residue of the function

$$f(z) = \sin\frac{1}{z-1}$$

at the point $z_0 = 1$.

Solution We saw in Example 7.10 that $z_0 = 1$ is an essential singularity of this function, and we obtained the Laurent series for it about that point:

$$\sin\frac{1}{z-1} = \frac{1}{z-1} - \frac{1}{3!}\cdot\frac{1}{(z-1)^3} + \frac{1}{5!}\cdot\frac{1}{(z-1)^5} - \cdots.$$

Having done all most of the work in that example, here it remains only to observe that the coefficient c_{-1} of $(z-1)^{-1}$ equals 1. Hence,

$$\operatorname{res}_1 f = c_{-1} = 1.$$

Now we turn to the notion of *residue* at the point *at infinity*. So let $z_0 = \infty$ be an isolated singularity of the function f, i.e. f is analytic in some punctured neighborhood

$$U = \{\, z : R < |z| < \infty \,\}$$

of ∞. Take a $\rho > R$ so that the circle $|z| = \rho$ lies within U, and make that circle a path by orienting it clockwise, i.e. so that in traversal U remains on the left; denote this path by Γ^-. We define the residue of f at ∞ by

$$\operatorname{res}_\infty f = \frac{1}{2\pi i} \int_{\Gamma^-} f(z)\, dz. \tag{7.21}$$

This is analogous to the definition (7.11) given for finite singularities; but we will see the relation between the residue and the coefficients of the Laurent series changes slightly at the point at infinity.

Theorem 7.22. *Suppose that the function f is analytic in a punctured neighborhood of ∞. The residue at infinity is equal to the coefficient c_{-1} of z^{-1} in the Laurent series expansion (7.8) of f about ∞, taken with the negative sign:*

$$\operatorname{res}_\infty f = -c_{-1}. \tag{7.22}$$

Proof. The both series in (7.8) converge uniformly on Γ^-, so we can substitute into the definition (7.21) and integrate termwise:

$$\operatorname{res}_\infty f = \frac{1}{2\pi i} \int_{\Gamma^-} f(z)\, dz = \frac{1}{2\pi i} \sum_{n=-\infty}^{\infty} c_n \int_{\Gamma^-} z^n\, dz,$$

where we have combined both sums of (7.21) into one. Invoking formula (5.6) once again, the right side reduces to just the term where $n = -1$:

$$\operatorname{res}_\infty f = \frac{1}{2\pi i} \cdot c_{-1} \int_{\Gamma^-} z^{-1}\, dz = \frac{1}{2\pi i} \cdot c_{-1}(-2\pi i),$$

where the negative sign is because the path is oriented clockwise. Then the desired formula (7.22) follows immediately. □

We should note one other difference from the case of finite singularities: in the case of $z_0 = \infty$, the coefficient c_{-1} belongs to the *regular* part of the Laurent series about $z_0 = \infty$, not the principal part. Therefore even if the function is analytic at ∞, the residue there may not be 0—unlike the case of finite points, where the residue is nonzero only if the point is a nonremovable singularity.

Example 7.23 Find the residue of the function

$$f(z) = 2 + \frac{3}{z}$$

at the point at infinity.

Solution This function is already in the form of its Laurent series expansion at $z_0 = \infty$, which consists of just these two terms. The coefficient of z^{-1} is 3, so

$$\operatorname{res}_\infty f = -3.$$

Note that this function is analytic at ∞, for the Laurent series has no principal part (the terms with z^n, $n > 0$, are absent).

Theorem 7.24 (about the sum of residues). *Suppose that the function f is analytic on the extended complex plane $\overline{\mathbb{C}}$, with the exception of finitely many singular points. Then the sum of the residues at all the singularities and at the point at infinity is zero.*

Proof. Let Γ be the circle $|z| = \rho$, where ρ is chosen sufficiently large that all finite singularities $z_1, z_2 \ldots, z_n$ of f lie in the interior of Γ. If Γ is oriented counterclockwise, then by Theorem 7.15

$$\frac{1}{2\pi i} \int_\Gamma f(z)\, dz = \sum_{k=1}^{n} \operatorname{res}_{z_k} f. \tag{7.23}$$

By the definition of the residue at the ∞,

$$-\frac{1}{2\pi i} \int_\Gamma f(z)\, dz = \frac{1}{2\pi i} \int_{\Gamma^-} f(z)\, dz = \operatorname{res}_\infty f. \tag{7.24}$$

Adding these two equations gives

$$0 = \operatorname{res}_\infty f + \sum_{k=1}^{n} \operatorname{res}_{z_k} f,$$

as required. □

This theorem is useful both for calculating integrals, as we will see in the next section, and for calculating the residue at the point at infinity.

Here is another convenient formula for $\operatorname{res}_\infty f$.

Property 7.25. *If f is analytic in a neighborhood of ∞,*

$$\operatorname{res}_\infty f = -\operatorname{res}_0 \left(\frac{1}{w^2} f\left(\frac{1}{w}\right) \right). \tag{7.25}$$

This formula reduces the calculation of the residue of $f(z)$ at $z_0 = \infty$, to the calculation of the residue of

$$g(w) = \frac{1}{w^2} f\left(\frac{1}{w}\right)$$

at $w_0 = 0$.

Proof. We start with the Laurent series expansion (7.8) of f in a punctured neighborhood $R < |z| < \infty$ of the point at infinity:

$$f(z) = \sum_{n=-\infty}^{\infty} c_n z^n,$$

where we have combined the regular and principal parts into one sum. Substituting $1/w$ for z, we get

$$f\left(\frac{1}{w}\right) = \sum_{n=-\infty}^{\infty} c_n \frac{1}{w^n} \quad \text{when} \quad R < \left|\frac{1}{w}\right| < \infty.$$

This can be rewritten as

$$g(w) = \frac{1}{w^2} f\left(\frac{1}{w}\right) = \sum_{n=-\infty}^{\infty} c_n w^{-n-2} \quad \text{when} \quad 0 < |w| < \frac{1}{R},$$

giving us the Laurent series expansion of g at $w_0 = 0$. Then $\text{res}_0\, g$ is the coefficient of w^{-1}, which is where $-n-2=-1$, or $n=-1$. So $\text{res}_0\, g = c_{-1}$. On the other hand, by Theorem 7.22, $\text{res}_\infty f = -c_{-1}$. Thus,

$$\text{res}_\infty f = -c_{-1} = -\text{res}_0 g,$$

and the desired equality (7.25) is proved. □

This property can also be proved by starting with the definitions of residues in terms of integrals, and applying the change of variable $w = 1/z$.

Example 7.26 Find the residue of the function

$$f(z) = \frac{z-1}{(z^2+1)(z+3)^2}$$

at the point at infinity.

Solution We will solve this in two different ways: first with Theorem 7.24. In Example 7.19 we saw that f has singularities at $z_1 = i$, $z_2 = -i$, and $z_3 = -3$, and that the residues there are

$$\text{res}_i f = 0.01(7+i), \quad \text{res}_{-i} f = 0.01(7-i), \quad \text{res}_{-3} f = -0.14.$$

Applying Theorem 7.24 we get

$$\text{res}_\infty f + 0.01(7+i) + 0.01(7-i) - 0.14 = 0,$$

so that $\text{res}_\infty f = 0$.

Now we will solve it using Property 7.25. First find $g(z)$:

$$g(w) = \frac{1}{w^2} f\left(\frac{1}{w}\right) = \frac{1}{w^2} \frac{\frac{1}{w} - 1}{\left(\frac{1}{w^2}+1\right)\left(\frac{1}{w}+3\right)^2} = \frac{(1-w)w}{(1+w^2)(1+3w)^2}$$

(we obtained an analogous result in Example 7.12). This last equation is valid when $w \neq 0$; but we see that the singularity at 0 is removable: by setting $g(0) = 0$, g becomes analytic at 0. Therefore by the formula (7.25),

$$\operatorname{res}_\infty f = -\operatorname{res}_0 g = 0.$$

Problems

1. Find the residues at the finite singular points of the function

$$f(z) = \frac{z^4}{(z^2+4)(z+1)^2}.$$

2. Find the residue of the function $f(z) = \dfrac{2z}{z^2+1}$ at the point $z_0 = i$ in two ways: a) Using the Laurent expansion of f in the neighborhood of the point z_0; b) By formula (7.15).

3. Find the residues of the following functions at the given points:

a) $f(z) = e^{\frac{z}{z-2}}, \quad z_0 = 2;$

b) $f(z) = (z-3)\sin\dfrac{2}{z-3}, \quad z_0 = 3;$

c) $f(z) = \dfrac{1}{z}\cos\dfrac{3}{z^2}, \quad z_0 = 0;$

d) $f(z) = (z+2)^2 e^{z/(z+2)}, \quad z_0 = -2;$

e) $f(z) = \dfrac{\cos z}{e^{2z}-1}, \quad z_0 = 0;$

f) $f(z) = \dfrac{e^x}{\tan z}, \quad z_0 = 0.$

4. Find the residue of the function $f(z) = (z+2)e^{-1/z}$ at the point at infinity.

5. Find the residue of the function f at the point at infinity.

a) $f(z) = (z+3)\sin\dfrac{1}{z};$

b) $f(z) = (z-1)\cos\dfrac{2}{z};$

c) $f(z) = e^{-2/z} + z^3 + z - 2;$

d) $f(z) = z^3\cos\dfrac{1}{z};$

e) $f(z) = \dfrac{5}{z} + \operatorname{Log}\left(1 + \dfrac{1}{z}\right).$

7.3 Computing Integrals with Residues

One of the most important uses of residues is in computing integrals: knowledge of a single coefficient from the Laurent series expansions at one or several points is often enough to compute an entire path integral.

7.3.1 Integrals over closed curves

Suppose that Γ is a closed curve, and that the function f is analytic on Γ and its interior, with the exception of isolated singularities. Then the contour integral of f along Γ can be calculated using the residue Theorem 7.15: find the residues at each singularity inside Γ, then sum them and multiply by $2\pi i$.

Example 7.27 Calculate the integral

$$\int\limits_{|z|=2} \frac{z-1}{(z^2+1)(z+3)^2}\, dz.$$

Solution Denote by $f(z)$ the integrand. From Example 7.19 we know that the only singularities of f which lie within the circle $|z| = 2$, are $z_1 = i$ and $z_2 = -i$; the third singularity, $z_3 = -3$ lies outside of it. Using the values of the residues at z_1 and z_2 found earlier,

$$\operatorname{res}_i f = \frac{7+i}{100} \quad \text{and} \quad \operatorname{res}_{-i} f = \frac{7-i}{100},$$

and applying formula (7.23) from Theorem 7.15, we get

$$\int\limits_{|z|=2} f(z)\, dz = \int\limits_{|z|=2} \frac{z-1}{(z^2+1)(z+3)^2}\, dz = 2\pi i \cdot \left(\frac{7+i}{100} + \frac{7-i}{100}\right) = 0.28\pi i.$$

If f has only isolated singularities in the extended complex plane $\overline{\mathbb{C}}$, then instead of computing the sum of the residues at the finite singularities, it may easier to find the residue at the point at infinity, and then apply Theorem 7.24 about the sum of the residues.

Example 7.28 Calculate the integral

$$\int\limits_{|z|=2} \frac{z^{15}}{(z^8+1)^2}\, dz.$$

Solution Denote by $f(z)$ the integrand; then $f(z)$ has 8 singularities, namely the roots of the equation $z^8 + 1 = 0$. Each of these points z_k is a pole of order 2, because in a punctured neighborhood of z_k the function $f(z)$ can be written in the form

$$f(z) = \frac{h(z)}{(z-z_k)^2},$$

where h is analytic in a neighborhood of z_k, and $h(z_k) \neq 0$. All these singularities lie inside the circle $|z| = 2$. Calculating the residues at all of these points would be rather tedious. So instead we will apply Theorem 7.24 to this function:

$$\operatorname{res}_\infty f + \sum_{k=1}^{8} \operatorname{res}_{z_k} f = 0.$$

Thus it suffices to find the residue at the point at infinity in order to find the sum of other residues. Using Property 7.25, we get

$$g(w) = \frac{1}{w^2} f\left(\frac{1}{w}\right) = \frac{1}{w^2} \cdot \frac{w^{-15}}{(w^{-8}+1)^2} = \frac{1}{w(1+w^8)^2},$$

whose residue at $w_0 = 0$ will be equal to the negative of the one we want. To find the residue at $w_0 = 0$, we write g in the form

$$g(z) = \frac{h_1(w)}{w}, \quad \text{where } h_1(w) = \frac{1}{(1+w^8)^2}.$$

Since $h_1(z)$ is analytic in a neighborhood of $w_0 = 0$ and $h_1(0) \neq 0$, we may apply formula (7.17): $\operatorname{res}_0 g = h_1(0) = 1$.

Putting everything together, we have

$$\int\limits_{|z|=2} \frac{z^{15}}{(z^8+1)^2}\, dz = 2\pi i \sum_{k=1}^{8} \operatorname{res}_{z_k} f = -2\pi i \operatorname{res}_\infty f = 2\pi i \operatorname{res}_0 g = 2\pi i.$$

7.3.2 Real integrals of the form $\int_0^{2\pi} R(\cos\phi, \sin\phi)\, d\phi$, where R is a rational function of $\cos\phi$ and $\sin\phi$

Integrals of this type arise in many settings, for example when solving boundary problems. They can be reduced to the type considered in the previous section, by means of the change of variable

$$z = e^{i\phi}, \quad dz = e^{i\phi} i\, d\phi = zi\, d\phi.$$

Then

$$d\phi = \frac{dz}{iz} = -\frac{idz}{z},$$

$$\cos\phi = \frac{1}{2}(e^{i\phi} + e^{-i\phi}) = \frac{1}{2}\left(z + \frac{1}{z}\right),$$

$$\sin\phi = \frac{1}{2i}(e^{i\phi} - e^{-i\phi}) = \frac{1}{2i}\left(z - \frac{1}{z}\right),$$

from formulas (4.28). As ϕ goes from 0 to 2π, the variable z describes the circle $|z| = 1$. Therefore after the change to the variable z, we obtain an integral around the unit circle; and the integrand is a ratio of two polynomials of z; such functions are called *rational functions.*

Example 7.29 Calculate the integral

$$\int_0^{2\pi} \frac{1}{1-2a\cos\phi+a^2}\,d\phi, \quad \text{where} \quad |a|<1.$$

Solution Making the above substitutions, we get the equivalent integral

$$\int_{|z|=1} \frac{i}{-z\left(1-2a\frac{1}{2}\left(z+\frac{1}{z}\right)+a^2\right)}\,dz = \int_{|z|=1} \frac{i}{az^2-(a^2+1)z+a}\,dz.$$

To factor the denominator, we must find its roots:

$$\begin{aligned} z &= \frac{a^2+1\pm\sqrt{(a^2+1)^2-4a^2}}{2a} = \frac{a^2+1\pm\sqrt{a^4-2a^2+1}}{2a} \\ &= \frac{a^2+1\pm(a^2-1)}{2a} = a, \;\; \frac{1}{a}. \end{aligned}$$

Denoting the integrand by $f(z)$ we now have

$$f(z) = \frac{i}{a(z-a)(z-\frac{1}{a})}.$$

So f has two singularities, at $z_1 = a$ and $z_2 = 1/a$, both of which are simple poles. Since $|a|<1$, z_1 lies inside the circle, while z_2 lies outside it. So by the residue Theorem 7.15,

$$\int_{|z|=1} f(z)\,dz = 2\pi i \operatorname{res}_a f.$$

To calculate the residue at $z_1 = a$ we could use any of the formulas from Theorem 7.17. For example, using formula (7.16),

$$h(z)=i, \quad \psi(z)=az^2-(a^2+1)z+a, \quad \psi'(z)=2az-(a^2+1),$$

$$\operatorname{res}_a f = \frac{h(a)}{\psi'(a)} = \frac{i}{2a^2-(a^2+1)} = \frac{i}{a^2-1}.$$

This yields

$$\int_0^{2\pi} \frac{1}{1-2a\cos\phi+a^2}\,d\phi = 2\pi i\frac{i}{a^2-1} = \frac{2\pi}{1-a^2}.$$

7.3.3 Improper integrals

Here we consider integrals over the entire real line defined by

$$\int_{-\infty}^{\infty} f(x)\,dx = \lim_{R\to\infty} \int_{-R}^{R} f(x)\,dx. \tag{7.26}$$

The integral defined by this equality, is called *improper integral in the sense of the principal value.* If the limit in (7.26) exists, the integral is said to converge; if not, it is said to diverge.

If each of the integrals

$$\int_{-\infty}^{0} f(x)\,dx = \lim_{R\to\infty} \int_{-R}^{0} f(x)\,dx,$$

$$\int_{0}^{\infty} f(x)\,dx = \lim_{R\to\infty} \int_{0}^{R} f(x)\,dx$$

converges, that is each limit exists, then the integral (7.26) along the real line also converges, and is equal to the sum of these two. However, the converse is not true: it is possible that (7.26) converges, while the integrals along half the real line diverge. For example, the integral

$$\int_{-\infty}^{\infty} \frac{x}{1+x^2}\,dx,$$

whose integrand is odd, converges to zero in the sense of the principal value, since for any $R > 0$,

$$\int_{-R}^{R} \frac{x}{1+x^2}\,dx = \frac{1}{2}\ln(1+x^2)\bigg|_{-R}^{R} = 0.$$

But both integrals

$$\int_{-\infty}^{0} \frac{x}{1+x^2}\,dx \quad \text{and} \quad \int_{0}^{\infty} \frac{x}{1+x^2}\,dx$$

diverge.

It is possible to calculate the values of many improper integrals of the form (7.26) (in the sense of the principal value) via the following theorem.

Theorem 7.30. *Suppose that the function $f(x)$ is defined on the real line, and satisfies the following two conditions.*

1. *The extension $f(z)$ of $f(x)$ to the complex plane has only isolated singularities, and none lying on the real line.*
2. *If $\gamma(R)$ denotes the semicircle in the upper (respectively, lower) half-plane which is centered at the origin and of radius R, then*

$$\lim_{R\to\infty} \int_{\gamma(R)} f(z)\, dz = 0. \tag{7.27}$$

Then the integral

$$\int_{-\infty}^{\infty} f(x)\, dx$$

is equal to the sum of the residues of the function $f(z)$ which lie in the upper (respectively, lower) half-plane, multiplied by $2\pi i$ (respectively, $-2\pi i$).

Proof. First we consider the case when the semicircle $\gamma(R)$ lie in the upper half-plane.

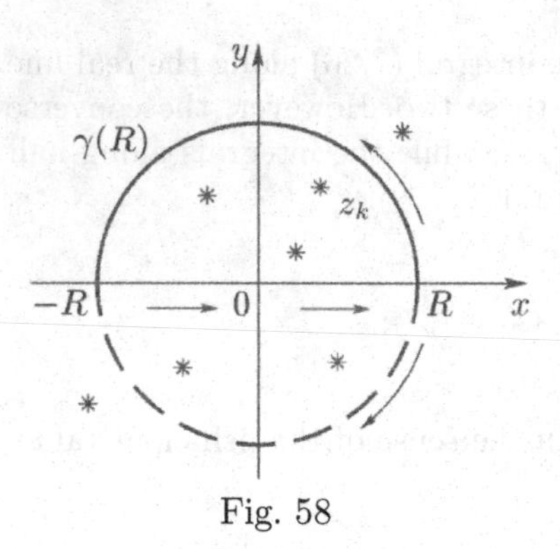

Fig. 58

Define a closed path Γ(R) formed by traversing the interval $[-R, R]$ from left to right, and the semicircle $\gamma(R)$ counterclockwise—Fig. 58. Then by the residue Theorem 7.15,

$$\int_{-R}^{R} f(x)\, dx + \int_{\gamma(r)} f(z)\, dz$$

$$= \int_{\Gamma(R)} f(z)\, dz = 2\pi i \sum \operatorname{res}_{z_k} f,$$

where the sum is taken over all singularities z_k which lie in the interior of $\Gamma(R)$. Taking the limit as $R \to \infty$, by definition (7.26) and assumption (7.27) we get

$$\int_{-\infty}^{\infty} f(x)\, dx = 2\pi i \sum \operatorname{res}_{z_k} f, \tag{7.28}$$

as desired; now the sum is over all singularities in the upper half-plane.

The proof when the $\gamma(R)$ lie in the lower half-plane is analogous, except that now $\gamma(R)$ must be oriented clockwise when combined with $[-R, R]$ to form $\Gamma(R)$, because the interval $[-R, R]$ is traversing from left to right in both cases. That means the $\Gamma(R)$ must also be oriented clockwise, and thus the signs of the residues must be reversed. □

Example 7.31 Evaluate the integral

$$\int_{-\infty}^{\infty} \frac{x^2+4}{(x^2+9)^2}\, dx.$$

Solution Here $f(z) = \frac{z^2+4}{(z^2+9)^2}$; we just replaced the real variable x in the integrand with the complex variable z. Clearly this function satisfies the first condition of Theorem 7.30, having only two singularities, at $z_1 = 3i$ and $z_2 = -3i$. To verify condition (7.27), we write $f(z)$ as

$$f(z) = \frac{z^2\left(1+\frac{4}{z^2}\right)}{z^4\left(1+\frac{9}{z^2}\right)^2} = \frac{1}{z^2}\cdot\frac{1+\frac{4}{z^2}}{\left(1+\frac{9}{z^2}\right)^2} = \frac{1}{z^2}h(z), \quad \text{where } h(z) = \frac{1+\frac{4}{z^2}}{\left(1+\frac{9}{z^2}\right)^2}.$$

Note that $h(z) \to 1$ as $z \to \infty$, and hence $|h(z)| < 2$ for sufficiently large $|z|$. This is helpful for checking the condition (7.27), because

$$|f(z)| = \frac{|h(z)|}{|z|^2} < \frac{2}{|z|^2}.$$

Therefore, since the arc length of $\gamma(R)$ is πR,

$$\left|\int_{\gamma(R)} f(z)\, dz\right| \le \frac{2}{R^2}\cdot \pi R = \frac{2\pi}{R} \to 0 \text{ as } R \to \infty.$$

Similar estimates are valid for both the upper and lower semicircles. Therefore, we can choose any of them as $\gamma(R)$.

Consider the case when $\gamma(R)$ is in the upper half-plane. Writing

$$f(z) = \frac{z^2+4}{(z^2+9)^2} = \frac{z^2+4}{(z+3i)^2(z-3i)^2},$$

we see the singularities $z_1 = 3i$ and $z_2 = -3i$ are poles of order 2. Of these, only the point $z_1 = 3i$ is in the upper half-plane. To find the residue at this point we apply formula (7.18) with $n = 2$:

$$(z-z_1)^2 f(z) = \frac{(z-3i)^2(z^2+4)}{(z+3i)^2(z-3i)^2} = \frac{z^2+4}{(z+3i)^2};$$

$$\frac{d}{dz}\left((z-z_1)^2 f(z)\right) = \frac{2z(z+3i)^2 - 2(z+3i)(z^2+4)}{(z+3i)^4}$$

$$= \frac{2z(z+3i) - 2(z^2+4)}{(z+3i)^3} = \frac{2(3iz-4)}{(z+3i)^3};$$

$$\text{res}_{3i} f = \frac{1}{1!}\lim_{z\to 3i} \frac{2(3iz-4)}{(z+3i)^3} = \frac{2(3i\cdot 3i - 4)}{(3i+3i)^3} = \frac{13}{108i}.$$

Finally, by formula (7.28) we get

$$\int_{-\infty}^{\infty} \frac{x^2+4}{(x^2+9)^2}\,dx = 2\pi i \operatorname{res}_{3i} f = 2\pi i\,\frac{13}{108i} = \frac{13\pi}{54}.$$

Although it would have been possible to evaluate this integral by finding the antiderivative of $f(x)$, the residue method used here is significantly easier.

Remark 7.32 The method we used to check the second condition (7.27) of the Theorem 7.30 can be applied when f is any rational function, if the degree of the polynomial in the denominator exceeds that of the numerator by at least 2; then condition (7.28) will be satisfied. In our example, the numerator was of degree 2, the denominator of degree 4.

The following theorem shows that this condition is satisfied by another class of functions, integrals of which arise, for example, in calculus and in the theory of the Laplace transform—see Chapter 9.

Theorem 7.33 (Jordan's lemma). *Suppose that $F(z)$ is a function analytic in the half-plane $\operatorname{Im} z \geq -a$, with the exception of at most finitely many isolated singularities, and that $F(z) \to 0$ as $z \to \infty$. Let $\gamma(R)$ be the arc of the circle $|z| = R$ above $\operatorname{Im} z = -a$, and let $t > 0$ be any positive number. Then*

$$\lim_{R\to\infty} \int_{\gamma(R)} e^{itz} F(z)\,dz = 0. \tag{7.29}$$

Proof. First we consider the case when $a > 0$. We will break $\gamma(R)$ up into three pieces, $\gamma_1(R)$, $\gamma_2(R)$, and $\gamma_3(R)$ (see Fig. 59), where $\gamma_3(R)$ is the part of $\gamma(R)$ in upper half-plane $\operatorname{Im} z \geq 0$, and $\gamma_1(R)$ and $\gamma_2(R)$ are the parts in the lower half-plane $\operatorname{Im} z < 0$. Obviously, the integral over $\gamma(R)$ is equal to the sum of integrals over these three arcs. Let us estimate each of these integrals.

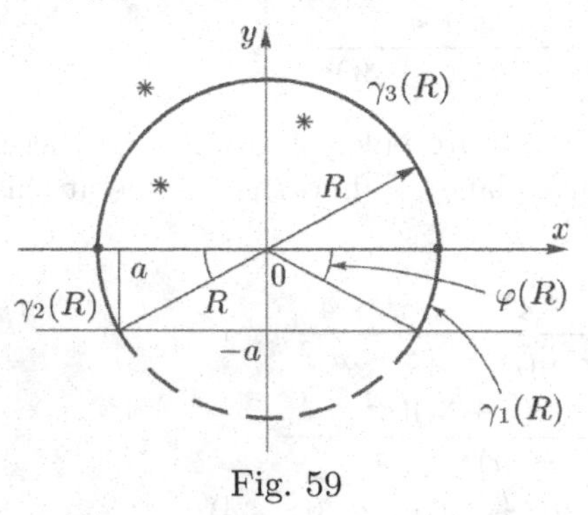

Fig. 59

For a point $z = x + iy$ on $\gamma_1(R)$, the imaginary part satisfies $y \geq -a$, or $-y \leq a$. Therefore

$$|e^{itz}| = |e^{it(x+iy)}| = |e^{itx}e^{-ty}| = e^{-ty} \leq e^{ta}.$$

Denote by $l(R)$ the arc length of $\gamma_1(R)$ and $\gamma_2(R)$, and by $\varphi(R)$ the central angle (from the origin) subtended by these arcs. It is easy to see (Fig. 59) that $\sin\varphi(R) = \frac{a}{R}$, so that $\varphi(R) = \sin^{-1}\left(\frac{a}{R}\right)$. Hence,

$$l(R) = R\varphi(R) = R\sin^{-1}\left(\frac{a}{R}\right).$$

Let $M(R)$ be the maximum of $|F(z)|$ on $\gamma(R)$. Since $F(z) \to 0$ as $z \to \infty$, so also $M(R) \to 0$ as $R \to \infty$. Then we get the estimate

$$\left|\int_{\gamma_1(R)} e^{itz}F(z)\,dz\right| \le e^{ta}M(R)l(R) = e^{ta}M(R)R\sin^{-1}\left(\frac{a}{R}\right).$$

For example, using the L'Hospital's rule, we can evaluate the limit

$$\lim_{R\to\infty} R\sin^{-1}\left(\frac{a}{R}\right) = a\lim_{R\to\infty}\left[\sin^{-1}\left(\frac{a}{R}\right)\Big/\frac{a}{R}\right] = a.$$

Therefore we get

$$\lim_{R\to\infty}\left|\int_{\gamma_1(R)} e^{itz}F(z)\,dz\right| \le \lim_{R\to\infty}\left[e^{ta}M(R)R\sin^{-1}\left(\frac{a}{R}\right)\right] = e^{ta}\cdot 0\cdot a = 0.$$

Exactly the same estimates are valid for $\gamma_2(R)$. Now we consider $\gamma_3(R)$.

For points $z \in \gamma_3(R)$, let $\phi = \operatorname{Arg} z$; then $0 \le \phi \le \pi$, and

$$z = z(\phi) = R(\cos\phi + i\sin\phi), \quad dz = iRe^{i\phi}d\phi,$$

$$\left|e^{itz(\phi)}\right| = \left|e^{itR(\cos\phi + i\sin\phi)}\right| = \left|e^{itR\cos\phi}e^{-tR\sin\phi}\right| = e^{-tR\sin\phi}.$$

We use this parametrization to estimate the integral:

$$\left|\int_{\gamma_3(R)} e^{itz}F(z)\,dz\right| \le M(R)\cdot R\int_0^{\pi} e^{-tR\sin\phi}\,d\phi.$$

We want to show that this integral approaches 0 as $R \to \infty$. Note that

$$\int_0^{\pi/2} e^{-tR\sin\phi}\,d\phi = \int_{\pi/2}^{\pi} e^{-tR\sin\phi}\,d\phi;$$

this equality is easy to prove, for example, by the change of variable $\alpha = \pi - \phi$. So,

$$\left|\int_{\gamma_3(R)} e^{itz}F(z)\,dz\right| \le 2M(R)\cdot R\int_0^{\pi/2} e^{-tR\sin\phi}\,d\phi.$$

When $0 \le \phi \le \frac{\pi}{2}$, the graph of $\sin\phi$ lies above the line segment connecting the points (0,0) and $(\frac{\pi}{2}, 1)$ of this graph. Hence, $\sin\phi \ge \frac{2}{\pi}\phi$ as $0 \le \phi \le \frac{\pi}{2}$. Therefore

$$e^{-tR\sin\phi} \le e^{-tR2\phi/\pi},$$

and

$$\left|\int_{\gamma_3(R)} e^{itz}F(z)\,dz\right| \le 2M(R)\cdot R\int_0^{\pi/2} e^{-tR2\phi/\pi}\,d\phi$$

$$= -M(R)\frac{\pi}{t}\,e^{-tR2\phi/\pi}\Big|_0^{\pi/2} = M(R)\cdot\frac{\pi}{t}\cdot(1-e^{-tR}).$$

Hence,

$$\lim_{R\to\infty}\left|\int_{\gamma_3(R)} e^{itz}F(z)\,dz\right| = 0\cdot\frac{\pi}{t}\cdot 1 = 0.$$

So we have shown that the integrals over all three pieces of $\gamma(R)$ approach 0 as $R\to\infty$, which finishes the case $a > 0$.

In the case when $a \le 0$, the arc $\gamma(R)$ lies in the closed upper half-plane $\operatorname{Im} z \ge 0$ and is a part of $\gamma_3(R)$; the parts $\gamma_1(R)$ and $\gamma_2(R)$ are absent. For $\gamma(R)$, the arguments given above for $\gamma_3(R)$ are valid, and the theorem is proved. □

The power of this theorem is that, however slowly the function $F(z)$ may tend to 0 as $z\to\infty$, multiplying by the factor e^{itz} causes the integral over $\gamma(R)$ to approach 0. Compare this with Example 7.29, where the integral only approached 0 because the integrand decreased sufficiently quickly, like $|z|^{-2}$.

A version of Theorem 7.33 holds for the case when $t < 0$, with the change that $\gamma(R)$ is now the part of the circle in the half-plane $\operatorname{Im} z \le -a$ (the dashed arc in Fig. 59). The proof would be analogous, or it could be derived from Theorem 7.33 by a change of variable $\zeta = -z$. In the case $t = 0$ the theorem is incorrect—see problem 6.

Example 7.34 Evaluate the integrals

$$\int_{-\infty}^{\infty}\frac{x\cos 2x}{x^2+9}\,dx \quad\text{and}\quad \int_{-\infty}^{\infty}\frac{x\sin 2x}{x^2+9}\,dx.$$

Solution Consider the function

$$f(z) = \frac{ze^{i2z}}{z^2+9};$$

then when $z = x$ is real,

$$f(x) = \frac{x(\cos 2x + i \sin 2x)}{x^2+9} = \frac{x\cos 2x}{x^2+9} + i\frac{x\sin 2x}{x^2+9}.$$

So the real and imaginary parts of $f(x)$ correspond to the two functions we wish to integrate. Therefore we can kill two birds with one stone by integrating the function $f(x)$, and then taking the real and imaginary parts of the result.

By defining the function F to the same as f, but without the factor e^{i2z}, i.e.

$$F(z) = \frac{z}{z^2+9},$$

then F satisfies the conditions of the previous Theorem 7.33: it has only two singularities, at $\pm 3i$, and

$$\lim_{z\to\infty} \frac{z}{z^2+9} = 0.$$

Therefore if $\gamma(R)$ is the part of the circle of radius R centered at the origin which lies in the upper half-plane $\operatorname{Im} z \geq 0$, then taking $t = 2$,

$$\lim_{R\to\infty} \int\limits_{\gamma(R)} \frac{ze^{i2z}}{z^2+9}\,dz = 0.$$

That we means we may now apply Theorem 7.30, according to which the integral of $f(x)$ along the real line is equal to the sum of the residues of $f(z)$ in the upper half-plane $\operatorname{Im} z > 0$ times $2\pi i$. The only such singularity is at $z_1 = 3i$. Writing

$$f(z) = \frac{ze^{i2z}}{(z-3i)(z+3i)},$$

we see that this is a simple pole, so to find the residue we can use any of the formulas in Theorem 7.17. Applying formula (7.17) with

$$h(z) = \frac{ze^{i2z}}{z+3i},$$

we get

$$\operatorname{res}_{3i} f = h(3i) = \frac{3i \cdot e^{i2\cdot 3i}}{2\cdot 3i} = \frac{1}{2}e^{-6}.$$

So for the integral we get

$$\int\limits_{-\infty}^{\infty} \frac{xe^{i2z}}{x^2+9}\,dx = 2\pi i\frac{1}{2}e^{-6} = i\frac{\pi}{e^6}.$$

Setting the real and imaginary parts of this value to first and second of the integrals, respectively, that we started with, we finally obtain

$$\int\limits_{-\infty}^{\infty} \frac{x\cos 2x}{x^2+9}\,dx = 0 \quad \text{and} \quad \int\limits_{-\infty}^{\infty} \frac{x\sin 2x}{x^2+9}\,dx = \frac{\pi}{e^6}.$$

We could have predicted that the first integral would come out to 0, because the integrand is an odd function, symmetric about the origin. But there was no easier way to evaluate the second integral.

Problems

1. Calculate the value of the integral

$$\int_{|z-i|=2} \frac{z^4}{(z^2+4)(z+1)^2}\,dz.$$

2. Compute the integrals.

a) $\displaystyle\int_{|z+2i|=3} \frac{z^4+1}{z^2(z^2+9)}\,dz;$ b) $\displaystyle\int_{|z|=3} \frac{z^4}{(z^2+4)(z+1)^2}\,dz;$

c) $\displaystyle\int_{|z-1|=3} \frac{z^4}{(z^2-9)(z-i)}\,dz;$ d) $\displaystyle\int_{|z|=2} \frac{z^8}{(z^2-1)^2}\,dz;$

e) $\displaystyle\int_{|z|=5} \frac{e^z}{z^2(z-i\pi)}\,dz.$

3. Compute the integral

$$\int_0^{2\pi} \frac{1}{5-\sqrt{21}\sin t}\,dt.$$

4. Compute the integrals.

a) $\displaystyle\int_0^{2\pi} \frac{1}{3+\sqrt{5}\sin t}\,dt$ b) $\displaystyle\int_0^{2\pi} \frac{1}{3\cos t+5}\,dt$

5. Calculate the value of the improper integral

$$\int_{-\infty}^{\infty} \frac{x^2-4}{(x^2+6x+13)(x^2+16)}\,dx.$$

6. Let $\gamma(R)$ be the arc of the circle $|z|=R$ above $\operatorname{Im} z=0$ oriented counterclockwise. Prove that

$$\lim_{R\to\infty} \int_{\gamma(R)} F(z)\,dz = 2\pi i \operatorname{res}_i F = i\pi \neq 0, \quad \text{where } F(z)=\frac{z}{z^2+1}.$$

Compare this equality with (7.29).

7. Compute the improper integrals.

a) $\displaystyle\int_{-\infty}^{\infty} \frac{x^2+5}{(x^2+2x+2)(x^2+1)}\,dx$; b) $\displaystyle\int_{-\infty}^{\infty} \frac{x^2+1}{(x^2+4x+13)^2}\,dx$;

c) $\displaystyle\int_{-\infty}^{\infty} \frac{x^2+4}{(x^2+9)^2}\,dx$; d) $\displaystyle\int_{-\infty}^{\infty} \frac{(x+4)\sin x}{x^2+4}\,dx$.

8. Prove that $\displaystyle\int_{0}^{\infty} \frac{\sin x}{x}\,dx = \frac{\pi}{2}$.

7.4 Logarithmic Residues and the Argument Principle

Consider the multivalued function

$$\begin{aligned}\log f(z) &= \ln|f(z)| + i\arg f(z)\\ &= \ln|f(z)| + i(\operatorname{Arg} f(z) + 2\pi k) \quad \text{where} \quad k = 0, \pm1, \pm2, \ldots.\end{aligned}$$

At all points z at which $f(z)$ is analytic and nonzero, $\log f(z)$ is a multivalued analytic function. Each of its regular branches, obtained by a particular choice of k, is a single-valued analytic function in some neighborhood of z. Since these branches differ only by an additive constant, they all have the same derivative, namely

$$\frac{d}{dz}\log f(z) = \frac{1}{f(z)}\cdot f'(z) = \frac{f'(z)}{f(z)}.$$

This is called the *logarithmic derivative of the function* f; it is a single-valued analytic function everywhere that $f(z)$ was both analytic and nonzero, in other words everywhere except at singularities and zeros of f. The residue of the logarithmic derivative of f at a point z_0 (that is the residue of the function $f'(z)/f(z)$) is called the *logarithmic residue* of f at z_0.

Theorem 7.35. *If z_0 is a zero of order n of an analytic function f, then the logarithmic residue of f at z_0 is n; if z_0 is a pole of f of order p, then the logarithmic residue of f at z_0 is $-p$.*

Proof. Suppose that z_0 is a zero of f of order n. Then, in some neighborhood of z_0, $f(z)$ can be written in the form

$$f(z) = (z-z_0)^n\phi(z),$$

where ϕ is an analytic function and $\phi(z_0) \neq 0$. Then

$$\begin{aligned}f'(z) &= n(z-z_0)^{n-1}\phi(z) + (z-z_0)^n\phi'(z),\\ \frac{f'(z)}{f(z)} &= \frac{n(z-z_0)^{n-1}\phi(z) + (z-z_0)^n\phi'(z)}{(z-z_0)^n\phi(z)} = \frac{n}{z-z_0} + \frac{\phi'(z)}{\phi(z)}.\end{aligned}$$

Since $\phi(z_0) \neq 0$, the function $\phi'(z)/\phi(z)$ is analytic in some neighborhood of z_0, and therefore has a power series expansion there. Therefore the principal part of the Laurent series expansion of $f'(z)/f(z)$ consists of just the single term $n/(z-z_0)$, and so the residue, which is equal to the coefficient c_{-1} of the $(z-z_0)^{-1}$ term (see (7.14)) is n.

Suppose now that z_0 is a pole of f of order p. We can prove the other assertion in the similar way, writing $f(z)$ in the form

$$f(z) = \frac{\phi(z)}{(z-z_0)^p},$$

where ϕ is analytic in some neighborhood of z_0, and $\phi(z_0) \neq 0$. But the following arguments are shorter. If z_0 is a pole of f of order p, then z_0 is a zero of order p of the function $g(z) = 1/f(z)$. Therefore, by the first part of the theorem,

$$\operatorname{res}_{z_0}\left(\frac{g'}{g}\right) = p.$$

Then using the identity

$$\frac{g'}{g} = (\log g)' = -(\log f)' = -\frac{f'}{f},$$

we get

$$\operatorname{res}_{z_0}\left(\frac{f'}{f}\right) = -\operatorname{res}_{z_0}\left(\frac{g'}{g}\right) = -p,$$

as desired. □

In the following theorems we will explore the connection between the number of zeros and poles of a function in the interior of a domain, and the behavior of the function on the boundary of the domain. In counting the number of zeros or poles, we will follow the convention that each zero or pole of order n will be counted n times. For example, if a domain contains a single pole, which is of order 3, and two zeros, one of order 1 and one of order 4, then we will say that the pole count $P = 3$ and the zero count $N = 5$.

Theorem 7.36 (Logarithmic residue theorem). *Suppose that f is a function analytic in some domain, Γ is a closed curve lying within that domain, and Γ contains no zeros of f. Let f be analytic in the interior of Γ with the possible exception a finite number of poles. If Γ is oriented so that points in its interior always lie on the left during traversal, then*

$$\frac{1}{2\pi i}\int_{\Gamma} \frac{f'(z)}{f(z)}\,dz = N - P, \tag{7.30}$$

where N is the zero count of f in the interior of Γ, and P is the pole count of f there.

Proof. Denote the logarithmic derivative of f by G, i.e. $G(z) = f'(z)/f(z)$. Because Γ contains neither zeros nor poles of f, this function G is analytic in a neighborhood of Γ. We are assuming that within the interior of Γ, f has only finitely many poles. Moreover, we may assume it has only finitely many zeros, otherwise the zeros would form a set of uniqueness, and the function would be identically zero, including the points of Γ.

So let $a_1, a_2, \ldots, a_l$ be the zeros of f, which are also the poles of G; and let $b_1, b_2, \ldots, b_m$ be the poles of f, which are also the zeros of G. And let n_k be the order of the zero of f at a_k, and let p_k be the order of the pole of f at b_k. Obviously, G is analytic at all other points inside Γ, and we can apply the residue Theorem 7.15 to this integral, according to which

$$\frac{1}{2\pi i}\int_{\Gamma}\frac{f'(z)}{f(z)}\,dz = \sum_{k=1}^{l}\operatorname{res}_{a_k}\frac{f'}{f} + \sum_{k=1}^{m}\operatorname{res}_{b_k}\frac{f'}{f}. \tag{7.31}$$

Now applying the previous Theorem 7.35, we have

$$\frac{1}{2\pi i}\int_{\Gamma}\frac{f'(z)}{f(z)}\,dz = \sum_{k=1}^{l} n_k - \sum_{k=1}^{m} p_k = N - P,$$

as desired. □

The integral on the left of equation (7.30) is called the logarithmic residue of f with respect to Γ, which explains the name of this theorem. We will show that this value has a geometric meaning, and consequently this theorem expresses certain geometric properties of the mapping $w = f(z)$.

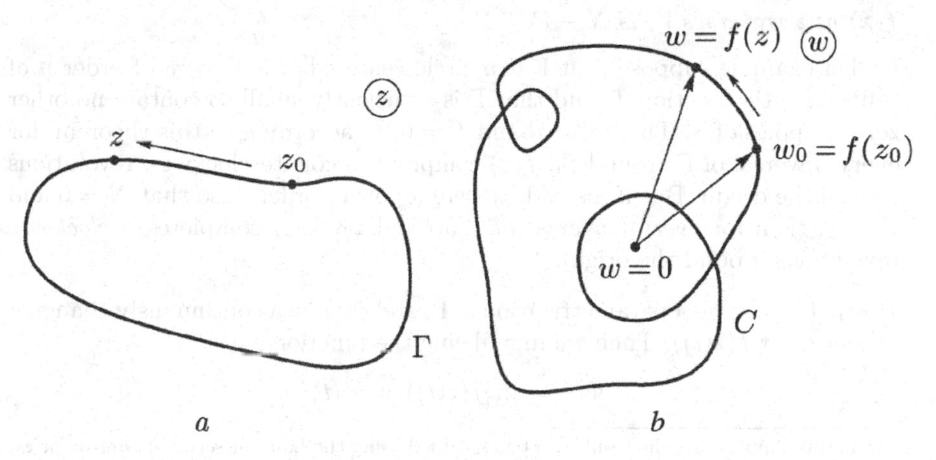

Fig. 60

In Fig. 60 *a*, we see a point z_0 on the path Γ; this point is mapped under $w = f(z)$ to a point $w_0 = f(z_0)$ in the w-plane, Fig. 60 *b*. If z traverses

Γ starting from z_0, then the corresponding point $w = f(z)$ will describe a trajectory C, starting at w_0, in the w-plane. As z returns along Γ to its starting point z_0, so w will return to w_0; therefore the path C will also be closed, although it may intersect itself.

Suppose that $z(t)$ is a parametrization of Γ on $[0, 1]$, with $z(0) = z_0 = z(1)$. We wish to define a continuous function $\phi(t)$ to represent $\arg f(z(t))$. At $t = 0$, we can simply take $\phi(0)$ to be the principal value $\operatorname{Arg} f(z(0))$. As t goes from 0 to 1,

$$\arg f(z(t)) = \operatorname{Arg} f(z(t)) + 2\pi k,$$

where k is an integer; we can always pick a suitable k so that the argument $\arg f(z(t))$ changes continuously. This value of $\arg f(z(t))$ we denote by $\phi(t)$. However, $\phi(0)$ may not equal $\phi(1)$; in general, when z traverses Γ starting from z_0 and returns to z_0, the argument of $f(z)$ will get some increment. Define

$$\Delta_\Gamma \arg f = \phi(1) - \phi(0).$$

This difference represents the number of times $f(z(t))$ has revolved around $w = 0$ when traversing C, times 2π; in Fig. 60, the number of revolutions is 2.

Theorem 7.37 (Argument principle)**.** *Suppose that f is a function analytic in some domain, Γ is a closed curve lying within that domain, and Γ contains no zeros of f. Let f be analytic in the interior of Γ with the possible exception of finitely many poles. If Γ is oriented so that points in its interior always lie on the left during traversal, then*

$$\Delta_\Gamma \arg f = 2\pi(N - P), \tag{7.32}$$

where N and P are the zero number and pole number, respectively, of f in the interior of Γ. In other words, the number of revolutions about $w = 0$ made by $f(z)$ as z traverses Γ, is $N - P$.

For example suppose that Γ is a circle centered at a zero z_0 of order n of some analytic function f, and that Γ is sufficiently small to contain no other zeros or poles of f. Then $N = n$ and $P = 0$, so according to this theorem, for every traversal of Γ around z_0, $f(z)$ completes n counterclockwise revolutions around the origin. But if instead z_0 were a pole of order p, so that $N = 0$ and $P = p$, then for every traversal of Γ around z_0, $f(z)$ completes p *clockwise* revolutions around the origin. [3]

Proof. Let $z(t)$ be a parametrization of Γ, and $\phi(t)$ be a continuously changing argument of $f(z(t))$. Then we may define the function

$$\Phi(t) = \ln|f(z(t))| + i\phi(t),$$

[3] In this simple case, the result can be explained using the Laurent series expansion at z_0. For if z_0 is a zero, then $f(z) = (z - z_0)^n \phi(z)$, where ϕ is analytic and nonzero at z_0, so $f(z) \approx \phi(z_0)(z - z_0)^n = C(z - z_0)^n$ at points near z_0 which makes n counterclockwise revolutions in the w-plane per each in the z-plane. And if z_0 is a pole, then $f(z) = \phi(z)(z - z_0)^{-p}$, and $f(z) \approx \phi(z_0)(z - z_0)^{-p} = C(z - z_0)^{-p}$, which makes p clockwise revolutions in the w-plane per each in the z-plane.

so that Φ is continuous, and $\Phi(t) = \log f(z(t))$ in a neighborhood of any $t \in [0,1]$, where log is a regular branch of logarithm. Then

$$\Phi'(t) = \frac{d}{dt} \log f(z(t)) = \frac{f'(z(t))}{f(z(t))} z'(t).$$

By the fundamental theorem,

$$\int_\Gamma \frac{f'(z)}{f(z)}\, dz = \int_0^1 \frac{f'(z(t))}{f(z(t))} z'(t)\, dt = \int_0^1 \Phi'(t)\, dt = \Phi(1) - \Phi(0).$$

Since $|f(z(1))| = |f(z(0))|$,

$$\Phi(1) - \Phi(0) = i(\phi(1) - \phi(0)) = i\Delta_\Gamma \arg f.$$

Therefore,

$$\Delta_\Gamma \arg f = \frac{1}{i} \int_\Gamma \frac{f'(z)}{f(z)}\, dz = \frac{1}{i} 2\pi i(N - P) = 2\pi(N - P),$$

where the last step is by the logarithmic residue Theorem 7.36. □

The quantity $N - P$ is often referred to as the *winding number* of Γ under f, since it represents the number of times that the image of Γ under f winds around the origin in the plane of $w = f(z)$. Theorem 7.36 means that the winding number is equal to the sum of the logarithmic residues.

The next theorem is a nice application of the argument principle.

Theorem 7.38 (Rouché's[4] theorem). *Let the functions f and g be analytic on the closed contour Γ and in its interior, and suppose that $|f(z)| > |g(z)|$ on Γ. Then within Γ, the function $F(z) = f(z) + g(z)$ has the same zero number, i.e. number of zeros counting multiplicities, as does f.*

Proof. First we show that neither f nor F have zeros on Γ. By assumption, on Γ

$$|f(z)| > |g(z)| \geq 0,$$

so f is never 0 there. And if $F(z) = 0$ for some z on Γ, then $f(z) + g(z) = 0$, implying $f(z) = -g(z)$, and $|f(z)| = |g(z)|$, which is a contradiction; hence F also has no zeros on Γ.

So we may apply the argument principle to these functions. Since $f(z)$ is never 0 on Γ, we may write

$$f(z) + g(z) = f(z)\left(1 + \frac{g(z)}{f(z)}\right)$$

there. Recall that multiplication of complex numbers results in the addition of their arguments. Therefore

$$\Delta_\Gamma \arg(f + g) = \Delta_\Gamma \arg f + \Delta_\Gamma \arg\left(1 + \frac{g}{f}\right).$$

[4] Eugène Rouché (1832–1910) was a French mathematician.

Now, by assumption $|g(z)/f(z)| < 1$ on Γ. Let

$$w(z) = 1 + \frac{g(z)}{f(z)};$$

then

$$|w - 1| = \left|\frac{g(z)}{f(z)}\right| < 1$$

for $z \in \Gamma$.

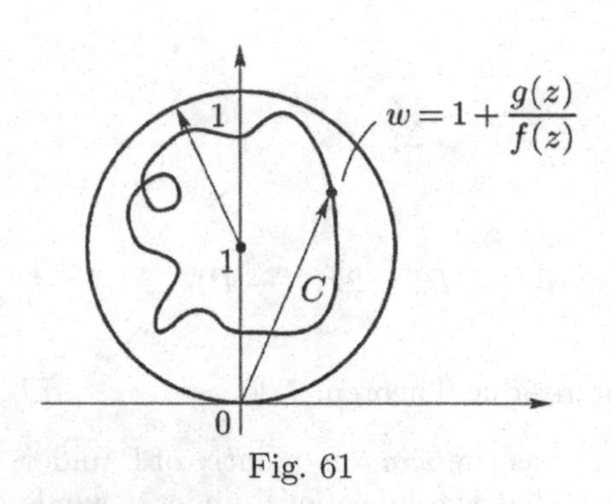

Fig. 61

This inequality means that as z traverses Γ, the point w remains within the disk $|w-1| < 1$, which lies on the upper side of the complex plane—see Fig. 61. Therefore the path of w never encircles the origin $w = 0$, and so the argument of w has the same value at the end of the traversal as it did at the beginning. In other words

$$\Delta_\Gamma \arg\left(1 + \frac{g}{f}\right) = 0.$$

Thus,

$$\Delta_\Gamma \arg(f + g) = \Delta_\Gamma \arg f,$$

and so by the previous theorem, the quantity $N - P$ is the same for f and F. Since neither function has any poles, $P = 0$ in both cases, and hence f and F have the same value of N, as desired. □

An important use of this theorem is to aid in the determination of the zero number of a function in a given domain.

Example 7.39 Determine the number of roots of the equation

$$z^8 - 4z^5 + z^2 - 1 = 0$$

which lie in the disk $|z| < 1$.

Solution Let us take Γ to be the circle $|z| = 1$, and

$$f(z) = -4z^5, \quad g(z) = z^8 + z^2 - 1,$$

so that $F(z) = f(z) + g(z)$ is the left side of the equation above. And at points on Γ,

$$|f(z)| = |-4z^5| = 4,$$
$$|g(z)| = |z^8 + z^2 - 1| \le |z^8| + |z^2| + |-1| = 3,$$

so $|f(z)| > |g(z)|$ on Γ. Therefore we may apply Rouché's theorem, according to which F has the same zero number as f. But $f(z) = -4z^5$ clearly has only one zero at $z = 0$, which is a zero of order 5, so $N = 5$. Therefore the original equation also has 5 roots in the disk $|z| < 1$, counting multiplicities.

Rouché's theorem also leads to an easy proof of the fundamental theorem of algebra.

Theorem 7.40 (Fundamental theorem of algebra). *Every polynomial*

$$P(z) = a_n z^n + a_{n-1}z^{n-1} + \cdots + a_0, \quad a_n \neq 0,$$

of degree n has exactly n roots, counting multiplicities, in the complex plane.

Proof. Let

$$f(z) = a_n z^n, \quad g(z) = a_{n-1}z^{n-1} + a_{n-2}z^{n-2} + \cdots + a_0,$$

so that $P(z) = f(z) + g(z)$. Then

$$\begin{aligned} \lim_{z\to\infty} \frac{f(z)}{g(z)} &= \lim_{z\to\infty} \frac{a_n z^n}{a_{n-1}z^{n-1} + a_{n-2}z^{n-2} + \cdots + a_0} \\ &= \lim_{z\to\infty} \frac{a_n}{\frac{a_{n-1}}{z} + \frac{a_{n-2}}{z^2} \cdots + \frac{a_0}{z^n}} = \infty, \end{aligned}$$

so there must exist some real $R_0 > 0$ such that $|f(z)/g(z)| \geq 2$ when $|z| \geq R_0$. Then for any $R > R_0$, the points z on the circle $|z| = R$ satisfy $|f(z)| > |g(z)|$, so we may apply Rouché's theorem. The power function $f(z)$ has n zeros, i.e. $z = 0$ with multiplicity n, in the interior of the circle; therefore by Rouché's theorem $P(z)$ also has exactly n roots in the disk $|z| < R$. Since this holds for any $R > R_0$, there must in fact be exactly n zeros of $P(z)$ in the entire complex plane. □

A *real* differentiable function $f(x)$ of one variable which is 1-1 on an interval, must be either monotone increasing or monotone decreasing on that interval. However, that does not guarantee that $f'(x) \neq 0$ on this interval—for example consider $f(x) = x^3$ on $(-1, 1)$. But using Rouché's theorem we see that the analogous property holds for complex analytic functions.

Lemma 7.41. *If a function $f(z)$ is analytic and 1-1 on the domain D then $f'(z) \neq 0$ on D.*

Proof. We will prove the equivalent (contrapositive) statement: if f is analytic on a domain D, and $f'(z_0) = 0$ at some $z_0 \in D$, then f is not 1-1. If f is constant, we are done; so assume f is not constant. Now suppose that z_0 is a root of order q of $f'(z)$, so that

$$f'(z_0) = f''(z_0) = \cdots = f^{(q)}(z_0) = 0, \quad \text{and} \quad f^{(q+1)}(z_0) \neq 0, \quad q \geq 1.$$

By the Uniqueness Theorem, since f is not constant on D, there must be some punctured disk $E = \{0 < |z - z_0| \leq r\}$ in D such that $f'(z) \neq 0$ and $f(z) \neq f(z_0)$ in E; let Γ be the circle $\{|z - z_0| = r\}$. Setting $w_0 = f(z_0)$, let μ be the minimal value of $|f(z) - w_0|$ on Γ. Then $\mu > 0$. Fix a point $w_1 \neq w_0$ for which $|w_1 - w_0| < \mu$, and consider two functions: $f_0(z) = f(z) - w_0$ and the constant function $g(z) = w_0 - w_1$. Note that f_0 has a zero of order $q + 1$ at z_0 and does not have other zeros within Γ. We have, for $z \in \Gamma$,

$$|g(z)| = |w_0 - w_1| < \mu \leq |f(z) - w_0| = |f_0(z)|.$$

By Rouché's theorem the function $F(z) = f_0(z) + g(z)$ has the same number of zeros counting multiplicities as f_0, which is $q + 1$.

Thus F has $q + 1$ zeros, which must lie in the punctured neighborhood E. Since $F'(z) = f'(z)$, the condition $f'(z) \neq 0$ in E implies that the multiplicity of each of these zeros is 1, hence they are at distinct points. Notice now that $F(z) = f(z) - w_1$, which means that at these $q+1$ zeros in E, $f(z) - w_1 = 0$, or equivalently $f(z) = w_1$. Therefore, $f(z)$ is not 1-1, so the lemma is proved. □

Another contrast with real functions comes from considering the converse of this lemma: "if $f'(z) \neq 0$ on D then $f(z)$ is 1-1". This *holds* for real differentiable functions (apply Rolle's theorem), but *fails* for complex analytic ones! For instance consider $f(z) = e^z$ in a domain D containing the points 0 and $2\pi i$. But a *local* version of the converse statement is valid—see problem 2.

As an application of Theorem 7.37 and Lemma 7.41 we prove Property 4.19. For the convenience of the reader we reformulate this property as Theorem 7.42.

Theorem 7.42. *Let D and D' be simply-connected domains which are bounded by closed Jordan curves Γ and Γ' in $\mathbb{C}$. Also let $w = f(z)$ be a function analytic on D and continuous on $\overline{D}$, which maps bijectively Γ onto Γ' and preserves orientation in the mapping from Γ onto Γ'. Then the function $w = f(z)$ carries out a conformal mapping of domain D onto domain D'.*

Proof. It suffices to prove that the mapping $w = f(z)$ is 1-1 on D; for then Lemma 7.41 would imply that $f'(z) \neq 0$ on D. But if a function $f(z)$ is 1-1 and analytic on the domain D and $f'(z) \neq 0$ on D, then this function performs a conformal mapping on D—see the beginning of Chapter 4. So, the theorem would be proved.

To prove that the mapping $w = f(z)$ is 1-1, we have to establish that

(a) for every point w in D', there is a unique preimage $z \in D$, i.e. $w = f(z)$;

(b) for every point z in D, the image $f(z)$ is in D'.

To prove (a), fix a point $w_0 \in D'$. Since f maps Γ bijectively onto Γ' and $w_0 \notin \Gamma'$, we must have $f(z) \neq w_0$ for $z \in \Gamma$. Also, by assumption f is continuous on $\overline{D}$; hence $f(z) \neq w_0$ also for z in D which are sufficiently close to Γ.

Again because f maps Γ bijectively onto Γ', the point $w = f(z)$ traverses Γ' exactly once as z traverses Γ. By assumption Γ' is a closed Jordan curve and so does not intersect itself (unlike the curve in Fig. 60 (b)). Hence the vector $f(z) - w_0$ revolves exactly once around the origin as z traverses Γ (see Fig. 62); that is $\Delta_\Gamma \arg(f(z) - w_0) = 2\pi i$, or

$$\frac{1}{2\pi i}\Delta_\Gamma \arg(f(z) - w_0) = 1. \tag{7.33}$$

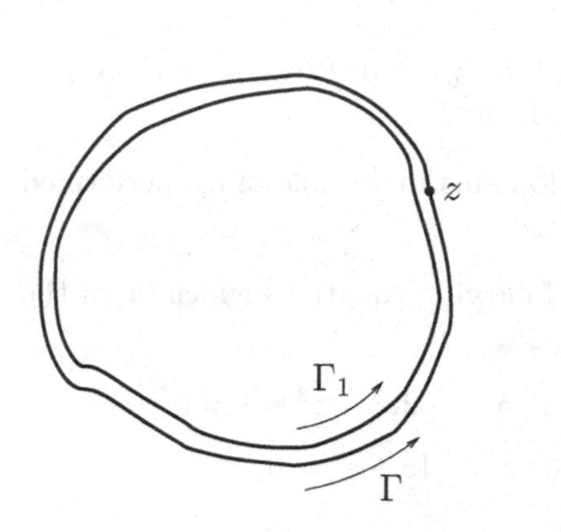

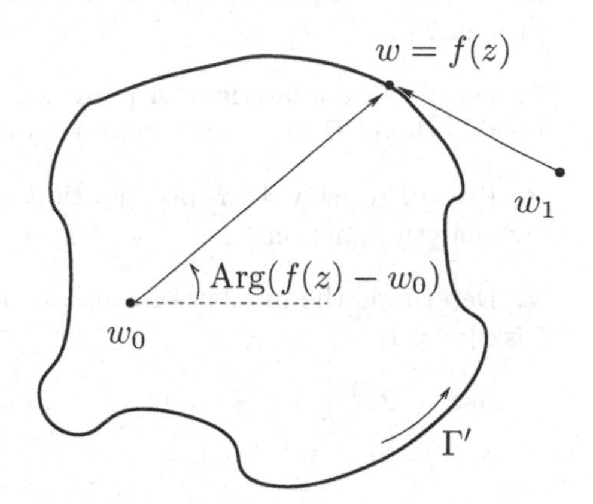

Fig. 62

This would suffice to prove (a), except that f is not assumed to be analytic on Γ, and so we cannot yet apply Theorem 7.37. The remedy is continuity: since f is continuous on $\overline{D}$, the change in the value of (7.33) is small under small deformations of Γ. Hence we can find Jordan curves Γ_1 in D which when substituted for Γ in (7.33) would take values arbitrarily close to 1. But this value can be only a natural number 1,2,...; therefore there is a closed Jordan curve Γ_1 in D for which (7.33) assumes the value 1, and moreover $f(z) \neq w_0$ for all z in D located between Γ and Γ_1. Applying (7.32) to the function $f(z) - w_0$ and to the closed curve Γ_1 we have

$$N = \frac{1}{2\pi i}\Delta_{\Gamma_1} \arg(f(z) - w_0) = 1.$$

Hence, there is the only one point z_0 within Γ_1 such that $f(z_0) = w_0$. Since there are no such points in $\overline{D}$ outside and on Γ_1, we have proved (a).

To prove (b), we first show that for $z \in D$, $f(z)$ must lie in $\overline{D'}$. Let $w_1 \notin \overline{D'}$. Then the vector $f(z) - w_1$ revolves zero times as z traverses Γ (see Fig. 62). In the same way as above we see that

$$N = \frac{1}{2\pi i}\Delta_\Gamma \arg(f(z) - w_1) = 0.$$

Therefore there are no points z in D for which $f(z) = w_1$.

Furthermore, $f(z)$ cannot lie on Γ'. For f is analytic on D, and therefore the image of D is an open set (Property 4.16); so if that image meets Γ', it would also contain points outside of $\overline{D'}$, which we have just seen is impossible. Therefore f maps D onto D'. This completes the proof. □

Problems

1. Let $f(z)$ be analytic at a point z_0, and let $f'(z)$ has zero of order q at z_0 (if $f'(z_0) \neq 0$ then $q = 0$). Prove that there are punctured neighborhoods $V = \{0 < |z - z_0| < r\}$ and $W = \{0 < |w - w_0| < \mu\}$ of the points z_0 and $w_0 = f(z_0)$, respectively, such that every point w_1 in W has exactly $q + 1$ preimages in V.

2. Let $f(z)$ be analytic at a point z_0, and $f'(z_0) \neq 0$. Prove that there is a neighborhood U of z_0, such that $f(z)$ is 1-1 on U.

3. Prove Property 4.16 (preservation of domains under mappings performed by analytic functions).

4. Determine the number of roots of the following equations which lie in the disk $|z| < 1$.

a) $z^4 - 2z^3 + 5z^2 - 1 = 0$;

b) $5z^5 - 2z^3 + z^2 - 1 = 0$;

c) $z^2 - 2z + 4e^{z+1} = 0$;

d) $z^2 - 4z + e^z = 0$.

5. How many zeros does the function $f(z)$ have in the annulus $1 < |z| < 2$?

a) $f(z) = z^3 + 3z - 1$;

b) $f(z) = z^3 + z - 3$.

8

Applications

In this chapter we consider some applications of complex variables to problems in hydrodynamics, electrostatics, and thermodynamics. Since most of these applications will be based on conformal mappings, we must first develop some further techniques for the construction of such mappings.

8.1 The Schwarz-Christoffel Transformation

The Schwarz-Christoffel transformation is a mapping of the upper half-plane onto a polygon. First, we consider a bounded polygon whose nonadjacent sides do not intersect except the case when adjacent sides form exterior angles of $-\pi$ (see Fig. 63). Then later we will consider the case when one or more vertices are at infinity.

Let $w_1, w_2, \ldots, w_n$ be consecutive vertices of the polygon, and let $\pi\alpha_1, \pi\alpha_2, \ldots, \pi\alpha_n$ be the corresponding *exterior* angles; in Fig. 63, $n = 8$. Each angle $\pi\alpha_k$ measures the change of direction of a point moving along the boundary of the polygon as the point passes through the vertex w_k.

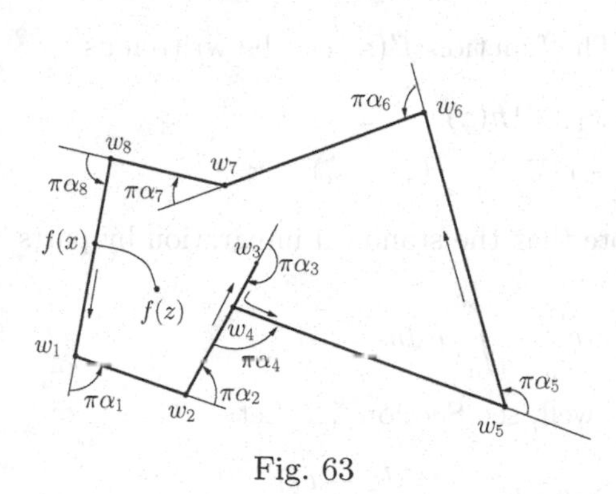

Fig. 63

The direction of motion is such that the interior of the polygon is on the left. We restrict each α_k to $-1 \leq \alpha_k < 1$ for $k = 1, \ldots, n$. When a point traverses the boundary, the total change of directions is 2π. Hence,

$$\alpha_1 + \alpha_2 + \cdots + \alpha_n = 2. \quad (8.1)$$

The existence of a conformal mapping of the upper half-plane onto a polygon follows from the Riemann Mapping Theorem 4.15. The following theorem gives the form of this mapping.

DOI: 10.1201/9780367810283-8

Theorem 8.1 (Schwarz-Christoffel[1]). *Let P be a polygon with exterior angles $\pi\alpha_1, \pi\alpha_2, \ldots, \pi\alpha_n$, $-1 \le \alpha_k < 1$ for $k = 1, \ldots, n$, and let z_0 be a fixed point in the upper half-plane Π. Then there exist real numbers $x_1, x_2, \ldots, x_n$ with $x_1 < x_2 < \cdots < x_n$, and complex constants A, B such that the function*

$$f(z) = A \int_{z_0}^{z} \frac{d\zeta}{(\zeta - x_1)^{\alpha_1}(\zeta - x_2)^{\alpha_2} \ldots (\zeta - x_n)^{\alpha_n}} + B \tag{8.2}$$

performs a conformal mapping of Π onto P. The integral in (8.2) is taken along any path joining z_0 to z and lying in Π.

We know that the power functions are multivalued in general. By $(z - x_k)^{\alpha_k}$ we understand the branch of the power function with a vertical cut below x_k. More precisely, if $\phi_k \in \arg(z - x_k)$ is chosen so that $-\frac{\pi}{2} < \phi_k \le \frac{3\pi}{2}$, then

$$(z - x_k)^{\alpha_k} = |z - x_k|^{\alpha_k} \exp(i\alpha_k \phi_k). \tag{8.3}$$

Therefore, the derivative

$$f'(z) = A(z - x_1)^{-\alpha_1}(z - x_2)^{-\alpha_2} \ldots (z - x_n)^{-\alpha_n}$$

is analytic on $\overline{\Pi}$ except the points $x_1, \ldots, x_n$. Hence, the same property holds for $f(z)$.

A rigorous proof of Theorem 8.1 is not given in this book; we refer the interested reader, for example, to the book [7] by R. A. Silverman. But we are going to discuss the formula (8.2), explain why $f(z)$ maps Π conformally onto the interior of a polygon, and give examples demonstrating how to use the formula.

(i) First we prove that *$f(z)$ is continuous on $\overline{\Pi}$ including the points x_k*; that is

$$\lim_{z \to x_k} f(z) = f(x_k) \quad \text{for} \quad z \in \overline{\Pi}.$$

Consider for instance the point x_1. The function $f'(z)$ can be written as

$$f'(z) = (z - x_1)^{-\alpha_1} h(z)$$
$$\text{where} \quad h(z) = A(z - x_2)^{-\alpha_2} \ldots (z - x_n)^{-\alpha_n};$$

obviously, $h(z)$ is analytic at x_1. Note that the standard integration by parts formula

$$\int_a^b u\, dv = uv\Big|_a^b - \int_a^b v\, du$$

is valid for analytic functions u, v as well; see Section 5.2. Let

$$u = h(\zeta), \quad dv = (\zeta - x_1)^{-\alpha_1}\, d\zeta, \quad v = \frac{(\zeta - x_1)^{1-\alpha_1}}{1 - \alpha_1}.$$

[1] German mathematicians Hermann A. Schwarz (1843–1921) and Elwin B. Christoffel (1829–1900) proved this theorem independently.

Then we have

$$\begin{aligned} f(z) &= \int_{z_0}^{z} (\zeta - x_1)^{-\alpha_1} h(\zeta)\, d\zeta \\ &= h(\zeta)\frac{(\zeta - x_1)^{1-\alpha_1}}{1-\alpha_1}\Big|_{z_0}^{z} - \frac{1}{1-\alpha_1}\int_{z_0}^{z} (\zeta - x_1)^{1-\alpha_1} h'(\zeta)\, d\zeta \\ &= h(z)\frac{(z - x_1)^{1-\alpha_1}}{1-\alpha_1} - h(z_0)\frac{(z_0 - x_1)^{1-\alpha_1}}{1-\alpha_1} \\ &\quad - \frac{1}{1-\alpha_1}\int_{z_0}^{z} (\zeta - x_1)^{1-\alpha_1} h'(\zeta)\, d\zeta. \end{aligned}$$

Since $1 - \alpha_1 > 0$, we see that $\lim (z - x_1)^{1-\alpha_1} = 0$ as $z \to x_k$, $z \in \overline{\Pi}$. Hence, the functions $(z - x_1)^{1-\alpha_1} h(z)$, $(\zeta - x_1)^{1-\alpha_1} h'(\zeta)$, and therefore the function $\int_{z_0}^{z} (\zeta - x_1)^{1-\alpha_1} h'(\zeta)\, d\zeta$ are continuous at x_1. So, the continuity of $f(z)$ at x_1 is proved, and $f(x_1)$ is finite. Obviously, the same arguments work for every x_k, $k = 1, \dots, n$. The continuity we have just established, allows us to select as z_0 any point in $\overline{\Pi}$ including points x_k.

Note in addition that there is a finite limit $f(x) \to w_0$ as $z \to \infty$ for $z \in \overline{\Pi}$; this follows from the estimate

$$|f'(z)| < \frac{M}{|z|^2} \quad \text{as} \quad |z| > R, \; z \in \overline{\Pi}, \tag{8.4}$$

where M and R are sufficiently large positive numbers. This estimate easily follows from (8.1). We leave the details for the reader.

(ii) Next we prove that *for every choice of real numbers $x_1, \dots, x_n$ and $\alpha_1, \dots, \alpha_n$ satisfying* (8.1) *and such that $-1 \leq \alpha_n < 1$, the function $f(z)$ maps Π onto the interior of a polygon with exterior angles $\pi\alpha_k$; the images w_k of points x_k are vertices of the polygon.*

According to (i), $f(z)$ maps the real axis onto a continuous bounded curve Γ. Consider a point x moving along the real axis in the positive direction. The function $f(z)$ is continuous at x and $f'(x) \neq 0$ except the points $x_1, \dots, x_n$. We know (see Section 3.3.1) that $\arg f'(x)$ is equal to the angle through which the tangent to a curve through x point is rotated under the mapping $f(z)$. Denote by $\mathbf{a}$ the unit tangent vector of the trajectory at x, that is $\mathbf{a} = (1, 0)$. Let us find the direction of the tangent vector to Γ at the point $f(x)$. It is obtained by turning the vector $\mathbf{a}$ through an angle $\arg f'(x)$. We have

$$\begin{aligned} \arg f'(z) &= \arg A + \arg(z - x_1)^{-\alpha_1} + \cdots + \arg(z - x_n)^{-\alpha_n} \\ &= \arg A - \alpha_1 \arg(z - x_1) - \cdots - \alpha_n \arg(z - x_n), \end{aligned} \tag{8.5}$$

where branches of $\arg(z - x_k)$ are indicated in (8.3).

Let $x \in (-\infty, x_1)$. Then $\arg(x - x_k) = \pi$ for all $k = 1, \dots, n$, and we have

$$\arg f'(x) = \arg A - \alpha_1 \pi - \alpha_2 \pi - \cdots - \alpha_n \pi, \quad x \in (-\infty, x_1).$$

Therefore, $\arg f'(z)$ is the same for all $x \in (-\infty, x_1)$. Hence, tangent vectors at all points of the corresponding piece of Γ have the same directions. We conclude that the image of the interval $(-\infty, x_1)$ is a line segment (w_0, w_1).

Consider now $x \in (x_1, x_2)$. Obviously, $\arg(x - x_1) = 0$, and $\arg(x - x_k) = \pi$ for $k = 2, \ldots, n$. Therefore,

$$\arg f'(x) = \arg A - \alpha_2\pi - \cdots - \alpha_n\pi, \quad x \in (x_1, x_2).$$

The same arguments as above show that the image of the interval (x_1, x_2) is a line segment (w_1, w_2). The difference between directions of the line segments (w_0, w_1) and (w_1, w_2) is $\pi\alpha_1$. Hence, these two segments form an angle with vertex at w_1, and the exterior angle is $\pi\alpha_1$. Continuing in the same way we see that Γ is a polygon with vertices $w_k = f(x_k)$, because the total change of directions is $\pi\alpha_1 + \pi\alpha_2 + \cdots + \pi\alpha_n = 2\pi$; the point w_0 is in the side (w_n, w_1).

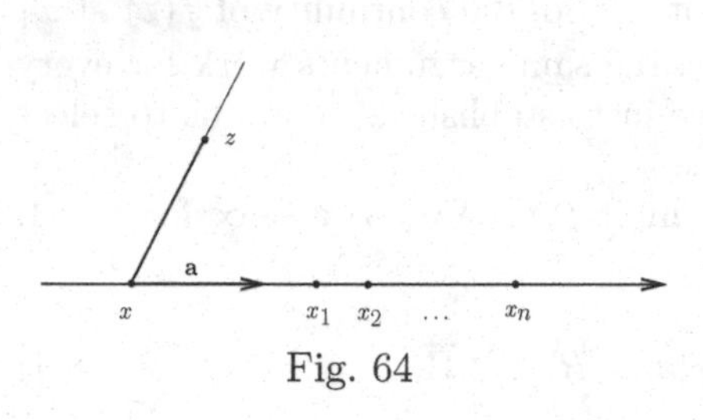

Fig. 64

Fix a point z in Π , and let x be a point on the real axis, $x \neq x_1, \ldots, x_n$. Obviously, the angle between vectors **a** and $\overline{xz}$ is less than π. The mapping $f(z)$ is conformal at x. Hence, the angle at the point $w = f(z)$ between Γ and the image of the line segment (x, z) is also less than π—see Fig. 64. It means that the image of Π is on the left of every side (w_{k-1}, w_k).

Moreover, we see that distinct line segments (w_{k-1}, w_k) cannot cross each other transversely, for then there would be points $w = f(z)$ on both sides of the segments. But possibly one of the line segments could overlap with an adjacent one—this would happen if some $\alpha_k = -1$. According to Property 4.19, $f(z)$ performs a conformal mapping of Π onto a polygon.

(iii) Now we explain why *for* **any** *polygon P there are points $x_1, \ldots, x_n$ on the real axis, and complex numbers A, B, such that the function $f(z)$ defined in* (8.2) *maps* Π *onto the given polygon P.*

It is sufficient to set $A = 1$, $B = 0$, and find $f(z)$ which maps Π onto a polygon P' which is similar to P and has the same orientation of the boundary; that is P' has the same angles in the same order when paths along the boundaries keep the polygons on the left, and side lengths of P' are proportional to the corresponding side lengths of P. In this case P can be obtained from P' by dilation, rotation, and translation. These transformations can be easily performed by choice of constants A (responsible for dilation and rotation) and B (responsible for translation). Notice that the exterior angles $\alpha_1, \ldots, \alpha_n$ are already fixed, and we have to choose only the points $x_1, \ldots, x_n$.

Two triangles are similar and have the same orientation if their corresponding angles are the same; side lengths will be proportional "automatically". Hence, to find a mapping of Π onto a triangle, *the points $x_1 < x_2 < x_3$ can be chosen arbitrarily.*

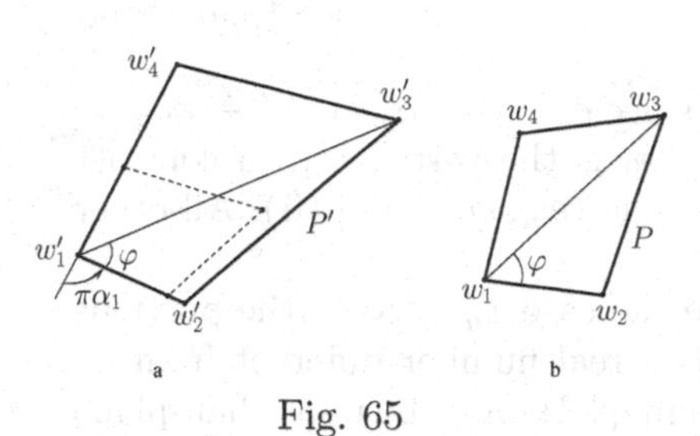

Fig. 65

For $n > 3$ the situation is different: the equality of angles does not imply similarity of P' and P—see Fig. 65a for the case $n = 4$. But the polygons P' and P are similar if in addition to the equality of angles, P' and P have the same angles φ between a side and diagonal. Fix x_1, x_2, x_4. If x_3 approaches x_2, then w_3' approaches w_2', and $\varphi \to 0$.

If x_3 approaches x_4, then $w_3' \to w_4'$, and φ approaches maximal possible value $\pi(1 - \alpha_1)$. Hence there is a value of x_3 when φ is exactly the same as in the given polygon P. Thus, there exists the desired mapping of Π onto P of the form (8.2).

In general, for $n > 3$ two polygons are similar if in addition to the equality of angles, the angles between a side and a diagonal, and between the diagonals coming from the same point, are the same. One can easily see that there are $n - 3$ such angles; in Fig. 66, $n = 6$.

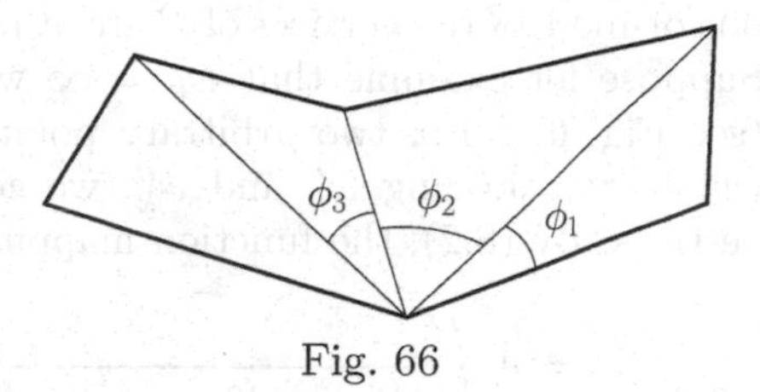

Fig. 66

If we fix three points x_k, we will have $n - 3$ other "free" parameters x_k to satisfy $n-3$ conditions for the angles φ_j. We will not prove the existence and uniqueness of a solution in this general case, and at this point our arguments are not rigorous. But we can explain why it is that *three points among the* x_k *can be chosen arbitrarily.* Note that this conclusion agrees with Riemann's mapping theorem (Theorem 4.15) which says that the mapping has three free real parameters. Finding the remaining $n-3$ points x_k is the main difficulty in applying the Schwarz-Christoffel formula. Before moving on to the examples, we make two useful remarks.

(iv) *One of the points* x_k *can be chosen at infinity. In this case the formula has the same form as* (8.2), *but the corresponding factor* $(\zeta - x_k)^{\alpha_k}$ *is absent.* For example, if $x_n = \infty$ then

$$f(z) = A \int_{z_0}^{z} \frac{d\zeta}{(\zeta - x_1)^{\alpha_1}(\zeta - x_2)^{\alpha_2} \dots (\zeta - x_{n-1})^{\alpha_{n-1}}} + B. \qquad (8.6)$$

For explanation we may just repeat the arguments given above. The existence of a finite limit w_0 of $f(z)$ as $z \to \infty$ follows from the estimate

$$|f'(z)| < \frac{M}{|z|^{\alpha_1 + \dots + \alpha_{n-1}}} = \frac{M}{|z|^{2-\alpha_n}}, \quad |z| > R, \; z \in \overline{\Pi}.$$

Here is essential that $2 - \alpha_n > 1$, because $\alpha_n < 1$. But the exterior angle between images of the rays $(-\infty, x_1)$ and (x_{n-1}, ∞), namely between the line segments (w_0, w_1) and (w_{n-1}, w_0), is not 0, since $\alpha_1 + \cdots + \alpha_{n-1} = 2 - \alpha_n \neq 0$. This exterior angle equals $\pi\alpha_n$. Therefore, w_0 is the vertex w_n, in contrast with the previous case when w_0 was on the side (w_n, w_1)—see **(ii)**. All other arguments given above are valid.

Another way to verify (8.6) is to reduce the case $x_n = \infty$ to the previous one by the mapping $\zeta = -\frac{1}{z-a}$, where a is a real number different from all $x_1, \ldots, x_{n-1}$ (possibly $a = 0$). This function maps Π onto the upper half-plane of another variable ζ, and maps points $x_1, \ldots, x_{n-1}, \infty$ onto n finite points $x'_1, \ldots, x'_{n-1}, x'_n$ with $x'_n = 0$. Substituting $\zeta = -\frac{1}{z-a}$ and $x'_k = -\frac{1}{x_k - a}$ instead of x_k in (8.2), we come to (8.6) (with z instead of ζ in the denominator and another constant A). We leave the details for the reader.

(v) Now we consider a case which is very important for applications: when one or more of the vertices of P are at infinity—a so-called *degenerate* polygon. Suppose for example that $w_k = \infty$ while the other vertices of P are finite (see Fig. 67). Fix two arbitrary points w'_k, w''_k on the rays (w_{k-1}, ∞) and (w_{k+1}, ∞). Joining w'_k and w''_k, we get a bounded polygon P' with $n+1$ vertices. By (8.2), the function mapping Π onto P' is

$$f(z) = A \int_{z_0}^{z} \frac{d\zeta}{(\zeta - x_1)^{\alpha_1} \ldots (\zeta - x'_k)^{\alpha'_k} (\zeta - x''_k)^{\alpha''_k} \ldots (\zeta - x_n)^{\alpha_n}} + B.$$

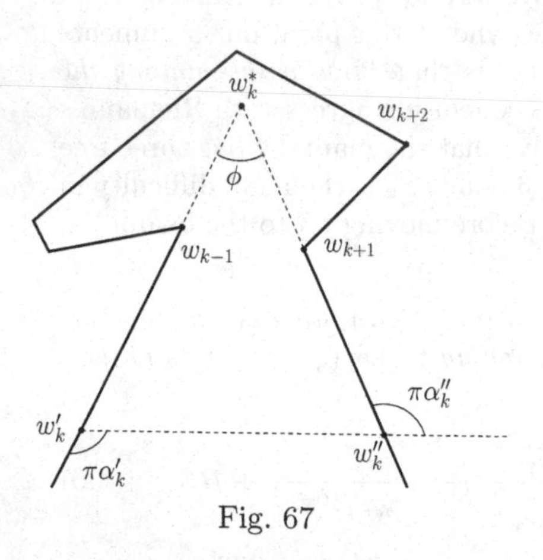

Fig. 67

Now, let the segment (w'_k, w''_k) approach infinity while remaining parallel to itself. Then the points x'_k and x''_k merge into a single point x_k. (Indeed, if x'_k, x''_k approach distinct points then the polygon is bounded—see **(ii)**). This point x_k corresponds to the vertex $w_k = \infty$. Passing to the limit we have

$$(\zeta - x'_k)^{\alpha'_k} (\zeta - x''_k)^{\alpha''_k} \to (\zeta - x_k)^{\alpha'_k + \alpha''_k}.$$

The sum of internal angles of the triangle $w'_k w''_k w^*_k$ equals π, and we have $(\pi - \pi\alpha'_k) + (\pi - \pi\alpha''_k) + \varphi = \pi$, that is $\pi\alpha'_k + \pi\alpha''_k = \pi + \varphi$—see Fig. 67.

Recall that we defined the angle between two lines at infinity as the angle at 0 between their images under the mapping $\frac{1}{z}$. One can easily see that if φ is a directed angle between rays at their finite point of intersection, then the angle between these rays at ∞ equals $-\varphi$, and the exterior angle at ∞ is $\pi\alpha_k = \pi - (-\varphi) = \pi + \varphi$. Therefore, $\alpha'_k + \alpha''_k = \alpha_k$, and our formula again has the form (8.2). If the rays (w_{k-1}, ∞) and (w_{k+1}, ∞) are parallel, we have $\varphi = 0$. Therefore, *the Schwarz-Christoffel formula* (8.2) *is valid also for polygons with* *one or more vertices at ∞, if the angle between two lines at ∞ is defined as the angle at the finite point of their intersection taken with the opposite sign. If the lines are parallel then the angle is 0, and the exterior angle is π, so that* $\alpha_k = 1$.

(vi) We conclude with useful remarks concerning finding the coefficients A, B. For $z = z_0$ the equalities (8.2), (8.6) imply $B = f(z_0)$. So, if $f(z_0)$ is known, we get B. Now we explain how to determine $\arg A$. Consider a point x on the real line such that $x > x_n$ if x_n is finite, and $x > x_{n-1}$ if $x_n = \infty$. Then $\arg(x - x_k) = 0$ for all points $x_k < x$ and for the selected branch of argument. Hence, by (8.5), $\arg f'(x) = \arg A$. Therefore, $\arg A$ *equals the argument of the side of the polygon starting from the vertex w_n if x_n is finite, and from w_{n-1} if x_n is infinite. In the other words,* $\arg A$ *is the angle between this side of the polygon and the u-axis in the w-plane, taking into account the bypass direction.*

Finally we note that the length of the side (w_k, w_{k+1}) of a polygon is given by the formula

$$|w_{k+1} - w_k| = \int_{x_k}^{x_{k+1}} |f'(x)|\, dx. \tag{8.7}$$

Indeed, if we write $f(x)$ in the form $f(x) = u(x) + iv(x)$, then $f'(x) = u'(x) + iv'(x)$, $|f'(x)| = \sqrt{(u'(x))^2 + (v'(x))^2}$, and (8.7) becomes the formula for arc length of a parametric curve—see Section 5.1.

Examples of the Schwarz-Christoffel transformation

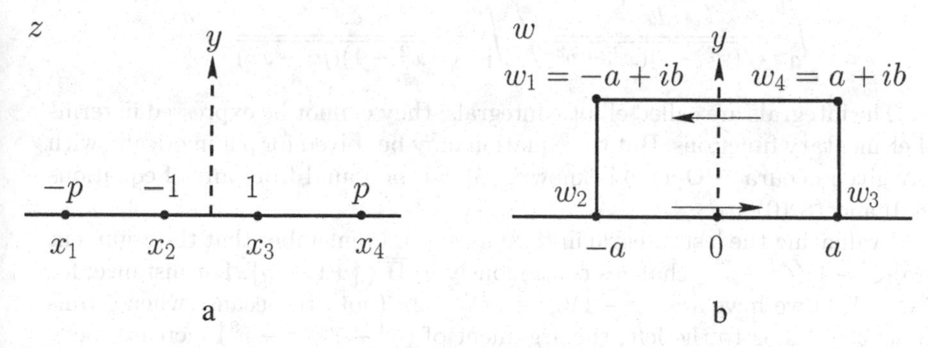

Fig. 68

Example 8.2 Map the upper half-plane Π onto the rectangle shown in Fig. 68 b.

Solution In this case $\alpha_1 = \cdots = \alpha_4 = \frac{1}{2}$. Since the values of f at three boundary points of Π may be chosen arbitrarily, we define $f(-1) = -a$, $f(0) = 0$, $f(1) = a$, and denote $x_2 = -1$, $x_3 = 1$. The remaining points x_1 and x_4 have to be determined.

k	w_k	α_k	x_k
1	$-a + ib$	1/2	$-p$
2	$-a$	1/2	-1
3	a	1/2	1
4	$a + ib$	1/2	p

Let $z_0 = 0$ and let $x_4 = p$, $p > 1$. Then by the symmetry, $x_1 = -x_4 = -p$. The data is shown in the table.

According to (8.2), the mapping is given by the formula

$$\begin{aligned} f(z) &= A \int_0^z \frac{d\zeta}{(\zeta+p)^{1/2}(\zeta+1)^{1/2}(\zeta-1)^{1/2}(\zeta-p)^{1/2}} + B \\ &= A \int_0^z \frac{d\zeta}{\sqrt{(\zeta^2-1)(\zeta^2-p^2)}} + B. \end{aligned} \tag{8.8}$$

To find B, we plug $z = 0$ into (8.8). Since $f(0) = 0$, we have $B = 0$. To find $\arg A$, we use the remark in **(vi)**. For $x > x_4$, $f(x)$ moves from w_4 toward w_1. Hence $\arg A = \pi$. Thus, A is real and negative: $A = -|A|$. For finding $|A|$ and p we use the formula (8.7) for the length of intervals. Since the length of $(0, w_3)$ equals a,

$$|A| \int_0^1 \frac{dx}{\sqrt{(x^2-1)(x^2-p^2)}} = a. \tag{8.9}$$

Evaluating the length of (w_3, w_4) we obtain the equality

$$|A| \int_1^p \frac{dx}{\sqrt{(x^2-1)(p^2-x^2)}} = b; \tag{8.10}$$

in (8.9), (8.10) we select the "usual" nonnegative values of square roots. Thus, we have two equations for two unknown parameters $|A|$ and p. Divide the first equation by the second one:

$$\int_0^1 \frac{dx}{\sqrt{(x^2-1)(x^2-p^2)}} \Big/ \int_1^p \frac{dx}{\sqrt{(x^2-1)(p^2-x^2)}} = \frac{a}{b}.$$

The integrals are called elliptic integrals; they cannot be expressed in terms of elementary functions. But the equation may be solved for p numerically with any given accuracy. Once p is known, $|A|$ can be found from any of equations (8.9) and (8.10).

Evaluating the last integral in (8.8) we should remember that the argument $\arg[(\zeta^2-1)(\zeta^2-p^2)]$ changes continuously in $\overline{\Pi} \setminus \{\pm 1, \pm p\}$. For instance, for $\zeta \in (-1, 1)$ we have $\arg[(\zeta^2-1)(\zeta^2-p^2)] = 2\pi$ (not 0!), because when ζ runs along the x axis to the left, the argument of $(\zeta^2-1)(\zeta^2-p^2)$ increases on π every time when ζ passes points p and 1—see **(ii)**. Therefore,

$$f'(x) = \frac{A}{\sqrt{(x^2-1)(p^2-x^2)}} = \frac{A}{e^{i\pi}|\sqrt{(x^2-1)(p^2-x^2)}|} > 0, \quad x \in (-1, 1).$$

It agrees with the direction of motion from w_2 to w_3.

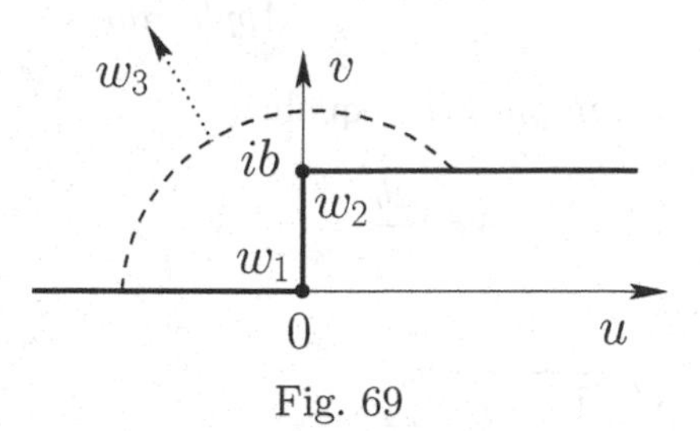

Fig. 69

Example 8.3 Map the upper half-plane Π onto the domain above the polygonal line shown in Fig. 69.

Solution The domain is a degenerate triangle with the vertex w_3 at infinity; the sides of this angle are horizontal rays. The angle between them equals π. Hence, the angle at ∞ equals $-\pi$, and the external angle is $\pi - (-\pi) = 2\pi$. So, $\alpha_3 = 2$. Because the point z_0 and boundary values $f(x_k)$ at three points x_k might be chosen arbitrarily, we set $z_0 = 0$ and define x_k and $f(x_k)$ according to the table. Note that $\sum_{k=1}^{\infty} \alpha_k = 2$. By (8.2) we have

k	w_k	α_k	x_k
1	0	1/2	0
2	ib	$-1/2$	1
3	∞	2	∞

$$f(z) = A \int_0^z \frac{d\zeta}{\zeta^{1/2}(\zeta - 1)^{-1/2}} + B = A \int_0^z \frac{\sqrt{\zeta - 1}}{\sqrt{\zeta}}\, d\zeta + B;$$

the factor corresponding to x_3 is absent—see **(iv)**. Since $f(0) = 0$, we have $B = 0$. To evaluate the integral, we may consider an interval in the real axis and use all methods of integration of real-valued functions of one real variable to evaluate an antiderivative $f(x)$. Since all differentiation formulas for functions of complex variable are the same as for real-valued functions of real variable, the obtained function $f(x)$ also is an antiderivative for the branch $f(\xi)$ which is analytic in Π and coincides with $f(x)$ on the given interval. It is convenient to choose the interval $(0, 1)$ and the branch of $\sqrt{\zeta}$ which is positive as $\zeta \in (0, \infty)$. Then $\sqrt{x - 1} = i\sqrt{1 - x}$ as $x \in (0, 1)$. We start with the substitution $u = \sqrt{x}$, so that $x = u^2$, $dx = 2u\, du$, and

$$\int \frac{\sqrt{x - 1}}{\sqrt{x}}\, dx = i \int \frac{\sqrt{1 - x}}{\sqrt{x}}\, dx = 2i \int \sqrt{1 - u^2}\, du.$$

Now we apply the standard trig substitution $u = \sin\theta$. Then $du = \cos\theta\, d\theta$, and the last integral is equal to

$$\begin{aligned} 2i \int \cos^2\theta\, d\theta &= 2i\frac{1}{2}\int (1 + \cos 2\theta)\, d\theta = i(\theta + \frac{1}{2}\sin 2\theta) + C \\ &= i(\sin^{-1} u + u\sqrt{1 - u^2}) + C = i(\sin^{-1}\sqrt{x} + \sqrt{x}\sqrt{1 - x}) + C. \end{aligned}$$

Now we replace x with the complex variable ζ, and interpret $\sin^{-1}\sqrt{\zeta}$ as the branch of the corresponding multivalued function such that $\sin^{-1} 1 = \pi/2$. Therefore,

$$f(z) = iA\left(\sin^{-1}\sqrt{\zeta} + \sqrt{\zeta(1 - \zeta)}\right)\Bigg|_0^z = iA\left(\sin^{-1}\sqrt{z} + \sqrt{z(1 - z)}\right).$$

To find A we use the condition $f(1) = ib$ which implies the equalities

$$ib = f(1) = iA(\sin^{-1} 1 + 0) = iA\frac{\pi}{2}, \quad A = \frac{2b}{\pi}.$$

Finally,

$$f(z) = \frac{2ib}{\pi}(\sin^{-1}\sqrt{z} + \sqrt{z(1-z)}).$$

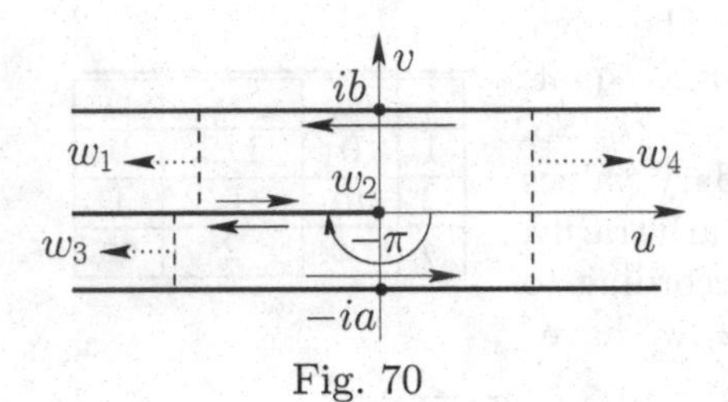

Fig. 70

Example 8.4 Map the upper half-plane Π onto the strip $-a < \operatorname{Im} w < b$ cut along the negative real axis, shown in Fig. 70.

Solution We regard the domain as a degenerate quadrangle with three vertices w_1, w_3, and w_4 at infinity. We choose $z_0 = 0$ and define x_k and $f(x_k)$ according to the table; the value $x_1 = -p$, $p > 0$, has to be determined. Note that the ray $(-\infty, 0)$ is the lower side of the angle with vertex w_1, and at the same time it is the upper side of the angle with vertex w_3; these sides are coincide. One may think about these sides as "the upper" and "the lower" sides of the ray. The arrows near the boundary indicate the direction of motion of the point $w = f(x)$ when x runs along the real axis from $-\infty$ to ∞. The exterior angles at w_k, $k = 1, 3, 4$ equal 0, and $\alpha_k = 1$, $k = 1, 3, 4$ – see **(v)** for a more detailed explanation. The exterior angle at w_2 equals $-\pi$, and $\alpha_2 = -1$. By (8.6) we have

k	w_k	α_k	x_k
1	∞	1	$-p$
2	0	-1	0
3	∞	1	1
4	∞	1	∞

$$f(z) = A\int_0^z \frac{\zeta\, d\zeta}{(\zeta - 1)(\zeta + p)} + B.$$

Since $f(0) = 0$ (see the above table), we have $B = 0$. To evaluate the integral we apply the partial fraction decomposition and obtain the formula

$$\begin{aligned} f(z) &= \frac{A}{p+1}(\log(\zeta - 1) + p\log(\zeta + p))\Big|_0^z \\ &= C(\log(z-1) + p\log(z+p)) - C(\log(-1) + p\log p), \end{aligned}$$

where $C = \frac{A}{p+1}$, and the branch of logarithm is indicated in (8.3), namely $\log z = \ln|z| + i\arg z$, $-\frac{\pi}{2} < \arg z \le \frac{3\pi}{2}$.

Hence,

$$f(z) = C(\log(z-1) - \log(-1) + p(\log(z+p) - \log p))$$
$$= C\left(\log(z-1) - i\pi + p\log\frac{z+p}{p}\right). \tag{8.11}$$

Note here that $\log(z+p) - \log p = \log\frac{z+p}{p}$ since $\arg p = 0$, although this would not be true for arbitrary p values.[2]

It remains to find C and p. If $z = x \in \mathbb{R}$ and x runs from 1 to ∞, then the corresponding point $w = f(x)$ is on the lower side $\operatorname{Im} f(x) = -a$ of the strip (see Fig. 70 and the table in the previous page), and runs from w_3 to w_4. The angle between this side of the strip and the real axis is 0. Hence, $\operatorname{Arg} A = 0$ (see **(vi)**). Therefore A and C are real positive numbers. In this interval $\operatorname{Arg}(x-1) = 0$ and $\operatorname{Arg}(\frac{x}{p}+1) = 0$. Comparing the imaginary parts of $f(x)$ and of the right hand side of (8.11) we have

$$\operatorname{Im} f(x) = \operatorname{Im} C(-i\pi) = -a.$$

Hence, $C\pi = a$, $C = \frac{a}{\pi}$.

To find p we assume that $z = x \in (-\infty, -p)$. Then the point $w = f(x)$ is on the upper side $\operatorname{Im} f(x) = b$ of the strip, and $\operatorname{Arg}(x-1) = \pi$, $\operatorname{Arg}(\frac{x}{p}+1) = \pi$. From (8.11) we obtain

$$\operatorname{Im} f(x) = \frac{a}{\pi}(\pi - \pi + p\pi) = b.$$

Therefore, $p = \frac{b}{a}$, and the desired mapping is

$$f(z) = \frac{a}{\pi}\log(z-1) + \frac{b}{\pi}\log\left(\frac{az}{b}+1\right) - ia. \tag{8.12}$$

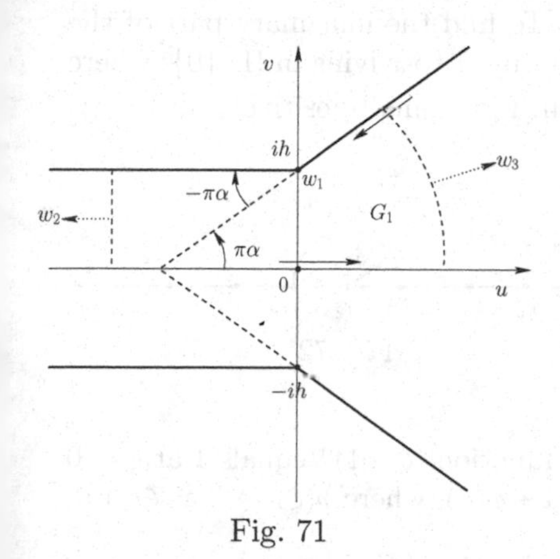

Fig. 71

Example 8.5 Map the strip $-\pi < \operatorname{Im} z < \pi$ onto the polygonal symmetric domain shown in Fig. 71.

[2]Indeed, if we were to take $p = -1$ and $z = 2$, we would get the peculiar equality $-\log(-1) = \log(-1)$. It occurs because the branches of logarithms on the left and on the right are different. The last equality is correct if we consider it as the equality of sets of all values of the multivalued function $\log z$.

Solution The domain is a degenerate quadrangle. We may map the given strip onto the upper half-plane, and then apply the Schwarz-Christoffel transformation. But it is much easier to make use of the symmetry principle (Theorem 5.28) and to work only with the upper half G_1 of the given domain which is a "degenerate triangle". Our plan is the following.

(1) Map the upper half-plane Π onto G_1 by the Schwarz-Christoffel transformation $w = f(z)$.

(2) Denote by D_1 the upper half $0 < \operatorname{Im}\chi < \pi$ of the strip $-\pi < \operatorname{Im}\chi < \pi$ in the χ-plane, and map D_1 onto Π by the function $z = e^{\chi}$. Then the function $w = g(\chi) = f(e^{\chi})$ maps D_1 onto G_1.

(3) Check the conditions of Theorem 5.28 to make sure that the obtained function $w = g(\chi)$ maps the whole strip $-\pi < \operatorname{Im}\chi < \pi$ onto the given polygonal domain.

So we start with the mapping of Π onto G_1. Here the angle with vertex w_3 at ∞ equals $-\pi\alpha$, and the exterior angle at ∞ is $\pi-(-\pi\alpha) = \pi(1+\alpha)$; therefore $\alpha_3 = 1 + \alpha$ (see **(v)** for details). The formula (8.6) yields

k	w_k	α_k	x_k
1	ih	$-\alpha$	-1
2	∞	1	0
3	∞	$1+\alpha$	∞

$$f(z) = A\int_{z_0}^{z} \frac{(\zeta+1)^{\alpha}}{\zeta}\, d\zeta + B. \qquad (8.13)$$

We choose $z_0 = -1$ because $f(-1) = ih$, and then for $z = -1$ we have $B = ih$.

To find A we use the same idea as in the previous example. Namely, we consider an appropriate value of x and compare the imaginary parts of $f(x)$ and of the right hand side in (8.13). However, here the situation is more complicated because the integral is not expressed in terms of elementary functions.

Consider $z = x \in (0, \infty)$. Then the point $w = f(x)$ is on the real axis; it runs to the right when x increases. Hence $\operatorname{Im} f(x) = 0$ and $\arg A = 0$ – see **(vi)**. Therefore, A is real and positive. To find the imaginary part of the integral, we may integrate along any path from -1 to x lying in $\overline{\Pi}\setminus\{0\}$, where the integrand $\frac{(\zeta+1)^{\alpha}}{\zeta}$ is an analytic function. Fix a small positive δ and consider a path consisting of the two intervals $[-1, -\delta]$, $[\delta, x]$, and a semicircle γ_δ of radius δ centered at $x = 0$—see Fig. 72. On both these intervals the integrand is real, and hence the imaginary parts of the integral over the intervals equal zero.

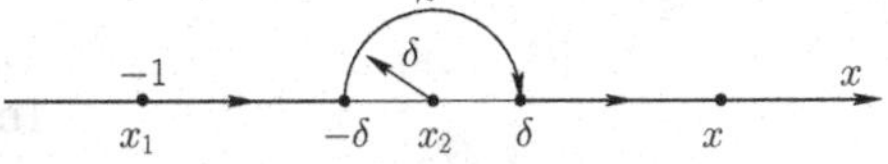

Fig. 72

Consider now the integral over γ_δ. The function $(\zeta+1)^{\alpha}$ equals 1 at $\zeta = 0$ and is continuous there. Hence, $(\zeta+1)^{\alpha} = 1 + \phi(\zeta)$, where $\phi(\zeta) \to 0$ as $\zeta \to 0$, and we have

$$\int_{\gamma_\delta} \frac{(\zeta+1)^{\alpha}}{\zeta}\, d\zeta = \int_{\gamma_\delta} \frac{1}{\zeta}\, d\zeta + \int_{\gamma_\delta} \frac{\phi(\zeta)}{\zeta}\, d\zeta.$$

The first integral on the right equals $-i\pi$ because we integrate in the clockwise direction – see Example 5.1. The integral of $\phi(\zeta)/\zeta$ tends to 0 as $\delta \to 0$ because $|\zeta| = \delta$ in this integral, the length of γ_δ equals $\pi\delta$, and $\phi(\zeta)$ approaches zero as $\delta \to 0$. Therefore,

$$\operatorname{Im} \int_{-1}^{x} \frac{(\zeta+1)^\alpha}{\zeta}\, d\zeta = -\pi \quad \text{as} \quad x \in (0,\infty).$$

From (8.13) we get

$$\operatorname{Im} f(x) = A(-\pi) + h = 0.$$

Therefore, $A = h/\pi$. Now (8.13) takes the form

$$f(z) = \frac{h}{\pi} \int_{-1}^{z} \frac{(\zeta+1)^\alpha}{\zeta}\, d\zeta + ih. \tag{8.14}$$

The next step is to use the mapping $z = e^\chi$. We obtain the function

$$w = g(\chi) = f(e^\chi) = \frac{h}{\pi} \int_{-1}^{e^\chi} \frac{(\zeta+1)^\alpha}{\zeta}\, d\zeta + ih,$$

which maps D_1 onto G_1. This expression can be simplified if we use the substitution $\zeta = e^\xi$. If $\zeta = -1$ then $\xi = i\pi$; if $\zeta = e^\chi$ then $\xi = \chi$. Hence,

$$w = g(\chi) = \frac{h}{\pi} \int_{i\pi}^{\chi} (e^\xi + 1)^\alpha\, d\xi + ih.$$

Now we check the assumptions of Theorem 5.28. The function $z = e^\chi$ maps the real axis in the χ-plane onto the interval $(0,\infty)$ in the z-plane; then our function $f(z)$ maps this interval onto the real axis in the w-plane. Therefore all conditions of Theorem 5.28 are satisfied. By the principle of symmetry, the same function $g(\chi)$ maps the whole strip $-\pi < \operatorname{Im}\chi < \pi$ onto the whole polygonal domain. Rewriting the function now using z instead of χ, we get

$$w = g(z) = \frac{h}{\pi} \int_{i\pi}^{z} (e^\xi + 1)^\alpha\, d\xi + ih. \tag{8.15}$$

Here z is a point in the strip $-\pi < \operatorname{Im}\chi < \pi$ (not in the half-lane Π).

Problems

1. This problem illustrates how the choice of points $x_1, x_2, \ldots$ may affect the function $f(z)$ (the choice of z_0 affects only B). Notice that if $f(z)$ and $g(z)$ are two conformal mappings of the upper half plane Π onto the same domain, then there is a Möbius transformation $\mu(z)$ of Π onto itself such that $g(z) = f(\mu(z))$.

Use the Schwarz-Christoffel transformation to map the upper half plane onto the strip $0 < \operatorname{Im} z < \pi$ in two ways:

(a) choose $x_1 = 0$, $x_2 = \infty$, with $f(x_k) = \infty$, $k = 1, 2$, and $z_0 = 1$. Compare your result with properties of the logarithmic function (Section 4.3.2).

(b) choose $x_1 = -1$, $x_2 = 1$, $z_0 = 0$.

2. Use the Schwarz-Christoffel transformation to find the mapping of the upper half plane onto the semi-infinite vertical strip

$$D = \{w = u + iv : -h < u < h,\ 0 < v < \infty\}$$

such that $f(-1) = -h$, $f(1) = h$, $f(\infty) = \infty$, $z_0 = 0$.

3. Find the mapping $f(z)$ of the upper half-plane $\operatorname{Im} z > 0$ onto the polygonal domain shown in Fig. 73 such that $f(-a^2) = h_1 + ih_2$, $f(0) = \infty$, $f(1) = 0$, $f(\infty) = \infty$, $z_0 = 1$, where h_1, h_2 are given numbers and a has to be determined.

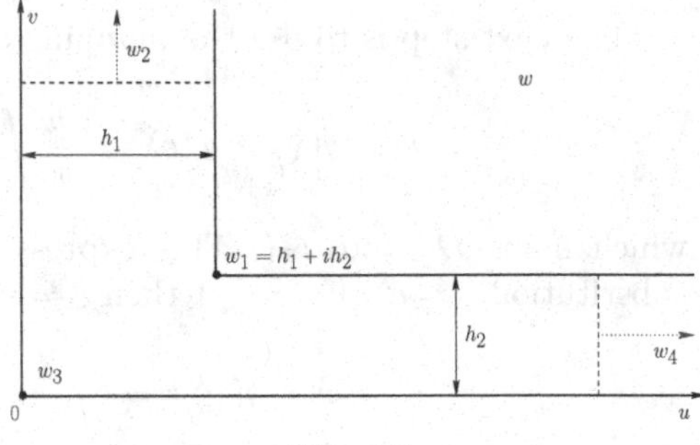

Fig. 73

(i) Determine α_k and construct the table as in the examples.

(ii) Write the Schwarz-Christoffel integral and determine B.

(iii) Determine A and a. It's even easier to find them now, before evaluation of the integral. Take $z = x > 1$ and conclude that A is real and positive.

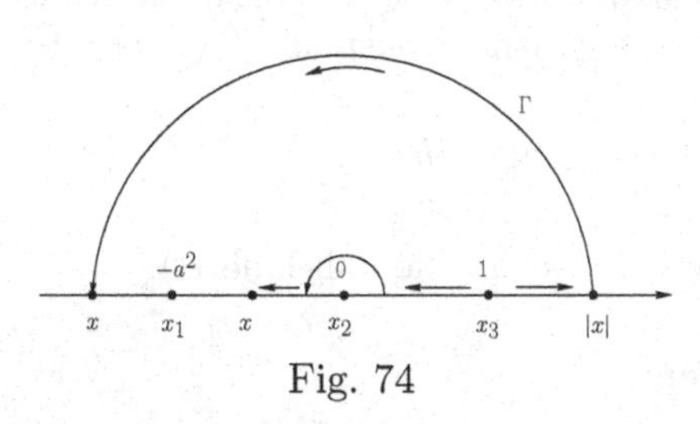

Fig. 74

Take $z = x \in (-a^2, 0)$ and compare the real parts of the left and right sides of the formula for f from part (ii). Then take $z = x < -a^2$ with a big absolute value $|x| > 1$, and compare the imaginary parts. The paths of integration are shown on Fig. 74. Notice that in both cases we are integrating over arcs whose images join the sides of angles with a vertex at infinity.

(iv) Evaluate the integral from part (ii). Use the substitution $t = \sqrt{\zeta + a^2}/\sqrt{\zeta - 1}$ and the partial fraction decomposition.

Here we consider the branch of the function which coincides with the corresponding real-valued function when $z = x > 1$—see the comments in the solution to the Example 8.3.

For example, for $z \in (0,1)$ we have $\arg t^2 = -\pi$ because the argument of t changes continuously when z moves from the interval $(1,\infty)$ to $(0,1)$ along an arc in Π. Hence, $t = -ir$, $a < r < \infty$ as $z = x \in (0,1)$, and $r \to a^+$ as $x \to 0^+$, while $r \to \infty$ as $x \to 1^-$. In this case,

$$\log \frac{t - ia}{t + ia} = \log \frac{r + a}{r - a}$$

tends to ∞ as $x \to 0^+$, and tends to 0 as $x \to 1^-$. Moreover,

$$\log \frac{t+1}{t-1} = \log \frac{-ir+1}{-ir-1} = i(\pi + 2\arg(-ir+1))$$

(to check the last equality, sketch a diagram with the points ± 1 and $-ir$). Hence, $\log \frac{t+1}{t-1}$ tends to $i(\pi + 2\arg(-ia+1))$ as $x \to 0^+$, and to 0 as $x \to 1^-$. Therefore, from (9.58) we see that the point $f(x)$ runs from ∞ to 0 along the imaginary axis as x runs from 0 to 1. This is consistent with Figs. 73 and 74. You may check the other intervals for x in the same way and see that $f(x)$ runs along the boundary of the domain in Fig. 73 as x changes from $-\infty$ to $+\infty$.

4. Find the mapping of the upper half-plane $\operatorname{Im} z > 0$ onto the same polygonal domain shown in Fig. 73 such that $f(-1) = \infty$, $f(0) = 0$, $f(a^2) = \infty$, $f(\infty) = h_1 + ih_2$, $z_0 = 0$, where h_1, h_2 are given numbers and $a > 1$ has to be determined (you will want to renumber the vertices in Fig. 73).

5. Let z_0, a be points of the upper half-plane Π, and let the real numbers x_k, α_k, $k = 1, \ldots, n$, be such that $\alpha_1 + \alpha_2 + \cdots + \alpha_n = 2$, and $-1 \leq \alpha_k < 1$. Let the function $f(x)$ be defined by

$$f(z) = A \int_{z_0}^{z} \frac{(\zeta - x_1)^{\alpha_1}(\zeta - x_2)^{\alpha_2} \ldots (\zeta - x_n)^{\alpha_n}}{(\zeta - a)^2(\zeta - \overline{a})^2}\, d\zeta + B,$$

where A, B are complex constants. Prove that $f(z)$ maps Π conformally onto the *exterior* of a polygon with external angles $\pi\alpha_k$; the images of points x_k are vertices of the polygon, and the image of a is ∞.

Hint. To prove the conformity of f at a, use the Laurent expansion for $f'(z)$ at a and write $f(z)$ in the form $f(z) = \frac{C}{z-a} + \phi(z)$, where $\phi(z)$ is analytic at a. Then use the same arguments as in the proof of conformity of Möbius transformations (Theorem 4.3).

6. Let D be the upper half of the w-plane cut along the vertical ray $\Gamma = \{w = re^{i\pi/2} : r \geq 1\}$. Let $w_1 = i$, and let $w_2 = \infty$ be the vertex of the angle between the rays Γ and $(-\infty, 0)$, and $w_3 = \infty$ be the vertex of the angle between the rays $(0, \infty)$ and Γ. Find the mapping $f(z)$ of the upper half-plane onto D such that:

(i) $f(-1) = w_1,\ f(0) = w_2,\ f(\infty) = w_3$;

(ii) $f(-1) = w_3,\ f(0) = w_1,\ f(1) = w_2$.

7. Consider the generalization of problem 6(i) where Γ is the ray $\Gamma = \{w = re^{i\phi} : r \geq 1\}$, and ϕ is a given angle, $0 < \phi < \pi/2$.

8.2 Hydrodynamics. Simply-connected Domains

In this section we consider the *steady planar flow* of a fluid. In this model, the flow is essentially two-dimensional (hence "planar"); that is, all particles of the fluid move parallel to the same plane, and particles lying along the same line which is perpendicular to the plane move with the same speed and direction, i.e. they have the same velocity vector—see Fig. 75, where the plane is horizontal.

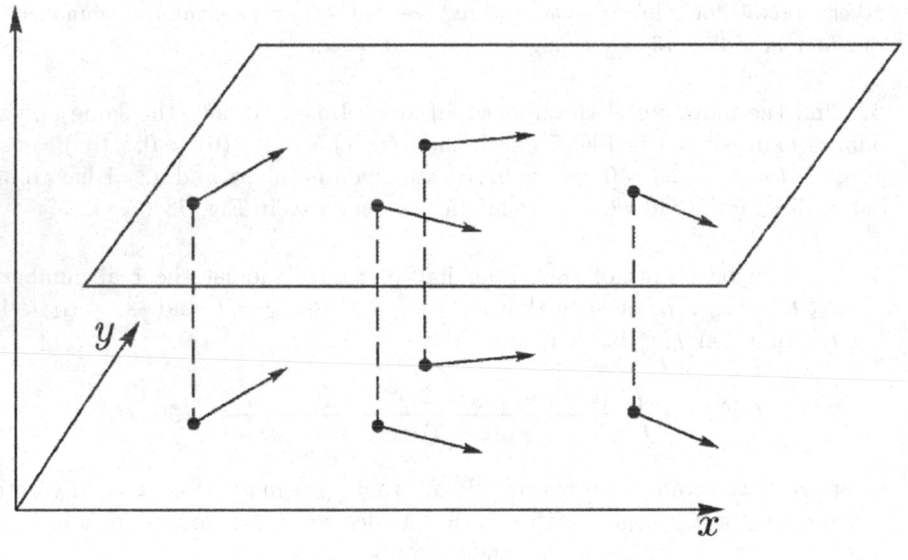

Fig. 75

We also assume that the flow is stable, in the sense of not varying with time (hence "steady"). Of course the flow may vary with position in the plane. Thus the velocity vector of a particle is independent of time and depends only on the projection (x, y) of the particle onto the plane. So we will represent the velocity by a vector function $\mathbf{V}$of the position (x, y) in the plane. We will denote the components of the vector $\mathbf{V}$ by X and Y. Then

$$\mathbf{V}(x, y) = X(x, y) + iY(x, y).$$

8.2.1 Complex potential of a vector field

We will assume that the functions X and Y have continuous first and second partial derivatives at all points of some domain D. Further, we assume that the vector field $\mathbf{V}$ satisfies the following two conditions.

1) $\mathbf{V}$ is a *potential field*, i.e. there exists some real-valued function u defined on D for which

$$X = \frac{\partial u}{\partial x}, \quad Y = \frac{\partial u}{\partial y}. \tag{8.16}$$

If this is the case, then from the equality of the mixed partial derivatives we get

$$\frac{\partial Y}{\partial x} = \frac{\partial^2 u}{\partial y \partial x} = \frac{\partial^2 u}{\partial x \partial y} = \frac{\partial X}{\partial y},$$

which can be written

$$\frac{\partial Y}{\partial x} - \frac{\partial X}{\partial y} = 0. \tag{8.17}$$

Conversely, if D is simply-connected then (8.17) implies (8.16). Moreover, in this case (8.17) and (8.16) are equivalent to the following property: $\int_\Gamma X\,dx + Y\,dy = 0$ for every closed contour Γ situated in D. Physically, this condition means that the fluid flow is free from circular motion, or eddies. Such fields are also called *irrotational.*

2) The vector field $\mathbf{V}$ is *solenoidal:*

$$\frac{\partial X}{\partial x} + \frac{\partial Y}{\partial y} = 0. \tag{8.18}$$

In a simply-connected domain D this condition is equivalent to the property that the vector field

$$i\mathbf{V} = -Y + iX$$

also has a potential function, i.e. some real function v such that

$$\frac{\partial v}{\partial x} = -Y, \quad \frac{\partial v}{\partial y} = X. \tag{8.19}$$

To interpret this condition in terms of fluid flow consider a smooth curve Γ in a simply-connected domain D. Imagine a vertical fence of unit height is built along Γ, but through which the fluid can flow unimpeded. Then $\int_\Gamma X\,dy - Y\,dx$ is equal to the amount of fluid that passes through the fence in one unit of time. This quantity is called the *flow through the curve* Γ or the *flux through* Γ. The property (8.18) is equivalent to requiring that the contour integral

$$\int_\Gamma X\,dy - Y\,dx = 0$$

for any closed contour Γ in D. The physical meaning of this property is that the quantity of the fluid entering Γ is equal to quantity leaving Γ, that is there are no sources or sinks inside Γ and therefore in the domain D. In other words, streamlines (the trajectories of the particles of the fluid) neither start nor stop anywhere in D. Sometimes the solenoidal property is called *incompressibility*, for it also means that the number of particles flowing to a point always equals the number flowing out from it.

The function v has its own physical meaning. Take any two points (x_1, y_1) and (x_2, y_2) in D, and connect them by some curve $\Gamma \subset D$. Then

$$\int_\Gamma X\,dy - Y\,dx = v(x_1, y_1) - v(x_2, y_2).$$

Hence the difference $v(x_1, y_1) - v(x_2, y_2)$ is equal to the flow through the curve Γ, and by this equality it depends only on the start and end points of the curve, and not on its shape.

Now let us consider the level curves of the function v, that is, sets of points $(x, y) \in D$ on which $v(x, y) = C$, where C is a constant. Fix C and find the tangent vector to the curve $v(x, y) = C$ at any one of its points (x, y). It's known that the gradient

$$\nabla v = \left(\frac{\partial v}{\partial x}, \frac{\partial v}{\partial y}\right)$$

is orthogonal to the level curve $v(x, y) = C$. According to (8.19),

$$\left(\frac{\partial v}{\partial x}, \frac{\partial v}{\partial y}\right) = (-Y, X).$$

Vector $(-Y, X)$ is orthogonal to (X, Y) since their dot product equals 0. So, the velocity vector $\mathbf{V} = (X, Y)$ is tangent to to the level curve $v(x, y) = C$. Therefore, the vector field $\mathbf{V}$ is tangent to the level curves of v at every point. That means: *the level curves $v(x, y) = C$ are streamlines of the flow of the fluid.* Therefore the function v is called the *stream function.*

We can now form the function $f(z) = u(x, y) + iv(x, y)$, where $z = x + iy$, making this a function of a complex variable. This is called the *complex potential* of the vector field $\mathbf{V}$. From equations (8.16) and (8.19) it follows that f satisfies the Cauchy-Riemann conditions (3.4), which means it is analytic in D. Thus, *the complex potential which arises from a planar potential and solenoidal steady fluid flow in a simply-connected domain, is an analytic function.*

Since $f(z)$ is analytic, its derivative is independent of direction, so we can calculate it by differentiating with respect to x; then applying the Cauchy-Riemann conditions, we get

$$f'(z) = \frac{\partial u}{\partial x} + i\frac{\partial v}{\partial x} = \frac{\partial u}{\partial x} - i\frac{\partial u}{\partial y} = X - iY = \overline{\mathbf{V}}. \tag{8.20}$$

So we see that *the derivative of the complex potential is the complex conjugate of the velocity of the fluid flow.*

The equality $f'(z_0) = 0$ means that the velocity $\mathbf{V}$ at z_0 is zero. Such points are called *stagnation points.*

The function u is called the *potential* of the vector field, and the level curves $u(x, y) = C$ are called *equipotentials.* The curves $u = C_1$, $v = C_2$ are orthogonal in the w-plane. If $f'(z) \neq 0$, then the mapping $f(z)$ preserves the angles at z. Therefore, the curves $u(x, y) = C_1$, $v(x, y) = C_2$ are orthogonal, that is *equipotentials are orthogonal to streamlines* except at stagnation points.

Now we will show that *any function $f(z) = u + iv$, analytic on D, can be interpreted as the complex potential of a vector field $\mathbf{V} = \frac{\partial u}{\partial x} - i\frac{\partial v}{\partial x}$, which is solenoidal as well as potential.* This is where the hydrodynamic interpretation of analytic functions lies.

So given such a function f, let us define a vector field $\mathbf{V} = X + iY$ by $X = \frac{\partial u}{\partial x}$ and $Y = -\frac{\partial v}{\partial x} = \frac{\partial u}{\partial y}$. Then our property 1 above, that $\mathbf{V}$ arises from a potential, is automatically satisfied. By virtue of the Cauchy-Riemann conditions, the equalities (8.19) are satisfied. Then differentiating the first of these equalities w.r.t. y, and the second w.r.t. x, we obtain the solenoidal condition (8.18). So $\mathbf{V}$ is both a solenoidal and potential field.

If a domain D is multiply-connected, the complex potential $f(z)$ can be constructed in a simply connected subdomain. An analytic continuation of $f(z)$ to D in such a situation will generally be multivalued. But the derivative $f'(z) = X - iY$ is a single-valued analytic function in D.

Interpreting an analytic function as a complex potential of a vector field enables us to solve a number of important applied problems. We consider some problems in hydrodynamics, electrostatics, and thermodynamics.

8.2.2 Simply-connected domains

Suppose that a flow $\mathbf{V}$ is potential and solenoidal in a simply-connected domain D. To study the vector flow for this scenario, we begin by finding the complex potential $f(z)$. Let $w = f(z) = u + iv$. We will exploit the fact that the streamlines of flow of the fluid are of the form $v(x, y) = C$. In the w-plane, the equations $v = C$ define the horizontal lines; so the function f will have to map the streamlines of flow in D to a family of horizontal lines in the w-plane. We make two assumptions.

(a) The boundary of D is one or two streamlines (that is a fluid flows along the boundary). Then $f(z)$ maps the boundary to horizontal lines in the w-plane.

(b) The total flux of the flow is given. We know that $v(z_2) - v(z_1)$ is the flux through a curve in D with endpoints z_1, z_2. So, the total flux is the maximal distance between the horizontal lines containing images of streamlines in the w-plane.

Let $f(z)$ be a conformal mapping of D onto a half-plane $\operatorname{Im} w > a$ if the total flux is infinite, and onto a strip $a < \operatorname{Im} w < b$ if the total flux is finite and equals $b - a$. Then the images of the horizontal lines $\operatorname{Im} w = C$ under the inverse mapping $f^{-1}(w)$ are streamlines of the flow with the given flux and satisfying (a). Moreover, we will be able to find conformal mappings satisfying additional conditions for specific problems. Since an additive constant to $f(z)$

would have no effect on the flow $X + iY = \overline{f'(z)}$, we can assume for simplicity that $a = 0$. Therefore, *to construct a complex potential* $f(z)$*, it's sufficient to find the conformal mapping of* D *onto either a half-plane* $\operatorname{Im} w > 0$*, or a strip* $0 < \operatorname{Im} w < b$*, satisfying some additional conditions depending on the specific problem.* To determine which of these cases takes place, we see whether the total flux is bounded or not. Two typical cases are considered below.

Once $f(z)$ is known, we can find the streamlines in two ways:

1. consider the equation $v(x, y) = \operatorname{Im} f(x, y) = C$ as an implicit equation of streamlines;

2. consider the equality $f(x, y) = u + iC$ as a system of equations with variables u, x, and y, exclude (if possible) u from the system, and obtain an explicit equations of streamlines $y = g(x, C)$.

Remark. In our discussion of the Schwarz-Christoffel mapping in the previous Section 8.1, we used the standard notation $w = f(z)$ for a mapping of the half-plane Π and a strip onto a polygonal domain P. Now we use the same notation $w = f(z)$ for a mapping of D (in particular, of P) onto Π or onto a strip, that is we assume that D is in the z-plane, and the image G is in the w-plane. So, *the complex potential* $f(z)$ *is inverse to the mappings considered in the previous section.* If the inverse $z = f^{-1}(w)$ to the complex potential (that is a mapping of Π in the w-plane onto D in the z-plane) is known, the streamlines have the equations $x = \operatorname{Re} f^{-1}(u, C)$, $y = \operatorname{Im} f^{-1}(u, C)$, which we may consider as parametric equations with the parameter u; here as before $z = x + iy$, $w = u + iv$. If possible, we exclude u from the system, and again obtain an explicit equation $y = g(x, C)$.

(a) Flow in a curvilinear half-plane. Consider a domain D bounded by a Jordan curve Γ with endpoints at infinity such that the angle at ∞ between the left and the right parts of Γ equals π, as seen in Fig. 76. Note that the figure shows the vertical cross section of the flow which is parallel to the (x, y)-plane. We assume that there is a limit $\lim_{z\to\infty} \mathbf{V}(z)$ which we denote by $\mathbf{V}_\infty$, and that $|\mathbf{V}_\infty| = V_\infty$ is a given positive number.

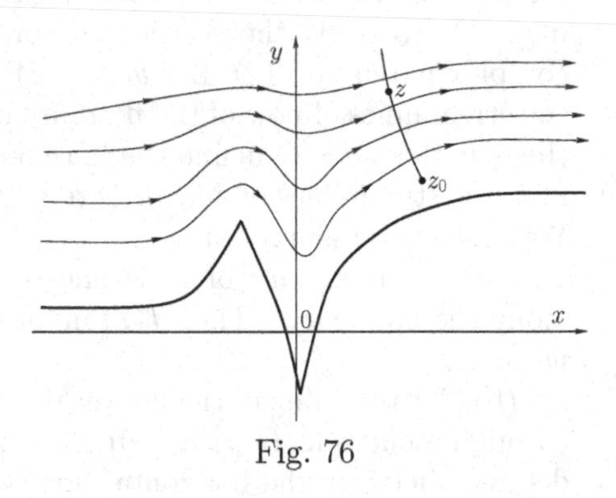

Fig. 76

The physical meaning of this assumption is that the flow is almost uniform at points which are far away from obstacles and other points where the shape of D changes. To determine whether the image of D is a strip or a half-plane, fix a point z_0 in D such that $\mathbf{V}(z)$ is close to $\mathbf{V}_\infty$ when $|z| \geq |z_0|$. Consider the curve in $|z| \geq |z_0|$ with endpoint at z_0, which is orthogonal to the streamlines (that is a part of an equipotential). This line does not end in D, and all normal vectors are almost parallel to $\mathbf{V}_\infty$. Hence this curve goes to infinity. Let z be a point on this curve. Then the flux through the part of the curve between z_0 and z approaches ∞ as $z \to \infty$. But the flux is equal to $v(z) - v(z_0)$; therefore $v(z) \to \infty$. So the image of D must be a half-plane $\operatorname{Im} z > a$, and we may choose $a = 0$. Moreover, we see that $f(\infty) = \infty$. By (8.20), $f'(z) = \overline{\mathbf{V}}$. Taking the limit as $z \to \infty$, we get $|f'(\infty)| = V_\infty$. Therefore, *the problem of finding the complex potential $f(z)$ is reduced to that of finding the conformal mapping from D onto the upper half-plane such that*

$$f(\infty) = \infty, \quad |f'(\infty)| = V_\infty. \tag{8.21}$$

Example 8.6 We will look at an infinitely deep planar flow around an obstacle of height h—see Fig. 77*a*. Here the domain D will be the upper half-plane with a segment cut from the bottom of the positive y-axis (the obstacle). We assume that the velocity vector $\mathbf{V}(z)$ approaches a fixed horizontal constant vector $\mathbf{V}_\infty$ with $|\mathbf{V}_\infty| = V_\infty$ as $|z| \to \infty$.

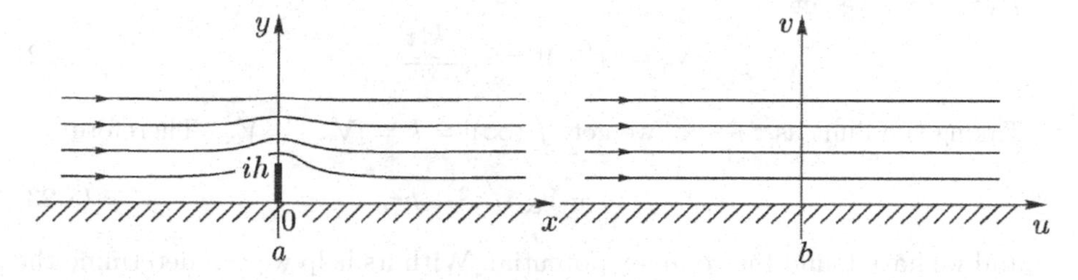

Fig. 77

To correlate Fig. 75 with Fig. 77*a*, imagine an endless vertical board and a horizontal flow. Then Fig. 77*a* is a horizontal cross section of the flow. As we saw, here G is just the upper half-plane Π, $\operatorname{Im} z > 0$. We consider two ways to find the conformal mapping of D onto Π satisfying (8.21). The first one does not require the Schwarz-Christoffel transformation, it is based only on properties of known functions; such an approach is not restricted to polygonal domains. The second method is based on the Schwarz-Christoffel transformations and allows us to consider a more general case.

First method. We will compose such a mapping in three steps.

1. The function $w_1 = z^2$ maps D to the complex plane minus the ray $\operatorname{Re} w_1 \geq -h^2$. The reader can verify this by considering the images of each of the four pieces of the boundary of D (including both sides of the "obstacle", $[0, ih]$).

2. The function $w_2 = w_1 + h^2$ slides the previous domain to the right h^2 units, producing the complex plane minus the positive real axis.

3. To this domain we apply a regular branch of the function $w = \sqrt{w_2}$, defined by
$$w = \sqrt{|w_2|}e^{i\phi/2},$$
$$\text{where } w_2 = |w_2|e^{i\phi} \quad \text{with} \quad 0 < \phi < 2\pi.$$

Under the last mapping $w = \sqrt{w_2}$, the upper edge of the cut, i.e. $\phi = 0$, goes to the same ray in the w-plane, while the lower edge of the cut, $\phi = 2\pi$, goes to the negative real axis $\operatorname{Arg} w = \pi$. Therefore the cut plane from step 2 goes to the upper half-plane in step 3.

The composition of these three functions gives the mapping we sought:

$$w = \sqrt{w_2} = \sqrt{w_1 + h^2} = \sqrt{z^2 + h^2}.$$

This mapping is not unique: a Möbius transform of the upper half plane onto itself could be added on as a step 4, but all we need for present purposes is a real multiplicative constant $k > 0$, to form $w = k\sqrt{z^2 + h^2}$.[3] We can determine the value of the constant k from the velocity V_∞: since by (8.20) $f'(z) = \overline{\mathbf{V}}$, we get

$$|\mathbf{V}| = |f'(z)| = \frac{k|z|}{\sqrt{|z^2 + h^2|}}. \tag{8.22}$$

Taking the limit as $z \to \infty$, we get $|f'(\infty)| = k = |\mathbf{V}_\infty| = V_\infty$. Therefore

$$f(z) = V_\infty \sqrt{z^2 + h^2}, \tag{8.23}$$

and we have found the complex potential. With its help we can determine the most important characteristics of the flow: the magnitude and direction of the velocity vector $\mathbf{V}$, the equation for the streamlines of flow, rate of outflow, etc. For example, from (8.22) we see that near the base of the obstruction ($z = 0$) the velocity is close to zero (*dead zone*) and $z = 0$ is a stagnation points, while near its top ($z = ih$) the velocity approaches infinity (*spike effect*).

Now let us find the equation for the streamlines of flow, which are curves defined by the equalities $v(x, y) = C$ for different positive constants C. So we can either find the imaginary part of the function $f(z) = u + iv$ defined in (8.23), or we can regard (8.23) as a system of equations with variables u, x, and y, and eliminate u. We will follow the second strategy: from (8.23) we get

$$\begin{aligned} (u + iv)^2 &= V_\infty^2((x + iy)^2 + h^2), \\ \text{or} \quad u^2 - v^2 + i\,2uv &= V_\infty^2(x^2 - y^2 + h^2) + i\,2V_\infty^2 xy. \end{aligned}$$

Equating the real and imaginary parts, we arrive at the system of equations

$$\begin{cases} u^2 - v^2 = V_\infty^2(x^2 - y^2 + h^2) \\ uv = V_\infty^2 xy. \end{cases}$$

[3] More precisely, the condition that ∞ be fixed by the mapping would restrict these Möbius transforms to linear ones $kw + b$. The constant b would not affect the flow, so the only extra parameter needed for our example is k.

We now solve for u in the second equation, and then substitute the resulting expression into the first, thus eliminating u. For convenience we will define $c = v/V_\infty$; since $v = C$ is constant, so is c. After simplifying algebraically we finally arrive at the equation

$$y = c\sqrt{1 + \frac{h^2}{x^2 + c^2}},$$

which corresponds to the lines of flow seen in Fig. 77a.

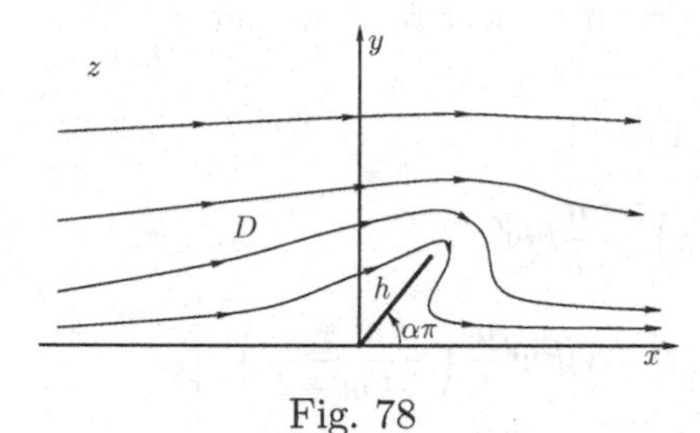

Fig. 78

Second method. We consider the more general problem when the obstacle is not necessarily perpendicular to the bottom—see Fig. 78. We will use the result of Example 8.4, but change the notations of variables, and map the upper half of the w-plane onto the domain D in the z-plane shown in the figure.

Choose a, b such that

$$\frac{a}{a+b} = \alpha; \text{ then } \frac{b}{a+b} = 1 - \alpha, \quad \frac{a}{b} = \frac{\alpha}{1-\alpha}.$$

Our function will be a composition of the following functions.

(1) The function from Example 8.4,

$$z_1 = \frac{a}{\pi}\log(w-1) + \frac{b}{\pi}\log\left(\frac{aw}{b} + 1\right) - ia,$$

maps the upper half-plane $\operatorname{Im} w > 0$ into the strip $-a < \operatorname{Im} z_1 < b$ cut along the negative real axis—see (8.12).

(2) The function $z_2 = z_1 + ia$ slides the previous domain a units upward.

(3) The function $z_3 = \frac{\pi}{a+b} z_2$ maps our strip onto the strip $0 < \operatorname{Im} z_3 < \pi$ cut along the horizontal ray $z_3 = x + i\frac{a\pi}{a+b} = x + i\pi\alpha$, $-\infty < x \le 0$.

(4) The function $z_4 = e^{z_3}$ maps the strip onto the upper half-plane $\operatorname{Im} z_4 > 0$, and maps the cut onto the interval $z_4 = re^{i\pi\alpha}$, $0 < r \le 1$.

(5) The last function $z = hz_4$ gives us the desired mapping onto domain D. Therefore (we write e^z as $\exp(z)$),

$$\begin{aligned} z = he^{z_3} &= h\exp\left(\frac{\pi}{a+b} z_2\right) \\ &= h\exp\left(\frac{\pi}{a+b}\left(\frac{a}{\pi}\log(w-1) + \frac{b}{\pi}\log\left(\frac{aw}{b} + 1\right)\right)\right) \\ &= h(w-1)^\alpha\left(\frac{aw}{b} + 1\right)^{1-\alpha}. \end{aligned}$$

Obviously, $z = \infty$ if and only if $w = \infty$. To satisfy the second equality in (8.21) we remark that the function

$$\begin{aligned} z = f^{-1}(w) &= h(kw-1)^{\alpha}\left(\frac{a}{b}kw+1\right)^{1-\alpha} \\ &= h(kw-1)^{\alpha}\left(\frac{\alpha}{1-\alpha}kw+1\right)^{1-\alpha} \end{aligned} \tag{8.24}$$

also maps the upper w-plane onto D and sends ∞ to ∞ for every $k > 0$. Differentiating equation (8.24) implicitly with respect to z, we get

$$\begin{aligned} 1 &= h\alpha(kw-1)^{\alpha-1}kw'\left(\frac{a}{b}kw+1\right)^{1-\alpha} \\ &\quad + h(kw-1)^{\alpha}(1-\alpha)\left(\frac{a}{b}kw+1\right)^{-\alpha}\frac{a}{b}kw' \\ &= h\left[\alpha kw'\left(\frac{\frac{a}{b}kw+1}{kw-1}\right)^{1-\alpha} + (1-\alpha)kw'\frac{a}{b}\left(\frac{kw-1}{\frac{a}{b}kw+1}\right)^{\alpha}\right]. \end{aligned}$$

Then taking the limit as w approach ∞, and recalling that $w' = \overline{\mathbf{V}} \to V_\infty$, we get

$$1 = h\left[\alpha kV_\infty\left(\frac{a}{b}\right)^{1-\alpha} + (1-\alpha)kV_\infty\frac{a}{b}\left(\frac{b}{a}\right)^{\alpha}\right] = h\left(\frac{a}{b}\right)^{1-\alpha}kV_\infty\,,$$

so that we obtain k:

$$k = \frac{1}{hV_\infty}\left(\frac{b}{a}\right)^{1-\alpha} = \frac{1}{hV_\infty}\left(\frac{1-\alpha}{\alpha}\right)^{1-\alpha}. \tag{8.25}$$

Therefore, the function inverse to the complex potential is given by (8.24) with k indicated in (8.25). From (8.24) we obtain the equations of streamlines

$$x + iy = h(k(u+iC)-1)^{\alpha}\left(\frac{\alpha}{1-\alpha}k(u+iC)+1\right)^{1-\alpha},$$

which can be viewed as parametric equations with parameter u.

In particular, for $\alpha = \frac{1}{2}$ the equation (8.24) has the form

$$z = h\sqrt{\left(\frac{w}{hV_\infty}\right)^2 - 1} = \sqrt{\left(\frac{w}{V_\infty}\right)^2 - h^2}, \quad \text{that is} \quad w = V_\infty\sqrt{z^2+h^2},$$

which coincides with (8.23).

(b) Flow in a curvilinear strip. Let Γ_1, Γ_2 be two Jordan curves with joint points only at infinity, and let D be a domain between these curves.

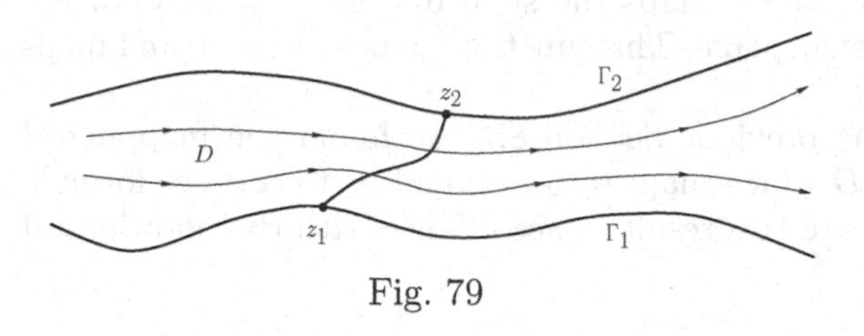

Fig. 79

We assume that the flow is potential and solenoidal in D, and the stream flows along Γ_1, Γ_2. This means that the function $v(z) = \operatorname{Im} f(z)$ is constant as $z \in \Gamma_1$ and as $z \in \Gamma_2$, where $f(z)$ is a complex potential.

Hence, the difference $v(z_2) - v(z_1)$ is the same for any points $z_1 \in \Gamma_1$ and $z_2 \in \Gamma_2$. The physical interpretation of this fact is that for any points $z_1 \in \Gamma_1$, $z_2 \in \Gamma_2$, and for any curve in D joining z_1, z_2, the flux through the curve is the same—see Fig. 79. We assume that the value (that is the total flux) $H = v(z_2) - v(z_1)$ is given. Then $f(z)$ maps D onto the strip $v_0 < \operatorname{Im} w < v_0 + H$; we may take $v_0 = 0$. In our case the streamlines tend to ∞ in both directions. For the images of the streamlines, namely for the horizontal lines $v = C$ in the plane of $w = u + iv$, we have $u \to +\infty$ as z tends to ∞ along a streamline in the direction of flow, and $u \to -\infty$ as z tends to ∞ in the opposite direction—it follows from the correspondence of boundaries under conformal mappings. We write this relation in the form

$$f(\pm\infty) = \pm\infty. \tag{8.26}$$

Therefore, *to determine the complex potential of a flow in a curvilinear strip* D, *we have to find the conformal mapping of* D *onto the strip* $0 < \operatorname{Im} w < H$ *satisfying* (8.26).

Example 8.7

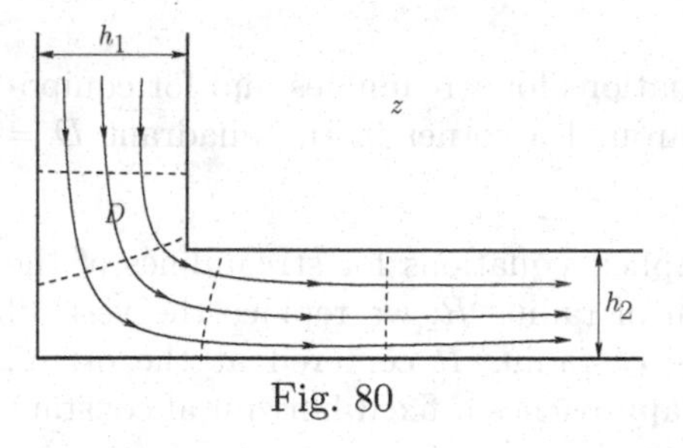

Fig. 80

Find parametric equations of the form $z = g(u, C)$ for the streamlines of the flow with total flux H shown in Fig. 80.

Solution As in the second method in the previous example, we construct the function $z = f^{-1}(w)$ such that $f^{-1}(w)$ is inverse to the complex potential $w = f(z)$, $f^{-1}(w)$ maps the strip $0 < \operatorname{Im} w < H$ onto the domain D, and

$$f^{-1}(\pm\infty) = \pm\infty \tag{8.27}$$

(clearly, (8.26) and (8.27) are equivalent). We find such a mapping in two steps.

(1) The function $w_1 = e^{\pi w/H}$ maps the strip $0 < \operatorname{Im} w < H$ onto the half-plane $\operatorname{Im} w_1 > 0$ in the w_1-plane. This function maps $-\infty$ to 0, and maps ∞ to ∞.

(2) In problem 3 in the previous Section 8.1, we found the mapping of the upper half-plane onto D which maps 0 to $-\infty$ and ∞ to ∞—see formula (9.58) in Answers. We rewrite this result in accordance with the notations of the current problem:

$$z = g_1(w_1) = \frac{h_2}{\pi}\left(ia\log\frac{t-ia}{t+ia} + \log\frac{t+1}{t-1}\right), \quad t = \left(\frac{w_1 + a^2}{w_1 - 1}\right)^{1/2}.$$

Thus, the composition

$$w = f^{-1}(w) = g_1(e^{\pi w/H})$$

maps the strip onto D and satisfies (8.27). To find the parametric equations we just write $u + iC$ for w, where $0 < C < H$:

$$z = g(u, C) = \frac{h_2}{\pi}\left(ia\log\frac{t-ia}{t+ia} + \log\frac{t+1}{t-1}\right), \quad t = \left(\frac{e^{\pi(u+iC)/H} + a^2}{e^{\pi(u+iC)/H} - 1}\right)^{1/2}.$$

Dotted lines in Fig. 80 indicate equipotentials. Another example of a similar kind will be given later—see Example 8.16 in Section 8.4.

Problems

1. Derive the formula (8.23) directly using the formula (8.2) for the Schwarz-Christoffel transformation.

2. Find the complex potential and equations for streamlines and for equipotentials of the flow with infinite flux around a corner in the quadrant $D = \{z = x + iy : x \geq 0,\ y \geq 0\}$.

3. Find the complex potential and implicit equations for streamlines of the flow around a semicircular obstruction of radius R; we replace the vertical obstacle in Fig. 77*a* by the semicircle of radius R centered at the origin. Assume that the velocity vector $\mathbf{V}(z)$ approaches a fixed horizontal constant vector $\mathbf{V}_\infty$ with $|\mathbf{V}_\infty| = V_\infty$ as $|z| \to \infty$.

4. (i) Find the complex potential $f(z)$ of the flow with total flux H in the upper half-plane $\operatorname{Im} z > 0$ cut along the vertical ray $\{z = re^{i\pi/2} : r \geq 1\}$. You may use the result of problem 6(i) in Section 8.1.

(ii) Find the velocity vector $\mathbf{V}(z)$ of the flow at $z = iy$, $0 < y < 1$.

8.3 Sources and Sinks. Flow around Obstacles

8.3.1 Sources and sinks

Recall that the flux of a vector field $\mathbf{V} = X + iY$ through a curve Γ is defined by the formula $\int_\Gamma X\,dy - Y\,dx$—see Section 8.2.1. We say that a point z_0 is an isolated *sourse of strength* N, if for every sufficiently small contour Γ enclosing z_0, the flux through Γ equals N. If we interpret $\mathbf{V}$ as a flow of fluid, it means that the point z_0 produces N units of fluid in a unit of time.

Let us consider the vector field generated by a single source of strength N.[4] We may assume that the source is located at the origin. In this case the direction of the vector $\mathbf{V}(z)$ coincides with the direction of the vector Oz for every $z \neq 0$, and $|\mathbf{V}|$ depends only on $r = |z|$—see Fig. 81. Hence, $\mathbf{V}$ has the form $\mathbf{V} = \phi(r)z$, where $\phi(r)$ is a positive function of r. From the equality (5.3) with $u = X$, $v = -Y$, we see that

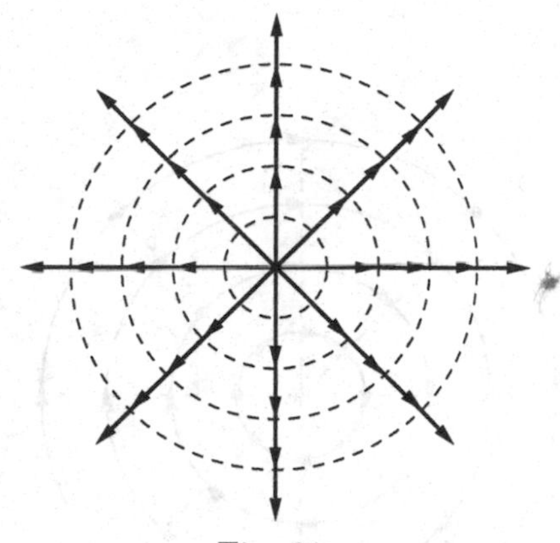

Fig. 81

$$\text{Flux through } \Gamma = \int_\Gamma X\,dy - Y\,dx = \operatorname{Im} \int_\Gamma \overline{\mathbf{V}(z)}\,dz. \tag{8.28}$$

For the source at the origin with strength N, the flux through any closed curve enclosing the origin equals N. Hence, we may take a circle $|z| = r$, $r > 0$, as Γ. We parametrize it in the standard way: $z = re^{it}$, $0 \le t < 2\pi$; then $dz = ire^{it}\,dt$. From (8.28) we have

$$\begin{aligned} N &= \operatorname{Im} \int_\Gamma \overline{\mathbf{V}(z)}\,dz = \operatorname{Im} \int_\Gamma \phi(r)\bar{z}\,dz \\ &= \operatorname{Im} \int_0^{2\pi} \phi(r)re^{-it}ire^{it}\,dt = \phi(r)r^2 2\pi. \end{aligned} \tag{8.29}$$

Therefore, $\phi(r) = N/(2\pi r^2)$, and we have

$$\mathbf{V} = \phi(r)z = \frac{N}{2\pi|z|^2}z = \frac{N}{2\pi}\cdot\frac{z}{z\bar{z}} = \frac{N}{2\pi}\cdot\frac{1}{\bar{z}} = \frac{Q}{\bar{z}}, \quad Q = \frac{N}{2\pi}. \tag{8.30}$$

[4] Recall that our plane vector field is a cross-section of a three-dimensional vector field whose vectors are parallel to the same plane. In three dimensions, we must think of the source as a line that is perpendicular to the plane and such that the entire line produces fluid evenly in all radial directions.

Now we may find the complex potential $f(z)$. Since $f'(z) = \overline{\mathbf{V}}$, we obtain the equalities

$$f'(z) = \frac{Q}{z}, \quad f(z) = Q \log z = Q \ln |z| + iQ \arg z \tag{8.31}$$

(we drop an additive constant). Therefore, the complex potential is a multi-valued function. The equipotentials $u = Q \ln |z| = C$, that is $|z| = c > 0$, and streamlines $v = Q \arg z = C$ are shown in Fig. 81. Obviously, if a source is at the point z_0, the complex potential is $f(z) = Q \log(z - z_0)$.

If N and Q are negative, that is if a point absorbs $|N|$ units of fluid in a unit of time, we say that the point is a *sink of strength* $|N|$. The equipotentials and streamlines are the same curves as in Fig. 81, just with the opposite direction of flow.

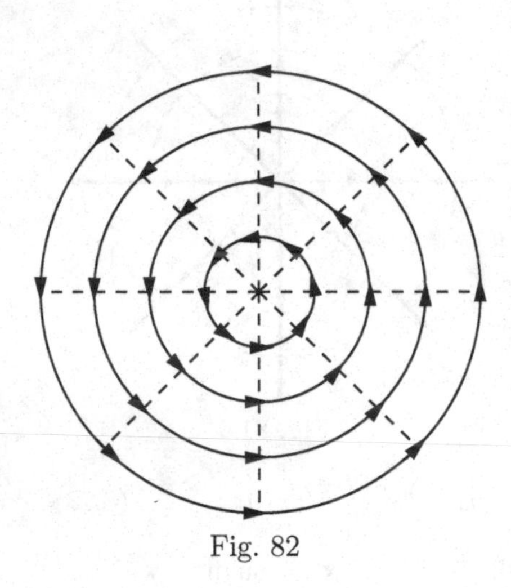

Fig. 82

8.3.2 Vortices

Recall that the circulation of a vector field $\mathbf{V} = X + iY$ around a contour Γ is defined by the formula $\int_\Gamma X\,dx + Y\,dy$. We say that z_0 is an isolated *vortex* point of $\mathbf{V}$ *of strength* κ, if for every sufficiently small contour Γ enclosing z_0, the circulation along Γ equals κ. Let us consider the vector field generated by a single vortex of strength κ located at the origin. In this case the vectors $\mathbf{V}(z)$ are tangent to circles centered at 0, the moduli $|\mathbf{V}(z)|$ depend only on $r = |z|$, that is $\mathbf{V}(z) = i\phi(r)z$, and a circulation of $\mathbf{V}$ around any contour enclosing 0 equals κ. In Fig. 82 the directions of streamlines are shown for $\kappa > 0$. Note that the complex potential cannot be analytic at the source, sinks, and vortex points since the corresponding integrals over Γ are not equal to 0. In the same way as for sources, we may show that the vector field $\mathbf{V}$ and its complex potential $f(z)$ are defined by the formulas (see problem 1)

$$\kappa = \operatorname{Re} \int_\Gamma \overline{\mathbf{V}(z)}\,dz, \quad \mathbf{V} = \frac{i\kappa}{2\pi} \cdot \frac{1}{\bar{z}}, \quad f(z) = -\frac{i\kappa}{2\pi} \log z. \tag{8.32}$$

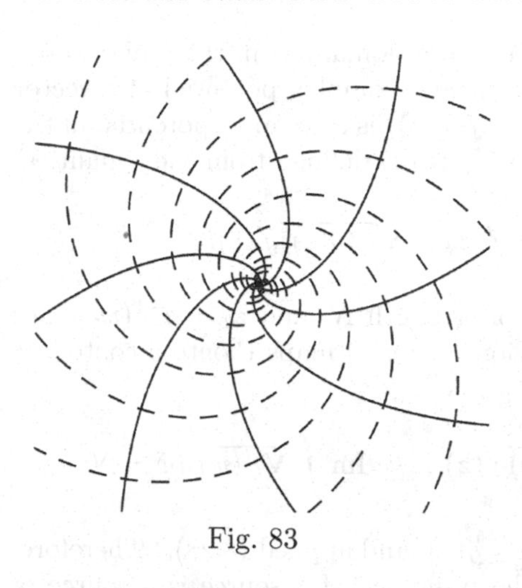

Fig. 83

We say that 0 is a *vortex-source point* of a vector field $\mathbf{V}$ if it's a vortex point and a source at the same time. In this case the complex potential of $\mathbf{V}$ is

$$f(z) = \frac{N - i\kappa}{2\pi} \log z;$$

the equipotentials and streamlines are spirals – see Fig. 83. Equations of the spirals are derived in problem 2. Therefore, if we interpret an analytic function as a complex potential of a vector field, then we may interpret isolated logarithmic branch points as vortex-source points.

Example 8.8 Suppose that $z = 1$ is a source and $z = -1$ is a sink of the same strength N. Find equations for equipotentials and streamlines for the system.

Solution The vector field of the system is the sum of the vector fields of the source and the sink. Hence, the complex potential is the sum of the corresponding complex potentials for these points. Setting $Q = \frac{N}{2\pi}$, we get

$$f(z) = Q\log(z-1) - Q\log(z+1) = Q\log\frac{z-1}{z+1} = u + iv,$$

$$u = Q\ln\left|\frac{z-1}{z+1}\right|, \quad v = Q\arg\frac{z-1}{z+1}.$$

Here we may take any branch of logarithm because an additive constant is unimportant. The equations $u =$ const for equipotentials and $v =$ const for streamlines are equivalent to

$$\left|\frac{z-1}{z+1}\right| = c,\ c > 0, \text{ and } \arg\frac{z-1}{z+1} = C.$$

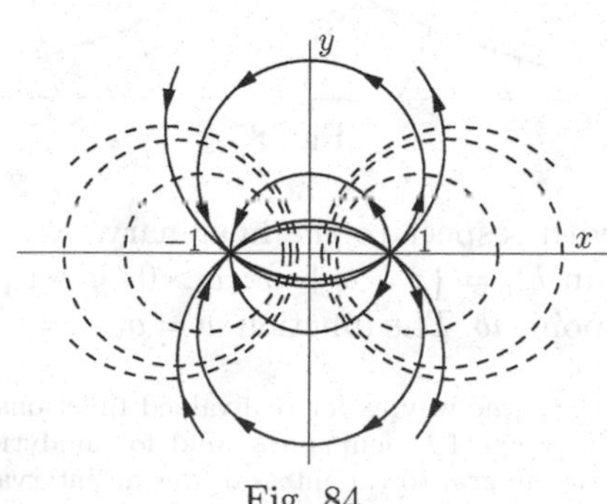

Fig. 84

Consider the Möbius transformation $\xi = \dfrac{z-1}{z+1}$. In the ξ-plane the lines $|\xi| = c$, $c > 0$, are circles, and $\arg\xi = C$ are rays. The inverse function $z = \dfrac{1+\xi}{1-\xi}$ also is a Möbius transformation, and according to Theorem 4.4, it maps circles onto circles or lines, and maps rays onto arcs of circles or rays. Therefore, equipotentials are either circles or lines, and streamlines are either arcs of circles or rays—see Fig. 84.

Let $w = g(z)$ be a conformal mapping of a domain D in the z-plane onto a domain G in the w-plane, and let $f(w)$ be a complex potential of a vector field $\mathbf{V}_G$ in G. Then the function $f_1(z) = f(g(z))$ is a complex potential in the z-plane. The corresponding vector field $\mathbf{V}_D(z)$ is defined from the equalities

$$\overline{\mathbf{V}_D(z)} = f_1'(z) = \frac{df(w)}{dw} g'(z) = \overline{\mathbf{V}_G(g(z))} g'(z).$$

Suppose now that $w_0 \in G$ is a source of strength N, and $z_0 = g^{-1}(w_0)$. Let Γ be a contour enclosing z_0. The function $w = g(z)$ maps Γ onto a contour γ enclosing w_0. Hence,

$$\operatorname{Im} \int_\Gamma \overline{\mathbf{V}_D(z)}\, dz = \operatorname{Im} \int_\Gamma \overline{\mathbf{V}_G(g(z))} g'(z)\, dz = \operatorname{Im} \int_\gamma \overline{\mathbf{V}_G(w)}\, dw = N;$$

where we have used the substitution $w = g(z)$, and applied (8.28).[5] Therefore, z_0 is a source of strength N. Thus, the preimage of a source is a source of the same strength. Obviously, the same arguments work when z_0 is a sink. In the same way we may consider the inverse mapping $z = g^{-1}(w)$ and see that *under a conformal transformation, a source or sink at a given point corresponds to a source or sink of the same strength at the image of that point.* The analogous statement holds for vortex points—see problem 3. Also this conclusion is valid if a source (sink) is located at the boundary of a domain. The proof is essentially the same; instead of a contour enclosing the point, we consider a curve with endpoints on the boundary of a domain on opposite sides of the point.

We start with an example where a source is inside the domain.

Example 8.9 A source of strength N is located at the point ib, $b > 0$, of the upper half-plane $D = \{\operatorname{Im} z > 0\}$. Find the complex potential and the corresponding vector field in D, derive the equations for streamlines and equipotentials, and draw some of these lines.

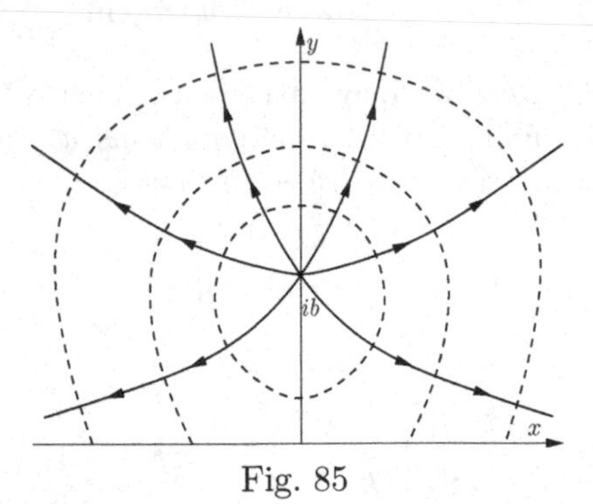

Fig. 85

Solution The problem is symmetric with respect to the imaginary axis – see Fig. 85. So, we may consider the domain $D_1 = \{z = x + iy : x > 0,\ y > 0\}$, and a source of strength $N/2$ at the same point ib. The function $w = g(z) = z^2$

[5]The substitution rule is justified exactly in the same way as for real-valued functions, because the chain rule and the Fundamental Theorem of Calculus are valid for analytic function; also one may prove it reducing a contour integral to the integral over an interval in the real axis.

maps D_1 onto the upper half-plane G in the w-plane. The image of ib is $-b^2$. We use symmetry once again (now with respect to the u-axis in the w-plane) and see that the problem in the entire w-plane with the source of strength N and the problem in the upper half-plane G (and in the lower half-plane) with the source of strength $N/2$ are the same. Hence, the complex potential in G is

$$f(w) = \frac{N}{2\pi}\log(w + b^2)$$

(see (8.31); we substitute $w - w_0 = w - (-b^2)$ instead of z); we may choose any branch of logarithm which is analytic in G. Thus, in D_1 we have

$$f_1(z) = f(g(z)) = \frac{N}{2\pi}\log(z^2 + b^2). \tag{8.33}$$

The same formula is valid in the entire half-plane D. Now we may find the vector field:

$$\overline{\mathbf{V}(z)} = f_1'(z) = \frac{N}{\pi}\cdot\frac{z}{z^2+b^2}; \quad \mathbf{V}(z) = \frac{N}{\pi}\cdot\frac{\overline{z}}{\overline{z}^2+b^2}.$$

In particular, we see that $\mathbf{V}(0) = 0$, so $z = 0$ is a stagnation point; while $\mathbf{V}(z) \to 0$ as $z \to 0$ and $z \to \infty$, and $\mathbf{V}(z) \to \infty$ as $z \to ib$.

We find the streamlines from the equalities

$$\operatorname{Im}\frac{N}{2\pi}\log(z^2+b^2) = \frac{N}{2\pi}\arg(z^2+b^2) = C.$$

Since C is an arbitrary constant, we may incorporate into it the factor $\frac{N}{2\pi}$. Taking into account that

$$z^2 + b^2 = (x+iy)^2 + b^2 = x^2 - y^2 + b^2 + i2xy,$$

we may write the equations for streamlines in the equivalent form

$$\tan C = \frac{\operatorname{Im}(z^2+b^2)}{\operatorname{Re}(z^2+b^2)} = \frac{2xy}{x^2-y^2+b^2}, \quad \text{or}$$

$$Kx^2 - Ky^2 - 2xy + Kb^2 = 0,$$

where $K = \tan C$. When $K = 0$, we have equations of lines $x = 0,\ y = 0$ (note that the line $y = 0$ is not in D); when $K \neq 0$, the streamlines are hyperbolas passing through the point ib—see Fig. 85.

The equipotentials are possibly even more remarkable curves. Their equations can be written in the following form:

$$\operatorname{Re}\log(z^2+b^2) = \ln|z^2+b^2| = C, \quad \text{or}$$

$$|z^2+b^2| = |z-ib|\cdot|z+ib| = c, \quad c > 0.$$

Therefore, equipotentials are the sets (or loci) of points such that the product of the distances to two fixed points, namely to $\pm ib$, is constant. Such curves are called Cassini ovals[6]; in Fig. 85 you see the upper half of them.

In fact any analogous problem with a single source (sink) at a point z_0 inside a simply connected unbounded domain D can be reduced to the problem in Example 8.9, if we find the conformal mapping $w = g(z)$ of D onto the upper half-plane such that $g(z_0) = ib$, $b > 0$, and $g(\infty) = \infty$. To obtain the complex potential, it's sufficient just to replace z by $g(z)$ in (8.33).

This approach works also for bounded domains D, but there must be a sink of the same strength at a boundary point ξ_0 of D, and such that $g(\xi_0) = \infty$.

Now we consider an example with a source at the boundary of a domain.

Example 8.10 A source of strength N is located at the point $h + id$ of the semi-infinite strip $D = \{z = x + iy : -h < x < h,\ 0 < y < \infty\}$—see Fig. 86. Find the equations for streamlines and the limit $\mathbf{V}(\infty)$ of the velocity of the flow as $z \to \infty$.

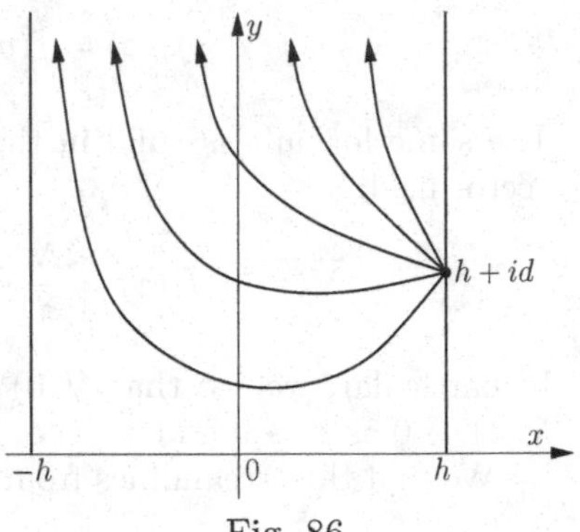

Fig. 86

Solution In the problem 2, Section 8.1, we found the conformal mapping of the upper half-plane onto the semi-infinite strip such that $f(\infty) = \infty$. In the context of the current problem, taking the half-strip D to be in the z-plane, and the half-plane G in the w-plane, the mapping is $z = \frac{2h}{\pi}\sin^{-1} w$. Therefore, the mapping of D onto G is given by the formula

$$w = g(z) = \sin\frac{\pi z}{2h}.$$

The image of the source $z_0 = h + id$ is

$$\begin{aligned} w_0 &= \sin\frac{\pi(h+id)}{2h} = \sin\left(\frac{\pi}{2} + \frac{i\pi d}{2h}\right) = \cos\frac{i\pi d}{2h} \\ &= \frac{1}{2}\left(e^{\pi d/(2h)} + e^{-\pi d/(2h)}\right) = \cosh\frac{\pi d}{2h} \end{aligned}$$

[6]Cassini ovals are named after the Italian astronomer Giovanni Domenico Cassini (1625–1712) who studied them in 1680. You may find more details in Wikipedia.

—see (4.33). The complex potential in the half-plane G is

$$f(w) = \frac{N}{\pi} \operatorname{Log}\left(w - \cosh\frac{\pi d}{2h}\right), \tag{8.34}$$

we may choose any branch of logarithm which is analytic in G. Here we use again (8.31), but write $\frac{N}{\pi}$ instead of $\frac{N}{2\pi}$ because whole fluid is coming into the upper half-plane; so, the equivalent source in the whole plane has double strength $2N$. Hence, the complex potential of the vector field in D is equal to

$$f_1(z) = f(g(z)) = \frac{N}{\pi} \operatorname{Log}\left(\sin\frac{\pi z}{2h} - \cosh\frac{\pi d}{2h}\right).$$

The streamlines have the implicit equations

$$\operatorname{Im} f_1(z) = \frac{N}{\pi} \operatorname{Arg}\left(\sin\frac{\pi z}{2h} - \cosh\frac{\pi d}{2h}\right) = C.$$

Let us find $\mathbf{V}(\infty)$. We have

$$\overline{\mathbf{V}(z)} = f_1'(z) = \frac{N}{\pi}\left(\sin\frac{\pi z}{2h} - \cosh\frac{\pi d}{2h}\right)^{-1} \frac{\pi}{2h} \cos\frac{\pi z}{2h}.$$

It's easy to see that, as $z \to \infty$ within D, we have $\operatorname{Im}\frac{\pi z}{2h} \to \infty$,

$$e^{-i\frac{\pi z}{2h}} \to \infty, \quad \frac{\sin\frac{\pi z}{2h}}{e^{-i\frac{\pi z}{2h}}} \to \frac{i}{2}, \quad \frac{\cos\frac{\pi z}{2h}}{e^{-i\frac{\pi z}{2h}}} \to \frac{1}{2}.$$

Now we divide $\cos\frac{\pi z}{2h}$ and $(\sin\frac{\pi z}{2h} - \cosh\frac{\pi d}{2h})$ by $e^{\frac{\pi z}{2h}}$ in the equality for $\overline{\mathbf{V}(z)}$ and pass to the limit:

$$\overline{\mathbf{V}(\infty)} = \lim_{z\to\infty \text{ within } D} \overline{\mathbf{V}(z)} = \frac{N}{2hi}, \quad \mathbf{V}(\infty) = \frac{Ni}{2h}.$$

In particular, it means that the velocity becomes more and more uniform, and streamlines are almost vertical as the distance from the source tends to infinity. Note that if we accept this intuitively obvious fact without proof, the limit can be found even without computation. Indeed, the source generates N units of fluid per a unit of time. Hence, the total flux through a cross-section of the strip perpendicular to streamlines is N. Since the width of the strip is $2h$, the magnitude of the velocity must be approximately $N/(2h)$.

8.3.3 Flow around obstacles

We now consider the flow around an object bounded by a piece-wise smooth closed Jordan curve Γ—see Fig. 87. In 3D the object is an infinite cylinder, and our plane is a cross section of the flow perpendicular to the cylinder. Let D be the unbounded domain in $\mathbb{C}$ exterior to Γ. Thus, ∞ is a point with a punctured neighborhood in D. Recall that a neighborhood of ∞ is a set $\{|z| > R\}$, that is the exterior of a disk.

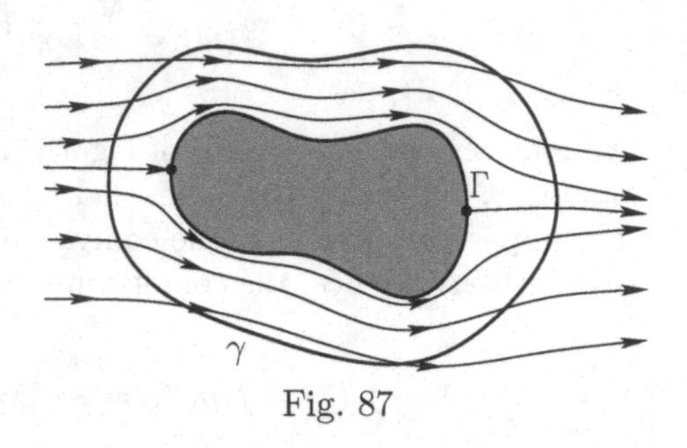

Fig. 87

We make the following assumptions regarding the vector field $\mathbf{V}$:

(a) $\mathbf{V}$ is potential and solenoidal in every simply connected subdomain of D.

(b) The limit of $\mathbf{V}(z)$ as $z \to \infty$ exists, that is the flow is almost uniform at a great distance from the object; this limit is denoted by $\mathbf{V}_\infty$.

(c) The curve Γ is a part of a streamline of the flow; in other words, the fluid flows around Γ.

(d) The circulation around any contour γ containing Γ in its interior, is a given value κ, that is

$$\int_\gamma X\,dx + Y\,dy = \operatorname{Re} \int_\gamma \overline{\mathbf{V}(z)}\,dz = \kappa. \tag{8.35}$$

The first equality follows from (5.3); and note that if Γ is outside γ, then, according to (a), the first integral in (8.35) equals zero. But this is not necessarily so if Γ is inside γ. In this case (a) implies only that the circulation is the same around any contour enclosing Γ—see problem 4.

The complex potential is defined in the same way as before. It is a single-valued analytic function in any simply connected subdomain of D, but now it is not necessarily single-valued in D. We prove that *under the assumptions (a)–(d), there is $R > 0$ such that the complex potential, up to an additive constant, can be written*

$$w = f(z) = \overline{\mathbf{V}_\infty} z + \frac{\kappa}{2\pi i} \log z - \frac{c_{-2}}{z} - \frac{c_{-3}}{2z^2} - \dots, \quad |z| > R, \tag{8.36}$$

where the coefficients $c_{-2}, c_{-3}, \dots$ are uniquely determined. Hence, the vector field satisfying (a)–(d), is unique. We give a proof of this statement because it is a nice example of the usefulness, for important applied problems, of the theory developed in the preceding sections. Also, the proof brings out more properties of $f(z)$.

Proof. From the equality $f'(z) = \overline{\mathbf{V}(z)}$ we see that $f'(z)$ is single-valued in D, and according to (b), there is a limit $\lim_{z\to\infty} f'(z) = \overline{\mathbf{V}_\infty}$. Hence $z_0 = \infty$ is a removable singularity of $f'(z)$, and by Theorem 7.11, the Laurent expansion of $f'(z)$ about infinity has the form

$$f'(z) = \overline{\mathbf{V}_\infty} + \frac{c_{-1}}{z} + \frac{c_{-2}}{z^2} + \frac{c_{-3}}{z^3} - \ldots \tag{8.37}$$

From this equality we get

$$\int_\gamma f'(z)\,dz = 2\pi i c_{-1}$$

for any contour γ enclosing Γ. On the other hand, from (8.29) and (8.35) we see that

$$\int_\gamma f'(z)\,dz = \int_\gamma \overline{\mathbf{V}(z)}\,dz = \kappa + iN,$$

where N is the flow through γ. Since sources and sinks are absent, $N = 0$ and therefore $c_{-1} = \kappa/(2\pi i)$. Then integrating both sides of (8.37), we obtain (8.36).

Let us prove that the coefficients $c_{-2}, c_{-3}, \ldots$ are uniquely determined. The function $f(z)$ maps streamlines onto horizontal lines. Since Γ is a part of a streamline (assumption (c)), the image of Γ is a horizontal interval, that is $\operatorname{Im} f(z)$ is constant when $z \in \Gamma$. Let f_1, f_2 be two complex potentials of vector fields satisfying (a)–(d), and let $F = f_1 - f_2$. From (8.36) we obtain the equality (up to an additive constant)

$$F(z) = -\frac{d_{-2}}{z} - \frac{d_{-3}}{2z^2} - \ldots, \qquad |z| > R,$$

where d_{-k} is the difference of the corresponding coefficients c_{-k} for f_1, f_2. Therefore, $F(z)$ is a single-valued analytic function in $|z| > R$, and hence in D. Thus, the function $v(z) = \operatorname{Im} F(z)$ is harmonic in D, and $v(z)$ is constant for $z \in \Gamma$. If the maximum or minimum value of $v(z)$ is attained at a point inside D, then $v(z)$ is constant in D by Corollary 6.30. Suppose that such a value is attained at ∞. Let $w = \frac{1}{z}$. The function $F_1(w) = F(\frac{1}{w}) = -d_{-2}w - \frac{d_{-3}}{2}w^2 - \ldots$ is analytic in $|w| < \frac{1}{R}$. Hence, the function $v_1(w) = \operatorname{Im} F_1(w) = v(\frac{1}{w})$ is harmonic and attains its maximum or minimum value at $w_0 = 0$. Again by Corollary 6.30, v is constant. Finally, if both these values are taken on Γ, then again $v(z) = C$ because its maximum and minimum values are the same. Thus, $v(z)$ is constant in D, and by the Cauchy-Riemann equations, $F(z) = f_1(z) - f_2(z) = C$. Hence, $f_1(z) = f_2(z) + C$, and the uniqueness (up to an additive constant) is proved. □

Now we consider the important example.

Example 8.11 Find the flow around a circular cylinder of radius R satisfying the assumptions (a)–(d)—see Fig. 88.

Solution Here D is the exterior of a disk of radius R, that is $D = \{|z| > R\}$, and Γ is the circle $|z| = R$. Denote for simplicity $|\mathbf{V}_\infty| = V_\infty$. We start with the case when $\mathbf{V}_\infty$ is a positive real number, so that $\mathbf{V}_\infty = |\mathbf{V}_\infty|$.

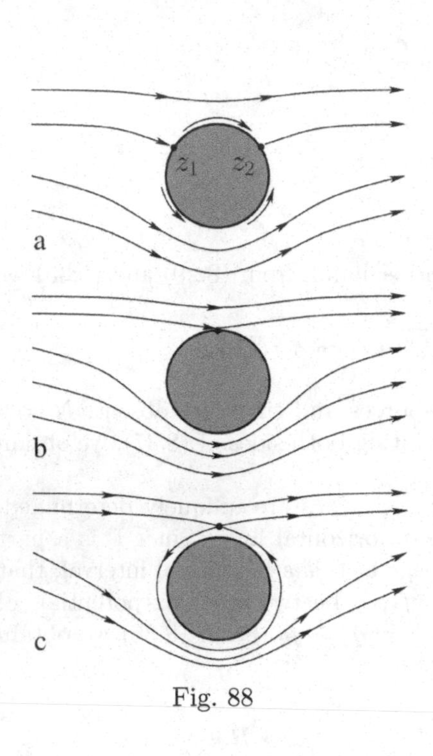

Fig. 88

We know that a complex potential $f(z)$ maps Γ onto a horizontal interval, and $f(\infty) = \infty$. A mapping with such properties has already been considered in Section 4.4.4—this is the Zhukovsky transform $w = \frac{1}{2}(z + \frac{1}{z})$, which maps conformally the exterior of the unit disk onto the exterior of the interval $[-1, 1]$. To map the exterior of a disk of radius R, we first map this disk onto the unit disk by the function $z_1 = z/R$, and then apply the Zhukovsky transform $w = \frac{1}{2}(z_1 + \frac{1}{z_1})$, obtaining $w = \frac{1}{2}(\frac{z}{R} + \frac{R}{z})$.

Note that the function $f(z) = k(\frac{z}{R} + \frac{R}{z})$ also maps the disk of radius R onto a horizontal interval for every real k, and therefore satisfies the condition (c). Since $f'(z) = \frac{k}{R} - \frac{kR}{z^2}$, we have $f'(\infty) = k/R$. So, the condition (c) will be fulfilled if $k/R = V_\infty$, that is $k = RV_\infty$, and

$$f(z) = V_\infty\left(z + \frac{R^2}{z}\right).$$

But (d) does not yet hold, because

$$\int_\gamma \overline{\mathbf{V}(z)}\, dz = \int_\gamma f'(z)\, dz = V_\infty \int_\gamma \left(1 - \frac{R^2}{z^2}\right) dz = 0$$

for every closed curve γ in D. To satisfy (d), we add the purely circulatory stream $\frac{\kappa}{2\pi i}\log z$ which also flows around the circle $|z| = R$:

$$f(z) = V_\infty z + \frac{V_\infty R^2}{z} + \frac{\kappa}{2\pi i}\log z. \tag{8.38}$$

Now all conditions (a)–(d) are satisfied. Indeed, just like the Zhukovsky transform, any branch of the function $\frac{\kappa}{2\pi i}\log z$ maps the circle $|z| = R$ onto a horizontal interval, namely onto

$$\frac{\kappa}{2\pi}(\arg z - i\ln R), \quad \theta_0 < \arg z \le \theta_0 + 2\pi,$$

and therefore (c) holds; (a),(b) are obvious, and (d) now holds since

$$\int_\gamma \overline{\mathbf{V}(z)}\,dz = \int_\gamma f'(z)\,dz = V_\infty \int_\gamma \left(1 - \frac{R^2}{z^2}\right) dz + \frac{\kappa}{2\pi i}\int_\gamma \frac{1}{z} = \kappa.$$

By the uniqueness (up to an additive constant) we established earlier, the function defined in (8.38) is the desired complex potential. Note that it has the form (8.36) with $c_{-2} = -V_\infty R^2$, $c_{-3} = c_{-4} = \cdots = 0$.

The general case $\mathbf{V}_\infty = V_\infty e^{i\alpha}$ can easily be reduced to the previous one by rotating the vector field by the angle $-\alpha$. First we apply the mapping $z_1 = ze^{-i\alpha}$, and then the transform (8.38) with z_1 instead of z. We get

$$\begin{aligned} f(z) &= V_\infty z e^{-i\alpha} + \frac{V_\infty R^2}{ze^{-i\alpha}} + \frac{\kappa}{2\pi i}\log(ze^{-i\alpha}) \\ &= \overline{\mathbf{V}_\infty} z + \frac{\mathbf{V}_\infty R^2}{z} + \frac{\kappa}{2\pi i}\log z - \frac{\kappa\alpha}{2\pi}, \quad \mathbf{V}_\infty = V_\infty e^{i\alpha}; \end{aligned} \tag{8.39}$$

the constant $-\frac{\kappa\alpha}{2\pi}$ may be dropped. For the vector field $\mathbf{V}(z)$ we have

$$\overline{\mathbf{V}(z)} = f'(z) = \overline{\mathbf{V}_\infty} - \frac{\mathbf{V}_\infty R^2}{z^2} + \frac{\kappa}{2\pi i z}.$$

Let us find the stagnation points of the flow. The equality $\mathbf{V}(z) = 0$ is equivalent to the quadratic equation

$$\overline{\mathbf{V}_\infty} z^2 + \frac{\kappa}{2\pi i} z - \mathbf{V}_\infty R^2 = 0 \tag{8.40}$$

with solutions

$$\begin{aligned} z_{1,2} &= \left(-\frac{\kappa}{2\pi i} \pm \sqrt{\left(\frac{\kappa}{2\pi i}\right)^2 + 4V_\infty^2 R^2}\right)\frac{1}{2\overline{\mathbf{V}_\infty}} \\ &= \frac{1}{4\pi\overline{\mathbf{V}_\infty}}\left(i\kappa \pm \sqrt{16\pi^2 V_\infty^2 R^2 - \kappa^2}\right). \end{aligned}$$

If $16\pi^2 V_\infty^2 R^2 - \kappa^2 > 0$, that is $|\kappa| < 4\pi V_\infty R$, then a straightforward calculation of the modulus shows that $|z_1| = |z_2| = R$. Therefore, both stagnation points are on the circle $|z| = R$—see Fig. 88 a; Fig. 88 illustrates the case when $\alpha = 0$ and $\kappa > 0$. At z_1 the streamline bifurcates in two; one part flows along the upper arc of the circle, and the other along the lower arc. At z_2 these two streamlines merge again. In the case $\kappa = 0$ when circulation is absent, z_1, z_2 are located at the endpoints of the diameter of Γ. When κ increases, the points z_1, z_2 approach each other, and when $\kappa = 4\pi V_\infty R$, these two points coincide—Fig. 88 b. For $\kappa > 4\pi V_\infty R$, there is the only one stagnation point with $|z_1| > R$—Fig. 88 c.

Suppose now that Γ is an arbitrary piece-wise smooth closed Jordan curve (a cross section of an infinite cylinder). To find the flow over Γ satisfying the conditions (a)–(d), we reduce the problem to the previous one. Let $z_1 = g(z)$ be the conformal mapping of the exterior D of Γ onto the exterior of the unit circle $|w| = 1$ such that

$$g(\infty) = \infty, \quad \operatorname{Arg} g'(\infty) = 0, \text{ where } g'(\infty) = \lim_{z\to\infty} g'(z) > 0. \tag{8.41}$$

Such a function $g(z)$ is unique. According to (8.39), for any complex number $\mathbb{A}$ the function

$$f(z_1) = \overline{\mathbb{A}} z_1 + \frac{\mathbb{A}}{z_1} + \frac{\kappa}{2\pi i} \log z_1$$

is a complex potential of a vector field around the circle $|z_1| = 1$ with $f'(\infty) = \overline{\mathbb{A}}$. Hence, the composition $f_1(z) = f(g(z))$ is a complex potential of a flow around Γ. To find $\mathbb{A}$, we use the condition (b) which is equivalent to the equality $f_1'(\infty) = \overline{\mathbf{V}_\infty}$. By the chain rule we have

$$\overline{\mathbf{V}_\infty} = f_1'(\infty) = f'(\infty) g'(\infty) = \overline{\mathbb{A}} g'(\infty), \quad \text{that is } \mathbb{A} = \frac{\mathbf{V}_\infty}{g'(\infty)} \tag{8.42}$$

(note that according to (8.41), $g'(\infty)$ is real). Therefore,

$$f_1(z) = f(g(z)) = \frac{\overline{\mathbf{V}_\infty}}{g'(\infty)} g(z) + \frac{\mathbf{V}_\infty}{g'(\infty) g(z)} + \frac{\kappa}{2\pi i} \log g(z). \tag{8.43}$$

We find the corresponding vector field $\mathbf{V}(z)$ in D from the equality

$$\overline{\mathbf{V}(z)} = f_1'(z) = \frac{\overline{\mathbf{V}_\infty}}{g'(\infty)} g'(z) - \frac{\mathbf{V}_\infty g'(z)}{g'(\infty) g^2(z)} + \frac{\kappa g'(z)}{2\pi i g(z)}. \tag{8.44}$$

It's easy to verify directly that $\mathbf{V}$ satisfies all conditions (a)–(d) (problem 5). Therefore, $f_1(z)$ is the desired complex potential.

Example 8.12 Find the complex potential and the stagnation points of the flow around the interval $[-h, h]$, if the velocity at infinity is $\mathbf{V}_\infty = V_\infty e^{i\alpha}$, $0 \le \alpha \le \pi$, and the circulation $\kappa = 0$— see Fig. 89.

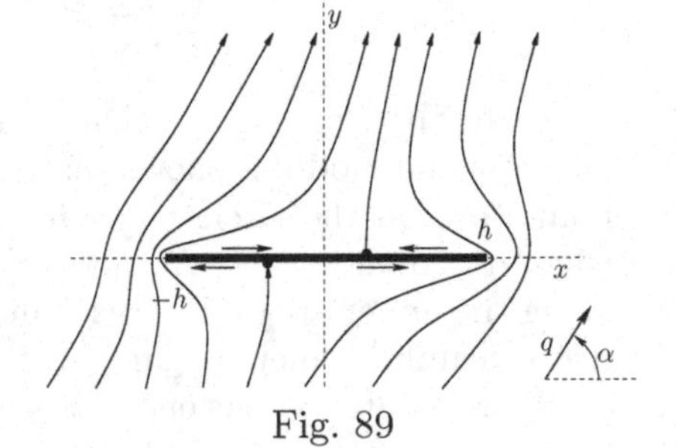

Fig. 89

Solution We know that the Zhukovsky function maps the exterior and the interior of the unit disk conformally onto the exterior of the interval $[-1, 1]$, and both these mappings are one-to-one. Let D be the exterior of the interval

$[-h, h]$. Thus, to find the conformal mapping $w = g(z)$ of D onto the exterior of the unit circle $|w| = 1$, we need two steps:

1. Map D onto the exterior of $[-1, 1]$ by the function $z_1 = z/h$.

2. Map the exterior of $[-1, 1]$ in the z_1-plane onto the exterior of the unit circle $|w| = 1$ by the corresponding branch of the function which is inverse to the Zhukovsky function $z_1 = \frac{1}{2}(w + \frac{1}{w})$.

The last equality is equivalent to the equality $w^2 - 2z_1 w + 1 = 0$. Solving this quadratic equation for w, we have $w = z_1 + \sqrt{z_1^2 - 1}$. So,

$$w = g(z) = \frac{1}{h}\left(z + \sqrt{z^2 - h^2}\right).$$

We define a branch of the square root function by taking

$$\arg(z^2 - h^2) = \operatorname{Arg}(z - h) + \operatorname{Arg}(z + h);$$

then

$$\sqrt{z^2 - h^2} = \exp\left(\frac{1}{2}\log(z^2 - h^2)\right)$$

is single-valued and continuous in D (why?[7]), and hence the function $g(z)$ is single-valued and continuous in D. Since $g(2h) > 1$, the function $g(z)$ maps D onto the exterior (but not onto the interior!) of the unit circle. Hence, our choice of the branch of the square root is correct.

Now we evaluate

$$g'(\infty) = \frac{1}{h}\lim_{z\to\infty}\left(1 + \frac{z}{\sqrt{z^2 - h^2}}\right) = \frac{2}{h}.$$

Therefore, $g(z)$ satisfies (8.41), and (8.43) yields the complex potential:

$$f_1(z) = \frac{\overline{\mathbf{V}_\infty}}{2}\left(z + \sqrt{z^2 - h^2}\right) + \frac{\mathbf{V}_\infty}{2}\cdot\frac{h^2}{z + \sqrt{z^2 - h^2}}, \quad \mathbf{V}_\infty = V_\infty e^{i\alpha}.$$

We can simplify this expression essentially using the identity

$$\left(z + \sqrt{z^2 - h^2}\right)\left(z - \sqrt{z^2 - h^2}\right) = h^2, \quad \text{that is}$$

$$\frac{h^2}{z + \sqrt{z^2 - h^2}} = z - \sqrt{z^2 - h^2}.$$

We have

$$\begin{aligned} f_1(z) &= \frac{V_\infty}{2}e^{-i\alpha}\left(z + \sqrt{z^2 - h^2}\right) + \frac{V_\infty}{2}e^{i\alpha}\left(z - \sqrt{z^2 - h^2}\right) \\ &= V_\infty z\,\frac{e^{i\alpha} + e^{-i\alpha}}{2} - iV_\infty\sqrt{z^2 - h^2}\,\frac{e^{i\alpha} - e^{-i\alpha}}{2i} \\ &= V_\infty\left(z\cos\alpha - i\sqrt{z^2 - h^2}\,\sin\alpha\right). \end{aligned}$$

[7] To prove this property show that for every real $x < -h$, the limit $\lim_{z\to x}\sqrt{z^2 - h^2} = -\sqrt{x^2 - h^2}$; it is the same as z approaches x from the upper or from the lower half-plane.

The vector field is defined by the equality

$$\overline{\mathbf{V}(z)} = f_1'(z) = V_\infty\left(\cos\alpha - \frac{iz}{\sqrt{z^2-h^2}}\sin\alpha\right).$$

To find the stagnation points of the flow we have to solve the equation $\mathbf{V}(z) = 0$, that is $\sqrt{z^2-h^2}\cos\alpha = iz\sin\alpha$. Squaring both parts of this equality, we have

$$(z^2-h^2)\cos^2\alpha = -z^2\sin^2\alpha, \quad z^2(\sin^2\alpha+\cos^2\alpha) = h^2\cos^2\alpha.$$

Therefore, $z^2 = h^2\cos^2\alpha$, and $z = \pm h\cos\alpha$. From Fig. 89 we see that the stagnation point on the upper side of the interval $[-h, h]$ is $h\cos\alpha$, and on the lower side $-h\cos\alpha$.

8.3.4 The Zhukovsky airfoils

Let γ be a circle with radius R centered at z_0, which intersects the x-axis at the point $a > 0$ and contains the point $-a$ in its interior—see Fig. 90. Let Γ be the image of γ under the mapping

$$w = \phi(z) = \frac{1}{2}\left(z + \frac{a^2}{z}\right). \tag{8.45}$$

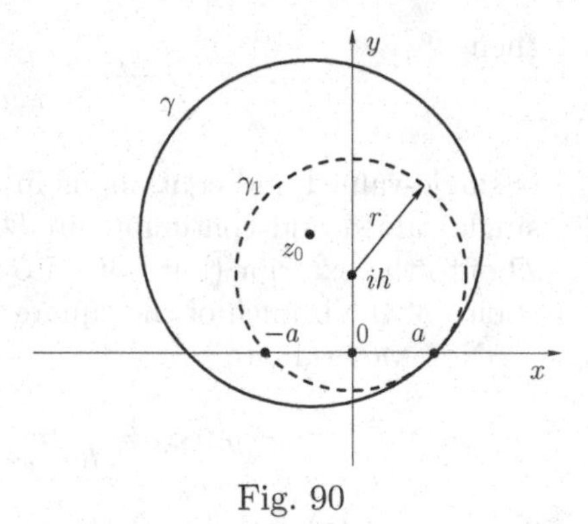

Fig. 90

Let us show that Γ has a cusp at the point $w = \phi(a) = a$. The unit tangent vector to Γ at $w = a$ is given by the limit

$$\lim_{z\to a}\frac{\phi(z)-a}{|\phi(z)-a|} = \lim_{z\to a}\frac{|z|}{z}\,\frac{z^2-2az+a^2}{|z^2-2az+a^2|} = \lim_{z\to a}\frac{(z-a)^2}{|z-a|^2}.$$

When z approaches a along γ from the opposite sides, the limits of the arguments of $(z-a)$ differ by π; hence, the limits of the arguments of $(z-a)^2$ differ by 2π. This means that the unit tangent vectors to the “upper” and “lower” sides of Γ at $w = a$ coincide. Therefore, Γ forms a sharp cusp at a—see Fig. 91.

The curve Γ is called the *Zhukovsky airfoil*. It resembles the cross-section of a wing, and Zhukovsky first investigated this class of curves to develop methods for calculating the wing of an aircraft. We will derive the formula for the lifting force in the subsequent sections.

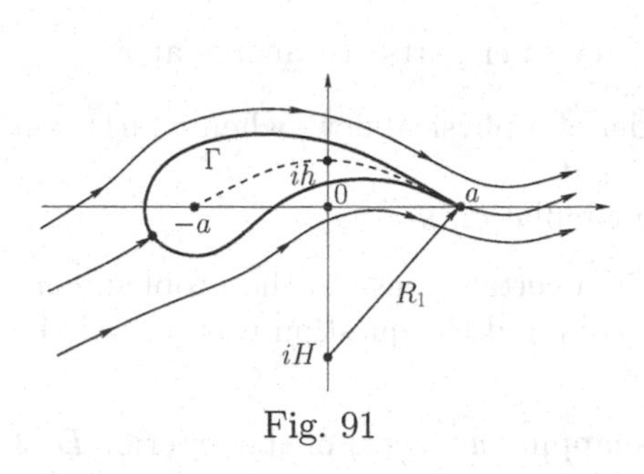

Fig. 91

To imagine the shape of Zhukovsky airfoils, it is useful to consider the auxiliary circle γ_1, which is centered at a point ih, $h > 0$, of the imaginary axis and is tangent to γ at a (drawn by the dotted line in Fig. 90). Let r be the radius of γ_1. The function $\phi(z)$ maps this circle onto the arc of the circle of radius $R_1 = \frac{r^2}{2h}$ centered at $i\frac{2h^2-r^2}{2h}$ (drawn by the dotted line in Fig. 91). The arc has endpoints at $\pm a$ and passes through ih — see problem 6. We denote this arc by A. It has the same unit tangent vector at a as Γ.

Therefore, the shape of a Zhukovsky airfoil depends on three real parameters: a characterizes width of a wing, h characterizes its curvature, and the distance $|z_0 - ih| = R - r$ between centers of the circles γ and γ_1 characterizes its thickness.

The important property of $\phi(z)$ is that ϕ *conformally maps the exterior of the circle* γ *onto the exterior of the airfoil* Γ. To this end we note, that the conformality of $\phi(z)$ at every point $z \neq 0$ and $z \neq \pm a$ is obvious[8] because $\phi(z)$ is differentiable and $\phi'(z) \neq 0$. The next step is to prove that $\phi(z)$ is one-to-one outside γ. In fact a stronger statement holds: in problems 7 and 8 we prove that ϕ maps the exterior of γ_1 (and also the interior of γ_1) bijectively onto the exterior of A.

Now we are ready to prove that $\phi(z)$ conformally maps the exterior of γ onto the exterior of Γ. We will start from the result of problem 8, that ϕ conformally maps the exterior of γ_1 onto the exterior of A; so here we need to show only that the domain and range of ϕ are as desired. Denote by E the exterior of γ, and by D the exterior of Γ. First we show that $\phi(z) \in D$ for every $z \in E$. Suppose that this statement is incorrect. Then there exists a point $z_1 \in E$ such that $\phi(z_1) \notin D$, that is $\phi(z_1)$ is inside Γ. Since $\phi(\infty) = \infty$, there is a point $z_2 \in E$ such that $w_2 = \phi(z_2) \in D$. Consider a curve γ_2 lying in E with endpoints z_1, z_2. Its image $\phi(\gamma_2)$ is a continuous curve with endpoints w_1, w_2 which are on opposite sides of Γ. Hence, γ_2 intersects Γ at some point w_3. The preimage z_3 of w_3 is a point in γ_2, and $z_3 \in \gamma$. Hence, γ_2 does not lie in E, and we come to a contradiction. Essentially the same arguments show that $\phi^{-1}(w) \in E$ for every $w \in D$ (just consider ϕ^{-1} instead of ϕ).

Example 8.13 Let Γ be a Zhukovsky airfoil with given z_0 and a, and let D be the exterior of Γ. Suppose that $\mathbf{V}$ is the vector field which is

1. potential and solenoidal in every simply connected subdomain of D,

2. $\lim_{z\to\infty} \mathbf{V}(z) = \mathbf{V}_\infty$, where $\mathbf{V}_\infty = V_\infty e^{i\alpha}$ is given,

[8] In fact ϕ is conformal at $z = 0$ and $z = \infty$ as well.

3. $\mathbf{V}$ flows around Γ (that is Γ is a streamline of the flow), and
4. the streamlines over the upper and the lower parts of Γ merge at a.

See Fig. 91; this is an approximate model of a physical flow, when whirls at a and other effects are not taken into account.

Find the complex potential and the circulation κ.

Notice that this problem is inverse in a certain sense to the problem considered in Example 8.12, where κ was given and the question was to find the stagnation points.

Solution Let us find the conformal mapping $w = g(z)$ of the exterior D of the airfoil Γ onto the exterior of the circle $|w| = R$. We assume that Γ is in the z-plane and the circle is in the w-plane, as in the formulas (8.43) and (8.44). It was proved earlier that the function $z = \frac{1}{2}(z_1 + \frac{a^2}{z_1})$ conformally maps the exterior of the circle $\gamma = \{|z_1 - z_0| = R\}$ in the z_1-plane onto D.

1. First we map D onto the exterior of γ by the branch of the function which is inverse to the function $z = \frac{1}{2}(z_1 + \frac{a^2}{z_1})$ and which conformally maps D onto the exterior of γ. To find z_1 from the last equality, we solve the quadratic equation $z_1^2 - 2z_1 z + a^2 = 0$ for z_1 and get

$$z_1 = z + \sqrt{z^2 - a^2}.$$

We select the branch of the square root as in the previous example. Namely, we define

$$\arg(z^2 - a^2) = \operatorname{Arg}(z - a) + \operatorname{Arg}(z + a)$$

and

$$\sqrt{z^2 - a^2} = \exp\left(\frac{1}{2}\log(z^2 - a^2)\right).$$

We may plug a test point $z = 2a$ and verify that $z_1 = a(2 + \sqrt{3})$ is outside γ indeed.

2. The next step is to map γ onto the circle $|w| = R$. It can be done easily by the translation $w = z_1 - z_0$. Therefore,

$$w = g(z) = z - z_0 + \sqrt{z^2 - a^2}.$$

It's not difficult to see that $g'(\infty) = 2$. To find κ, we again reduce the problem to the case of the flow around the circle $|w| = R$. The function $w = g(z)$ carries the streamlines around Γ onto the streamlines of the vector field around the circle whose velocity at infinity equals $\mathbf{V}_\infty / g'(\infty) = \mathbf{V}_\infty/2$—see (8.42). The point a on Γ goes to the point $g(a) = a - z_0$ on the circle $|w| = R$. Therefore, we have to find κ under condition that the streamlines over the circle merge at the point $a - z_0$. This is a stagnation point on the circle. Hence we may use the equality (8.40) with $\mathbf{V}_\infty/2$ instead of $\mathbf{V}_\infty$ and $z = a - z_0$:

$$\frac{1}{2}\left(\overline{\mathbf{V}_\infty}(a-z_0)^2 + \frac{\kappa}{\pi i}(a-z_0) - \mathbf{V}_\infty R^2\right) = 0,$$

$$\kappa = \pi i\left(\frac{\mathbf{V}_\infty R^2}{a-z_0} - \overline{\mathbf{V}_\infty}(a-z_0)\right).$$

We may simplify this expression using the elementary formulas

$$R^2 = |a-z_0|^2 = (a-z_0)\overline{(a-z_0)},$$

$$\overline{\mathbf{V}_\infty}(a-z_0) = \overline{\mathbf{V}_\infty\overline{(a-z_0)}}, \text{ and } z - \overline{z} = 2i\operatorname{Im} z$$

for every $z \in \mathbb{C}$. We get

$$\kappa = \pi i\left(\mathbf{V}_\infty\overline{(a-z_0)} - \overline{\mathbf{V}_\infty\overline{(a-z_0)}}\right) = -2\pi\operatorname{Im}\left(\mathbf{V}_\infty\overline{(a-z_0)}\right).$$

We can also write κ in terms of the angle α, the angle of the predominant flow, and $\theta = \operatorname{Arg}(a - z_0)$:

$$\kappa = -2\pi V_\infty R\sin(\alpha - \theta), \quad a - z_0 = Re^{i\theta}. \tag{8.46}$$

Now we may find the complex potential from the equality (8.43) with $\frac{1}{R}g(z)$ instead of $g(z)$ to have a mapping of D onto the unit circle.

In conclusion we remark that we cannot find κ directly from (8.44), by in plugging $z = a$ and setting the right hand side to zero, because $g'(a) = \infty$. Moreover, one can prove in general that in (8.44) the limit of $\overline{\mathbf{V}(z)}$ as $z \to a$ is *never* 0.

8.3.5 Lifting force

In this section we find the force exerted by a vector field on a unit length of a cross section of an infinite cylinder. The sum of these forces over the whole cross section is called a *lifting force*, although the force is not necessarily vertical. As before, we assume that the vector field $\mathbf{V} = \mathbf{V}(z)$ satisfies the assumptions (a)–(d) in Section 8.3.3.

Let $P = P(z)$ be a pressure at a point z of the flow $\mathbf{V}(z)$, and let ρ be the (uniform) density of the fluid. Bernoulli's Law[9] states that

$$P(z) = C - \frac{1}{2}\rho|\mathbf{V}(z)|^2, \tag{8.47}$$

where C is some constant (we do not take the gravitational force into account).[10]

[9]Daniel Bernoulli (1700–1782) was a Swiss mathematician and physicist.

[10]For a derivation of Bernoulli's Law, see for example the book by S. D. Fisher [4], p. 251.

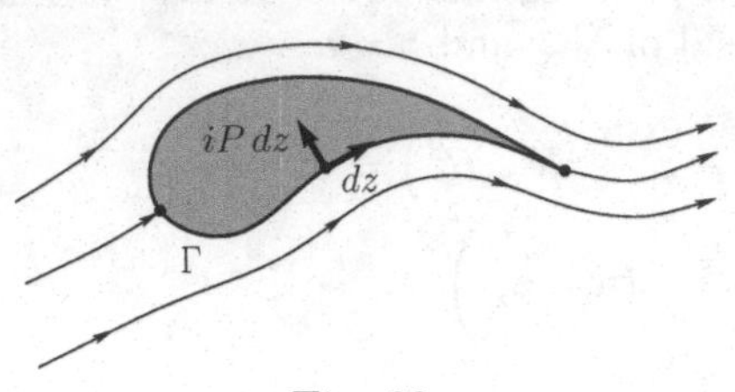

Fig. 92

We consider a cross section Γ perpendicular to the cylinder. We assume that Γ is piece-wise smooth, and we consider an element dz of Γ at a point z where Γ is smooth. Since the pressure acting on dz is normal to Γ and directed toward its interior (Fig. 92), the force acting on dz is

$$iP(z)\,dz = iC\,dz - i\frac{1}{2}\rho|\mathbf{V}(z)|^2\,dz.$$

Hence, the total force $\mathbf{F}$ acting on Γ is given by the integral

$$\mathbf{F} = \int_\Gamma iP(z)\,dz = iC\int_\Gamma dz - i\frac{1}{2}\rho\int_\Gamma |\mathbf{V}(z)|^2\,dz.$$

Since $\int_\Gamma dz = 0$ (why?), we get

$$\mathbf{F} = -i\frac{1}{2}\rho\int_\Gamma |\mathbf{V}(z)|^2\,dz. \tag{8.48}$$

It is convenient to express $\mathbf{F}$ via the complex potential $f(z)$. Let $\mathbf{V}(z) = |\mathbf{V}(z)|e^{i\theta}$. Since Γ is a streamline, the vector $\mathbf{V}(z)$ is tangent to Γ. Hence, $\theta = \arg\mathbf{V}(z) = \arg(dz)$, and $\mathbf{V}(z) = \overline{f'(z)}$, $z \in \Gamma$. Therefore,

$$|\mathbf{V}(z)| = \mathbf{V}(z)e^{-i\theta} = \overline{f'(z)}e^{-i\theta},\quad dz = |dz|e^{i\theta},\quad \overline{dz} = |dz|e^{-i\theta}.$$

From (8.48) we have

$$\begin{aligned}\mathbf{F} &= -i\frac{1}{2}\rho\int_\Gamma (\overline{f'(z)})^2 e^{-i2\theta}e^{i\theta}\,|dz| \\ &= -i\frac{1}{2}\rho\int_\Gamma (\overline{f'(z)})^2\,\overline{dz} = \overline{i\frac{1}{2}\rho\int_\Gamma (f'(z))^2\,dz},\end{aligned}$$

and we obtain the formula[11]

$$\overline{\mathbf{F}} = i\frac{1}{2}\rho\int_\Gamma (f'(z))^2\,dz. \tag{8.49}$$

Instead of Γ we may integrate over any piece-wise smooth curve surrounding Γ, since $(f'(z))^2$ is analytic outside Γ.

Now it's easy to obtain a formula for $\mathbf{F}$ in terms of the circulation κ. Let $|z| = r$ be a circle surrounding Γ. Applying (8.36) and (8.49) we get

$$\begin{aligned}\overline{\mathbf{F}} &= i\frac{1}{2}\rho\int_\Gamma \left(\overline{\mathbf{V}_\infty} + \frac{\kappa}{2\pi i z} + \frac{c_{-2}}{z^2} + \frac{c_{-3}}{z^3} + \dots\right)^2 dz \\ &= i\frac{1}{2}\rho\frac{2\overline{\mathbf{V}_\infty}\kappa}{2\pi i}\int_\Gamma \frac{dz}{z} = i\frac{1}{2}\rho\frac{2\overline{\mathbf{V}_\infty}\kappa}{2\pi i}2\pi i = i\rho\kappa\overline{\mathbf{V}_\infty},\end{aligned}$$

[11] This formula was obtained by Russian mechanic and mathematician S. A. Chaplygin in 1910, and by German physicist P. R. Blasius in 1911.

because the integrals of other terms are zero. Finally,

$$\mathbf{F} = -i\rho\kappa\mathbf{V}_\infty. \tag{8.50}$$

This is the celebrated *Kutta-Zhukovsky theorem.*[12] According to this theorem, the magnitude of the force exerted on Γ equals $\rho\kappa|\mathbf{V}_\infty|$; the force is perpendicular to $\mathbf{V}_\infty$, and its direction is obtained by rotating the vector $\mathbf{V}_\infty$ through 90° counterclockwise if $\kappa < 0$, and clockwise if $\kappa > 0$. In other words, $\mathbf{V}_\infty$ is rotating in the direction opposite to the circulation around Γ. For example, if $\alpha = 0$, that is the wind is horizontal and directed to the right, and $\kappa < 0$, then $\mathbf{F}$ will be directed upward.

Example 8.14 Let Γ be the Zhukovsky airfoil considered in Example 8.13; we use the notation from that Section 8.3.4. We have proved that in this case

$$\kappa = -2\pi V_\infty R \sin(\alpha - \theta)$$

(equation (8.46)). Hence,

$$\mathbf{F} = i2\rho\pi V_\infty R \sin(\alpha - \theta)\mathbf{V}_\infty, \quad \mathbf{V}_\infty = V_\infty e^{i\alpha}.$$

So, the magnitude of the lifting force is $2\rho\pi V_\infty^2 R \sin(\alpha - \theta)$. If the wind is horizontal and directed to the right, then the lifting force is directed upward.

Problems

1. Prove the formulas (8.32) for the vector field with vortex of strength κ at the origin.

2. Suppose that the origin is a vortex point of strength $\kappa \neq 0$ and a source of strength $N \neq 0$ at the same time.

(i) Derive equations of the equipotentials and streamlines in polar coordinates.

(ii) Determine how the signs of κ and N affect the streamlines, that is when the streamlines spiral into the origin clockwise, and when they spiral counterclockwise.

3. Prove that under a conformal transformation, a vortex point corresponds to a vortex point of the same strength.

4. Let D be the domain exterior to a closed Jordan curve Γ, and let a vector field $\mathbf{V}$ be potential and solenoidal in every simply connected subdomain of D. Prove that the circulation around any contour enclosing Γ is the same.

[12] M W Kutta (1867–1944) was a German mathematician.

5. Suppose that Γ is an arbitrary piecewise smooth closed Jordan curve and $w = g(z)$ is the conformal mapping of the exterior of Γ onto the exterior of the unit circle $|w| = 1$ satisfying (8.41). Verify that the vector field defined by the equality (8.44), satisfies all conditions (a)–(d).

6. Prove that the function $\phi(z) = \frac{1}{2}(z + \frac{a^2}{z})$ maps a circle γ_1 of radius r, centered at a point ih, $h > 0$, and passing through $\pm a$, onto the arc of the circle of radius $R_1 = \frac{r^2}{2h}$ centered at $iH = i\frac{2h^2 - r^2}{2h}$. The arc has endpoints at $\pm a$ and lies in the upper half-plane.

7. Let G_e be the exterior and G_i be the interior of γ_1; we also use the notation of Section 8.3.4. Prove that if $\phi(z_1) = \phi(z_2)$, $z_1 \neq z_2$, then $z_1 \in G_e$ if and only if $z_2 \in G_i$.

8. In the notation of Section 8.3.4 and problem 7, prove that $\phi(z)$ maps the sets G_e, G_i bijectively, and hence conformally, onto the exterior of the arc A.

9. Consider again the air foil from Example 8.13. Suppose that V_∞ is fixed, and that the wind is horizontal and toward the right. Show that the lifting force depends only on $\operatorname{Im} z_0$.

10. A source of strength N is located at the point h, $h > 0$ on the side of the angle $D = \{z = x + iy : x \geq 0,\ y \geq 0\}$. Find the equations for streamlines and the limit $\mathbf{V}(\infty)$ of the velocity of the flow as $z \to \infty$.

11. A source of strength N is located at the point h, $h > 0$ on the side of the angle $D = \{z = x + iy : x \geq 0,\ y \geq 0\}$, and a sink of the same strength N is located at the point ih. Find the equations for streamlines and the limit $\mathbf{V}(\infty)$ of the velocity of the flow as $z \to \infty$.

8.4 Other Interpretations of Vector Fields

Fundamentally, the possibility of applying complex analysis to the problems of hydrodynamics arises because the vector field of velocities is potential and solenoidal outside obstacles, sources, and sinks. In this section we consider two more very important vector fields with such properties, namely the electric field and heat flow.

8.4.1 Electrostatics

An *electric field* $\mathbf{E}(z)$ is the force experienced by a unit positive charge at a point z in a given region. We assume that the electric field is independent of time (that is steady, or static), and is generated by charged

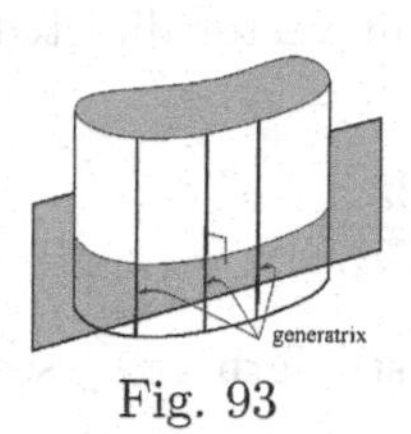

Fig. 93

infinitely long cylindrical surfaces or wires which are ideal conductors. The charge they carry is uniform along any *generatrix* of the cylinder, the line whose movement forms its surface (Fig. 93). So, the vectors $\mathbf{E}(z)$ are orthogonal to a generatrix, and the electric field is the same in any plane perpendicular to the generatrix. Thus, again we have a plane vector field, making it possible to use the methods of complex analysis.

We will make use of *Coulomb's Law* according to which two particles of charges q_1 and q_2, exert a force on one another whose magnitude is

$$|\mathbf{F}| = C\frac{q_1 q_2}{s^2},$$

where s is the distance between the particles. The force is directed along the line through the two points and is attractive if q_1 and q_2 have opposite signs and repulsive if the signs of q_1 and q_2 are the same. The constant C depends on the choice of units; we shall take $C = 1$.

Example 8.15 Find the electric field generated by an infinite wire perpendicular to the complex plane, passing through a point z_0 and carrying a uniform charge of λ coulombs per unit length—see Fig. 94.

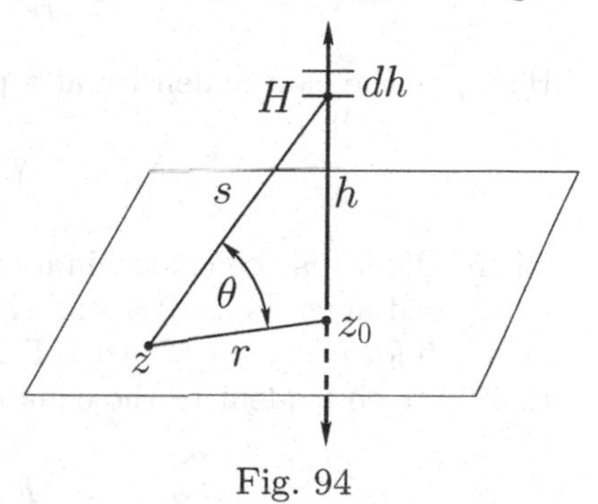

Fig. 94

Solution Suppose that a unit charge is at the point $z \neq z_0$. Fix a point H on the wire and consider an element dh of the wire (Fig. 94). This element carries a charge $\lambda\, dh$, and according to Coulomb's Law, the magnitude of the force between this element and the unit charge is

$$|d\mathbf{E}| = \frac{\lambda\, dh}{s^2} = \frac{\lambda\, dh}{r^2 + h^2}.$$

Since the force $d\mathbf{E}$ is directed along the line through z and H, the magnitude of its projection onto the z-plane equals

$$|d\mathbf{E}|\cos\theta = \frac{\lambda\, dh}{s^2}\cdot\frac{r}{s} = \frac{\lambda r\, dh}{(r^2 + h^2)^{3/2}}.$$

Due to the symmetry about the plane, the total force $\mathbf{E}$ has zero vertical component. Hence, $|\mathbf{E}|$ is equal to the sum of the projections of $|d\mathbf{E}|$. To evaluate the integral of the last expression, we use the substitution $h = r\tan\theta$, so that $dh = r\sec^2\theta\, d\theta$, and $r^2 + h^2 = r^2\sec^2\theta$. We have

$$|\mathbf{E}| = \int_{-\infty}^{\infty} \frac{\lambda r\, dh}{(r^2 + h^2)^{3/2}} = \lambda\int_{-\pi/2}^{\pi/2} \frac{r^2\sec^2\theta\, d\theta}{r^3\sec^3\theta} = \frac{\lambda}{r}\int_{-\pi/2}^{\pi/2}\cos\theta\, d\theta = \frac{2\lambda}{r}.$$

Now we take into account the direction of $\mathbf{E}$. Since the unit vector directed from z_0 to z is $(z - z_0)/r$, we have

$$\mathbf{E} = \frac{2\lambda}{r} \cdot \frac{\overline{z - z_0}}{r} = 2\lambda \cdot \frac{\overline{z - z_0}}{(z - z_0)\overline{(z - z_0)}} = \frac{2\lambda}{\overline{z - z_0}}.$$

Note that $\mathbf{E}$ coincides with the vector field for a source of strength $4\pi\lambda$—see (8.30).

The general static two-dimensional electric field has many properties which are similar to those of the fluid flow field. Let $\mathbf{E}(x, y) = E_x(x, y) + iE_y(x, y)$. In this case the general Maxwell equations have the form

$$\frac{\partial E_y}{\partial x} - \frac{\partial E_x}{\partial y} = 0, \tag{8.51}$$

$$\frac{\partial E_x}{\partial x} + \frac{\partial E_y}{\partial y} = 4\pi\rho. \tag{8.52}$$

Here ρ is the charge density at a point $z = (x, y)$, that is

$$\rho(z) = \lim_{r \to 0} \frac{Q(z, r)}{\pi r^2},$$

where $Q(z, r)$ is the charge in a cylinder of unit high with the base of radius r centered at z. Thus, if a simply-connected domain D is charge-free, that is $\rho(z) = 0$ for every $z \in D$, then $\mathbf{E}$ is potential and solenoidal in D, and (8.51), (8.52) are equivalent to the equalities

$$\int_\Gamma E_x\,dx + E_y\,dy = 0, \tag{8.53}$$

$$\int_\Gamma -E_y\,dx + E_x\,dy = 0, \tag{8.54}$$

respectively; here Γ is any closed contour in D. It's interesting to note that (8.53) holds even for closed contours in multiply-connected domains, that is the circulation along any closed contour is zero (in contrast with the vector fields $\mathbf{V}$ in hydrodynamics). The following physical arguments explain this property. For any (not necessarily closed) curve Γ the integral $\int_\Gamma E_x\,dx + E_y\,dy$ is equal to the work of the force $\mathbf{E}$ to carry the charge along Γ. Note that no additional energy required to maintain the steady electric field. If we assume that (8.53) does not hold, then traversing the closed curve Γ more and more times, we get an infinite source of energy, that is perpetual motion!

The equalities (8.53), (8.54) imply the existence of real functions u, v such that

$$E_x = -\frac{\partial v}{\partial x}, \quad E_y = -\frac{\partial v}{\partial y},$$

$$E_y = -\frac{\partial u}{\partial x}, \quad E_x = \frac{\partial u}{\partial y}.$$

(we introduce the functions $u\,,v$ in the way accepted in electrostatics, which is slightly different from those in Section 8.2; some explanation for the difference is given below). Since (8.53) holds even in multiply-connected domains, the potential $v(x,y)$ of $\mathbf{E}$ is single-valued in any domain (in contrast with the vector field $\mathbf{V}$); the function u is single-valued in simply-connected domains, but might be multi-valued in multiply-connected ones. We see that

$$\frac{\partial u}{\partial x}=\frac{\partial v}{\partial y},\quad \frac{\partial u}{\partial y}=-\frac{\partial v}{\partial x}.$$

Hence, $u\,,v$ satisfy the Cauchy-Riemann equations, and therefore the function

$$f(z)=u(x,y)+iv(x,y)$$

is analytic in D. It is called the *complex potential* of $\mathbf{E}$. We get

$$f'(z)=\frac{\partial u}{\partial x}+i\frac{\partial v}{\partial x}=-E_y-iE_x=-i(E_x-iE_y)=-i\overline{\mathbf{E}(z)}.$$

Therefore,

$$\mathbf{E}(z)=-i\overline{f'(z)}. \tag{8.55}$$

The vector $(E_x,E_y)=(\frac{\partial u}{\partial y},-\frac{\partial u}{\partial x})$ is perpendicular to $\nabla u=(\frac{\partial u}{\partial x},\frac{\partial u}{\partial y})$, and ∇u at a point (x,y) is the normal vector to the level curve $u(x,y)=C$ through (x,y). Hence, the vector $\mathbf{E}=(E_x,E_y)$ is a tangent vector to this curve. This is why the curves $u(x,y)=C$ are called *lines of force.* The orthogonal curves $v(x,y)=C$ are called *equipotentials.*

The equipotentials in electrostatics are analogous to the flow lines in fluid flow, in that both correspond to the level curves $v(x,y)=C$. A more essential parallel is in their behavior around boundaries of the domain. Let a charged cylindrical conducting surface intersects the complex plane along the curve Γ. Then Γ must be a part of an equipotential. Indeed, if the potential $v(x,y)$ is not constant on Γ, the field produces a movement of charges along Γ, and hence the field is not static. Thus, in electrostatics the equipotentials include the boundary of the conductor, analogous to the way the boundary of a domain is a flow line in hydrodynamics.

The problem of finding complex potentials is essentially the same as those that we considered earlier. For example, instead of a flow in a curvilinear strip bounded by curves Γ_1, Γ_2, considered in Section 8.2.2 **(b)**, now we have a condenser with plates Γ_1, Γ_2 (more precisely, the condenser with cylindrical plates intersecting the complex plane along the curves Γ_1, Γ_2). Then we apply conformal mappings exactly as before. To determine the width H of the strip on which we map D, again we need the value $v(z_2)-v(z_1)$, where $z_1\in\Gamma_1$, $z_2\in\Gamma_2$. In the flow problem this difference is the total flux; now it is a difference of potentials on Γ_1, Γ_2. For example, if we interpret the channel in Example 8.7 as a condenser—see Fig. 80, then the solid lines will be equipotentials, and the dotted lines will be the lines of force.

Let us consider one more example.

Example 8.16 Find the electric field near the edges of a semi-infinite parallel-plate condenser, if the distance between plates is $2h$, and potentials at the lower and upper plates are $-V$ and V, respectively—see Fig. 95.

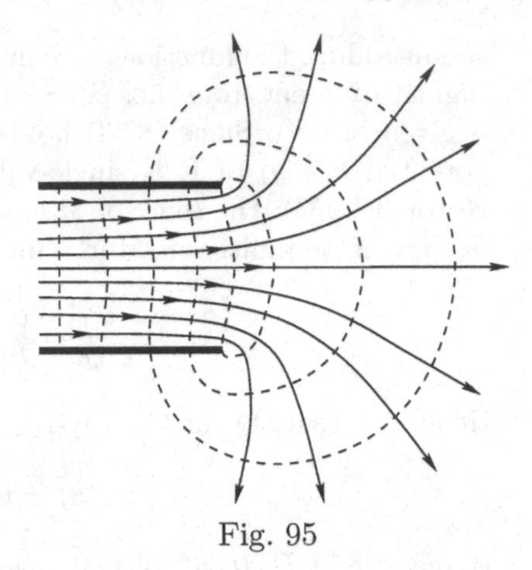

Fig. 95

Solution In this problem D is the complement of two rays—the plates of the condenser. The complex potential $w = f(z)$ maps D onto the strip $-V < \operatorname{Im} w < V$ in such a way that the lower and upper plates of the condenser go to the lower and upper boundaries of the strip, respectively. As in Section 8.2.2, we find the inverse mapping $z = f^{-1}(w)$ of the strip onto D. Even a more general case has been already considered in Example 8.5, and the following result was obtained: the function

$$z = \frac{h}{\pi} \int_{i\pi}^{w_1} (e^{\xi} + 1)^{\alpha} \, d\xi + ih \tag{8.56}$$

maps the strip $-\pi < \operatorname{Im} w_1 < \pi$ in the w_1-plane onto the polygonal domain in the z-plane, shown in Fig. 71; here we use a different notation for the variables. The lower and the upper boundaries of the strip go to lower and upper boundaries of the polygonal domain, respectively. Our domain D is the special case when $\alpha = 1$. To apply (8.56), we map our strip $-V < \operatorname{Im} w < V$ onto the strip $-\pi < \operatorname{Im} w_1 < \pi$ by the function $w_1 = \frac{\pi}{V} w$, and plug in $\alpha = 1$:

$$z = \frac{h}{\pi} \int_{i\pi}^{\pi w/V} (e^{\xi} + 1) \, d\xi + ih = \frac{h}{\pi} (e^{\xi} + \xi) \Big|_{i\pi}^{\pi w/V} + ih.$$

Hence,

$$z = f^{-1}(w) = \frac{h}{\pi} \left(e^{\pi w/V} + \frac{\pi w}{V} + 1 \right).$$

Taking real and imaginary parts of both sides of this equality, we get two equations

$$x = \frac{h}{\pi} \left(e^{\pi u/V} \cos \frac{\pi v}{V} + \frac{\pi u}{V} + 1 \right),$$

$$y = \frac{h}{\pi} \left(e^{\pi u/V} \sin \frac{\pi v}{V} + \frac{\pi v}{V} \right),$$

$$z = x + iy, \quad w = u + iv.$$

If we fix v by setting $v = C$, then we obtain the parametric equations with parameter u for equipotentials (solid lines in Fig. 32); fixing u, we obtain the parametric equations with parameter v for force lines (dotted lines).

To find the electric field $\mathbf{E}$, we use (8.55). It's convenient to write $\mathbf{E}$ via w:

$$\mathbf{E} = -i\overline{\left(\frac{dw}{dz}\right)} = -i\frac{1}{\overline{\left(\frac{dz}{dw}\right)}} = -i\frac{1}{\frac{h}{\pi}\frac{\pi}{V}\left(\overline{e^{\pi w/V}}+1\right)} = -i\frac{V}{h}\frac{1}{e^{\pi\overline{w}/V}+1}.$$

If w tends to $-\infty$ inside the strip $-V < \operatorname{Im} w < V$ (that is $u \to -\infty$), then $e^{\pi w/V}$ and $e^{\pi\overline{w}/V}$ tend to 0. Hence z tends to $-\infty$ inside the condenser and $\mathbf{E}$ tends to $-iV/h$ (compare with the answer to problem 1). If $w \to \pm iV$, then $e^{\pi w/V}$ and $e^{\pi\overline{w}/V}$ tend to -1. In this case z approaches the ends $\pm ih$ of the condenser and $\mathbf{E}$ becomes infinitely large.

8.4.2 Heat flow

We briefly consider one more interpretation of vector fields and related analytic functions. Namely, we consider a steady (that is independent of time) distribution of temperature in a plane domain D. Let $u = u(x, y)$ be a temperature at a point $z = (x, y)$. It is well known that u satisfies the Laplace[13] equation

$$\frac{\partial^2 u}{\partial x^2} + \frac{\partial^2 u}{\partial y^2} = 0.$$

Hence, u is harmonic in D. Let v be harmonically conjugate to u. If D is simply-connected, then v is single-valued. In multiply connected domains v might be mulivalued. The analytic function

$$f(z) = u(x, y) + iv(x, y)$$

is called *the complex potential.* Let us clarify the physical meaning of v. Let k be the coefficient of thermal conductivity of the material. The vector

$$\mathbf{Q} = -k\nabla u$$

is called *the vector of heat flow*; the minus sign appears because the heat flows from high temperatures to the lower ones. By a general property of gradients, $\mathbf{Q}(x_0, y_0)$ is orthogonal to the isotherm $u(x, y) = C$ passing through some point (x_0, y_0). Since the curves $u(x, y) = C$ and $v(x, y) = C$ are orthogonal, $\mathbf{Q}(x_0, y_0)$ is a tangent vector to the curve $v(x, y) = C$ through the point (x_0, y_0). Therefore, $v(x, y) = C$ are *heat streamlines.* If $\mathbf{Q}$ and dz are orthogonal at some point, then $|\mathbf{Q}|\,|dz|$ is the amount of heat flowing through dz per unit of time. If the angle between dz and $\mathbf{Q}$ is θ, this amount is

$$|\mathbf{Q}|\,|dz|\sin\theta = k\frac{\partial u}{\partial y}\,dx - k\frac{\partial u}{\partial x}\,dy = -k\left(\frac{\partial v}{\partial x}\,dx + \frac{\partial v}{\partial y}\,dy\right);$$

[13] Pierre-Simon, marquis de Laplace (1749–1827) was a great French mathematician.

in the last equality we used the Cauchy-Riemann equations. Thus, for a curve γ in D with endpoints z_1, z_2, the amount of heat flowing through γ per unit time is

$$\begin{aligned} -k\int_\gamma \frac{\partial u}{\partial x}\,dy - \frac{\partial u}{\partial y}\,dx &= -k\int_\gamma \frac{\partial v}{\partial x}\,dx + \frac{\partial v}{\partial y}\,dy \\ &= -k\int_\gamma dv = k(v(z_1) - v(z_2)), \end{aligned} \tag{8.57}$$

which up to the multiplicative constant $-k$ coincides with the first equality in (8.28). Moreover,

$$\mathbf{Q} = -k\left(\frac{\partial u}{\partial x} + i\frac{\partial u}{\partial y}\right) = -k\left(\frac{\partial u}{\partial x} - i\frac{\partial v}{\partial x}\right) = -k\overline{f'(z)}.$$

So, there is a direct analogy between the fluid flow and the heat flow.

Example 8.17 We may interpret the source-sink system in Example 8.8 as a heat source and heat sink of strength N, located at the points 1 and -1 respectively, and we consider the vector field $\mathbf{Q}$ instead of $\mathbf{V}$. The complex potential in this case is

$$f(z) = \frac{N}{2\pi k}\log\frac{z+1}{z-1}$$

—see problem 3. So the temperature $u(z) = \operatorname{Re} f(z)$ increases to ∞ as z approaches 1, and $u(z) \to -\infty$ as $z \to -1$. The solid lines in Fig. 84 indicate the heat streamlines, and the dotted lines are isotherms.

We may consider analogs of other problems on fluid flow in the similar way. The analog of flux is the heat flow defined in (8.57).

8.4.3 Remarks on boundary value problems

The classical topic in the theory of partial differential equations is the following *Dirichlet*[14] *problem*: find a function that is harmonic in a domain D and has prescribed values along the boundary γ of D. If γ is continuous and piecewise smooth, and the boundary values are piecewise continuous on γ, then the solution to the Dirichlet problem exists and unique. To show how useful conformal mappings are for this task, we start with a special case.

Suppose that the boundary γ consists of two parts γ_1 and γ_2, the temperature at each of these parts is constant and equals T_1 and T_2, respectively. We know that isotherms have equations $u(x, y) = C$. Thus, to find the complex potential $w = f(z)$ of the heat flow in D, it is sufficient to find a conformal mapping of D onto the vertical strip $T_1 < \operatorname{Re} w < T_2$, and which maps γ_1 and γ_2 onto the vertical lines $\operatorname{Re} w = T_1$ and $\operatorname{Re} w = T_2$, respectively. Then the temperature $u(x, y)$ in D equals $\operatorname{Re} f(z)$.

[14] Johann Peter Dirichlet (1805–1859) was a famous German mathematician.

Example 8.18 Find the temperature distribution in the upper half-plane D if the temperature at the boundary is $u(x, 0) = T_1$ as $x < 0$, and $u(x, 0) = T_2$ as $x > 0$, $T_1 < T_2$.

Solution (1) The function $w_1 = \operatorname{Log} z$ maps D onto the horizontal strip $0 < \operatorname{Im} w_1 < \pi$ in the plane of variable $w_1 = u_1 + v_1$; the image of the ray $x > 0$ is the line $v_1 = 0$, and the image of $x < 0$ is the line $v_1 = \pi$.

(2) The next step is to map the obtained strip onto the vertical strip $T_1 < \operatorname{Re} w < T_2$, $w = u + iv$, in such a way that $v_1 = 0$ goes to $u = T_2$, and $v_1 = \pi$ goes to $u = T_1$. In turn, it requires several steps.

(a) The thickness of our strip is π, but we need the strip of thickness $T_2 - T_1$. So we apply the mapping $w_2 = \frac{T_2 - T_1}{\pi} w_1$.

(b) To have a vertical strip with the boundary values T_1, T_2, we rotate the strip obtained in (a) by $\pi/2$: $w_3 = i w_2$. Now we have the strip $-(T_2 - T_1) < \operatorname{Re} w_3 < 0$.

(c) The last step is to translate the strip to the right by adding T_2. Finally we have

$$\begin{aligned} w = u + iv = f(z) &= T_2 + i\frac{T_2 - T_1}{\pi} \operatorname{Log} z \\ &= T_2 - \frac{T_2 - T_1}{\pi} \operatorname{Arg} z + i\frac{T_2 - T_1}{\pi} \ln |z|. \end{aligned}$$

Therefore, $u(x, y) = T_2 - \frac{T_2 - T_1}{\pi} \operatorname{Arg} z$. We see that isotherms $u(x, y) = C$ are the rays emanating from the origin, and the heat streamlines $v(x, y) = C$ are the semicircles centered at the origin.

What do we do if the boundary values are more complicated? We will not consider general methods of solving the problem in this book.[15] But we shall show how conformal mappings may reduce a difficult problem to an easier one. We prove a simple but important fact.

Theorem 8.19. *Let $w = g(z)$ be a conformal mapping of a domain D onto a domain G, and let $H(w)$ be a harmonic function in G. Then the function $H(g(z))$ is harmonic in D.*

Proof. Choose a point $z_0 \in D$, and let $w_0 = g(z_0)$. Denote by I a harmonic conjugate to H in some neighborhood U of w_0. Then the function $F(w) = H(w) + iI(w)$ is analytic in U. Hence, $F(g(z))$ is analytic in a neighborhood of z_0 because the superposition of analytic functions is analytic. Therefore, $\operatorname{Re} F(g(z)) = H(g(z))$ is harmonic near z_0. Since z_0 is any point in D, the function $H(g(z))$ is harmonic in D, as required to prove. □

Suppose now that we have to solve a Dirichlet problem for a harmonic function $h(z)$ in a domain D with given boundary values on the boundary γ of D. The mapping $w = g(z)$ can be viewed as carrying those boundary values

[15] The reader may find more details, for example, in the book by R. V. Churchill [2].

to Γ of G, via $H(w) = h(g^{-1}(w))$. Therefore, the Dirichlet problem for h in D becomes the corresponding Dirichlet problem for the harmonic function H in G. This problem in G may be easier than the initial one. After finding $H(w)$ we have $h(z) = H(g(z))$.

Note that the approach we used above for the special case of boundary values T_1, T_2, may be viewed as a special case of this method when $H(w) = \operatorname{Re} w$ and the complex potential f equals g.

Problems

1. Find the complex potential and the electric field inside an infinite condenser with plates $\operatorname{Im} z = \pm ih$, if the potentials at the lower and upper plates are $-V$ and V, respectively. Sketch equipotentials and force lines.

2. Suppose the boundary of the unit disk D consists of two conductors, insulated from each other at the points ± 1. The potentials at the lower and upper plates are 0 and V, respectively. Find the complex potential and the electric field inside the disk.

3. Prove that if the origin is a heat source of strength N, then the complex potential $f(z)$ equals

$$f(z) = -\frac{N}{2\pi k} \log z,$$

where k is the coefficient of thermal conductivity of the material.

4. Find the complex potential of the heat flow and the temperature at any point (x, y) of the strip $a < \operatorname{Im} z < b$, if the temperature at points of the lines $\operatorname{Im} z = a$, $\operatorname{Im} z = b$ is constant and equals T_1 and T_2, respectively. Sketch the isotherms and heat streamlines.

5. Find the complex potential of the heat flow and the temperature at any point (x, y) of the following domains D:

(a) D is the upper half of the unit disk $|z| < 1$; the temperature at points of the diameter $(-1, 1)$ is 0, and at points of the upper semi-circle is T.

(b) D is the sector $0 < \operatorname{Arg} z < \pi/2$; the temperature on sides $\operatorname{Arg} z = 0$ and $\operatorname{Arg} z = \pi/2$ equals 0 and T, respectively.

6. Consider two parallel half-planes with distance $2h$ between them, as the plates of the condenser in Fig. 95. Suppose that the lower and the upper plates have temperature $-T$ and T, respectively. Find the inverse of the complex potential. Sketch the isotherms and heat streamlines.

9

The Laplace Transform

We will work with a certain class of functions $f(t)$, of a real variable t, sufficiently wide for practical applications. We then introduce a rule by which each function of that class is associated to a unique function $F(p)$ of a complex variable p. This rule is called a *transformation* or *operator*, and $F(p)$ is called the *transform* of the original function $f(t)$. To each alteration of the original function there corresponds an alteration of the transform. What makes this correspondence useful is the crucial feature that operations carried out on the transforms tend to be simpler than the corresponding ones performed on the original functions. For example, differentiation of the original function corresponds to multiplication of the transform by the variable p, and integration to division by p.

Suppose we have such a transformation. Then to solve a differential equation for a function $y(t)$, for example, we might use the following strategy.

1. Instead of the original function $y(t)$ consider its transform $Y(p)$.
2. Transform the differential equation, so that the operations of differentiation of $y(t)$ become simpler operations on the transform $Y(p)$: the original differential equation for $y(t)$ becomes a linear algebraic equation for $Y(p)$.
3. Solve the algebraic equation, obtaining a transform $Y(p)$.
4. Restore the function $y(t)$ those transform is $Y(p)$, that is perform the inverse transformation from $Y(p)$ to $y(t)$. This function $y(t)$ solves the given differential equation.

Similar so-called symbolic calculus had been developed in the 19th century, in which taking the nth derivative of a function $f(t)$ was regarded as applying the formal symbol p^n to f (in the modern sense p is the argument of the transform $F(p)$). This approach proved to be quite convenient for solving problems associated with linear differential equations. The English engineer and physicist Oliver Heaviside (1850–1925) contributed to the development and popularization of this method, successfully applying symbolic calculus to problems in electromagnetism. But the rules for the symbolic operations remained without formal mathematical foundations until the work of a series of 20th century mathematicians established the necessary theory.

DOI: 10.1201/9780367810283-9

9.1 The Laplace Transform

Let $f(t)$ be a function of the real variable t, for all $t \in \mathbb{R}$; the values of $f(t)$ may be either real or complex, although in our applications they will be real. The function f is said to be *piecewise differentiable* if, for every finite interval I, f fails to be differentiable at only finitely many points of I, and all its points of discontinuity are jumps (i.e. there are right and left limits of the function at these points).

We now introduce a class of functions for which the transformation will be defined. We assume that the following three conditions are satisfied:

(1) $f(t) = 0$ when $t < 0$;
(2) f is piecewise differentiable;
(3) there exist real numbers M and σ such that

$$|f(t)| \leq Me^{\sigma t} \quad \text{for all} \quad t \in \mathbb{R}. \tag{9.1}$$

As a rule, condition (1) does not limit the possibilities for applications. For example, $f(t)$ might describe some physical process that starts at a particular moment $t = 0$, but then evolves into the indefinite future. In this case only the values of $f(t)$ for $t \geq 0$ are significant; the values for $t \leq 0$ can be chosen arbitrary, in particular equal to zero.

Condition (3) says that the growth of f is at most exponential, i.e. $f(t)$ does not grow too fast as $t \to \infty$. In practice, this does not greatly limit the class of functions, since all the elementary functions we have seen in this book (for example, $\sin t$, $\cos t$, t^n, e^{at}, and so on) satisfy this condition for $t \geq 0$. A function violating this condition would have super-exponential growth, like $f(t) = e^{t^2}$.

Besides the function which is identically zero, the simplest member of this class of functions is the *Heaviside function*

$$h(t) = \begin{cases} 0, & t < 0 \\ 1, & t \geq 0. \end{cases} \tag{9.2}$$

Like any bounded function, this satisfies condition (9.1) with $\sigma = 0$. When the Heaviside function is multiplied into any other function $\phi(t)$, the resulting function is identical to ϕ on $t \geq 0$, but identically zero on $t < 0$. Therefore if ϕ fails to satisfy condition (1), but does satisfy conditions (2) and (3), then the function $f(t) = h(t)\phi(t)$ will satisfy all three conditions. For example,

$$h(t) \sin \omega t, \quad h(t)t^n, \quad \text{and} \quad h(t)e^{at}$$

are all satisfy conditions (1)–(3). For simplicity we will usually omit the factor $h(t)$, taking it as understood that when we are going to find the transform of a function we redefine its values for negative t to be 0; so we will usually just write 1 instead of $h(t)$, $\sin \omega t$ instead of $h(t) \sin \omega t$, and so on.

The Heaviside function allows us to write the definition of functions defined by different formulas on different intervals (that is piecewise defined functions) in a more convenient form. Such functions appear in various applications.

Let $f(t)$ be a function on the interval $t \geq 0$, and let $f_1(t)$ be a "piece" of $f(t)$ on the interval $[a, b)$, $a \geq 0$, that is $f_1(t) = f(t)$ when $t \in [a, b)$, and $f_1(t) = 0$ otherwise. To set the value of $f_1(t)$ to zero for $t < a$, we multiply $f(t)$ by $h(t - a)$. To get zero for $t \geq b$, we can subtract from $f(t)$ the values $f(t)$ as $t \geq b$, that is subtract $h(t - b)f(t)$. Thus,

$$f_1(t) = h(t-a)f(t) - h(t-b)f(t) = [h(t-a) - h(t-b)]f(t). \qquad (9.3)$$

Example 9.1 Using the Heaviside function, write down the piecewise definition of the function

$$f(t) = \begin{cases} 0, & 0 \leq t < 2, \\ 3t, & 2 \leq t < 4, \\ 2, & t \geq 4, \end{cases}$$

in one line.

Solution The function $f(t)$ is the sum $f_1(t) + f_2(t)$, where $f_1(t)$ is the "piece" of the function $f(t) = 3t$ on the interval $[2, 4)$, and $f_2(t)$ is the "piece" of the function $f(t) = 2$ on the interval $[4, \infty)$. For f_2 we need not to subtract $h(t - b)f(t)$. Applying (9.3) we have

$$\begin{aligned} f_1(t) &= h(t-2)3t - h(t-4)3t; \quad f_2(t) = h(t-4)2; \\ f(t) &= f_1(t) + f_2(t) = h(t-2)3t - h(t-4)3t + h(t-4)2 \\ &= h(t-2)3t - h(t-4)(3t-2). \end{aligned}$$

If the function f satisfies the growth condition (9.1) for some $M > 0$ and σ_1, then the condition will still be satisfied with that same M and any $\sigma > \sigma_1$. On the other hand, if for some σ_2 condition (9.1) fails for any $M > 0$, then for all $\sigma < \sigma_2$ the condition will still fail for any $M > 0$. Therefore the real number line is divided into two groups, forming the two rays $(-\infty, \sigma_0)$ and (σ_0, ∞): for any σ in the first, the condition (9.1) fails for all $M > 0$, while for any σ in the second, the condition is satisfied for some $M > 0$. The number σ_0, which separates the two sets, is called the *growth index* or just the *index* of the function f. When $\sigma = \sigma_0$, condition (9.1) may or may not be satisfied, depending on the function. As examples:

For the functions 1, $\sin \omega t$, and $\cos \omega t$, the growth index is $\sigma_0 = 0$, and when $\sigma = \sigma_0 = 0$ the condition is satisfied when $M = 1$.

For t^n with $n > 0$, the index is still $\sigma_0 = 0$. Indeed, for any $\sigma > 0$,

$$\lim_{t \to \infty} \frac{t^n}{e^{\sigma t}} = 0,$$

and therefore condition (9.1) holds with some $M > 0$. But at $\sigma = \sigma_0 = 0$ the condition fails for every $M > 0$, since $|t^n| > Me^0$ when $t > \sqrt[n]{M}$.

We still need a few more definitions before getting to the transformation alluded to at the start of the chapter. Let ϕ be a piecewise continuous function defined for $t \geq 0$. Recall that the improper integral of ϕ over $[0, \infty)$ is defined by

$$\int_0^\infty \phi(t)\, dt = \lim_{R\to\infty} \int_0^R \phi(t)\, dt, \tag{9.4}$$

if the limit exists, in which case we say the integral converges and ϕ is *integrable* over $[0, \infty)$; if the limit does not exist, we say the integral diverges. We will say that ϕ is *absolutely integrable* over $[0, \infty)$ if the integral of $|\phi|$ converges, i.e. if the limit

$$\int_0^\infty |\phi(t)|\, dt = \lim_{R\to\infty} \int_0^R |\phi(t)|\, dt \tag{9.5}$$

exists (and is finite).

The following theorem is analogous to Theorem 6.2 for convergence of series.

Theorem 9.2. *If a function is absolutely integrable over* $[0, \infty)$ *then it is integrable over that set.*

Since we will not be using this theorem directly, we omit the proof; the reader can derive Theorem 9.2 from Theorem 6.2. As with Theorem 6.2, the converse is false: integrability does not imply absolute integrability.

Now suppose that ϕ is a function which depends on both a real variable t and a complex variable p, that is $\phi = \phi(t, p)$. If for some p the integral

$$\int_0^\infty \phi(t, p)\, dt \tag{9.6}$$

converges, we denote its value by $F(p)$. Thus, F is a function defined on the set of values of p for which the integral converges.

Analogous to the concept of uniform convergence of a series, we say that the integral (9.6) *converges uniformly on the domain D to the function $F(p)$* if for any $\epsilon > 0$ there is some R_ϵ depending on ϵ, for which

$$\left| \int_0^R \phi(t, p)\, dt - F(p) \right| < \epsilon$$

whenever $R > R_\epsilon$ and $p \in D$.

Uniform convergence is an essentially stronger condition than ordinary pointwise convergence: the difference is that in the definition, R_ϵ may depend only on ϵ, and *not* on p. So once again, uniform convergence of the integral implies convergence, but not conversely.

Now finally we are ready to define our transformation of functions satisfying the conditions (1)–(3).

The Laplace transformation of the function $f(t)$ is a rule defined by the formula

$$F(p) = \int_0^\infty f(t)e^{-pt}\, dt, \tag{9.7}$$

associating the function $f(t)$ of the real variable t with the function $F(p)$ of the complex variable p.

The function $F(p)$ is defined on the set of those p for which the integral (9.7) converges; it is called the *Laplace transform*, or just the *transform* of the function $f(t)$. We denote this relationship between f and F by

$$f(t) \xrightarrow{\mathcal{L}} F(p) \quad \text{and} \quad F(p) \xrightarrow{\mathfrak{I}} f(t).$$

Notice that in the literature, the term "transform" is often used for both the mapping $f(t)$ to $F(p)$ and the function $F(p)$.

In this text we will look only at the Laplace transformation, but other transformations can be used in operational calculus. Heaviside considered the transformation $\widetilde{F}(p) = pF(p)$; others studied include the Mellin and Fourier transformations, which are useful for certain applications. But the Laplace transformation is the most common for applications we are going to consider.

Theorem 9.3. *If $f(t)$ is a piecewise continuous function with growth index σ_0, then the integral (9.7) converges absolutely for all p in the half-plane* $\operatorname{Re} p > \sigma_0$*; and it converges uniformly on every half-plane* $\operatorname{Re} p > \sigma_1$ *where $\sigma_1 > \sigma_0$. The function $F(p)$ defined by formula (9.7) is analytic in the half-plane* $\operatorname{Re} p > \sigma_0$.

Proof. Fix a real number $\sigma_1 > \sigma_0$, and set

$$F_n(p) = \int_{n-1}^{n} f(t)e^{-pt}\, dt \quad \text{where} \quad n = 1, 2, \ldots.$$

We will first show that for each n, the function $F_n(p)$ is analytic on the entire complex plane of the variable p. Take any closed path Γ lying in $\mathbb{C}$; then

$$\int_\Gamma F_n(p)\, dp = \int_\Gamma \left(\int_{n-1}^{n} f(t)e^{-pt}\, dt \right) dp = \int_{n-1}^{n} \left(\int_\Gamma f(t)e^{-pt}\, dp \right) dt. \tag{9.8}$$

The change of the order of integration in the last equality can be justified as follows. Suppose that $p(\tau) = x(\tau) + iy(\tau)$ is a parametrization of Γ, where $\alpha \le \tau \le \beta$. As we saw in Section 5.1, integrals along Γ can be reduced to integrals with respect to the real variable τ; in which case the double integral is over the rectangle

$$\alpha \le \tau \le \beta, \quad n-1 \le t \le n.$$

You will recall from calculus that such a double integral over a rectangle in $\mathbb{R}^2$ can be taken in either order of the variables.

Resuming from (9.8), since $f(t)$ does not depend on p we may pull it out of the inner integral:

$$\int_{\Gamma} f(t)e^{-pt}\, dp = f(t)\int_{\Gamma} e^{-pt}\, dp = 0,$$

where the last equality is by the Cauchy-Goursat's Theorem 5.7. So we have now established that

$$\int_{\Gamma} F_n(p)\, dp = 0 \quad \text{for all} \quad n = 1, 2, \ldots.$$

From the continuity of e^{-pt} and the piecewise continuity of $f(t)$, we can easily conclude that the functions $F_n(p)$ are also continuous on $\mathbb{C}$. Therefore by Morera's Theorem 5.26, all the $F_n(p)$ are analytic on $\mathbb{C}$.

Fix a number σ_2 such that $\sigma_1 > \sigma_2 > \sigma_0$. Since the growth index of $f(t)$ equals σ_0 and $\sigma_2 > \sigma_0$, there exists M for which $|f(t)| \le Me^{\sigma_2 t}$. Let $\operatorname{Re} p = \sigma > \sigma_1$; then

$$\begin{aligned} |F_n(p)| &= \left| \int_{n-1}^{n} f(t)e^{-pt}\, dt \right| \le \int_{n-1}^{n} |f(t)||e^{-pt}|\, dt \\ &\le M \int_{n-1}^{n} e^{\sigma_2 t}e^{-\sigma t}\, dt < M \int_{n-1}^{n} e^{(\sigma_2-\sigma_1)t}\, dt \\ &= \frac{M}{\sigma_2-\sigma_1} e^{(\sigma_2-\sigma_1)t}\Big|_{n-1}^{n} < \frac{M}{\sigma_1-\sigma_2} e^{(\sigma_2-\sigma_1)(n-1)} = a_n, \end{aligned}$$

using a_n to denote this upper bound of $|F_n(p)|$. Note that these a_n form a geometric sequence with ratio

$$\frac{a_{n+1}}{a_n} = e^{\sigma_2-\sigma_1} < 1,$$

so that the series $\sum_{n=1}^{\infty} a_n$ converges. Since $|F_n(p)| < a_n$ for all p in the half-plane $\operatorname{Re} p > \sigma_1$, the series $\sum_{n=1}^{\infty} F_n(p)$ converges absolutely and uniformly, by Weierstrass' uniform convergence Theorem 6.5. Denote this sum by $S(p)$; by Theorem 6.9, $S(p)$ is analytic on the half-plane $\operatorname{Re} p > \sigma_1$.

Next we show that $S(p)$ is equal to the function $F(p)$ defined in formula (9.7). Denote by

$$S_n(p) = \sum_{k=1}^{n} F_k(p)$$

the partial sums of $S(p)$; we know that these partial sums converge to $S(p)$ on the half-plane. Also,

$$S_n(p) = \sum_{k=1}^{n} \int_{k-1}^{k} f(t)e^{-pt}\,dt = \int_{0}^{n} f(t)e^{-pt}\,dt.$$

Therefore

$$\lim_{n\to\infty} S_n(p) = S(p)$$

implies that

$$\lim_{n\to\infty} \int_{0}^{n} f(t)e^{-pt}\,dt = S(p).$$

Now we prove that also

$$\lim_{R\to\infty} \int_{0}^{R} f(t)e^{-pt}\,dt = S(p),$$

and that the convergence is uniform. Take any $R > 1$, and choose n so that $n \le R < n+1$. Then

$$\begin{aligned}
\left| \int_{0}^{R} f(t)e^{-pt}\,dt - S(p) \right| &= \left| \int_{0}^{n} f(t)e^{-pt}\,dt + \int_{n}^{R} f(t)e^{-pt}\,dt - S(p) \right| \\
&\le \left| \int_{0}^{n} f(t)e^{-pt}\,dt - S(p) \right| + \left| \int_{n}^{R} f(t)e^{-pt}\,dt \right| \\
&\le |S_n(p) - S(p)| + \int_{n}^{R} |f(t)e^{-pt}|\,dt \le \sum_{k=n+1}^{\infty} a_k + a_{n+1}.
\end{aligned}$$

Let us denote this last expression by ϵ_n; note that this depends only on n, and not on p. Furthermore, $\epsilon_n \to 0$ as $n \to \infty$. Therefore the integral

$$F(p) = \int_{0}^{\infty} f(t)e^{-pt}\,dt$$

converges to $S(p)$ uniformly on the domain $\mathrm{Re}\,p > \sigma_1$. Thus, $S(p) = F(p)$, which is analytic on the half-plane $\mathrm{Re}\,p > \sigma_1$.

Since our $\sigma_1 > \sigma_0$ was arbitrary, it now follows easily that $F(p)$ is in fact analytic on the half-plane $\mathrm{Re}\,p > \sigma_0$. For any p' with $\mathrm{Re}\,p' > \sigma_0$, we can always find some intermediate σ_1, i.e. such that $\mathrm{Re}\,p' > \sigma_1 > \sigma_0$; and then the argument just given shows that F is analytic at p'. □

Now let us calculate the Laplace transforms of some common functions. Recall our convention that these functions are considered to be identically zero for $t < 0$, or in other words that we always multiply these functions by the Heaviside function $h(t)$ before calculating the Laplace transform.

Example 9.4 Find the Laplace transform of the Heaviside function itself, defined in (9.2).

Solution Take any p for which $\operatorname{Re} p > 0$. From the definition (9.7) of the Laplace transformation,

$$F(p) = \int_0^\infty h(t)e^{-pt}\,dt = \int_0^\infty e^{-pt}\,dt = -\lim_{R\to\infty} \frac{e^{-pt}}{p}\bigg|_0^R = \frac{1}{p},$$

since

$$\lim_{R\to\infty} |e^{-pR}| = \lim_{R\to\infty} e^{-(\operatorname{Re} p)R} = 0.$$

Therefore

$$h(t) \xrightarrow{\mathcal{L}} \frac{1}{p} \quad \text{or} \quad 1 \xrightarrow{\mathcal{L}} \frac{1}{p}. \tag{9.9}$$

Example 9.5 Find the Laplace transform of the exponential function e^{at} (or $h(t)e^{at}$).

Solution Set $\sigma_0 = \operatorname{Re} a$, and take any p for which $\operatorname{Re} p > \sigma_0$. Then

$$F(p) = \int_0^\infty e^{at}e^{-pt}\,dt = \int_0^\infty e^{-(p-a)t}\,dt = -\lim_{R\to\infty} \frac{e^{-(p-a)t}}{p-a}\bigg|_0^R = \frac{1}{p-a}.$$

So

$$e^{at} \xrightarrow{\mathcal{L}} \frac{1}{p-a}. \tag{9.10}$$

Note in this example that the transform $F(p) = 1/(p-a)$ turns out to be analytic in the entire complex plane, apart from the singularity at $p = a$; this even though the integral that defines it converges only when $\operatorname{Re} p > \sigma_0$. In other words, $F(p)$ has an analytic continuation to a larger domain. As we will see in the following examples, this situation is typical: as a rule, the function $F(p)$ turns out to be defined and analytic on a significantly larger part of $\mathbb{C}$ than the half-plane $\operatorname{Re} p > \sigma_0$. According to Theorem 9.3, $F(p)$ can have no singularities in that half-plane: all its singularities lie on or to the left of the line $\operatorname{Re} p = \sigma_0$.

To each function $f(t)$ satisfying conditions (1)–(3), the Laplace transformation associates its transform $F(p)$. It turns out that distinct at some point of continuity functions $f(t)$ are associated with distinct transforms—the association is 1-1. Even better, there is a so-called *inversion formula* which can recover from a known transform $F(p)$, the original function $f(t)$.

Theorem 9.6 (Inversion theorem). *Suppose that a function $f(t)$ satisfies conditions (1)–(3) and has growth index σ_0, and $F(p)$ is its Laplace transform. Then at all points t at which f is continuous, the value $f(t)$ can be expressed in terms of F by the inversion formula*

$$f(t) = \frac{1}{2\pi i} \int_{\sigma - i\infty}^{\sigma + i\infty} e^{pt} F(p)\, dp, \tag{9.11}$$

where the integral is over any line of the form $\operatorname{Re} p = \sigma$ with $\sigma > \sigma_0$, and the principal value of the improper integral is used, i.e.

$$\int_{\sigma - i\infty}^{\sigma + i\infty} e^{pt} F(p)\, dp = \lim_{R \to \infty} \int_{\sigma - iR}^{\sigma + iR} e^{pt} F(p)\, dp.$$

The proof of Theorem 9.6 is omitted. Since this theorem will be used in the proof of Theorem 9.29, we give an informal justification of formula (9.11) in Appendix, without a rigorous proof.

In the examples above we calculated Laplace transforms directly from the definition (9.7). But in many cases it is much more convenient to make use of general properties of the transformation, and we turn to some of these next.

Problems

1. Determine whether the function $f(t)$ satisfies the conditions (1)–(3) stated at the beginning of Section 9.1. If yes, indicate the growth index σ_0. Assume that $f(t) = 0$ as $t < 0$.

a) $f(t) = te^{3t}$; b) $f(t) = e^t \sin t$; c) $f(t) = e^t \tan t$;

d) $f(t) = \ln(t+1)$; e) $f(t) = t^2 \cos t$; f) $f(t) = e^{e^t}$.

2. Using the Heaviside function, write down the piecewise definition of the function $f(t)$ in one line.

a) $f(t) = \begin{cases} 3t - 1, & 0 \le t < 3, \\ t + 2, & 3 \le t < 4, \\ 0, & t \ge 4; \end{cases}$ b) $f(t) = \begin{cases} 0, & 0 \le t < 1, \\ 2t + 1, & 1 \le t < 3, \\ 7, & t > 3; \end{cases}$

c) $f(t) = \begin{cases} \sin t, & 2\pi \le t < 3\pi, \\ 0, & \text{otherwise}; \end{cases}$ d) $f(t) = \begin{cases} t^2, & 0 \le t < 1, \\ 1, & t \ge 1. \end{cases}$

3. The following Comparison Test for improper integrals is valid (compare with the Comparison Test for series): *Suppose that $f(t)$ and $g(t)$ are continuous functions on an interval $[a, \infty)$, and $|g(t)| \le |f(t)|$. Then*

if $\int_0^\infty |f(t)|\,dt$ converges, then so does $\int_0^\infty |g(t)|\,dt$;
if $\int_0^\infty |g(t)|\,dt$ diverges, then so does $\int_0^\infty |f(t)|\,dt$.
Prove that the following integrals converge (do not evaluate the integrals):

a) $\displaystyle\int_0^\infty \frac{x^{10}}{e^{2x}}\,dx$; b) $\displaystyle\int_0^\infty \frac{\sin x}{x^2}\,dx$; c) $\displaystyle\int_0^\infty e^{-x^2}\,dx$.

4. Using definition (9.7), find the Laplace transform of the function $f(t)$ and indicate the half-plane of convergence of the integral. Assume that $f(t) = 0$ when $t < 0$.

a) $f(t) = te^{at}$; b) $f(t) = \begin{cases} 1, & 3 \le t < 5, \\ 0 & \text{otherwise}; \end{cases}$

c) $f(t) = \begin{cases} t, & 0 \le t < 2, \\ 0, & t \ge 2; \end{cases}$ d) $f(t) = \begin{cases} e^t, & 0 \le t < 1, \\ e, & t \ge 1; \end{cases}$

e) $f(t) = \begin{cases} \sin t, & 0 \le t < \pi, \\ 0, & t \ge \pi; \end{cases}$ f) $f(t) = t^2$.

9.2 Properties of the Laplace Transformation

In what follows $f(t)$, $g(t)$, ... denote functions satisfying conditions (1)–(3), and $F(p)$, $G(p)$, ... denote their respective Laplace transforms.

Property 9.7 (Linearity). *If $f(t) \xrightarrow{\mathcal{L}} F(p)$ and $g(t) \xrightarrow{\mathcal{L}} G(p)$, then for any complex numbers α and β,*

$$\alpha f(t) + \beta g(t) \xrightarrow{\mathcal{L}} \alpha F(p) + \beta G(p). \tag{9.12}$$

Proof. This follows immediately from the definition (9.7) and the linearity of the integral. The Laplace transform of $\alpha f(t) + \beta g(t)$ equals

$$\int_0^\infty (\alpha f(t)+\beta g(t))e^{-pt}\,dt = \alpha \int_0^\infty f(t)e^{-pt}\,dt + \beta \int_0^\infty g(t)e^{-pt}\,dt = \alpha F(p) + \beta G(p).$$

□

Note however that, since $F(p)$ and $G(p)$ will generally be defined on different half-planes $\operatorname{Re} p > \sigma_1$ and $\operatorname{Re} p > \sigma_2$, their sum will only be defined on the intersection of these, i.e. $\operatorname{Re} p > \sigma_0$, where $\sigma_0 = \max(\sigma_1, \sigma_2)$.

Example 9.8 Find the Laplace transforms of the functions $\sin \omega t$, $\cos \omega t$, $\sinh \omega t$, and $\cosh \omega t$.

Solution These functions can be expressed in terms of the exponential function using the formulas (4.28) and (4.33):

$$\begin{aligned} \sin z &= \frac{e^{iz} - e^{-iz}}{2i}, \qquad \cos z = \frac{e^{iz} + e^{-iz}}{2}, \\ \sinh z &= \frac{e^{z} - e^{-z}}{2}, \qquad \cosh z = \frac{e^{z} + e^{-z}}{2}. \end{aligned} \tag{9.13}$$

We saw in the last section (see (9.10)) that

$$e^{i\omega t} \xrightarrow{\mathcal{L}} \frac{1}{p - i\omega} \quad \text{and} \quad e^{-i\omega t} \xrightarrow{\mathcal{L}} \frac{1}{p + i\omega},$$

from which, using the linearity of the transformation,

$$\sin \omega t = \frac{1}{2i} e^{i\omega t} - \frac{1}{2i} e^{-i\omega t} \xrightarrow{\mathcal{L}} \frac{1}{2i} \frac{1}{p - i\omega} - \frac{1}{2i} \frac{1}{p + i\omega} = \frac{\omega}{p^2 + \omega^2}. \tag{9.14}$$

The reader can independently provide analogous arguments for the other formulas:

$$\cos \omega t \xrightarrow{\mathcal{L}} \frac{p}{p^2 + \omega^2}, \quad \sinh \omega t \xrightarrow{\mathcal{L}} \frac{\omega}{p^2 - \omega^2}, \quad \cosh \omega t \xrightarrow{\mathcal{L}} \frac{p}{p^2 - \omega^2}.$$

Property 9.9 (Behavior of $F(p)$ as $p \to \infty$). *The transform $F(p)$ of any function $f(t)$ satisfying (1)–(3), approaches 0 as $\operatorname{Re} p \to +\infty$.*

Proof. Let σ_0 be the growth index of f, and take some $\sigma_1 > \sigma_0$; also let σ denote $\operatorname{Re} p$. Then $|f(t)| \le M e^{\sigma_1 t}$, and for $\sigma > \sigma_1$ we have

$$\begin{aligned} |F(p)| &= \left| \int_0^\infty f(t) e^{-pt} \, dt \right| \le \int_0^\infty |f(t)| \left| e^{-pt} \right| dt \le M \int_0^\infty e^{\sigma_1 t} e^{-\sigma t} \, dt \\ &= M \int_0^\infty e^{-(\sigma - \sigma_1)t} \, dt = -\frac{M}{\sigma - \sigma_1} \lim_{R \to \infty} e^{-(\sigma - \sigma_1)t} \Big|_0^R = \frac{M}{\sigma - \sigma_1}. \end{aligned}$$

Here we see that $F(p) \to 0$ as $\sigma \to \infty$, as desired. □

Theorem 9.10. *If $f(t) \xrightarrow{\mathcal{L}} F(p)$, then for any $\lambda > 0$*

$$f(\lambda t) \xrightarrow{\mathcal{L}} \frac{1}{\lambda} F\left(\frac{p}{\lambda}\right). \tag{9.15}$$

Proof. We introduce a new variable $\tau = \lambda t$. Then $t = \frac{\tau}{\lambda}$, $dt = \frac{1}{\lambda} d\tau$. Substituting these into the integral, we get

$$\int_0^\infty f(\lambda t) e^{-pt}\, dt = \int_0^\infty f(\tau) e^{-p\frac{\tau}{\lambda}} \frac{1}{\lambda}\, d\tau = \frac{1}{\lambda} \int_0^\infty f(\tau) e^{-(\frac{p}{\lambda})\tau}\, d\tau = \frac{1}{\lambda} F\left(\frac{p}{\lambda}\right),$$

because of course the variable of integration, whether t or τ, makes no difference in the definition (9.7). □

In many applications $f(t)$ is a signal emitted over time t, so $f(t-\tau)$ would represent the same signal sent with a time delay τ; hence the name of the following theorem.

Theorem 9.11 (Time delay theorem). *If* $f(t) \xrightarrow{\mathcal{L}} F(p)$, *then for any* $\tau > 0$

$$f(t-\tau) = h(t-\tau) f(t-\tau) \xrightarrow{\mathcal{L}} e^{-p\tau} F(p); \tag{9.16}$$

here h *is the Heaviside function defined in (9.2).*

Proof. The first equality holds because $f(t) = 0$ as $t < 0$, and therefore $f(t-\tau) = 0$ as $t < \tau$. Hence,

$$\int_0^\infty f(t-\tau) e^{-pt}\, dt = \int_\tau^\infty f(t-\tau) e^{-pt}\, dt.$$

We introduce a new variable $\zeta = t - \tau$. Then $t = \zeta + \tau$, $dt = d\zeta$, and as t goes from τ to ∞, the new variable ζ goes from 0 to ∞. So computing the Laplace transform of $f(t-\tau)$ and changing the variable to ζ, we get

$$\int_0^\infty f(\zeta) e^{-p(\zeta+\tau)}\, d\zeta = e^{-p\tau} \int_0^\infty f(\zeta) e^{-p\zeta}\, d\zeta = e^{-p\tau} F(p),$$

as desired. □

The reader should be careful with using formula (9.16), because $F(p)$ is the Laplace transform of $f(t)$, but it is *not* the transform of $f(t-\tau)$.

Example 9.12 Using the relation $e^{3t} \xrightarrow{\mathcal{L}} \frac{1}{p-3}$—see formula (9.10)—find the Laplace transform of the function $h(t-2)e^{3t}$.

Solution Because we wish to apply the previous theorem, we write

$$h(t-2) f(t-\tau) = h(t-2) e^{3t};$$

hence $\tau = 2$ and $f(t-2) = e^{3t}$. To find $f(t)$, we introduce a new variable $x = t - 2$. Then $t = x + 2$, and

$$f(x) = e^{3(x+2)} = e^6 e^{3x}.$$

Of course, we can use the notation t for the argument of f again and write $f(t) = e^6 e^{3t}$—*not* just e^{3t}! Therefore, $F(p) = \frac{e^6}{p-3}$. By formula (9.16) we get

$$h(t-2)e^{3t} \xrightarrow{\mathcal{L}} e^{-2p}\frac{e^6}{p-3} = \frac{e^{-2(p-3)}}{p-3}.$$

Example 9.13 Find the Laplace transform of the rectangular impulse $f(t)$ of height 1 and width l, as shown in Fig. 96.

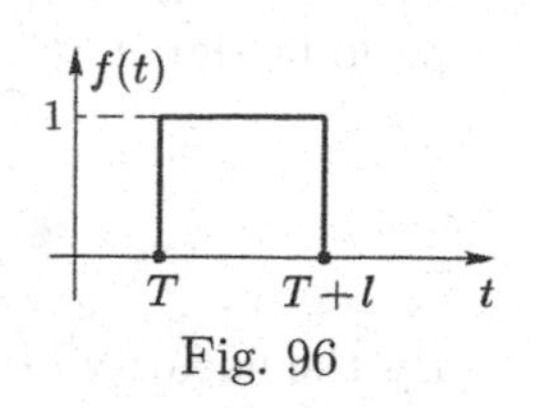

Fig. 96

Solution We could easily find the transform directly from the definition by computing the integral:

$$F(p) = \int_T^{T+l} 1 \cdot e^{-pt}\, dt = -\frac{1}{p}e^{-pt}\Big|_T^{T+l}$$

$$= \frac{1}{p}(e^{-pT} - e^{-p(T+l)}) = \frac{e^{-pT}}{p}(1 - e^{-pl}).$$

But other methods are possible. Let $h(t)$ be the Heaviside function, defined in (9.2).The functions $h(t-T)$ and $h(t-T-l)$ are obtained from it by shifting to the right by T and $T+l$ units, respectively. Hence the given impulse $f(t)$ can be written in the form

$$f(t) = h(t-T) - h(t-T-l).$$

Then we may apply the time delay theorem (Theorem 9.11) to get the transforms of $h(t-T)$ and $h(t-T-l)$:

$$h(t) \xrightarrow{\mathcal{L}} \frac{1}{p}, \quad h(t-T) \xrightarrow{\mathcal{L}} e^{-pT}\frac{1}{p}, \quad h(t-T-l) \xrightarrow{\mathcal{L}} e^{-p(T+l)}\frac{1}{p}.$$

Using the linearity of the Laplace transformation, we have

$$f(t) \xrightarrow{\mathcal{L}} e^{-pT}\frac{1}{p} - e^{-p(T+l)}\frac{1}{p} = \frac{e^{-pT}}{p}(1 - e^{-pl}).$$

The following example shows how to find the Laplace transform of a periodic signal if one knows the transform of a single cycle, or impulse.

Example 9.14 Find the Laplace transform of a series of rectangular impulses, having amplitude 1, width l, and period T (Fig. 97).

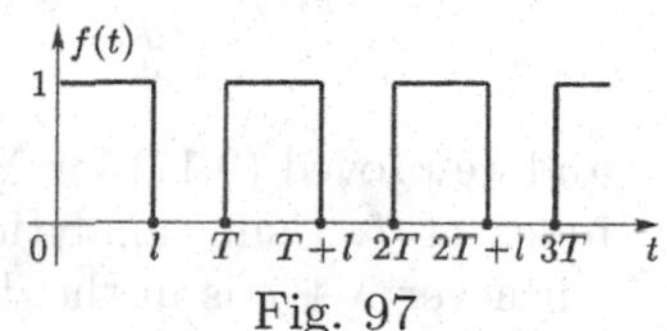

Fig. 97

Solution Let us denote by $f_n(t)$ the nth impulse, $n = 0, 1, 2, \ldots$, which starts at time nT and extends through l time units. From the preceding example we have that

$$f_n(t) \overset{\mathcal{L}}{\rightarrow} \frac{e^{-pnT}}{p}(1 - e^{-pl}),$$

where we have replaced T by nT. These f_n, when summed, yield the given periodic signal, denoted f:

$$f(t) = \sum_{n=0}^{\infty} f_n(t).$$

By the linearity of the Laplace transformation we get

$$\begin{aligned} F(p) &= \sum_{n=0}^{\infty} \frac{e^{-pnT}}{p}(1 - e^{-pl}) \\ &= \frac{1}{p}(1 - e^{-pl}) \sum_{n=0}^{\infty} \left(e^{-pT}\right)^n = \frac{1 - e^{-pl}}{p(1 - e^{-pT})}, \end{aligned}$$

using the formula for the sum of a geometric series with ratio $q = e^{-pT}$.

A slightly different method would have been to start with the transform $F_0(p)$ of $f_0(t)$, and then find the $f_n(t)$ using the time delay formula:

$$f_n(t) = f_0(t - nT) \overset{\mathcal{L}}{\rightarrow} e^{-pnT} F_0(p),$$

so that

$$F(p) = \sum_{n=0}^{\infty} e^{-pnT} F_0(p) = \frac{F_0(p)}{1 - e^{-pT}}.$$

According to Theorem 9.11, shifting $f(t)$ results in multiplying its Laplace transform by the exponent $e^{-p\tau}$. The following theorem states that multiplication of the original function by an exponential leads to a shift of its transform.

Theorem 9.15 (Shift theorem). *If $f(t) \overset{\mathcal{L}}{\rightarrow} F(p)$, then for any complex number λ,*

$$e^{-\lambda t} f(t) \overset{\mathcal{L}}{\rightarrow} F(p + \lambda). \tag{9.17}$$

Proof. Directly from the definition (9.7),

$$e^{-\lambda t} f(t) \overset{\mathcal{L}}{\rightarrow} \int_0^{\infty} e^{-\lambda t} f(t) e^{-pt}\, dt = \int_0^{\infty} f(t) e^{-(\lambda + p)t}\, dt = F(p + \lambda),$$

and we proved (9.17) for λ and p with $\mathrm{Re}(\lambda + p) > \sigma_0$, where σ_0 is the growth index of f. Using analytic continuation, we extend this relation to all λ, p whenever $\lambda + p$ is in the domain of analyticity of F. □

Example 9.16 Find the Laplace transforms of the functions $e^{at}\sin\omega t$ and $e^{at}\cos\omega t$.

Solution First take $f(t) = \sin\omega t$; according to (9.14), in this case

$$F(p) = \frac{\omega}{p^2+\omega^2}.$$

By the previous theorem, the Laplace transform of $e^{-\lambda t}f(t)$ will be $F(p+\lambda)$. So if we take $\lambda = -a$, we get

$$e^{at}\sin\omega t \xrightarrow{\mathcal{L}} F(p-a) = \frac{\omega}{(p-a)^2+\omega^2}. \tag{9.18}$$

Analogously, using the Laplace transform of $\cos\omega t$, we get

$$e^{at}\cos\omega t \xrightarrow{\mathcal{L}} \frac{p-a}{(p-a)^2+\omega^2}. \tag{9.19}$$

Theorem 9.17 (Laplace transform of the derivative). *Suppose that $f(t)$ is a continuous function for $t \geq 0$, and that $f(t)$ and $f'(t)$ satisfy the conditions (1)–(3). Let $F(p)$ denote the Laplace transform of f. Then*

$$f'(t) \xrightarrow{\mathcal{L}} pF(p) - f(0). \tag{9.20}$$

Furthermore, if $f^{(n-1)}(t)$ is continuous for $t \geq 0$, and functions $f^{(n-1)}(t)$, $f^{(n)}(t)$ satisfy the conditions (1)–(3), then

$$f^{(n)}(t) \xrightarrow{\mathcal{L}} p^nF(p) - p^{n-1}f(0) - p^{n-2}f'(0) - \cdots - f^{(n-1)}(0), \tag{9.21}$$

where $f^{(k)}(0)$ is understood as the right limit of $f^{(k)}(t)$ as $t \to 0^+$.

Proof. Let σ_0 be the growth index of f, let $\sigma = \operatorname{Re} p$, and choose some σ_1 in between them, i.e. $\sigma_0 < \sigma_1 < \sigma$. Applying the definition of the Laplace transformation to f', and then integrating by parts, we get

$$f'(t) \xrightarrow{\mathcal{L}} \int_0^\infty f'(t)e^{-pt}\,dt = \lim_{R\to\infty} f(t)\,e^{-pt}\Big|_0^R + p\int_0^\infty f(t)e^{-pt}\,dt, \tag{9.22}$$

where we have used $(e^{-pt})' = -pe^{-pt}$; integration by parts is possible due to the conditions on f. Using the growth condition (9.1),

$$|f(R)e^{-pR}| \leq Me^{\sigma_1 R}|e^{-pR}| = Me^{-(\sigma-\sigma_1)R} \to 0$$

as $R \to \infty$. Therefore the limit in (9.22) equals $-f(0)$, and since the last integral is $F(p)$, we obtain (9.20).

Now we simply repeat the application of formula (9.20): let F_1 be the transform of f'. By (9.20), $F_1(p) = pF(p) - f(0)$. Then

$$\begin{aligned} f''(t) = (f'(t))' &\overset{\mathcal{L}}{\to} pF_1(p) - f'(0) \\ &= p(pF(p) - f(0)) - f'(0) = p^2F(p) - pf(0) - f'(0). \end{aligned}$$

Similarly,

$$\begin{aligned} f'''(t) = (f''(t))' &\overset{\mathcal{L}}{\to} p(p^2F(p) - pf(0) - f'(0)) - f''(0) \\ &= p^3F(p) - p^2f(0) - pf'(0) - f''(0), \end{aligned}$$

etc. Since this can repeated any number of times, formula (9.21) is established. □

Note that if $f(0) = f'(0) = \cdots = f^{(n-1)}(0) = 0$, then this formula reduces to

$$f^{(n)}(t) \overset{\mathcal{L}}{\to} p^nF(p);$$

i.e. *in the presence of the zero initial conditions, differentiating it n times corresponds to multiplication of the transform by* p^n.

Theorem 9.18 (Laplace transform of the integral). *If f satisfies the conditions (1)–(3) given at the start of Section 9.1 and $f(t) \overset{\mathcal{L}}{\to} F(p)$, then the function $\phi(t) = \int_0^t f(\tau)\,d\tau$ also satisfies these conditions, and*

$$\int_0^t f(\tau)\,d\tau \overset{\mathcal{L}}{\to} \frac{F(p)}{p}. \tag{9.23}$$

Proof. We will show that ϕ satisfies the conditions (1)–(3). Since $f(t) = 0$ when $t < 0$, the same is true for ϕ, and in fact also $\phi(0) = 0$ even if $f(0) \neq 0$. Moreover ϕ is continuous and piecewise differentiable on the entire real line. Since f satisfies the growth condition (9.1) for some M and σ,

$$|\phi(t)| \le \int_0^t |f(\tau)|\,d\tau \le M\int_0^t e^{\sigma\tau}\,d\tau = \frac{M}{\sigma}(e^{\sigma t} - 1) < M_1e^{\sigma t}, \quad M_1 = \frac{M}{\sigma}.$$

Thus ϕ also satisfies the growth condition.

Let $\Phi(p)$ be the Laplace transform of $\phi(t)$. From the equalities $\phi'(t) = f(t)$, $\phi(0) = 0$, and (9.20) we get

$$f(t) = \phi'(t) \overset{\mathcal{L}}{\to} p\Phi(p) - \phi(0) = p\Phi(p).$$

Since the Laplace transform of a function is unique, we must have $p\Phi(p) = F(p)$, from which equation (9.23) follows immediately. □

Theorem 9.19 (Derivative of the transform). *If* $f(t) \xrightarrow{\mathcal{L}} F(p)$, *then*

$$-tf(t) \xrightarrow{\mathcal{L}} F'(p). \tag{9.24}$$

Proof. We will assume the following property of the integral: The derivative of

$$F(p) = \int_0^\infty f(t)e^{-pt}\, dt$$

can be calculated by differentiating under the integral sign, i.e.

$$F'(p) = \int_0^\infty \frac{d}{dp}\left(f(t)e^{-pt}\right)\, dt = \int_0^\infty -tf(t)e^{-pt}\, dt.$$

We do not provide a justification for this property. Under that assumption, equation (9.24) follows immediately from the definition of the Laplace transform of $-tf(t)$. □

Applying this theorem n times, we get

$$(-1)^n t^n f(t) \xrightarrow{\mathcal{L}} F^{(n)}(p). \tag{9.25}$$

Example 9.20 Find the Laplace transforms of the functions t^n, $t^n e^{at}$, $t\sin\omega t$, and $t\cos\omega t$.

Solution Since from Example 9.4 we know that $1 \xrightarrow{\mathcal{L}} 1/p$, then by (9.24) we get

$$-t \xrightarrow{\mathcal{L}} \frac{d}{dp}\frac{1}{p} = -\frac{1}{p^2}, \quad \text{or} \quad t \xrightarrow{\mathcal{L}} \frac{1}{p^2}.$$

Applying the theorem repeatedly, we obtain

$$-t^2 = -t \cdot t \xrightarrow{\mathcal{L}} \frac{d}{dp}\frac{1}{p^2} = -\frac{2}{p^3}, \qquad \text{or} \quad t^2 \xrightarrow{\mathcal{L}} \frac{2}{p^3},$$

$$-t^3 = -t \cdot t^2 \xrightarrow{\mathcal{L}} \frac{d}{dp}\frac{2}{p^3} = -\frac{2\cdot 3}{p^4}, \quad \text{or} \quad t^3 \xrightarrow{\mathcal{L}} \frac{2\cdot 3}{p^4},$$

and in general

$$t^n \xrightarrow{\mathcal{L}} \frac{n!}{p^{n+1}} \quad \text{for} \quad n = 1, 2 \ldots. \tag{9.26}$$

To find the Laplace transform of the function $t^n e^{at}$, we use the previous result together with the Shift Theorem 9.15, as in Example 9.16. Substituting $p + \lambda = p - a$ for p in equation (9.26), we get

$$t^n e^{at} \xrightarrow{\mathcal{L}} \frac{n!}{(p-a)^{n+1}}, \quad n = 1, 2, \ldots \tag{9.27}$$

The transforms of $t \sin \omega t$ and $t \cos \omega t$ can be found easily by combining our earlier results for $\sin \omega t$ and $\cos \omega t$ with Theorem 9.19. Relations (9.14), (9.2), and (9.24) imply

$$-t \sin \omega t \xrightarrow{\mathcal{L}} \frac{d}{dp} \frac{\omega}{p^2 + \omega^2} = -\frac{2p\omega}{(p^2 + \omega^2)^2},$$

$$-t \cos \omega t \xrightarrow{\mathcal{L}} \frac{d}{dp} \frac{p}{p^2 + \omega^2} = \frac{\omega^2 - p^2}{(p^2 + \omega^2)^2}.$$

Therefore

$$t \sin \omega \xrightarrow{\mathcal{L}} \frac{2p\omega}{(p^2 + \omega^2)^2}, \quad t \cos \omega t \xrightarrow{\mathcal{L}} \frac{p^2 - \omega^2}{(p^2 + \omega^2)^2}. \tag{9.28}$$

Theorem 9.21 (Integral of transform). *Suppose that both functions $f(t)$ and $f(t)/t$ satisfy the conditions (1)–(3) in Section 9.1, and that $f(t) \xrightarrow{\mathcal{L}} F(p)$. Then*

$$\frac{f(t)}{t} \xrightarrow{\mathcal{L}} \int_p^\infty F(s)\, ds. \tag{9.29}$$

Proof. Denote by $\Phi(p)$ the Laplace transform of $f(t)/t$, i.e.

$$\Phi(p) = \int_0^\infty \frac{f(t)}{t} e^{-pt}\, dt.$$

Let us find the derivative $\Phi'(p)$, which as in the proof of Theorem 9.19 we compute by differentiating under the integral sign:

$$\begin{aligned}\Phi'(p) &= \int_0^\infty \frac{d}{dp} \left(\frac{f(t)}{t} e^{-pt} \right) dt \\ &= \int_0^\infty -t \frac{f(t)}{t} e^{-pt}\, dt = -\int_0^\infty f(t) e^{-pt}\, dt = -F(p).\end{aligned}$$

So, $F(p) = -\Phi'(p)$. Therefore, by the Fundamental theorem,

$$\int_p^R F(s)\, ds = -\int_p^R \Phi'(s)\, ds = -(\Phi(R) - \Phi(p)) = \Phi(p) - \Phi(R).$$

Now by Property 9.9, $\Phi(R) \to 0$ as $R \to \infty$, so we get

$$\int_p^\infty F(s)\, ds = \lim_{R\to\infty} \int_p^R F(s)\, ds = \Phi(p) - \lim_{R\to\infty} \Phi(R) = \Phi(p),$$

which is equivalent to the desired equation (9.29). □

Example 9.22 Find the Laplace transform of the function $f(t) = \dfrac{\sin \omega t}{t}$.

Solution Note that $t = 0$ is a removable singularity of this function, since

$$\lim_{t \to 0+} \frac{\sin \omega t}{t} = \omega.$$

Therefore, $f(t)$ satisfies the conditions (1)–(3). Combining our previously calculated transform $\sin \omega t \xrightarrow{\mathcal{L}} \frac{\omega}{p^2+\omega^2}$ with the preceding theorem we have

$$\frac{\sin \omega t}{t} \xrightarrow{\mathcal{L}} \int_p^{\infty} \frac{\omega}{s^2 + \omega^2}\, ds = \lim_{R \to \infty} \omega \frac{1}{\omega} \tan^{-1} \frac{s}{\omega}\bigg|_p^R = \frac{\pi}{2} - \tan^{-1} \frac{p}{\omega}. \tag{9.30}$$

The formula obtained in this example allows us to calculate some interesting integrals, since it means that

$$\int_0^{\infty} \frac{\sin \omega t}{t} e^{-pt}\, dt = \frac{\pi}{2} - \tan^{-1} \frac{p}{\omega}.$$

For example, setting $p = 0$ gives us

$$\int_0^{\infty} \frac{\sin \omega t}{t}\, dt = \frac{\pi}{2}.$$

We should mention that before this calculation, we really should have verified that the improper integral converges; but here we omit the proof of this. Notice however that the proof of the convergence is provided by another method for calculating this integral given in Section 7.3, problem 8.

In order to state the next property, we need to define a new operation on functions. If the functions f and g are defined and piecewise continuous on $(-\infty, \infty)$, then for each value of t we can consider the integral

$$\int_{-\infty}^{\infty} f(\tau) g(t - \tau)\, d\tau,$$

which depends on t. If in addition f and g equal 0 on the interval $(-\infty, 0)$, then $f(\tau)g(t-\tau) = 0$ as $\tau < 0$ and $t - \tau < 0$. Therefore, in fact the integration is over the interval $[0, t]$. The *convolution of f with g* is the function of t defined by this integral, and is denoted $f * g$:

$$f * g(t) = \int_0^t f(\tau) g(t - \tau)\, d\tau. \tag{9.31}$$

Note that if $t < 0$, then $f * g(t) = 0$.

Property 9.23. *The convolution operation is commutative, i.e.*

$$f * g(t) = g * f(t). \tag{9.32}$$

Proof. We introduce a new variable $\zeta = t - \tau$. Then $\tau = t - \zeta$, $d\tau = -d\zeta$. When τ runs from 0 to t, the variable ζ changes from t to 0. Making these substitutions in the integral in (9.31) yields

$$f * g(t) = \int_0^t f(\tau) g(t-\tau)\, d\tau = -\int_t^0 f(t-\zeta) g(\zeta)\, d\zeta$$

$$= \int_0^t g(\tau) f(t-\tau)\, d\tau = g * f(t),$$

what was required. □

Property 9.24. *Suppose that functions f and g satisfy the conditions (1)–(3) in Section 9.1, and their growth indices are σ_0 and s_0 respectively. Then the convolution $f * g(t)$ also satisfies these conditions, and its growth index is no greater than* $\max(\sigma_0, s_0)$.

Proof. As noted above, $f * g(t) = 0$ when $t < 0$. The piecewise differentiability of $f * g$ follows from the corresponding property of f and g, though we omit the proof.

It remains to verify the growth condition (9.1). Consider any σ which is greater than both σ_0 and s_0, and choose σ_1 such that $\sigma > \sigma_1 > s_0$. By the growth condition (9.1), there exist positive numbers M_1 and M_2 such that

$$|f(t)| < M_1 e^{\sigma t} \quad \text{and} \quad |g(t)| < M_2 e^{\sigma_1 t}.$$

Then from formula (9.31),

$$|f * g(t)| = \left| \int_0^t f(\tau) g(t-\tau)\, d\tau \right| \le \int_0^t |f(\tau)|\, |g(t-\tau)|\, d\tau$$

$$\le M_1 M_2 \int_0^t e^{\sigma\tau} e^{\sigma_1(t-\tau)}\, d\tau = M_1 M_2 e^{\sigma_1 t} \int_0^t e^{(\sigma-\sigma_1)\tau}\, d\tau$$

$$= M_1 M_2 e^{\sigma_1 t} \frac{1}{\sigma - \sigma_1} e^{(\sigma-\sigma_1)\tau} \Big|_0^t = \frac{M_1 M_2}{\sigma - \sigma_1} e^{\sigma_1 t} (e^{(\sigma-\sigma_1)t} - 1)$$

$$< \frac{M_1 M_2}{\sigma - \sigma_1} e^{\sigma_1 t} e^{(\sigma-\sigma_1)t} = M e^{\sigma t}, \quad \text{where } M = \frac{M_1 M_2}{\sigma - \sigma_1}.$$

Thus the convolution $f * g$ satisfies the growth condition (9.1) for any $\sigma > \max(\sigma_0, s_0)$; and therefore its growth index is no greater than $\max(\sigma_0, s_0)$. □

Now we are ready to prove the following theorem about the relation of the Laplace transform to the convolution operation.

Theorem 9.25 (Product of Laplace transforms). *Suppose that* $f(t) \xrightarrow{\mathcal{L}} F(p)$ *and* $g(t) \xrightarrow{\mathcal{L}} G(p)$. *Then*

$$f * g(t) \xrightarrow{\mathcal{L}} F(p)G(p), \tag{9.33}$$

i.e. the transform of the convolution is the product of the transforms.

Proof. Substituting the formula (9.31) into the definition of the Laplace transform, we get

$$f * g(t) \xrightarrow{\mathcal{L}} \int_0^\infty \left(\int_0^t f(\tau)g(t-\tau)\,d\tau \right) e^{-pt}\,dt = \int_0^\infty \int_0^t f(\tau)g(t-\tau)e^{-pt}\,d\tau\,dt.$$

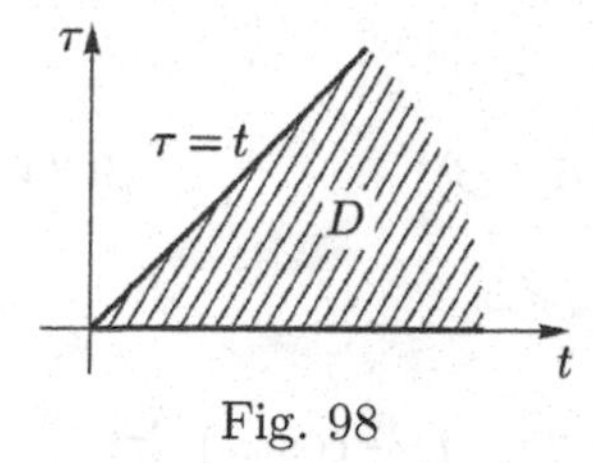

Fig. 98

Thus we obtain a double integral over the unbounded domain D, shown in Fig. 98. We wish now to change the order of integration. This is justifiable here because of the integral's absolute convergence when $\operatorname{Re} p > \max(\sigma_0, s_0)$, which holds because of Property 9.24; but we do not provide a detailed proof.

For fixed $\tau > 0$, the variable t changes from τ to ∞. So on exchanging the order of integration, we get

$$\int_0^\infty \int_0^t f(\tau)g(t-\tau)e^{-pt}\,d\tau\,dt = \int_0^\infty \int_\tau^\infty f(\tau)g(t-\tau)e^{-pt}\,dt\,d\tau$$
$$= \int_0^\infty f(\tau) \int_\tau^\infty g(t-\tau)e^{-pt}\,dt\,d\tau.$$

Next we change the variable of integration in the inner integral, from t to $\zeta = t - \tau$, so that $t = \zeta + \tau$, $dt = d\zeta$, and adjust the limits for the new variable, obtaining

$$\int_\tau^\infty g(t-\tau)e^{-pt}\,dt = \int_0^\infty g(\zeta)e^{-p(\zeta+\tau)}\,d\zeta = e^{-p\tau}\int_0^\infty g(\zeta)e^{-p\zeta}\,d\zeta = e^{-p\tau}G(p).$$

Finally we resume the computation of the transform of $f * g$:

$$f * g(t) \xrightarrow{\mathcal{L}} \int_0^\infty f(\tau)e^{-p\tau}G(p)\,d\tau = G(p)\int_0^\infty f(\tau)e^{-p\tau}\,d\tau = G(p)F(p),$$

as desired. □

This formula (9.33) is often used to determine the original function corresponding to a given transform, when the latter can be broken up into factors of known transforms.

Example 9.26 Find the original function corresponding to the transform

$$F(p) = \frac{\omega^2}{(p^2+\omega^2)^2}.$$

Solution The transform $F(p)$ can be written as

$$F(p) = \frac{\omega}{p^2+\omega^2} \cdot \frac{\omega}{p^2+\omega^2},$$

and we have seen that

$$\sin \omega t \xrightarrow{\mathcal{L}} \frac{\omega}{p^2+\omega^2}.$$

Therefore, using the trig identity

$$\sin\alpha \cdot \sin\beta = \frac{1}{2}(\cos(\alpha-\beta) - \cos(\alpha+\beta)) \tag{9.34}$$

with the previous theorem, we have

$$F(p) \xrightarrow{\mathcal{L}^{-1}} \int_0^t \sin\omega\tau \cdot \sin\omega(t-\tau)\, d\tau = \frac{1}{2}\int_0^t (\cos\omega(2\tau - t) - \cos\omega t)\, d\tau$$

$$= \frac{1}{4\omega}\sin\omega(2\tau - t)\Big|_0^t - \frac{1}{2}\tau\cos\omega t\Big|_0^t = \frac{1}{2\omega}\sin\omega t - \frac{1}{2}t\cos\omega t.$$

This result is easy to verify using the linearity of the Laplace transformation with the earlier results from Examples 9.8 and 9.20:

$$\frac{1}{2\omega}\sin\omega t - \frac{1}{2}t\cos\omega t \xrightarrow{\mathcal{L}} \frac{1}{2\omega}\frac{\omega}{p^2+\omega^2} - \frac{1}{2}\frac{p^2-\omega^2}{(p^2+\omega^2)^2}$$

$$= \frac{p^2+\omega^2-p^2+\omega^2}{2(p^2+\omega^2)^2} = \frac{\omega^2}{(p^2+\omega^2)^2}.$$

The application of Theorem 9.25 to solving differential equations will be given in Section 9.3.4.

The properties of the Laplace transformation provided above enable us to find the transforms for a number of other functions—see the examples at the end of this section.

For convenience, we assemble here in a single table all the Laplace transforms of elementary functions (lines 1–15), as well as basic properties of the transformation (lines 16–26). The last column refers to the text formula where the result was first presented; formulas 13 and 14 were not derived in the text, but can be found using the same method as in 11 and 12, i.e. as in Example 9.20.

	$f(t)$	$F(p)$	text
1.	1	$\frac{1}{p}$	(9.9)
2.	e^{at}	$\frac{1}{p-a}$	(9.10)
3.	$\sin \omega t$	$\frac{\omega}{p^2+\omega^2}$	(9.14)
4.	$\cos \omega t$	$\frac{p}{p^2+\omega^2}$	page 275
5.	$\sinh \omega t$	$\frac{\omega}{p^2-\omega^2}$	page 275
6.	$\cosh \omega t$	$\frac{p}{p^2-\omega^2}$	page 275
7.	$e^{at} \sin \omega t$	$\frac{\omega}{(p-a)^2+\omega^2}$	(9.18)
8.	$e^{at} \cos \omega t$	$\frac{p-a}{(p-a)^2+\omega^2}$	(9.19)
9.	t^n, $n = 1, 2, \ldots$	$\frac{n!}{p^{n+1}}$	(9.26)
10.	$t^n e^{at}$, $n = 1, 2, \ldots$	$\frac{n!}{(p-a)^{n+1}}$	(9.27)
11.	$t \sin \omega t$	$\frac{2p\omega}{(p^2+\omega^2)^2}$	(9.28)
12.	$t \cos \omega t$	$\frac{p^2-\omega^2}{(p^2+\omega^2)^2}$	(9.28)
13.	$t \sinh \omega t$	$\frac{2p\omega}{(p^2-\omega^2)^2}$	
14.	$t \cosh \omega t$	$\frac{p^2+\omega^2}{(p^2-\omega^2)^2}$	
15.	$\frac{\sin \omega t}{t}$	$\frac{\pi}{2} - \tan^{-1} \frac{p}{w}$	(9.30)
16.	$\alpha f(t) + \beta g(t)$	$\alpha F(p) + \beta G(p)$	(9.12)
17.	$f(\lambda t)$	$\frac{1}{\lambda} F\left(\frac{p}{\lambda}\right)$	(9.15)
18.	$h(t-\tau) f(t-\tau)$	$e^{-p\tau} F(p)$	(9.16)
19.	$e^{-\lambda t} f(t)$	$F(p+\lambda)$	(9.17)
20.	$f'(t)$	$pF(p) - f(0)$	(9.20)
21.	$f^{(n)}(t)$	$p^n F(p) - \sum_{k=0}^{n-1} p^{n-1-k} f^{(k)}(0)$	(9.21)
22.	$\int_0^t f(\tau)\, d\tau$	$\frac{F(p)}{p}$	(9.23)
23.	$-t f(t)$	$F'(p)$	(9.24)
24.	$\frac{f(t)}{t}$	$\int_p^\infty F(s)\, ds$	(9.29)
25.	$f * g(t)$	$F(p) G(p)$	(9.33)

Problems

1. Find the Laplace transform of the function $f(t)$ graphed in Fig. 99.

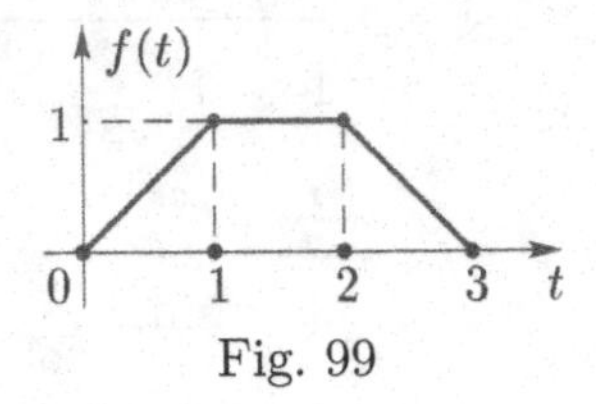

Fig. 99

2. Find the Laplace transform of the functions $f(t)$ defined in Section 9.1, problem 2.

3. Find the Laplace transform of the functions given by the diagrams.

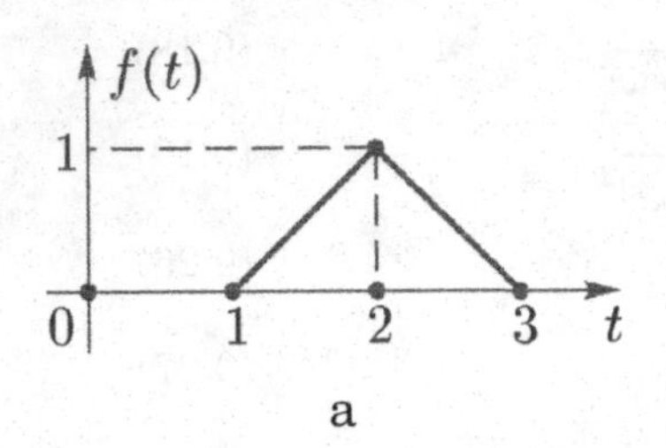

a

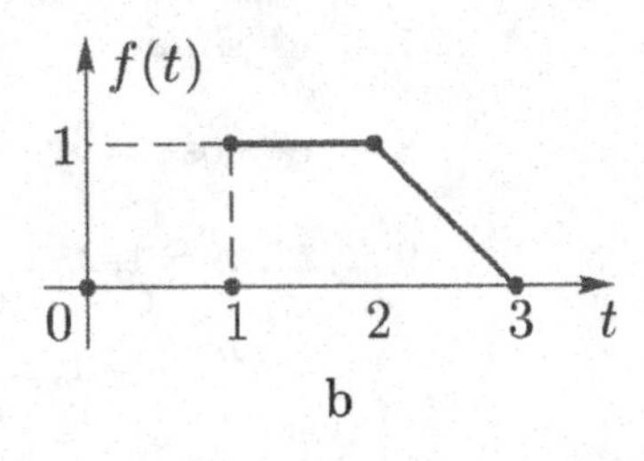

b

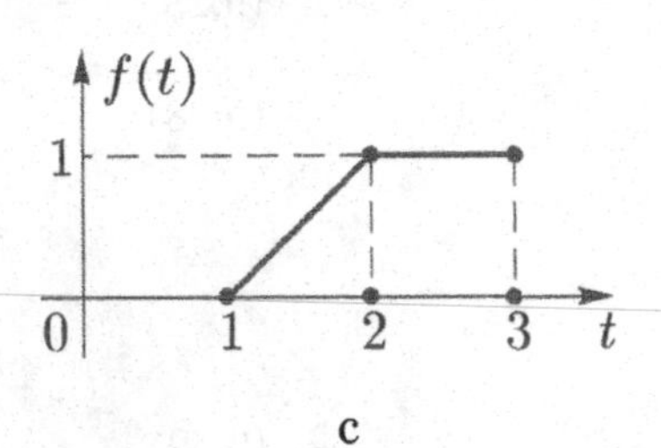

c

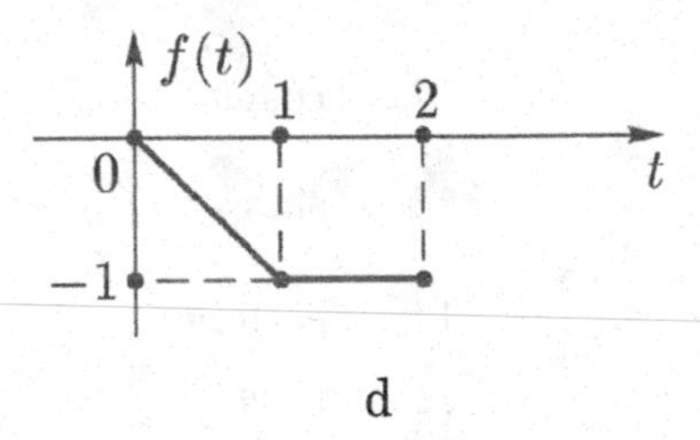

d

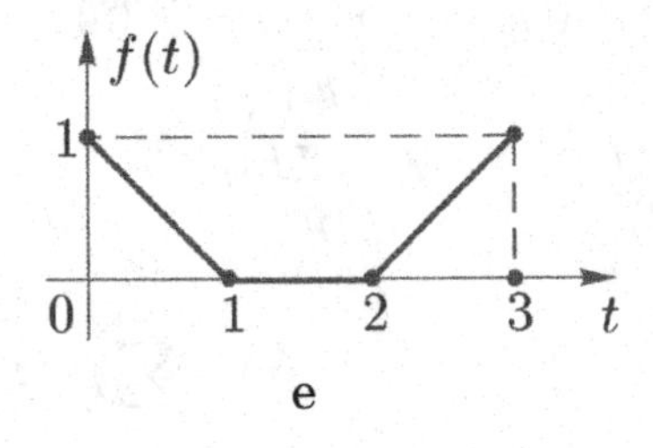

e

Fig. 100

4. Find the Laplace transform of an infinite series of repetitive signals:
1) with a period of 3, of which the first signal is shown in Fig. 100 a and b;
2) with a period of 2, of which the first signal is shown in Fig. 100 d.

5. Let $f(t)$ be an infinite series of repetitive signals with a period of T, of which the first signal is a given function $f_0(t)$, $0 \le t \le T$. Prove that

$$f(t) \xrightarrow{\mathcal{L}} \frac{F_0(p)}{1 - e^{-pT}}, \quad \text{where } F_0(p) = \int_0^T f_0(t) e^{-pt}\, dt.$$

6. Find the Laplace transform of the function $f(t)$ and indicate the half-plane of convergence of the integral in (9.7).

a) $f(t) = \dfrac{e^{2t} - 1}{t}$; b) $f(t) = \dfrac{1 - \cos \omega t}{t}$.

7. Find the original function corresponding to the Laplace transform

$$F(p) = \frac{1}{(p^2 + 4p + 13)^2},$$

and verify your answer by calculating the transform of the function you find.

8. Find the original function for the Laplace transform, then verify it by calculating the transform of the function you obtained.

a) $F(p) = \dfrac{1}{(p^2 + 2p + 5)^2}$ b) $F(p) = \dfrac{p^2}{(p^2 + 4)^2}$

9. Find the original function for this Laplace transform in two ways: using the convolution Theorem 9.25, and using formula 11 in the table.

$$F(p) = \frac{p}{(p^2 + 16)^2}$$

9.3 Applications to Differential Equations

9.3.1 Linear ODEs

One of the most important applications of the Laplace transformation is to the solution of linear differential equations with constant coefficients, that is equations of the form

$$a_n y^{(n)} + a_{n-1} y^{(n-1)} + \cdots + a_0 y = f(t), \tag{9.35}$$

where the a_k are constants, called the *coefficients of the equation*, $f(t)$ is a given function, and $y = y(t)$ is the unknown function which we hope to find. A *Cauchy problem* is to find the solution to such an equation which satisfies given initial conditions

$$y(0) = y_0, \quad y'(0) = y_0', \ \ldots, \ y^{(n-1)}(0) = y_0^{(n-1)}, \tag{9.36}$$

where the $y_0, y_0', \ldots, y_0^{(n-1)}$ are arbitrary given values.

We will assume in this section that the right side of the differential equation, $f(t)$, satisfies the conditions (1)–(3) in Section 9.1. It can be proved that under these assumptions, a solution $y(t)$ to the Cauchy problem exists uniquely, and also satisfies these conditions. Here we will explain the basic steps of the method of solution using the Laplace transformation, as sketched in the introduction to this chapter.

1. Find the Laplace transform $F(p)$ of $f(t)$; we will use $Y(p)$ to denote the transform of the function $y(t)$ which we seek.

2. Using the linearity of the Laplace transformation, and the formula (9.21) to transform each derivative $y^{(k)}$, we transform the differential equation (9.35) to one involving $F(p)$ and $Y(p)$ instead of $f(t)$ and $y(t)$:

$$\begin{aligned} &a_n(p^nY - p^{n-1}y_0 - p^{n-2}y_0' - \cdots - y_0^{(n-1)}) \\ &+a_{n-1}(p^{n-1}Y - p^{n-2}y_0 - p^{n-3}y_0' - \cdots - y_0^{(n-2)}) \\ &+\cdots + a_1(pY - y_0) + a_0Y = F(p). \end{aligned} \tag{9.37}$$

   This new equation is simpler than that original equation (9.35), because the original was a differential equation, while the new one is a linear algebraic equation for Y.

3. Solve the new algebraic equation for Y. We will keep the terms containing Y of the left, while bringing all other terms to the right of the equation:

$$\begin{aligned} &Y(a_np^n + a_{n-1}p^{n-1} + \cdots + a_0) \\ &= F(p) + a_n(p^{n-1}y_0 + p^{n-2}y_0' + \cdots + y_0^{(n-1)}) \\ &+ a_{n-1}(p^{n-2}y_0 + p^{n-3}y_0' + \cdots + y_0^{(n-2)}) + \cdots + a_1y_0. \end{aligned} \tag{9.38}$$

   We denote the parenthetical expression on the left by

$$A(p) = a_np^n + a_{n-1}p^{n-1} + \cdots + a_0;$$

   this is called the *characteristic polynomial* of the linear differential equation. The right side of (9.38), except for the term $F(p)$, is denoted by $B(p)$; clearly this is another polynomial in p, of order $n-1$. With this notation, equation (9.38) can put in the form

$$A(p)Y = F(p) + B(p),$$

   from which we get the solution

$$Y = Y(p) = \frac{F(p)}{A(p)} + \frac{B(p)}{A(p)}. \tag{9.39}$$

   Note that with *null initial conditions*, meaning that $y_0 = y_0' = \cdots = y_0^{(n-1)} = 0$, we have $B(p) = 0$.

4. The final step is to find the function $y(t)$ those transform is $Y(p)$; that function is the solution to the original Cauchy problem. In Section 9.3.4 we will say more about the functions those transforms are $\frac{F(p)}{A(p)}$ and $\frac{B(p)}{A(p)}$.

Example 9.27 Find the solution to the differential equation

$$y'' - 3y' + 2y = \sin 2t,$$

satisfying the initial conditions $y(0) = 0,\ y'(0) = 1$.

Solution According to formula 3 in the Laplace transform table,

$$\sin 2t \xrightarrow{\mathcal{L}} \frac{2}{p^2+4}.$$

Using properties 21 and 20 of the table, we can transform the differential equation to

$$(p^2Y - p\cdot 0 - 1) - 3(pY - 0) + 2Y = \frac{2}{p^2+4},$$

from which we get

$$(p^2 - 3p + 2)Y = \frac{2}{p^2+4} + 1 = \frac{p^2+6}{p^2+4},$$

$$Y = Y(p) = \frac{p^2+6}{(p^2+4)(p^2-3p+2)}.$$

This transform is a rational function in which the degree of the numerator is less than that of the denominator.

It remains for us to find the original function $y(t)$ corresponding to this transform. We defer this step till the next section, where we look in detail at methods of recovering an original function from its transform.

9.3.2 Finding the original function from its transform

We will look at two ways of doing this, i.e. inverting the Laplace transformation. The first method is to expand the rational function $Y(p)$ into a sum of simpler ones using the technique of partial fractions. After this, it may be possible to recognize each term as the transform of a simple function from the Laplace transform table. We assume the reader has learned the technique of partial fractions in an earlier calculus course, so we will just give a brief outline.

Initially the denominator of $Y(p)$ will be factored into terms of the form $(ap+b)^k$ or $(ap^2+bp+c)^k$. Each factor of the form $(ap+b)^k$ corresponds, in the expansion of $Y(p)$, to the sum

$$\frac{A_1}{ap+b} + \frac{A_2}{(ap+b)^2} + \cdots + \frac{A_k}{(ap+b)^k};$$

each factor of the form $(ap^2+bp+c)^k$ corresponds to the sum

$$\frac{B_1p+C_1}{ap^2+bp+c}+\frac{B_2p+C_2}{(ap^2+bp+c)^2}+\cdots+\frac{B_kp+C_k}{(ap^2+bp+c)^k}.$$

The coefficients A_i, B_i, C_i are defined uniquely; they can be found by setting the original $Y(p)$ equal to the expression for the expanded sum.[1] We will illustrate this in more detail in the next example, which finishes the one we started in the last section.

Example 9.28 Find the function $y(t)$ those transform is

$$Y(p)=\frac{p^2+6}{(p^2+4)(p^2-3p+2)}.$$

Solution First let us factor the denominator: the polynomial p^2-3p+2 has two real roots, $p_1=1$ and $p_2=2$; the polynomial p^2+4 has no real roots. So the denominator can be written

$$(p^2+4)(p^2-3p+2)=(p^2+4)(p-1)(p-2).$$

The factors $(p-1)$ and $(p-2)$ are of the first type, with $k=1$; so each corresponds to a single term of the form $A/(p-p_m)$. The factor (p^2+4) is of the second type with $k=1$, and corresponds to $(Bp+C)/(p^2+4)$. So setting the original $Y(p)$ equal to its expansion, we get

$$\frac{p^2+6}{(p^2+4)(p^2-3p+2)}=\frac{p^2+6}{(p^2+4)(p-1)(p-2)}=\frac{A_1}{p-1}+\frac{A_2}{p-2}+\frac{Bp+C}{p^2+4}.$$

Multiplying through by the denominator and canceling all common factors, we get

$$p^2+6=A_1(p-2)(p^2+4)+A_2(p-1)(p^2+4)+(Bp+C)(p-1)(p-2). \quad (9.40)$$

From this equation we can find the values of A_1, A_2, B, and C, either by plugging in a few concrete values for p, like the roots of the denominator; or by expanding the right side and collecting like terms, whose coefficients we set equal to those on the left side. In practice it is best to use a combination of these tricks.

So for example we set $p=1$ to get

$$1^2+6=A_1(1-2)(1^2+4)+A_2\cdot 0+(B\cdot 1+C)\cdot 0,$$

[1] When factoring the denominator into factors of the form $(ap+b)^k$ or $(ap^2+bp+c)^k$, it should be borne in mind that the zeros of these factors must be different. For example, the expansion of the denominator $(p^2-4)(p^2-3p+2)$ (slightly different from the expression in the next example) should be $(p+2)(p-1)(p-2)^2$ but not $(p-2)(p+2)(p-1)(p-2)$.

giving $7 = -5A_1$, or $A_1 = -\frac{7}{5}$. Then we set $p = 2$ to get

$$2^2 + 6 = A_2(2^2 + 4),$$

or $A_2 = \frac{5}{4}$. We see that by substituting a root of the denominator into (9.40), we immediately get one of the coefficients. Also we may plug other numbers into (9.40) and obtain equations for the remaining coefficients. For example setting $p = 0$ we get

$$6 = -8A_1 - 4A_2 + 2C = -8\Big(-\frac{7}{5}\Big) - 4\cdot\frac{5}{4} + 2C = \frac{56}{5} - 5 + 2C;$$

$$6 - \frac{56}{5} + 5 = 2C, \quad C = -\frac{1}{10}.$$

We could find B in a similar way substituting into (9.40) any other value of p, but it is actually easier now to switch to the other trick. Let us find the coefficient of p^3 on both sides of equation (9.40); on the left it is 0; on the right it is

$$A_1 + A_2 + B = -\frac{7}{5} + \frac{5}{4} + B = -\frac{3}{20} + B.$$

Hence $0 = -\frac{3}{20} + B$, and $B = \frac{3}{20}$.

So we have obtained the expansion of $Y(p)$ into partial fractions:

$$\frac{p^2+6}{(p^2+4)(p^2-3p+2)} = -\frac{7}{5}\cdot\frac{1}{p-1} + \frac{5}{4}\cdot\frac{1}{p-2} + \frac{3}{20}\cdot\frac{p}{p^2+4} - \frac{1}{10}\cdot\frac{1}{p^2+4}.$$

At last we are ready to give the solution to the Cauchy problem from Example 9.27. From formulas 2, 4, and 3 of the Laplace transform table, we see that

$$\sin 2t \xrightarrow{\mathcal{L}} \frac{2}{p^2+4} \quad \text{or} \quad \frac{1}{p^2+4} \risingdotseq \frac{1}{2}\sin 2t,$$

$$\frac{p}{p^2+4} \risingdotseq \cos 2t, \quad \frac{1}{p-1} \risingdotseq e^t, \quad \frac{1}{p-2} \risingdotseq e^{2t}.$$

Since the Laplace transformation is linear, the same property holds for its inverse. Therefore, the solution is

$$y(t) = -\frac{7}{5}e^t + \frac{5}{4}e^{2t} + \frac{3}{20}\cos 2t - \frac{1}{20}\sin 2t. \tag{9.41}$$

The second method for recovering the original function for a known transform, is to use the theory of residues.

Theorem 9.29 (Decomposition theorem). *Suppose that $y(t) \xrightarrow{\mathcal{L}} Y(p)$, and that $Y(p)$ is a rational function with poles at $p_1, p_2, \ldots, p_k$. Then*

$$y(t) = \sum_k \operatorname{res}_{p_k}(Y(p)e^{pt}), \tag{9.42}$$

i.e. the function $y(t)$ is equal to the sum of the residues of the function $Y(p)e^{pt}$ at all its poles.

Proof. Let σ_0 be the growth index of $y(t)$, and take any $\sigma > \sigma_0$. By Theorem 9.3, the function $Y(p)$ is analytic on $\operatorname{Re} p > \sigma_0$; therefore all its poles lie in the half plane $\operatorname{Re} p \leq \sigma_0 < \sigma$.

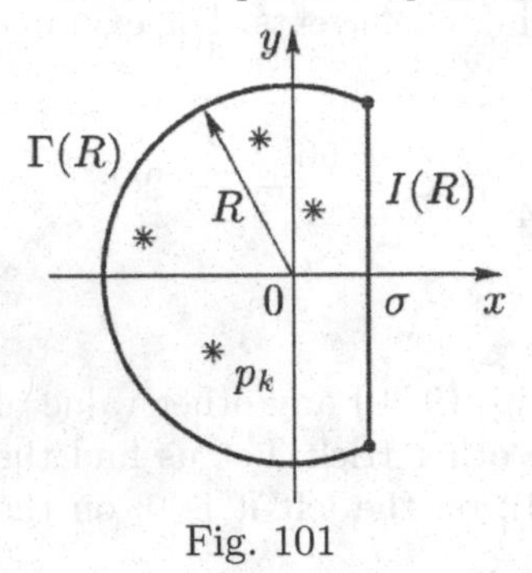

Fig. 101

Because Y is a rational function, there are only finitely many of these poles, so for any sufficiently large R, all the poles will lie inside the region within the circle $|p| = R$ and to the left of the line $\operatorname{Re} p = \sigma$ (Fig. 101). We will denote by $\Gamma(R)$ the arc of the circle that bounds this region, and by $I(R)$ the line segment that bounds it; then $I(R)$ is the set of points z such that

$$\operatorname{Re} z = \sigma \text{ and } -\sqrt{R^2 - \sigma^2} \leq \operatorname{Im} z \leq \sqrt{R^2 - \sigma^2}.$$

According to the residue Theorem 7.15, for any sufficiently large R,

$$\begin{aligned}\sum_k \operatorname{res}_{p_k}(Y(p)e^{pt}) &= \frac{1}{2\pi i} \int_{I(R)\cup\Gamma(R)} Y(p)e^{pt}\,dp \\ &= \frac{1}{2\pi i} \int_{I(R)} Y(p)e^{pt}\,dp + \frac{1}{2\pi i} \int_{\Gamma(R)} Y(p)e^{pt}\,dp;\end{aligned}$$

while the left side of this equation does not depend on R at all. Taking the limit as $R \to \infty$, we get

$$\sum_k \operatorname{res}_{p_k}(Y(p)e^{pt}) = \frac{1}{2\pi i} \lim_{R\to\infty} \int_{I(R)} Y(p)e^{pt}\,dp + \frac{1}{2\pi i} \lim_{R\to\infty} \int_{\Gamma(R)} Y(p)e^{pt}\,dp.$$

We claim that the second limit is equal to 0,

$$\lim_{R\to\infty} \int_{\Gamma(R)} Y(p)e^{pt}\,dp = 0. \tag{9.43}$$

To show this, we make a change of variable $z = -ip$ in the integral; then $p = iz$, $dp = idz$. In multiplying by $-i$, the p-plane is rotated $\frac{\pi}{2}$ radians clockwise; so the vertical line $\operatorname{Re} p = \sigma$ becomes the horizontal line $\operatorname{Im} z = -\sigma$, and the left half-plane $\operatorname{Re} p < \sigma$ becomes the upper half-plane $\operatorname{Im} p > -\sigma$. At the same time, the arc $\Gamma(R)$ becomes $\gamma(R)$, lying in that upper half-plane as in Fig. 59. The changes in the integral give us

$$\int_{\Gamma(R)} Y(p)e^{pt}\,dp = i \int_{\gamma(R)} Y(iz)e^{izt}\,dz.$$

We would like now to apply Jordan's Lemma 7.33, but first we must verify the condition that $Y(iz) \to 0$ as $z \to \infty$. We showed something weaker in Property 9.9: that $Y(p) \to 0$ as $\text{Re}\,p \to \infty$. However, in this context, when $Y(p)$ is assumed to be a rational function, that is good enough: for the only way a rational function could satisfy Property 9.9 is if the denominator is of strictly higher degree than the numerator; and in that case $Y(p) \to 0$ as $p \to \infty$ in any direction. Therefore also $Y(iz) \to 0$ as $z \to \infty$, and we may apply Jordan's lemma, which verifies the claim (9.43).

So,

$$\sum_k \text{res}_{p_k}(Y(p)e^{pt}) = \frac{1}{2\pi i} \lim_{R\to\infty} \int_{I(R)} Y(p)e^{pt}\,dp = \frac{1}{2\pi i} \int_{\sigma-i\infty}^{\sigma+i\infty} Y(p)e^{pt}\,dp.$$

And the formula on the right is exactly the inverse Laplace transform given in Theorem 9.6—see the inversion formula (9.11); hence it is equal to $y(t)$, and equation (9.42) is proved. □

Example 9.30 Find the inverse of the transform

$$Y(p) = \frac{p^2+6}{(p^2+4)(p^2-3p+2)},$$

using residues.

Solution Factoring the denominator completely over the complex numbers, we get

$$Y(p)e^{pt} = \frac{(p^2+6)e^{pt}}{(p-1)(p-2)(p-2i)(p+2i)}.$$

This function has four singularities, at 1, 2, $2i$, and $-2i$, all of which are simple poles. Therefore the residues at these points can be found using any of the formulas in Theorem 7.17. For comparison we will use formula (7.17) for $p_0 = 1$, and formula (7.15) for the others; the formula (7.16) of that theorem is not as convenient in this problem. So for the first pole we write

$$Y(p)e^{pt} = \frac{(p^2+6)e^{pt}}{(p-1)(p-2)(p^2+4)} = \frac{h(p)}{p-1}, \quad \text{where } h(p) = \frac{(p^2+6)e^{pt}}{(p-2)(p^2+4)};$$

$$\text{res}_1(Y(p)e^{pt}) = \frac{(1+6)e^t}{(1-2)(1+4)} = -\frac{7}{5}e^t.$$

At the other poles, using formula (7.15), we get

$$\text{res}_2(Y(p)e^{pt}) = \lim_{p\to 2}(p-2)Y(p)e^{pt} = \lim_{p\to 2} \frac{(p^2+6)e^{pt}}{(p-1)(p^2+4)}$$
$$= \frac{(4+6)e^{2t}}{(2-1)(4+4)} = \frac{5}{4}e^{2t},$$

and

$$\operatorname{res}_{2i}(Y(p)e^{pt}) = \lim_{p\to 2i}(p-2i)Y(p)e^{pt} = \lim_{p\to 2i}\frac{(p^2+6)e^{pt}}{(p^2-3p+2)(p+2i)}$$
$$= \frac{(-4+6)e^{2it}}{(-4-6i+2)(2i+2i)} = \frac{e^{2it}}{4(3-i)} = \frac{(3+i)e^{2it}}{40};$$

$$\operatorname{res}_{-2i}(Y(p)e^{pt}) = \lim_{p\to -2i}(p+2i)Y(p)e^{pt} = \lim_{p\to -2i}\frac{(p^2+6)e^{pt}}{(p^2-3p+2)(p-2i)}$$
$$= \frac{(-4+6)e^{-2it}}{(-4+6i+2)(-2i-2i)} = \frac{e^{-2it}}{4(3+i)} = \frac{(3-i)e^{-2it}}{40}.$$

Applying the previous theorem, we sum the residues to get the function $y(t)$:

$$y(t) = -\frac{7}{5}e^t + \frac{5}{4}e^{2t} + \frac{(3+i)e^{2it}}{40} + \frac{(3-i)e^{-2it}}{40}.$$

Note that last two terms are conjugates of one another, so their sum is a real number which we may rewrite as

$$\frac{3}{40}e^{2it} + \frac{i}{40}e^{2it} + \frac{3}{40}e^{-2it} - \frac{i}{40}e^{-2it} = \frac{3}{40}(e^{2it}+e^{-2it}) + \frac{i}{40}(e^{2it}-e^{-2it})$$
$$= \frac{3}{20}\cos 2t - \frac{1}{20}\sin 2t.$$

So our new solution agrees with (9.41) obtained in the previous example.

The solution $y(t)$, which we have calculated as the solution to the differential equation in Example 9.27, is only guaranteed to be valid when $t \geq 0$, since we have defined the Laplace transform only for functions that are identically 0 when $t < 0$. However, it is easy to verify that the function defined by equality (9.41) is the solution to the Cauchy problem in Example 9.27 on the whole real line.

9.3.3 Differential equations with piecewise defined right hand sides

In this section, we consider linear differential equation of the form (9.35) with piecewise defined functions $f(t)$. Such equations arise in various applications, for example, when $f(t)$ is a short-term signal or a short external influence on the system. In these cases the method outlined is especially effective.

We will need to restore original functions (that is find the inverse) of the Laplace transforms of the form

$$Y(p) = F(p)e^{-p\tau}.$$

For this we will apply formula 18 in the table from right to left in two steps:

1. Find the original function $f(t)$ of $F(p)$.
2. Substitute $(t-\tau)$ for t and multiply $f(t-\tau)$ by $h(t-\tau)$, where h is the Heaviside function.

Example 9.31 Find the inverse of the transform

$$Y(p) = \frac{e^{-ap}}{p^2(p^2+1)}.$$

Solution 1. First we find the original function $f(t)$ of

$$F(p) = \frac{1}{p^2(p^2+1)}.$$

We apply partial fractions and suggest the reader to find $f(t)$ using residues. We have

$$\begin{aligned}\frac{1}{p^2(p^2+1)} &= \frac{A}{p} + \frac{B}{p^2} + \frac{Cp+D}{p^2+1};\\ 1 &= Ap(p^2+1) + B(p^2+1) + (Cp+D)p^2\\ &= (A+C)p^3 + (B+D)p^2 + Ap + B.\end{aligned}$$

For $p = 0$ we get $B = 1$. To find A, C, and D, we find coefficients of p^3, p^2, and p on both sides of the last equation: we have $0 = A + C$, $0 = B + D$, and $0 = A$, respectively. Therefore, $C = 0$ and $D = -1$. So,

$$F(p) = \frac{1}{p^2(p^2+1)} = \frac{1}{p^2} - \frac{1}{p^2+1}.$$

Note that we could have found this representation more easily by

$$\frac{1}{p^2(p^2+1)} = \frac{1+p^2-p^2}{p^2(p^2+1)} = \frac{1+p^2}{p^2(p^2+1)} - \frac{p^2}{p^2(p^2+1)} = \frac{1}{p^2} - \frac{1}{p^2+1}.$$

According to formulas 9 and 3 of the table we obtain $f(t)$:

$$\frac{1}{p^2} - \frac{1}{p^2+1} \risingdotseq t - \sin t = f(t).$$

2. Now we substitute $(t-a)$ for t and multiply $f(t-a)$ by $h(t-a)$:

$$Y(p) = \frac{e^{-ap}}{p^2(p^2+1)} \risingdotseq h(t-a)[(t-a) - \sin(t-a)] = y(t). \tag{9.44}$$

Let's apply this technique to solving differential equations with piecewise defined right-hand sides. Note that the scheme for solving such equations is exactly the same as before.

Example 9.32 Find the solution to the differential equation

$$y'' + y = g(t), \quad \text{where} \quad g(t) = \begin{cases} t, & 0 \le t < \pi, \\ 0, & t \ge \pi, \end{cases}$$

satisfying the initial conditions $y(0) = 1,\ y'(0) = 2$.

Solution To find the Laplace transform of $g(t)$, we write $g(t)$ as

$$g(t) = t - h(t - \pi)t.$$

By formula 9 of the table, $t \xrightarrow{\mathcal{L}} \frac{1}{p^2}$. For the term $h(t-\pi)t$ we apply formula 18 with $\tau = \pi$ and $f(t-\pi) = t$. Hence $f(t) = t + \pi$ (we use the same arguments as in Example 9.12). Therefore,

$$t - h(t-\pi)t \xrightarrow{\mathcal{L}} \frac{1}{p^2} - e^{-p\pi}\left(\frac{1}{p^2} + \frac{\pi}{p}\right).$$

Using properties 21 and 20 of the table we transform the given differential equation to

$$p^2 Y - p - 2 + Y = \frac{1}{p^2} - e^{-p\pi}\left(\frac{1}{p^2} + \frac{\pi}{p}\right),$$

or

$$(p^2 + 1)Y = p + 2 + \frac{1}{p^2} - e^{-p\pi}\left(\frac{1}{p^2} + \frac{\pi}{p}\right),$$

$$Y = \frac{p}{p^2+1} + \frac{2}{p^2+1} + \frac{1}{p^2(p^2+1)} - e^{-p\pi}\left(\frac{1}{p^2(p^2+1)} + \frac{\pi}{p(p^2+1)}\right).$$

It remains to find the inverse Laplace transform. We have

$$\frac{1}{p^2(p^2+1)} = \frac{1}{p^2} - \frac{1}{p^2+1}, \qquad \frac{1}{p(p^2+1)} = \frac{1}{p} - \frac{p}{p^2+1};$$

the first equality was obtained in the previous example. Applying formulas 1, 3, 4, 18 of the table and relation (9.44), we get the desired solution of the given differential equation:

$$y(t) = \cos t + 2\sin t + t - \sin t - h(t-\pi)[(t-\pi) - \sin(t-\pi) + \pi - \pi\cos(t-\pi)].$$

Notice that $y_h(t) = \cos t + 2\sin t$ is the solution of the homogeneous equation $y'' + y = 0$ with initial conditions $y(0) = 1,\ y'(0) = 2$, and the rest of the expression for $y(t)$ is the solution of the initial value problem $y'' + y = g(t)$ with $y(0) = y'(0) = 0$. So, if our differential equation describes a vibration system, the resulting oscillation of the system is the sum of oscillations caused by the initial conditions (namely, by the initial displacement $y(0) = 1$ and the initial velocity $y'(0) = 2$), and oscillations caused by the external force $g(t)$.

We may simplify the answer using the identities $\sin(t - \pi) = -\sin t$ and $\cos(t - \pi) = -\cos t$:

$$y(t) = \cos t + \sin t + t - h(t - \pi)(t + \sin t + \pi \cos t)$$
$$= \begin{cases} \cos t + \sin t + t, & 0 \le t < \pi, \\ (1 - \pi)\cos t, & t \ge \pi. \end{cases}$$

Note that $y(t)$ is continuous and differentiable on the entire interval $(0, \infty)$ including $t = \pi$.

9.3.4 Application of the convolution operation to solving differential equations

According to the equality (9.39), the Laplace transform $Y(p)$ of a solution of a linear differential equation (9.35) with initial conditions (9.36) consists of two parts:

$$Y(p) = \frac{B(p)}{A(p)} + \frac{F(p)}{A(p)} = Y_h(p) + Y_{nh}(p);$$

here we introduce the notations $Y_h(p) = \frac{B(p)}{A(p)}$ and $Y_{nh}(p) = \frac{F(p)}{A(p)}$. Recall that $A(p)$ is the characteristic polynomial of the given differential equation, $F(p) \stackrel{\mathcal{L}}{\leftarrow} f(t)$, and $B(p)$ is a polynomial generated by initial conditions. Finding the inverses of $Y_h(p)$ and $Y_{nh}(p)$, we get

$$y(t) = y_h(t) + y_{nh}(t), \quad \text{where} \quad y_h(t) \stackrel{\mathcal{L}}{\rightarrow} Y_h(p), \quad y_{nh}(t) \stackrel{\mathcal{L}}{\rightarrow} Y_{nh}(p).$$

Therefore, the solution $y(t)$ also consists of two parts. If our equation is homogeneous, that is $f(t) = 0$, then $F(p) = 0$, and $Y_{nh}(p)$, $y_{nh}(t)$ also equal zero; therefore $y(t) = y_h(t)$. Hence, the function $y_h(t)$ is the solution of the homogeneous equation with initial conditions (9.36). On the other hand, if the initial conditions are zero, then $B(p) = 0$ and $y(t) = y_{nh}(t)$. Therefore, $y_{nh}(t)$ is the solution of the non-homogeneous equation (9.35) with zero initial conditions.

We see from (9.38) that $Y_h(p) = \frac{B(p)}{A(p)}$ is a ratio of polynomials $B(p)$ and $A(p)$. So, we know two methods of finding its inverse $y_h(t)$. Concerning the inverse of $Y_{nh}(p)$, we show that Theorem 9.25 on a product of Laplace transforms allows us to write the function $y_{nh}(t)$ as a convolution integral.

Let $g(t)$ be the inverse of $\frac{1}{A(p)}$, that is

$$g(t) \stackrel{\mathcal{L}}{\rightarrow} \frac{1}{A(p)}.$$

Since $f(t) \xrightarrow{\mathcal{L}} F(p)$ and

$$Y_{nh}(p) = F(p) \cdot \frac{1}{A(p)},$$

by Theorem 9.25 we have $y_{nh}(t) = f * g(t)$. So,

$$y(t) = y_h(t) + y_{nh}(t) = y_h(t) + \int_0^t f(\tau)g(t-\tau)\,d\tau. \tag{9.45}$$

This method of solving differential equations is applied in cases where it is difficult to calculate the transform $F(p)$ of $f(t)$ on the right side of the differential equation (9.35). Formula (9.45) is also convenient when you want to solve several differential equations with the same initial conditions and the same left-hand sides, but different right-hand sides. Then it suffices to find y_h and g only ones; solutions of the equations with other $f(t)$ can be found by the use of formula (9.45).

Example 9.33 Express the solution of the initial value problem

$$y'' + 4y' + 4y = f(t), \quad y(0) = -1, \quad y'(0) = 3,$$

in terms of a convolution integral.

Solution We apply the Laplace transformation to both parts of the differential equation and get

$$\begin{aligned}
p^2Y - p\cdot(-1) - 3 + 4(pY + 1) &= F(p),\\
Y(p^2 + 4p + 4) + p + 1 &= F(p),\\
Y &= -\frac{p+1}{p^2+4p+4} + \frac{F(p)}{p^2+4p+4}\\
&= Y_h + Y_{nh}.
\end{aligned}$$

To find $y_h(t)$ (the inverse of $Y_h(p)$), we may use any method. Let us apply residues. The function

$$Y_h(p)e^{pt} = -\frac{p+1}{p^2+4p+4}e^{pt} = -\frac{p+1}{(p+2)^2}e^{pt}$$

has a pole of the second order at $p = -2$. By Theorem 9.29 and formula (7.18),

$$\begin{aligned}
y_h(t) = \operatorname{res}_{-2}(Y_h(p)e^{pt}) &= \lim_{p\to-2}\frac{d}{dt}\left[-(p+2)^2\frac{p+1}{(p+2)^2}e^{pt}\right]\\
&= -\lim_{p\to-2}(e^{pt} + (p+1)te^{pt}) = -e^{-2t} + te^{-2t}.
\end{aligned}$$

Alternatively, we can find the partial fractions decomposition of Y_h without calculating the coefficients A, $B, \ldots$ as follows:

$$-\frac{p+1}{p^2+4p+4} = -\frac{p+2-1}{(p+2)^2} = -\frac{p+2}{(p+2)^2} + \frac{1}{(p+2)^2} = -\frac{1}{(p+2)} + \frac{1}{(p+2)^2}.$$

Then we apply formulas 2, 10 of the table with $a = -2$ and $n = 1$ and get the same result $y_h(t) = -e^{-2t} + te^{-2t}$.

To find $y_{nh}(t)$ (the inverse of $Y_{nh}(p)$), we need the inverse of

$$G(p) = \frac{1}{p^2 + 4p + 4} = \frac{1}{(p+2)^2}.$$

We may apply residues again, but it's easier to use formula 10 of the table:

$$g(t) = te^{-2t}.$$

Therefore, $y_{nh}(t) = f * g(t)$. Finally we have, from (9.45),

$$y(t) = y_h(t) + y_{nh}(t) = -e^{-2t} + te^{-2t} + \int_0^t f(\tau)(t-\tau)e^{-2(t-\tau)}\, d\tau. \qquad (9.46)$$

Example 9.34 Find the solution to the differential equation

$$y'' + 4y' + 4y = \frac{e^{-2t}}{t^2+1},$$

with the initial conditions $y(0) = -1, \quad y'(0) = 3.$

Solution The use of the convolution integral in this case is due in part to the absence anywhere, in our table of Laplace transforms, of the expression on the right-hand side. This initial value problem is a special case of the problem considered in the previous example with

$$f(t) = \frac{e^{-2t}}{t^2+1}.$$

So, we may plug this function into formula (9.46). We get

$$\int_0^t f(\tau)(t-\tau)e^{-2(t-\tau)}\, d\tau = \int_0^t \frac{e^{-2\tau}}{\tau^2+1}(t-\tau)e^{-2(t-\tau)}\, d\tau$$

$$= e^{-2t}\int_0^t \frac{t-\tau}{\tau^2+1}\, d\tau = e^{-2t}\left(\int_0^t \frac{t}{\tau^2+1}\, d\tau - \int_0^t \frac{\tau}{\tau^2+1}\, d\tau\right)$$

$$= e^{-2t}\left(t\tan^{-1}\tau\Big|_0^t - \frac{1}{2}\ln(\tau^2+1)\Big|_0^t\right) = e^{-2t}\left(t\tan^{-1}t - \frac{1}{2}\ln(t^2+1)\right).$$

Hence,

$$y(t) = -e^{-2t} + te^{-2t} + e^{-2t}\left(t\tan^{-1}t - \frac{1}{2}\ln(t^2+1)\right).$$

9.3.5 Systems of differential equations

Analogously we may apply the operational calculus to the solution of *systems* of linear differential equations with constant coefficients. For example, suppose we must solve a system of two differential equations, with initial conditions, for two unknown functions $x(t)$ and $y(t)$. We would first transform the equations to an algebraic system in unknowns $X(p)$ and $Y(p)$; that linear system could then be solved, to find X and Y; then we would try to recover the functions $x(t)$ and $y(t)$ from $X(p)$ and $Y(p)$.

Example 9.35 Solve the system of differential equations

$$\begin{cases} x' = -x + 3y + 1 \\ y' = x + y \end{cases} \quad \text{with} \quad x(0) = 1 \quad \text{and} \quad y(0) = 2.$$

Solution Using formulas 20 and 1 from the table, the original system transforms to

$$\begin{cases} pX - 1 = -X + 3Y + \dfrac{1}{p} \\ pY - 2 = X + Y \end{cases} \quad \text{or} \quad \begin{cases} (p+1)X - 3Y = \dfrac{1}{p} + 1 \\ -X + (p-1)Y = 2. \end{cases}$$

This system of linear equations in X and Y can be solved using any of the standard methods learned in a linear algebra course: substitution, elimination, row reduction, Cramer's rule, matrix inverses, etc. Let us apply the substitution method. Perhaps it is even slightly easier to apply Cramer's rule—see the solution to problem 12 f.

Here the simplest is probably to solve for X in the second equation:

$$X = (p-1)Y - 2,$$

and then substitute that result for X in the first equation, which we then solve for Y:

$$\begin{aligned} (p+1)((p-1)Y - 2) - 3Y &= \frac{1}{p} + 1, \\ (p^2 - 4)Y - 2(p+1) &= \frac{1}{p} + 1, \\ Y &= \frac{2p + 3 + \frac{1}{p}}{p^2 - 4} = \frac{2p^2 + 3p + 1}{p(p^2 - 4)}. \end{aligned}$$

Substituting this into the formula for X gives the rest of the solution:

$$\begin{aligned} X &= (p-1)Y - 2 = (p-1)\left(\frac{2p^2 + 3p + 1}{p(p^2 - 4)}\right) - 2 \\ &= \frac{2p^3 + p^2 - 2p - 1}{p(p^2 - 4)} - \frac{2p(p^2 - 4)}{p(p^2 - 4)} = \frac{p^2 + 6p - 1}{p(p^2 - 4)}. \end{aligned}$$

Now we must find the inverse transforms of $X(p)$ and $Y(p)$. To remind the reader of both ways of restoring original functions, we will do this using two different methods: partial fractions to find $x(t)$, and residues to find $y(t)$.

Expanding $X(p)$ using partial fractions we get

$$\frac{p^2+6p-1}{p(p-2)(p+2)} = \frac{A}{p} + \frac{B}{p-2} + \frac{C}{p+2}$$
$$= \frac{A(p-2)(p+2)+Bp(p+2)+Cp(p-2)}{p(p-2)(p+2)}$$

(each factor in the denominator has the form $(ap+b)^k$ with $k=1$). Setting the numerators equal,

$$p^2+6p-1 = A(p-2)(p+2)+Bp(p+2)+Cp(p-2),$$

and substituting values of p gives

$$\begin{aligned} &\text{when} \quad p=0 \quad \text{then} \quad -1=-4A \quad \text{and} \quad A=\tfrac{1}{4};\\ &\text{when} \quad p=2 \quad \text{then} \quad 15=8B \quad \text{and} \quad B=\tfrac{15}{8};\\ &\text{when} \quad p=-2 \quad \text{then} \quad -9=8C \quad \text{and} \quad C=-\tfrac{9}{8}. \end{aligned}$$

Therefore

$$X(p) = \frac{1}{4}\cdot\frac{1}{p} + \frac{15}{8}\cdot\frac{1}{p-2} - \frac{9}{8}\frac{1}{p+2}.$$

Using formulas 1 and 2 from the table, we get the inverse transform

$$x(t) = \frac{1}{4} + \frac{15}{8}e^{2t} - \frac{9}{8}e^{-2t}.$$

For Y we will use formula (9.42), with the function

$$Y(p)e^{pt} = \frac{(2p^2+3p+1)e^{pt}}{p(p^2-4)} = \frac{(2p^2+3p+1)e^{pt}}{p(p-2)(p+2)}.$$

This has simple poles at 0, 2 and -2; using formula (7.15) we find the residues to be

$$\begin{aligned} \operatorname{res}_0(Y(p)e^{pt}) &= \lim_{p\to 0}\frac{p(2p^2+3p+1)e^{pt}}{p(p-2)(p+2)}\\ &= \lim_{p\to 0}\frac{(2p^2+3p+1)e^{pt}}{(p-2)(p+2)} &&= -\frac{1}{4};\\ \operatorname{res}_2(Y(p)e^{pt}) &= \lim_{p\to 2}\frac{(2p^2+3p+1)e^{pt}}{p(p+2)} &&= \frac{15}{8}e^{2t};\\ \operatorname{res}_{-2}(Y(p)e^{pt}) &= \lim_{p\to -2}\frac{(2p^2+3p+1)e^{pt}}{p(p-2)} &&= \frac{3}{8}e^{-2t}. \end{aligned}$$

So we get

$$y(t) = -\frac{1}{4} + \frac{15}{8}e^{2t} + \frac{3}{8}e^{-2t}.$$

Problems

1. Using two methods, that of partial fractions and that of residues, find the inverse of the rational function

$$F(p) = \frac{2p+1}{(p+1)(p^2+4p+5)}.$$

2. Find the inverse of the Laplace transform in two ways: using partial fractions, and using residues.

a) $F(p) = \dfrac{p+4}{(p-2)(p^2+2p+2)}$; b) $F(p) = \dfrac{2}{p^2(p+1)}$;

c) $F(p) = \dfrac{p+10}{p^3-6p^2+10p}$; d) $F(p) = \dfrac{2p+1}{(p+2)(p-1)^2}$;

e) $F(p) = \dfrac{2p+3}{(p-1)(p^2+4)}$.

3. Find the solution to the initial value problem

$$y'' + y' - 2y = e^t, \quad y(0) = -1, \quad y'(0) = 0.$$

4. Find the solution of the differential equation satisfying the given initial conditions.

a) $y'' + y' - 2y = e^{-t}, \quad y(0) = -1,\ y'(0) = 0$;

b) $y'' + y = 6e^{-t}, \quad y(0) = 3,\ y'(0) = 1$;

c) $y'' - 3y' + 2y = 12e^{3t}, \quad y(0) = 2,\ y'(0) = 6$;

d) $y'' - 2y' - 3y = 2t, \quad y(0) = 1,\ y'(0) = 1$;

e) $y'' - 9y = \sin t - \cos t, \quad y(0) = -3,\ y'(0) = 2$.

5. Find the inverse of the Laplace transform. Use any method.

a) $F(p) = \dfrac{(p^2+6)e^{-p\pi}}{(p^2+4)(p^2-3p+2)}$; b) $F(p) = \dfrac{(p+4)e^{-2p}}{(p-2)(p^2+2p+2)}$;

c) $F(p) = \dfrac{2e^{-3p}}{p^2(p+1)}$; d) $F(p) = \dfrac{(9p+10)e^{-p\pi}}{p^3-6p^2+10p}$;

e) $F(p) = \dfrac{(2p+1)e^{-5p}}{(p+2)(p-1)^2}$; f) $F(p) = \dfrac{(2p+3)e^{-p\pi}}{(p-1)(p^2+4)}$.

6. Find the solution to the initial value problem for a first order differential equation

$$y' - 2y = f(t), \quad y(0) = 5,$$

where $f(t)$ is a function defined in problem 2 of Section 9.1.

7. Find the solution to the initial value problem

$$y' + 3y = f(t), \quad y(0) = 0,$$

where $f(t)$ is a function defined in Section 9.2, problem 3.

8. Solve the initial value problem

$$y'' - 3y' + 2y = f(t), \quad y(0) = 0, \quad y'(0) = 1, \quad \text{where}$$

a) $f(t) = \begin{cases} 1, & 0 \le t < 3, \\ 0, & t \ge 3. \end{cases}$; b) $f(t) = \begin{cases} \sin t, & 0 \le t < \pi, \\ 0, & t \ge \pi. \end{cases}$.

9. (i) Express the solution of the initial value problem in terms of a convolution integral. (ii) Solve the problem for the given function $f(t)$.

a) (i) $y'' - 2y' + y = f(t), \quad y(0) = -1,\ y'(0) = 2,$ (ii) $f(t) = \dfrac{e^t}{t^2+4}$;

b) (i) $y'' - 6y' + 9y = f(t), \quad y(0) = 2,\ y'(0) = -1,$ (ii) $f(t) = \dfrac{e^{3t}}{t^2+1}$;

c) (i) $y'' + 9y = f(t), \quad y(0) = 1,\ y'(0) = 0,$ (ii) $f(t) = \sin 3t$;

d) (i) $y'' - 4y' + 5y = f(t), \quad y(0) = 0,\ y'(0) = 0,$ (ii) $f(t) = e^{2t}$.

10. Solve the initial value problem

$$y' - 3y = \frac{e^{3t}}{t^2+4}, \quad y(0) = 1.$$

11. Using convolution, prove that the solution to the initial value problem

$$y' + ay = f(t), \quad y(0) = y_0, \quad \text{where } a \text{ is a given constant,}$$

has the form $y(t) = e^{-at}\left(y_0 + \int_0^t f(\tau)e^{a\tau}\, d\tau\right)$.

12. Find the solution of the system of differential equations satisfying the given initial conditions.

a) $\begin{cases} x' = x + 3y + 2, & x(0) = -1, \\ y' = x - y + 1, & y(0) = 2. \end{cases}$ b) $\begin{cases} x' = x + 4y, & x(0) = 1, \\ y' = 2x - y + 9, & y(0) = 0. \end{cases}$

c) $\begin{cases} x' = x + 2y + 1, & x(0) = 0, \\ y' = 4x - y, & y(0) = 1. \end{cases}$ d) $\begin{cases} x' = 2x + 5y, & x(0) = 1, \\ y' = x - 2y + 2, & y(0) = 1. \end{cases}$

e) $\begin{cases} x' = -2x + 5y + 1, & x(0) = 0, \\ y' = x + 2y + 1, & y(0) = 2. \end{cases}$ f) $\begin{cases} x' = 3x + y, & x(0) = 2, \\ y' = -5x - 3y + 2, & y(0) = 0. \end{cases}$

Solutions, Hints, and Answers to Selected Problems

Section 1.2

3. (a) *Hint.* Represent z_1, z_2 in trigonometric form:

$$z_1 = r_1(\cos\phi_1 + i\sin\phi_1), \quad z_2 = r_2(\cos\phi_2 + i\sin\phi_2);$$
$$z_1 - z_2 = (r_1\cos\phi_1 - r_2\cos\phi_2) + i(r_1\sin\phi_1 - r_2\sin\phi_2).$$

Then apply (1.1) and the addition formula (1.9) for cosine. The identity we prove is equivalent to the law of cosines (also known as the cosine rule) which generalizes the Pythagorean theorem.

(b) Use the same representation for z_1, z_2 as in (a). To interpret the identity geometrically, sketch a parallelogram with sides z_1, z_2, and its diagonals.

7. *Solution.* Recall that the algebraic, or Cartesian, form of a complex number is $z = x + iy$.

1. First we put the number $z = \sqrt{3} - i$ in trigonometric, or polar, form; for this we need the modulus r and argument ϕ of z:

$$r = \sqrt{\left(\sqrt{3}\right)^2 + (-1)^2} = \sqrt{3+1} = 2;$$
$$\cos\phi = \frac{x}{r} = \frac{\sqrt{3}}{2}; \quad \sin\phi = \frac{y}{r} = -\frac{1}{2}.$$

We see that z lies in the fourth quadrant, so $\phi = -\frac{\pi}{6} + 2\pi k, \quad k \in \mathbb{Z}$. We may take any of these values for ϕ, so take $k = 0$:

$$\sqrt{3} - i = 2\left(\cos\left(-\frac{\pi}{6}\right) + i\sin\left(-\frac{\pi}{6}\right)\right).$$

2. We find z^7 using de Moivre's formula (1.15):

$$\left(\sqrt{3} - i\right)^7 = 2^7\left(\cos\left(-\frac{7\pi}{6}\right) + i\sin\left(-\frac{7\pi}{6}\right)\right)$$
$$= 2^7\left(-\cos\frac{\pi}{6} + i\sin\frac{\pi}{6}\right) = 2^7\left(-\frac{\sqrt{3}}{2} + i\frac{1}{2}\right) = 64(-\sqrt{3} + i).$$

9. *Solution.* 1. Find the modulus and argument of the number $z = -\frac{1}{2} - i\frac{\sqrt{3}}{2}$ which stands under the root symbol:

$$r = \sqrt{\left(-\frac{1}{2}\right)^2 + \left(-\frac{\sqrt{3}}{2}\right)^2} = \sqrt{\frac{1}{4} + \frac{3}{4}} = 1;$$

$$\cos\phi = -\frac{1}{2}; \quad \sin\phi = -\frac{\sqrt{3}}{2}.$$

Since z lies in the third quadrant, $\phi = -\frac{2\pi}{3} + 2\pi k$, for $k \in \mathbb{Z}$.

2. Applying formula (1.17) for the n-th root, we get

$$\sqrt[4]{\frac{-1 - i\sqrt{3}}{2}} = \sqrt[4]{1}\left(\cos\frac{-\frac{2\pi}{3} + 2\pi k}{4} + i\sin\frac{-\frac{2\pi}{3} + 2\pi k}{4}\right)$$
$$= \cos\left(-\frac{\pi}{6} + \frac{\pi}{2}k\right) + i\sin\left(-\frac{\pi}{6} + \frac{\pi}{2}k\right), \quad \text{for} \quad k = 0, 1, 2, 3.$$

Substituting each of these values of k, we get the desired roots:

$$w_0 = \cos\left(-\frac{\pi}{6}\right) + i\sin\left(-\frac{\pi}{6}\right) = \frac{\sqrt{3}}{2} - i\frac{1}{2};$$
$$w_1 = \cos\frac{\pi}{3} + i\sin\frac{\pi}{3} = \frac{1}{2} + i\frac{\sqrt{3}}{2};$$
$$w_2 = \cos\frac{5\pi}{6} + i\sin\frac{5\pi}{6} = -\frac{\sqrt{3}}{2} + i\frac{1}{2};$$
$$w_3 = \cos\frac{4\pi}{3} + i\sin\frac{4\pi}{3} = -\frac{1}{2} - i\frac{\sqrt{3}}{2}.$$

No further distinct roots are produced by new values of k.

10. a) *Hint.* Since $-1 = \cos(\pi + 2\pi k) + i\sin(\pi + 2\pi k)$, we have

$$w_0 = \frac{1}{2} + i\frac{\sqrt{3}}{2}, \quad w_1 = -1, \quad \frac{1}{2} - i\frac{\sqrt{3}}{2};$$

other values of k repeat these roots.

b) *Hint.* The modulus and argument of $-\frac{1}{2} + i\frac{\sqrt{3}}{2}$ are 1 and $\frac{2\pi}{3}$ respectively. Hence,

$$\sqrt[4]{-\frac{1}{2} + i\frac{\sqrt{3}}{2}} = \cos\left(\frac{\pi}{6} + \frac{\pi}{2}k\right) + \sin\left(\frac{\pi}{6} + \frac{\pi}{2}k\right), \quad \text{where} \quad k \in \mathbb{Z}.$$

When $k = 0, 1, 2, 3$, we get

$$w_0 = \frac{\sqrt{3}}{2} + i\frac{1}{2}, \quad w_1 = -\frac{1}{2} + i\frac{\sqrt{3}}{2}, \quad w_2 = -\frac{\sqrt{3}}{2} - i\frac{1}{2}, \quad w_3 = \frac{1}{2} - i\frac{\sqrt{3}}{2};$$

other values of k repeat these roots.

11. *Solution.*

1. First we rewrite $z = 2 - 2i$, which is to be raised to the power 4, in trigonometric form:

$$r = |z| = \sqrt{2^2 + (-2)^2} = \sqrt{8} = 2\sqrt{2};$$
$$\cos\phi = \frac{2}{s\sqrt{2}} = \frac{1}{\sqrt{2}}, \quad \sin\phi = -\frac{1}{\sqrt{2}}.$$

Since z lies in the fourth quadrant, we may take $\phi = -\frac{\pi}{4}$, so that

$$2 - 2i = 2\sqrt{2}\left(\cos\left(-\frac{\pi}{4}\right) + i\sin\left(-\frac{\pi}{4}\right)\right).$$

2. Now we may use de Moivre's formula to raise this number to the power 4:

$$(2-2i)^4 = (2\sqrt{2})^4\left(\cos\left(-\frac{4\pi}{4}\right) + i\sin\left(-\frac{4\pi}{4}\right)\right) = 64(-1 + i0) = -64.$$

To find $(1 - 2i)^2$, it is easier to just expand algebraically:

$$(1-2i)^2 = 1 - 4i + (2i)^2 = 1 - 4i - 4 = -3 - 4i.$$

3. Expressing the sums in algebraic form gives:

$$(2-2i)^4 + 72 + 4i = -64 + 72 + 4i = 8 + 4i,$$
$$(1-2i)^2 + 5i = -3 - 4i + 5i = -3 + i.$$

4. Now we put the fraction is algebraic form. We do this by multiplying the numerator and denominator by the conjugate of the denominator:

$$\frac{8+4i}{-3+i} = \frac{(8+4i)(-3-i)}{(-3+i)(-3-i)} = \frac{-24 - 12i - 8i + 4}{(-3)^2 + 1} = \frac{-20 - 20i}{10} = -2 - 2i.$$

5. We will take the cube root by applying de Moivre's formula. First we must put the number $z = -2 - 2i$ in trigonometric form:

$$r = \sqrt{(-2)^2 + (-2)^2} = \sqrt{8} = 2\sqrt{2};$$
$$\cos\phi = -\frac{2}{2\sqrt{2}} = -\frac{1}{\sqrt{2}}, \quad \sin\phi = -\frac{1}{\sqrt{2}}.$$

Since z is in the third quadrant, we may take $\phi = -\frac{3\pi}{4}$ (or, just as easily, $\phi = \frac{5\pi}{4}$):

$$-2-2i = 2\sqrt{2}\left(\cos\left(-\frac{3\pi}{4}\right) + i\sin\left(-\frac{3\pi}{4}\right)\right).$$

By de Moivre's formula (1.17),

$$\sqrt[3]{-2-2i} = \sqrt[3]{2\sqrt{2}}\left(\cos\frac{-\frac{3\pi}{4}+2\pi k}{3} + i\sin\frac{-\frac{3\pi}{4}+2\pi k}{3}\right), \quad k = 0, 1, 2.$$

As $\sqrt[3]{2\sqrt{2}} = \sqrt{2}$, for $k = 0, 1, 2$ we get

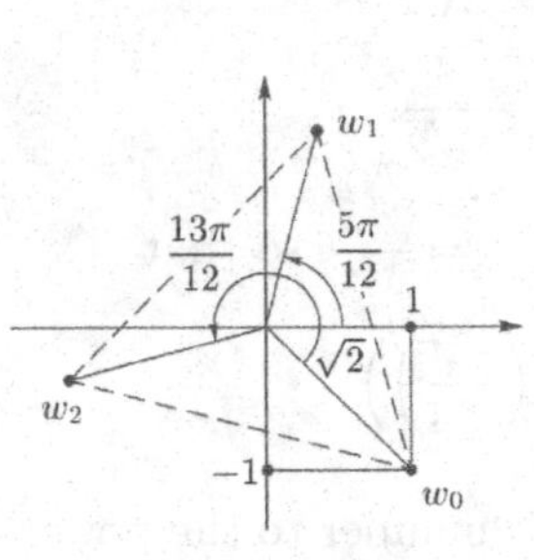

Fig. 102

$$w_0 = \sqrt{2}\left(\cos\left(-\frac{\pi}{4}\right) + i\sin\left(-\frac{\pi}{4}\right)\right) = 1 - i;$$

$$w_1 = \sqrt{2}\left(\cos\frac{5\pi}{12} + i\sin\frac{5\pi}{12}\right);$$

$$w_2 = \sqrt{2}\left(\cos\frac{13\pi}{12} + i\sin\frac{13\pi}{12}\right).$$

Finally, plotting these values in the complex plane yields Fig. 102. The points w_0, w_1, and w_2 are the vertices of an equilateral triangle centered at the origin.

13. *Solution.* By the fundamental theorem of algebra (Theorem 7.40), this equation has six roots, among which some may be repeated. We make the change of variable $t = z^3$, $t^2+28t+27 = 0$, changing the equation to quadratic form, and then

$$t = \frac{-28 \pm \sqrt{28^2 - 4\cdot 1\cdot 27}}{2\cdot 1} = \frac{-28 \pm 26}{2} = -1, -27.$$

Now we find the values of z. When $t = -27$,

$$z^3 = -27 = 27\left(\cos\pi + i\sin\pi\right)$$

in trigonometric form. Using de Moivre's formula (1.17), we get

$$z = \sqrt[3]{-27} = \sqrt[3]{27}\left(\cos\frac{\pi+2\pi k}{3} + i\sin\frac{\pi+2\pi k}{3}\right), \quad k = 0, 1, 2.$$

So

$$\text{when } k = 0, \quad z_1 = 3\left(\cos\frac{\pi}{3} + i\sin\frac{\pi}{3}\right) = \frac{3}{2} + i\frac{3\sqrt{3}}{2};$$

$$\text{when } k = 1, \quad z_2 = 3\left(\cos\pi + i\sin\pi\right) = -3;$$

$$\text{when } k = 2, \quad z_3 = 3\left(\cos\frac{5\pi}{3} + i\sin\frac{5\pi}{3}\right) = \frac{3}{2} - i\frac{3\sqrt{3}}{2};.$$

Similarly, when $t = -1 = \cos\pi + i\sin\pi$, we get

$$z_4 = \cos\frac{\pi}{3} + i\sin\frac{\pi}{3} = \frac{1}{2} + i\frac{\sqrt{3}}{2}; \quad z_5 = -1; \quad z_6 = \frac{1}{2} - i\frac{\sqrt{3}}{2}.$$

Thus all roots of the equation turn out to be distinct.

14. *Solution.* This equation has four roots. On making the change of variable $t = z^2$, the equation changes to quadratic form $t^2 - 2t + 4 = 0$, and

$$t = \frac{2 \pm \sqrt{(-2)^2 - 4\cdot 1\cdot 4}}{2\cdot 1} = \frac{2 \pm \sqrt{-12}}{2} = \frac{2 \pm i2\sqrt{3}}{2} = 1 \pm i\sqrt{3}.$$

Now we find the values of z. When

$$t = 1 + i\sqrt{3} = 2\left(\cos\frac{\pi}{3} + i\sin\frac{\pi}{3}\right)$$

we get by de Moivre's formula (1.17)

$$z = \sqrt{1 + i\sqrt{3}} = \sqrt{2}\left(\cos\frac{\frac{\pi}{3} + 2\pi k}{2} + i\sin\frac{\frac{\pi}{3} + 2\pi k}{2}\right), \quad k = 0, 1.$$

Then

$$\text{when } k = 0, \quad z_1 = \sqrt{2}\left(\cos\frac{\pi}{6} + i\sin\frac{\pi}{6}\right) = \sqrt{2}\left(\frac{\sqrt{3}}{2} + i\frac{1}{2}\right);$$

$$\text{when } k = 1, \quad z_2 = \sqrt{2}\left(\cos\frac{7\pi}{6} + i\sin\frac{7\pi}{6}\right) = -\sqrt{2}\left(\frac{\sqrt{3}}{2} + i\frac{1}{2}\right).$$

Similarly, when

$$t = 1 - i\sqrt{3} = 2\left(\cos\left(-\frac{\pi}{3}\right) + i\sin\left(-\frac{\pi}{3}\right)\right)$$

we get

$$z_3 = \sqrt{2}\left(\cos\left(-\frac{\pi}{6}\right) + i\sin\left(-\frac{\pi}{6}\right)\right) = \sqrt{2}\left(\frac{\sqrt{3}}{2} - i\frac{1}{2}\right);$$

$$z_4 = \sqrt{2}\left(\cos\frac{5\pi}{6} + i\sin\frac{5\pi}{6}\right) = \sqrt{2}\left(-\frac{\sqrt{3}}{2} + i\frac{1}{2}\right).$$

15. a) *Hint.* Setting $t = z^3$, the equation becomes $t^2 - 9t + 8 = 0$, so that $t = 1,\ 8$. Solving $t = z^3$ for each of these values gives

$$z = 1, \quad -\frac{1}{2} + i\frac{\sqrt{3}}{2}, \quad -\frac{1}{2} - i\frac{\sqrt{3}}{2}, \quad 2, \quad -1 + i\sqrt{3}, \quad -1 - i\sqrt{3}.$$

16. a) The inequality $|z - i| < 2$ represents the open disk centered at i with radius 2; while the inequality $\operatorname{Im} z > 2$ represents the open half-plane above $y = 2$. The desired set is the intersection of these sets—see Fig. 103.

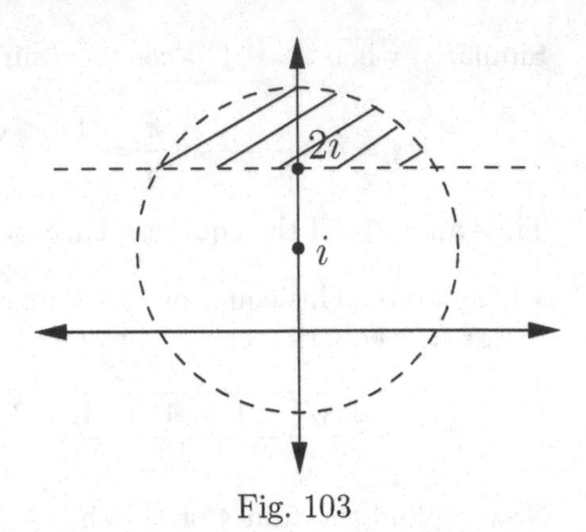

Fig. 103

c) The inequality $|z - 1 + i| < 1$ represents the open disk centered at $1 - i$ and with radius 1; the inequality $|\operatorname{Arg} z + \frac{\pi}{4}| < \frac{\pi}{6}$ represents the sector between the rays at angles $-\frac{\pi}{12}$ and $-\frac{5\pi}{12}$ to the x-axis. The desired set is shown in Fig. 104.

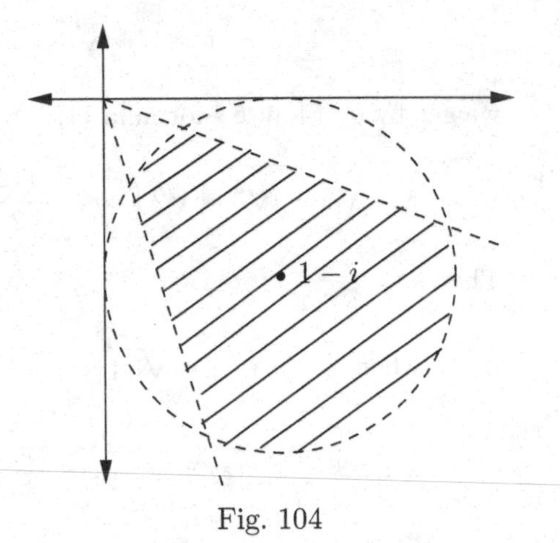

Fig. 104

e) The inequality $|z - 1 + i| < 2$ represents the open disk centered at $1 - i$ and with radius 2; the inequality $|\operatorname{Arg} z| < \frac{\pi}{3}$ represents the sector between the rays at angles $\pm\frac{\pi}{3}$ to the x-axis; and $\operatorname{Im} z > -1$ is the half-plane above $y = -1$. The intersection of these three sets is shown in Fig. 105.

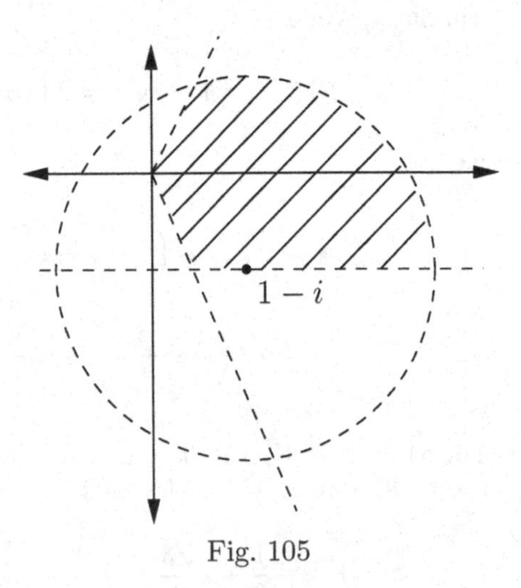

Fig. 105

17. a) $-\frac{\pi}{3} \le \operatorname{Arg} z \le \frac{\pi}{4}, \ |z - 3| \le 3.$

Section 2.1

1. *Hint.* The strip is an open connected set, that is a domain. Its boundary is connected since the lines $\operatorname{Im} z = 0$ and $\operatorname{Im} z = 1$ meet at the boundary point at ∞.

2. a) *Solution.* The inequality $|z - 2i| < 2$ says that the distance from z to $z_0 = 2i$ is less than 2; this set of points is the interior of the circle centered at $2i$ with radius 2.
The condition $1 < \operatorname{Im} z < 3$ says that the y-coordinate of the point z is between 1 and 3; the set of such z is the strip lying between the horizontal lines $y = 1$ and $y = 3$.
The desired set is the intersection of the disk and the strip, shown in Fig. 106; note that the boundary of the shaded region is not included in the set. The set is a simply connected domain.
b) The set is not a domain.

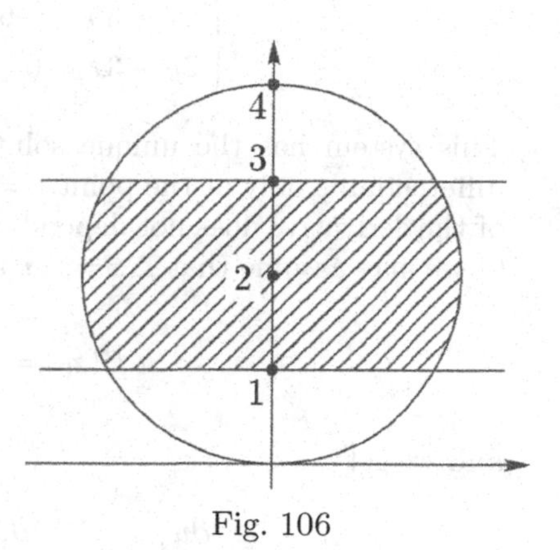

Fig. 106

4. *Answer.* All sets are closed simply connected domains.

Section 3.1

1. *Hint.* To prove that a differentiable function at a point z_0 has a derivative and $A = f'(z_0)$, use definitions (3.1) and (3.2).

To prove that if a function $w = f(z)$ has a derivative at the point z_0 then it is differentiable at z_0 and $A = f'(z_0)$, take

$$\alpha(\Delta z) = \frac{\Delta w}{\Delta z} - f'(z_0). \tag{9.47}$$

2. *Solution.* We first express the given function $w = f(z)$ in the form $f(z) = u(x, y) + iv(x, y)$. Writing $z = x + iy$, and using the equalities $\overline{z} = x - iy$ and $\operatorname{Im} z = y$, we separate the real and imaginary parts of the entire expression:

$$f(z) = (x + iy)(x - iy - 3y) = x^2 + y^2 - 3xy - i3y^2.$$

According to Theorem 3.2, the function f is differentiable at a point $z = (x, y)$ if and only if the Cauchy-Riemann conditions (3.4) are satisfied at that point. Taking the partial derivatives,

$$\frac{\partial u}{\partial x} = 2x - 3y, \quad \frac{\partial v}{\partial y} = -6y, \quad \frac{\partial u}{\partial y} = 2y - 3x, \quad \frac{\partial v}{\partial x} = 0,$$

we see that the Cauchy-Riemann conditions lead to the system of equations

$$\begin{cases} 2x - 3y = -6y \\ 2y - 3x = 0 \end{cases} \quad \text{or} \quad \begin{cases} 2x + 3y = 0 \\ 3x - 2y = 0. \end{cases}$$

This system has the unique solution $x = 0$, $y = 0$. Thus the function is differentiable only at the point $z = 0$. Because the limit in the definition (3.1) of the derivative does not depend on the direction along which Δz approaches 0, we may assume that $\Delta z = \Delta x$ and $\Delta y = 0$. Then

$$f'(z_0) = \left.\frac{\partial u}{\partial x}\right|_{z_0} + i\left.\frac{\partial v}{\partial x}\right|_{z_0},$$

and we get

$$f'(z_0) = \left.\frac{\partial u}{\partial x}\right|_{(0,0)} + i\left.\frac{\partial v}{\partial x}\right|_{(0,0)} = (2x - 3y)\Big|_{(0,0)} + i0 = 0.$$

Note that the given function f is not analytic anywhere, even at $z = 0$, because that would require the point to have a neighborhood in which the function is differentiable.

3. a) *Solution.* From $w = z \operatorname{Im} z = (x + iy)y = xy + iy^2$ we get $u(x, y) = xy$, $v(x, y) = y^2$,

$$u_x = y, \quad v_x = 0, \quad u_y = x, \quad v_y = 2y.$$

Now applying the Cauchy-Riemann conditions (3.4), we get

$$\begin{cases} u_x = v_y \\ u_y = -v_x, \end{cases} \quad \text{or} \quad \begin{cases} y = 2y \\ x = 0, \end{cases}, \quad \begin{cases} y = 0 \\ x = 0, \end{cases}$$

so that the function is differentiable only at the origin. At that point we have, since w' is independent of direction,

$$w' = \frac{\partial w}{\partial x} = u_x + iv_x = y = 0.$$

d) *Answer.* The function is differentiable only at the origin; $w'(0) = 1$.

Section 3.2

1. *Solution.*

1. First we check that the given function is in fact the real part of an analytic function, substituting u into the Laplace equation $\frac{\partial^2 u}{\partial x^2} + \frac{\partial^2 u}{\partial y^2} = 0$:

$$\frac{\partial u}{\partial x} = e^{-y}\cos x, \quad \frac{\partial^2 u}{\partial x^2} = -e^{-y}\sin x;$$
$$\frac{\partial u}{\partial y} = -e^{-y}\sin x + 1, \quad \frac{\partial^2 u}{\partial y^2} = e^{-y}\sin x;$$
$$\frac{\partial^2 u}{\partial x^2} + \frac{\partial^2 u}{\partial y^2} = -e^{-y}\sin x + e^{-y}\sin x = 0.$$

So the function u satisfies the Laplace equation everywhere in the plane.

2. Applying the Cauchy-Riemann conditions, we get

$$\frac{\partial v}{\partial y} = \frac{\partial u}{\partial x} = e^{-y}\cos x; \quad \frac{\partial v}{\partial x} = -\frac{\partial u}{\partial y} = e^{-y}\sin x - 1.$$

3. Integrating the second equation with respect to x, and treating y as a constant, we get

$$v(x,y) = \int (e^{-y}\sin x - 1)\,dx = e^{-y}\int \sin x\,dx - \int 1 \cdot dx$$
$$= -e^{-y}\cos x - x + C(y),$$

where $C(y)$ is some function of y.

4. Substituting this expression for $v(x,y)$ into the equation

$$\frac{\partial v}{\partial y} = e^{-y}\cos x,$$

we get

$$e^{-y}\cos x + C'(y) = e^{-y}\cos x,$$

hence $C'(y) = 0$ and $C(y)$ is some constant C_1.

5. Writing $f(z)$ in the form $f(z) = u(x,y) + iv(x,y)$, we find C_1 from the condition $f(0) = i$:

$$f(z) = e^{-y}\sin x + y + i(-e^{-y}\cos x - x + C_1);$$
$$i = f(0) = 0 + i(-1 + 0 + C_1);$$
$$i = -i + iC_1 \quad \text{so} \quad C_1 = 2.$$

6. Finally we get

$$f(z) = e^{-y} \sin x + y + i(-e^{-y} \cos x - x + 2).$$

After studying the exponential and trigonometric functions of a complex variable, we can write this expression in terms of z, replacing x by z and y by 0, as in Example 3.10:

$$\begin{aligned} f(z) &= e^0 \sin z + 0 + i(-e^0 \cos z - z + 2) \\ &= \sin z - i \cos z - iz + 2i \\ &= -ie^{iz} - iz + 2i; \end{aligned}$$

the last equality follows from (4.28).

Constructing an analytic function from its imaginary part can be carried out analogously—see Example 3.10.

2. b) *Solution.* First we check that the function $u = x^3 - 3xy^2 + 1$ satisfies the Laplace equation. Then, from the Cauchy-Riemann conditions we get

$$v_y = u_x = 3x^2 - 3y^2, \quad v_x = -u_y = 6xy.$$

Integrating the second equation with respect to x, we get

$$v = 3x^2 y + C(y),$$

where C is a function that depends only on y. Combining this with the first equation gives

$$v_y = 3x^2 + C'(y) = 3x^2 - 3y^2 \implies C'(y) = -3y^2,$$

so that $C(y) = -y^3 + D$, where D is a constant. This means that

$$f(z) = u + iv = x^3 - 3xy^2 + 1 + i(3x^2 y - y^3 + D) = z^3 + 1 + iD.$$

Finally, applying the condition $f(0) = 1$ gives $1 = 0^3 + 1 + iD$, $iD = 0$, so that $f(z) = z^3 + 1$.

3. a) *Solution.* First we check that the function $v = e^x \cos y$ satisfies the Laplace equation. Then, from the Cauchy-Riemann conditions we get

$$u_x = v_y = -e^x \sin y, \quad u_y = -v_x = -e^x \cos y.$$

Integrating the first equation with respect to x we have

$$u = -e^x \sin y + C(y),$$

where C is a function that depends only on y. Combining this with the second equation gives us

$$u_y = -e^x \cos y + C'(y) = -e^x \cos y \implies C'(y) = 0,$$

so that $C(y) = D$, where D is a constant. This means that

$$f(z) = u + iv = -e^x \sin y + D + ie^x \cos y = ie^z + D.$$

Applying the condition $f(0) = 1 + i$ gives $1 + i = ie^0 + D = i + D$, $D = 1$, so that $f(z) = ie^z + 1$.

Section 4.1

1. *Solution.* Triangle ABC can be transformed into triangle $A'B'C'$ through a combination of expansion, rotation, and translation. Therefore the transformation is a linear one, $w = az + b$.

1. Since the similarity ratio is 2, we start with the expansion $w_1 = 2z$. This takes the point $A = (-1, 0)$ to $A_1 = (-2, 0)$.

2. Next we rotate the resulting triangle counterclockwise about the origin through the angle $\pi/2$. This transformation can be written

$$w_2 = w_1 e^{i\pi/2} = iw_1.$$

   This takes A_1 to $A_2 = (0, -2)$.

3. Last, we translate the resulting triangle by the vector

$$\mathbf{A_2 A'} = (3, 2) = 3 + 2i.$$

   This transformation can be written $w = w_2 + 3 + 2i$.

The composition of these three transformations yields the desired function:

$$w = w_2 + 3 + 2i = iw_1 + 3 + 2i = 2iz + 3 + 2i.$$

4. *Solution.* First we determine the image of the boundary of the disk, i.e. the circle $|z + 2i| = 2$. Because Möbius transformations preserve circles (Theorem 4.4), this circle is mapped to another circle—which will be a line if its radius is ∞. In fact the point $z = 0$ lies on the circle $|z + 2i| = 2$, and this point is sent to $w = \infty$. Therefore the image has a radius $R = \infty$, and hence is a line. To graph the line, it suffices to find two points on it:

$$\text{when } z_1 = -4i \text{ we get } w_1 = \frac{1}{-4i} = \frac{i}{4};$$

$$\text{when } z_2 = 2 - 2i \text{ we get } w_2 = \frac{1}{2 - 2i} = \frac{1}{4} + \frac{i}{4}.$$

So the image of the circle $|z+2i|=2$ is the line through these two points, which is the horizontal line $y=\frac{1}{4}$. Therefore the image of the disk $|z+2i|<2$ is one of the half-planes bounded by this line, either the upper or the lower.

To find out which is the case, we need only check to which half-plane an arbitrary point in the interior of the disk is mapped. Using the center of the disk $z_0=-2i$, we see that

$$w_0=\frac{1}{z_0}=-\frac{1}{2i}=\frac{i}{2}.$$

Because this point lies above the line $y=\frac{1}{4}$, we see that the disk $|z+2i|<2$ is mapped to the half-plane $\operatorname{Im} w>\frac{1}{4}$.

The image of the boundary $|z+2i|=2$ may be found in other ways, as seen in the next example.

5. *Solution.* Again, we first determine the image of the circle $|z+i|=3$. Since $z=0$ does not lie on this circle, none of its points are sent to $w=\infty$. This means that the image is a circle of finite radius. To find the center and radius we write $z=x+iy$ and $w=u+iv$, then expand $|z+i|=3$ into

$$|x+i(y+1)|=3, \quad \text{or} \quad x^2+(y+1)^2=9, \quad x^2+y^2+2y-8=0.$$

The equality $w=1/z$ means that $z=1/w$, so that

$$x+iy=\frac{1}{u+iv}=\frac{u-iv}{u^2+v^2},$$

from which we get

$$x=\frac{u}{u^2+v^2} \quad \text{and} \quad y=-\frac{v}{u^2+v^2}.$$

Now we substitute these expressions into the equation of the circle:

$$\frac{1}{u^2+v^2}-\frac{2v}{u^2+v^2}-8=0,$$

$$8(u^2+v^2)+2v-1=0, \quad u^2+v^2+\frac{v}{4}-\frac{1}{8}=0.$$

Completing $v^2+\frac{v}{4}$ to a square gives

$$u^2+v^2+\frac{v}{4}+\left(\frac{1}{8}\right)^2-\frac{1}{8}=\left(\frac{1}{8}\right)^2, \quad u^2+\left(v+\frac{1}{8}\right)^2=\frac{9}{64}.$$

Thus the circle $|z+i|=3$ is mapped to the circle with center $w_0=(0,-\frac{1}{8})$ and radius $R=\frac{3}{8}$. The point $z=0$, lying in the interior of the disk $|z+i|<3$, goes to the point $w=\infty$. That means that under $w=\frac{1}{z}$ the disk $|z+i|<3$ is mapped to the exterior of the disk with center $w_0=-\frac{i}{8}$ and radius $R=\frac{3}{8}$. This set can defined by the inequality $\left|w+\frac{i}{8}\right|>\frac{3}{8}$.

7. *Hint.* Let Γ be the circle $|z| = 1$, and let Γ' be the image of Γ under the Möbius transformation f of the form (4.12). By Theorem 4.4, Γ' is a circle. The points z_0 and $1/\overline{z_0}$ are symmetric about Γ. By Theorem 4.6, their images $f(z_0) = 0$ and $f(1/\overline{z_0}) = \infty$ are symmetric about Γ'. It is possible if and only if Γ' is a circle of radius $R < \infty$ centered at the origin (why?). Since $|f(1)| = 1$, the radius $R = 1$.

Another solution is based on the direct verification of the equality $|f(e^{it})| = 1$ for all real t.

8. *Solution.* In Example 4.8 we saw that a Möbius transformation mapping the unit disk to itself must have the form (4.12):

$$w = e^{i\phi}\frac{z - z_0}{1 - \overline{z_0}z},$$

where z_0 is the point that goes to $w = 0$. In our case, the first condition means that $z_0 = -\frac{i}{2}$. It remains only to determine ϕ, which we will get from the second condition, after first calculating $w'(z_0)$:

$$w'(z) = e^{i\phi}\frac{1 \cdot (1 - \overline{z_0}z) - (-\overline{z_0})(z - z_0)}{(1 - \overline{z_0}z)^2} = e^{i\phi}\frac{1 - \overline{z_0}z_0}{(1 - \overline{z_0}z)^2}.$$

Using $\overline{z_0}z_0 = |z_0|^2$, when $z = z_0 = -\frac{i}{2}$ we get

$$w'(z_0) = e^{i\phi}\frac{1 - |z_0|^2}{(1 - |z_0|^2)^2} = e^{i\phi}\frac{1}{1 - |z_0|^2},$$

$$w'\left(-\frac{i}{2}\right) = e^{i\phi}\frac{1}{1 - \frac{1}{4}} = \frac{4}{3}e^{i\phi}.$$

Hence

$$\left|w'\left(-\frac{i}{2}\right)\right| = \frac{4}{3} \quad \text{and} \quad \operatorname{Arg} w'\left(-\frac{i}{2}\right) = \phi.$$

This means that the dilation coefficient at the point $-\frac{i}{2}$ is equal to $\frac{4}{3}$; and from the second condition given in the problem we get $\phi = \frac{\pi}{4}$. The desired transformation is the function

$$w = e^{i\pi/4}\frac{z + \frac{i}{2}}{1 - \overline{(-\frac{i}{2})}z} = e^{i\pi/4}\frac{z + \frac{i}{2}}{1 - \frac{i}{2}z}$$

(where we have used $\overline{i} = -i$).

9. *Solution.* From formula (4.12), any Möbius transformation of the disk to itself is of the form

$$w = f(z) = e^{i\phi}\frac{z - a}{1 - \overline{a}z}.$$

where a is the point which is mapped to the origin. In our case $a = \frac{1}{2}$, so

$$f(z) = e^{i\phi}\frac{z - \frac{1}{2}}{1 - \frac{1}{2}z} = e^{i\phi}\frac{2z - 1}{2 - z}.$$

Taking the derivative,

$$f'(z) = e^{i\phi}\frac{3}{(2-z)^2}; \quad f'\left(\frac{1}{2}\right) = e^{i\phi}\cdot\frac{3}{(\frac{3}{2})^2} = e^{i\phi}\cdot\frac{4}{3}.$$

The condition $\operatorname{Arg} w'(\frac{1}{2}) = \frac{\pi}{2}$ becomes $\frac{\pi}{2} = \operatorname{Arg}\left(e^{i\phi}\cdot\frac{4}{3}\right) = \phi$. Thus the transformation is

$$w = f(z) = e^{i\frac{\pi}{2}}\frac{2z-1}{2-z} = i\frac{2z-1}{2-z},$$

and the dilation coefficient at $z = \frac{1}{2}$ is $\left|f'\left(\frac{1}{2}\right)\right| = \frac{4}{3}$.

10. *Solution.* In Example 4.7 we saw that a conformal mapping from the upper half-plane to the unit disk must have the form (4.11):

$$w = e^{i\phi}\frac{z-z_0}{z-\overline{z_0}},$$

where z_0 is the point that goes to $w=0$. In our case, from the first condition, that point is $2i$. To determine ϕ we use the second condition, so we must first find $w'(z_0)$:

$$w'(z) = e^{i\phi}\frac{1\cdot(z-\overline{z_0}) - 1\cdot(z-z_0)}{(z-\overline{z_0})^2} = e^{i\phi}\frac{z_0-\overline{z_0}}{(z-\overline{z_0})^2},$$

$$w'(z_0) = e^{i\phi}\frac{z_0-\overline{z_0}}{(z_0-\overline{z_0})^2} = e^{i\phi}\frac{1}{z_0-\overline{z_0}}.$$

Therefore,

$$w'(2i) = \frac{e^{i\phi}}{2i+2i} = \frac{e^{i\phi}}{4i} = -\frac{ie^{i\phi}}{4} = \frac{1}{4}e^{i\pi}e^{i\frac{\pi}{2}}e^{i\phi} = \frac{1}{4}e^{i(\frac{3\pi}{2}+\phi)}$$

(where we have used $-1 = e^{i\pi}$ and $i = e^{i\frac{\pi}{2}}$). Then

$$|w'(2i)| = \frac{1}{4} \quad \text{and} \quad \operatorname{Arg} w'(2i) = \frac{3\pi}{2}+\phi.$$

This means that the dilation coefficient at the point $2i$ is $\frac{1}{4}$; and from the second condition given in the problem we get

$$\frac{3\pi}{2}+\phi = \pi \quad \text{or} \quad \phi = -\frac{\pi}{2}.$$

Therefore the desired transformation is

$$w = e^{-i\frac{\pi}{2}}\frac{z-2i}{z+2i} = -i\frac{z-2i}{z+2i}.$$

12. *Solution.* We may just find the function which is inverse to those in Example 4.7, solving the equation (4.11) for z and interchanging z and w. Let us find another solution which is independent of Example 4.7.

Let w_0, $\operatorname{Im} w_0 > 0$, be the image of 0 under the mapping $f(z)$, that is $f(0) = w_0$. Then $f(\infty) = \overline{w_0}$. For a Möbius transformation

$$f(z) = \frac{az+b}{cz+d} \tag{9.48}$$

these equalities imply that $b = dw_0$, $a = c\overline{w_0}$. Hence,

$$f(z) = \frac{c\overline{w_0}z + dw_0}{cz+d} = \frac{\overline{w_0}z + Aw_0}{z+A}, \quad A = \frac{d}{c}.$$

Since one point of the unit circle, say $e^{i\alpha}$, goes to infinity, $e^{i\alpha} + A = 0$. Therefore,

$$w = f(z) = \frac{\overline{w_0}z - e^{i\alpha}w_0}{z - e^{i\alpha}}, \quad \operatorname{Re} w_0 > 0. \tag{9.49}$$

Again our mapping depends on three real parameters, namely on two components of w_0 and α.

13. *Solution.* When we traverse the upper semicircle $|z| = 1$ counterclockwise, the point $w = f(z)$ moves from 0 to ∞ along the real axis $\operatorname{Im} w = 0$. Hence, $f(1) = 0$, $f(-1) = \infty$. We plug these values in (9.49) and get $e^{i\alpha} = -1$, $w_0 = ih$, where h is any positive number, say 1. Therefore,

$$w = f(z) = -ih\,\frac{z-1}{z+1}, \quad h > 0. \tag{9.50}$$

Another solution is to use (9.48) directly, taking into account that $f(z)$ is real and positive on the upper semicircle.

Section 4.2

1. *Answer.*

$$(a)\ f(z) = \sqrt[3]{r}\left(\cos\frac{\phi - 2\pi}{3} + i\sin\frac{\phi - 2\pi}{3}\right), \quad f(1) = -\frac{1}{2} - i\frac{\sqrt{3}}{2};$$

$$(b)\ f(z) = \sqrt[3]{r}\left(\cos\frac{\phi}{3} + i\sin\frac{\phi}{3}\right), \quad f(1) = 1.$$

2. *Hint.* Show that for every real $x < -1$, the limit of $f(z)$ as $z \to x$ is the same when z approaches x over the upper and the lower half-planes. Also consider the case $x \in (-1, 1)$.

3. *Answer.* The numbers 1 and $(-1)(-1)$ under the first and the second roots are on different sheets of the Riemann surface: their arguments are 0 and 2π, respectively. Therefore, the different branches of the function $\sqrt{z}$ are selected.

Section 4.3

2. *Answer.* (*a*) $z = i(\pi+2\pi k)$; (*b*) $z = 1+i2\pi k$; (*c*) $z = 1+i(-\frac{\pi}{2}+2\pi k)$, $k \in \mathbb{Z}$.

3. *Solution.* We break the solution into four steps.

1. Using a linear function $w = az + b$ we can map the given strip to the strip $-\infty < x < +\infty$, $0 < y < \pi$. The first strip has a width of $\sqrt{2}$, so to map it onto a strip of width π, we apply a similarity transform with coefficient $\frac{\pi}{\sqrt{2}}$; for this we can take

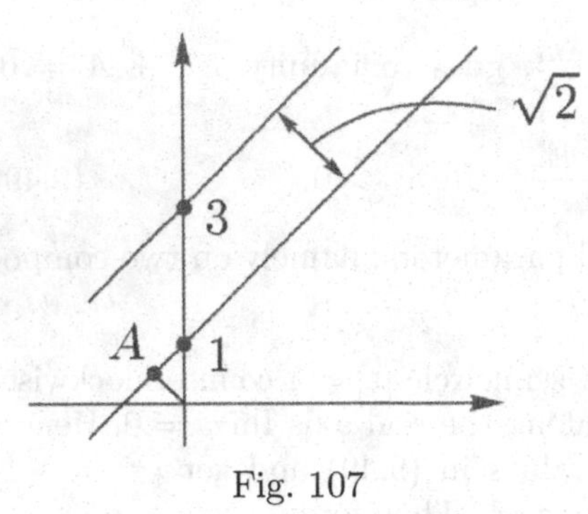

Fig. 107

$$w_1 = \frac{\pi}{\sqrt{2}}z.$$

Then the point $A = (-\frac{1}{2}, \frac{i}{2})$ goes to

$$A_1 = \left(-\frac{\pi}{2\sqrt{2}}, i\frac{\pi}{2\sqrt{2}}\right).$$

Then we must rotate the resulting strip clockwise through an angle of $\frac{\pi}{4}$, so that it becomes parallel to the x-axis. This transformation can be written in the form

$$w_2 = w_1 e^{-i\frac{\pi}{4}} = \frac{\pi z}{\sqrt{2}} e^{-i\frac{\pi}{4}}.$$

This takes the point A_1 to $A_2 = (0, \frac{\pi}{2})$.

Finally we apply a translation along the vector $\mathbf{A_2O}$:

$$w_3 = w_2 - \frac{i\pi}{2} = \frac{\pi z}{\sqrt{2}} e^{-i\frac{\pi}{4}} - \frac{i\pi}{2}.$$

Instead of taking A, to be $(-\frac{1}{2}, \frac{i}{2})$, we could have chosen any point on the line $y = x+1$ (e.g. $(0,1)$). Using the equality $e^{-i\frac{\pi}{4}} = \frac{1}{\sqrt{2}} - \frac{i}{\sqrt{2}}$, the transformation we have constructed may be written in the form

$$w_3 = \frac{\pi(1-i)}{2} z - \frac{i\pi}{2}.$$

2. Next, we map the strip just obtained to the upper half-plane $\operatorname{Im} z > 0$. In Section 4.3.1 we saw that the exponential function $w = e^z$ maps the strip $0 < \operatorname{Im} z < \pi$ onto the upper half-plane. So the mapping we need is $w_4 = e^{w_3}$.

3. Now we map the upper half-plane to the unit disk with the help of the Möbius transformation

$$w = e^{i\phi}\frac{z - z_0}{z - \overline{z_0}}, \qquad \operatorname{Im} z > 0.$$

Since no additional conditions were given in the problem, we may choose the two parameters ϕ and z_0 (with $\operatorname{Im} z_0 > 0$) arbitrarily. Taking, for example, $\phi = 0$ and $z_0 = i$, we would get

$$w = \frac{w_4 - i}{w_4 + i}.$$

4. The desired mapping is the composition of the three constructed above:

$$w = \frac{w_4 - i}{w_4 + i} = \frac{e^{w_3} - i}{e^{w_3} + i} = \frac{\exp\left(\frac{\pi(1-i)}{2}z - \frac{i\pi}{2}\right) - i}{\exp\left(\frac{\pi(1-i)}{2}z - \frac{i\pi}{2}\right) + i}.$$

4. b) *Solution.* We proceed in four steps.

1. Map the oblique strip onto a horizontal one bounded below by the x-axis:

$$w_1 = e^{-i\frac{\pi}{3}}z = \left(\frac{1}{2} - i\frac{\sqrt{3}}{2}\right)z.$$

Since $i\pi$ is on the upper boundary of the original strip, and $w_1(i\pi) = \frac{\sqrt{3}}{2}\pi + i\frac{\pi}{2}$, this maps the strip onto $0 < \operatorname{Im} z < \frac{\pi}{2}$.

2. Double the height of the strip, so it becomes $0 < \operatorname{Im} z < \pi$: $w_2 = 2w_1$.
3. Map the new strip onto the upper half-plane: $w_3 = e^{w_2}$.
4. Map the upper half-plane onto the unit disk. Any such Möbius transformation must have the form of formula (4.11):

$$w_4 = e^{i\phi}\frac{w_3 - a}{w_3 - \overline{a}},$$

where ϕ is real and a is a point in the upper half-plane which maps to 0. We may choose the two parameters ϕ and z_0 (with $\operatorname{Im} z_0 > 0$) arbitrarily, for example, $\phi = 0$ and $z_0 = i$.

Composing these functions gives

$$w_2 = 2w_1 = (1 - i\sqrt{3})z, \quad w_3 = e^{(1-i\sqrt{3})z},$$

$$w_4 = \frac{e^{(1-i\sqrt{3})z} - i}{e^{(1-i\sqrt{3})z} + i}.$$

10. *Answer.* $0 < \operatorname{Arg} z \leq \pi$.

Section 4.4

2. *Solution.* According to the definition of the general power function given in Section 4.4.1,

$$\left(-\sqrt{3}+i\right)^{-6i} = e^{-6i\log(-\sqrt{3}+i)}.$$

The values of the logarithm are found via formula (4.21):

$$\log\left(-\sqrt{3}+i\right) = \ln\left|-\sqrt{3}+i\right| + i\left(\operatorname{Arg}\left(-\sqrt{3}+i\right)+2\pi k\right);$$

$$\left|-\sqrt{3}+i\right| = \sqrt{\left(-\sqrt{3}\right)^2+1} = 2; \quad \cos\phi = -\frac{\sqrt{3}}{2}, \quad \sin\phi = \frac{1}{2},$$

where $\phi = \operatorname{Arg}\left(-\sqrt{3}+i\right)$. Since ϕ lies in the second quadrant, $\phi = \frac{5\pi}{6}$. This gives us

$$\begin{aligned}\left(-\sqrt{3}+i\right)^{-6i} &= \exp\left(-6i\left(\ln 2 + i\left(\frac{5\pi}{6}+2\pi k\right)\right)\right)\\ &= \exp(5\pi + 12\pi k - i6\ln 2)\\ &= [\exp(5\pi+12\pi k)](\cos(6\ln 2) - \sin(6\ln 2)), \quad \text{for} \quad k\in\mathbb{Z}.\end{aligned}$$

Our result is that the power $\left(-\sqrt{3}+i\right)^{-6i}$ has an infinite set of values. All these values lie on a single ray, at the angle $\theta = -6\ln 2$ to the positive x-axis. The moduli of these values form two geometric progressions, both starting at $e^{5\pi}$ (when $k=0$), with ratios $e^{12\pi}$ (when k is positive) and $e^{-12\pi}$ (when k is negative).

3. *Solution.* Since $1+i = \sqrt{2}\left(\frac{1}{\sqrt{2}}+i\frac{1}{\sqrt{2}}\right) = \sqrt{2}(\cos\frac{\pi}{4}+i\sin\frac{\pi}{4})$, then

$$\begin{aligned}(1+i)^i &= \exp\left(i(\ln\sqrt{2}+i(\tfrac{\pi}{4}+2\pi k))\right)\\ &= \left[\exp\left(-\tfrac{\pi}{4}-2\pi k\right)\right]\left(\cos(\tfrac{1}{2}\ln 2)+i\sin(\tfrac{1}{2}\ln 2)\right), \quad \text{for} \quad k\in\mathbb{Z}.\end{aligned}$$

4. a) *Hint.* See problem 3 in Section 4.2.
b) *Solution.*

$$z_1^a\cdot z_2^a = e^{(a\ln|z_1|+i\phi_1)}\cdot e^{(a\ln|z_2|+i\phi_2)} = e^{a(\ln(|z_1|\cdot|z_2|)+i(\phi_1+\phi_2))} = (z_1z_2)^a.$$

8. *Solution.* By formula (4.30),

$$\begin{aligned}\cos^{-1}(-5i) &= i\log\left(-5i+\sqrt{(-5i)^2-1}\right) = i\log\left(-5i+\sqrt{-26}\right)\\ &= i\log\left(-5i\pm i\sqrt{26}\right) = i\log\left(i\left(-5\pm\sqrt{26}\right)\right).\end{aligned}$$

We write $\pm$ before $\sqrt{26}$ because usually the root of a positive number is understood as its principal value, but the root in formula (4.30) is to be understood as two-valued. We will consider separately each of the signs of the root. First with $+$, making use of the equality $\left(\sqrt{26}-5\right)\left(\sqrt{26}+5\right)=1$, we get

$$\left|i\left(-5+\sqrt{26}\right)\right|=\sqrt{26}-5=\left(\sqrt{26}+5\right)^{-1}; \quad \operatorname{Arg}\left(i\left(-5+\sqrt{26}\right)\right)=\frac{\pi}{2}.$$

Then by the formula (4.21) for the log,

$$\begin{aligned} i\log\left(i\left(-5+\sqrt{26}\right)\right) &= i\ln\left(\sqrt{26}-5\right)-\left(\frac{\pi}{2}+2\pi k\right) \\ &= -i\ln\left(\sqrt{26}+5\right)-\left(\frac{\pi}{2}+2\pi k\right) \quad \text{for} \quad k\in\mathbb{Z}. \end{aligned}$$

Analogously, when the sign is $-$,

$$\begin{aligned} \left|i\left(-5-\sqrt{26}\right)\right| &= \sqrt{26}+5; \quad \operatorname{Arg}\left(-i\left(5+\sqrt{26}\right)\right)=-\frac{\pi}{2}; \\ i\log\left(-i\left(5+\sqrt{26}\right)\right) &= i\ln\left(\sqrt{26}+5\right)-\left(-\frac{\pi}{2}+2\pi k\right) \\ &= i\ln\left(\sqrt{26}+5\right)+\left(\frac{\pi}{2}+2\pi k\right) \quad \text{for} \quad k\in\mathbb{Z}. \end{aligned}$$

We have written $+2\pi k$ instead of $-2\pi k$, since k runs through the set of integers. Finally we get

$$\cos^{-1}(-5i)=\pm\left(\frac{\pi}{2}+2\pi k+i\ln\left(\sqrt{26}+5\right)\right), \quad k\in\mathbb{Z}.$$

Note that we could have obtained the second value from the first by applying a general remark made in Section 4.4.3, according to which a change in the sign of the root leads to a change in the sign of the expression $\log\left(z+\sqrt{z^2+1}\right)$.

9. *Solution.* By formula (4.31), when $z=\dfrac{-2\sqrt{3}+3i}{3}$ we have

$$\tan^{-1}z=\frac{1}{2i}\log\frac{i-z}{i+z}=\frac{1}{2i}\log\frac{2\sqrt{3}}{-2\sqrt{3}+6i};$$

$$\frac{2\sqrt{3}}{-2\sqrt{3}+6i}=\frac{1}{-1+i\sqrt{3}}=\frac{-1-i\sqrt{3}}{4};$$

$$\left|\frac{-1-i\sqrt{3}}{4}\right|=\frac{1}{4}\left|1+i\sqrt{3}\right|=\frac{1}{2}; \quad \phi=\operatorname{Arg}\frac{-1-i\sqrt{3}}{4}=-\frac{2\pi}{3},$$

since $\sin\phi=-\frac{\sqrt{3}}{2}$, $\cos\phi=-\frac{1}{2}$. This gives us

$$\begin{aligned} \tan^{-1}\frac{-2\sqrt{3}+3i}{3} &= \frac{1}{2i}\left(\ln\frac{1}{2}+i\left(-\frac{2\pi}{3}+2\pi k\right)\right) \\ &= -\frac{\pi}{3}+\pi k+\frac{1}{2i}\ln\frac{1}{2}=-\frac{\pi}{3}+\pi k+\frac{i}{2}\ln 2, \quad k\in\mathbb{Z}. \end{aligned}$$

10. d) *Solution.* All solutions of the equation $\tan z = a$ are given by the formula $z = \tan^{-1} a$. By (4.31),

$$\tan^{-1} a = \frac{1}{2i} \log \frac{i-a}{i+a}, \quad \text{where } a = \frac{2\sqrt{3}+3i}{7}.$$

Elementary computations give

$$\frac{i-a}{i+a} = \frac{1}{2}\left(\frac{1}{2} + i\frac{\sqrt{3}}{2}\right).$$

Combining this with the formula for $\tan^{-1}$, we get

$$z = \tan^{-1} \frac{2\sqrt{3}+3i}{7} = \frac{1}{2i}\left(\ln\frac{1}{2} + i\left(\frac{\pi}{3} + 2\pi k\right)\right) = \frac{\pi}{6} + \pi k + i\frac{\ln 2}{2}, \quad k \in \mathbb{Z}.$$

Section 5.1

1. *Solution.* We first write the curve in the form of a parametric equation,

$$z = z(t) = x(t) + iy(t), \quad \alpha \le t \le \beta.$$

Because this curve consists of two parts, the segments AB and BC, which will have different equations, we consider each one separately.

The part AB can be given as $y = x+1$ for $-1 \le x \le 0$. Identifying x with the paramter t, we may write $z(t) = t + i(t+1)$, $\quad -1 \le t \le 0$.

The other part BC can be given as $y = -x+1$ for $0 \le x \le 1$. Identifying x with the paramter t, we may write $z(t) = t + i(-t+1)$, $\quad 0 \le t \le 1$.

Now we compute the integral along Γ using the formula (5.5):

$$\int_\Gamma f(z)\, dz = \int_\alpha^\beta f(z(t)) z'(t)\, dt.$$

We will apply this to each part of the curve separately, and then take the sum of the two integrals.

For the segment AB, $z'(t) = 1 + i$, so that

$$\int_{AB} \overline{z}^2\, dz = \int_{AB} (x - iy)^2\, dz = \int_{-1}^{0} (t - i(t+1))^2 z'(t)\, dt$$

$$= \int_{-1}^{0} (t^2 - 2it(t+1) - (t+1)^2)(1+i)\, dt$$

$$= \int_{-1}^{0} (2t^2 - 1 + i(-2t^2 - 4t - 1))\, dt$$

$$= \frac{2t^3}{3} - t + i\left(-\frac{2t^3}{3} - 2t^2 - t\right)\Bigg|_{-1}^{0} = -\frac{1}{3} + \frac{i}{3}.$$

For the segment BC, $z'(t) = 1 - i$, so that

$$\int_{BC} \overline{z}^2\, dz = \int_{BC} (x - iy)^2\, dz = \int_{0}^{1} (t - i(-t+1))^2 z'(t)\, dt$$

$$= \int_{0}^{1} (t^2 - 2it(-t+1) - (-t+1)^2)(1-i)\, dt$$

$$= \int_{0}^{1} (2t^2 - 1 + i(2t^2 - 4t + 1))\, dt$$

$$= \frac{2t^3}{3} - t + i\left(\frac{2t^3}{3} - 2t^2 + t\right)\Bigg|_{0}^{1} = -\frac{1}{3} - \frac{i}{3}.$$

So the integral over Γ is

$$\int_{\Gamma} f(z)\, dz = \int_{AB} f(z)\, dz + \int_{BC} f(z)\, dz = -\frac{1}{3} + \frac{i}{3} - \frac{1}{3} - \frac{i}{3} = -\frac{2}{3}.$$

2. *Solution.* We may parametrize the upper unit semicircle traversed in the clockwise direction, with $z(t) = e^{it}$, where t changes from π to 0. Then

$$z'(t) = ie^{it}, \quad f(z(t)) = \left(\overline{e^{it}}\right)^2 = e^{-2it};$$

$$\int_{\Gamma} f(z)\, dz = \int_{\pi}^{0} e^{-2it}\, ie^{it}\, dt = i\int_{\pi}^{0} e^{-it}\, dt = -\frac{i}{i}\, e^{-it}\Bigg|_{\pi}^{0} = -(1 - (-1)) = -2.$$

4. b) *Solution.* The parabola $y = x^2$ on the interval $0 \le x \le 2$ can be parametrized by $z(t) = t + it^2$, where $t \in [0, 2]$. Then

$$z'(t) = 1 + i2t, \quad f(z(t)) = (t + it^2)\,\mathrm{Re}(t + it^2) = (t + it^2)t = t^2 + it^3;$$

$$\int_\Gamma f(z)\,dz = \int_0^2 (t^2 + it^3)(1 + i2t)\,dt = \int_0^2 (t^2 - 2t^4 + i3t^3)\,dt$$

$$= \frac{t^3}{3} - \frac{2t^5}{5} + i\frac{3t^4}{4}\bigg|_0^2 = \frac{8}{3} - \frac{64}{5} + 12i = -\frac{152}{15} + 12i.$$

5. a) *Solution.* For any complex numbers z_1, z_2 we have $|z_1 + z_2| \ge |z_1| - |z_2|$ —see problem 1 in Section 1.2. Hence, $|z^2 + 1| \ge |z^2| - 1 = 3$ as $|z| = 2$. Therefore,

$$|f(z)| = \frac{1}{|z^2 + 1|} \le \frac{1}{3} \quad \text{as } |z| = 2.$$

The length of the circle $|z| = 2$ equals 4π. Now the desired estimate follows from (5.9) with $M = \frac{1}{3}$ and $\ell = 4\pi$.

Section 5.2

2. a) *Solution.* We break the integral up into the three pieces AO, OB, and BC.

AO: We parametrize this segment by $z(t) = t$, $-1 \le t \le 0$. Then $z'(t) = 1$;

$$f(z(t)) = t^2; \quad \int_{AO} f(z)\,dz = \int_{-1}^0 t^2 \cdot 1\,dt = \frac{t^3}{3}\bigg|_{-1}^0 = \frac{1}{3}.$$

OB: We use the parametrization $z = it$, $0 \le t \le 1$. Then $z'(t) = i$;

$$f(z(t)) = \overline{it}^2 = (-it)^2 = -t^2;$$

$$\int_{OB} f(z)\,dz = \int_0^1 -t^2 \cdot i\,dt = -i\frac{t^3}{3}\bigg|_0^1 = -\frac{i}{3}.$$

BC: The part BC can be given as $y = 1 - x$ for $0 \le x \le 1$. Identifying x with the parameter t, we have $z(t) = t + i(1-t)$. Then $z'(t) = 1 - i$;

$$f(z(t)) = \overline{z(t)}^2 = (t + i(t-1))^2 = 2t - 1 + i\,2(t^2 - t);$$

$$\int\limits_{BC} f(z)\,dz = \int\limits_0^1 (2t - 1 + i\,2(t^2-t))(1-i)\,dt$$

$$= (1-i)\left[t^2 - t + 2i\left(\frac{t^3}{3} - \frac{t^2}{2}\right)\right]\Bigg|_0^1 = (1-i)\cdot 2i \cdot \left(-\frac{1}{6}\right) = -\frac{1}{3} - \frac{i}{3}.$$

Summing the three pieces, we get

$$\int\limits_{AOBC} f(z)\,dz = \frac{1}{3} - \frac{i}{3} - \frac{1}{3} - \frac{i}{3} = -\frac{2i}{3}.$$

The integrals along different paths from A to C may differ, because the function $f(z) = \overline{z}^2$ is not analytic in any domain (prove it).

4. b) *Solution.* Let us first estimate the denominator of the fraction as $|z| \ge 2$:

$$|z^4 + z - 2| = |z^4|\left|1 + \frac{1}{z^3} - \frac{2}{z^4}\right| \ge |z^4|\left(1 - \frac{1}{|z|^3} - \frac{2}{|z|^4}\right)$$
$$\ge |z^4|\left(1 - \frac{1}{2^3} - \frac{2}{2^4}\right) > |z^4| \cdot \frac{1}{2} \quad \text{as } |z| \ge 2.$$

Hence, the denominator is not zero as $|z| \ge 2$; therefore, the function $f(z) = \frac{z}{z^4+z-2}$ is analytic on $D = \{z : |z| \ge 2\}$. By Corollary 5.11,

$$\int\limits_{|z|=2} \frac{z\,dz}{z^4 + z - 2} = \int\limits_{|z|=R} \frac{z\,dz}{z^4 + z - 2} \quad \text{for any } R > 2.$$

Using the estimate for the denominator, we have

$$\left|\frac{z}{z^4 + z - 2}\right| < \frac{2|z|}{|z^4|} = \frac{2}{R^3} \quad \text{for } |z| = R > 2.$$

The length of the circle $|z| = R$ equals $2\pi R$. Applying (5.9) with $M = \frac{2}{R^3}$ and $\ell = 2\pi R$, we get

$$\left|\int\limits_{|z|=R} \frac{z\,dz}{z^4 + z - 2}\right| < \frac{2}{R^3} \cdot 2\pi R = \frac{4\pi}{R^2} \to 0 \quad \text{as } R \to \infty.$$

We see that the original integral over the circle $|z| = 2$ is a fixed number, the modulus of which is less than any positive number. Therefore, the integral must be zero.

Section 5.3

1. *Solution.* There are two ways to test f for analyticity. The first is to write it in the form $f(z) = u(x,y) + iv(x,y)$, that is with the real and imaginary parts separated; then to compute the partial derivatives of u and v and see if the Cauchy-Riemann conditions (3.4) are satisfied for all x and y. The second, and easier, method uses the properties of analytic functions (see Section 3.1.3).

The function $\sin z$ is analytic in $\mathbb{C}$, as shown in Section 4.4.2. The function $5z$ is obviously also analytic. Therefore the composition $\sin 5z$ is analytic in $\mathbb{C}$. It then follows that the function $f(z) = z \sin 5z$ will be analytic in $\mathbb{C}$, being the product of analytic functions.

Now note that the required antiderivative $F(z)$ is given by the formula

$$F(z) = \int_{\frac{\pi}{2}}^{z} \zeta \sin 5\zeta \, d\zeta,$$

where the integral is over any path from $\frac{\pi}{2}$ to z. Indeed, according to Corollary 5.14, the function F defined here will be an antiderivative of f, and the equality $F(\frac{\pi}{2}) = 0$ is obvious. This integral can be computed using formula (5.19), one of the forms of the fundamental theorem of calculus. Moreover, as noted in Remark 5.15, we may use all the usual formulas and techniques for integrating. In this case we will integrate by parts:

$$\int_a^b u \, dv = uv\Big|_a^b - \int_a^b v \, du.$$

In our problem

$$u = \zeta, \quad dv = \sin 5\zeta \, d\zeta, \quad du = d\zeta, \quad v = -\frac{1}{5}\cos 5\zeta.$$

Then

$$\int_{\frac{\pi}{2}}^{z} \zeta \sin 5\zeta \, d\zeta = -\frac{\zeta}{5}\cos 5\zeta\Big|_{\frac{\pi}{2}}^{z} - \int_{\frac{\pi}{2}}^{z} \left(-\frac{1}{5}\right)\cos 5\zeta \, d\zeta$$

$$= -\frac{\zeta}{5}\cos 5\zeta\Big|_{\frac{\pi}{2}}^{z} + \frac{1}{25}\sin 5\zeta\Big|_{\frac{\pi}{2}}^{z} = -\frac{z}{5}\cos 5z + 0 + \frac{1}{25}\sin 5z - \frac{1}{25}.$$

So,

$$F(z) = -\frac{z}{5}\cos 5z + \frac{1}{25}\sin 5z - \frac{1}{25}.$$

2. c) *Solution.* Both functions z^2+2 and e^z are analytic in the entire complex plane, and so their composition is also. Hence the function $f(z) = ze^{z^2+2}$ is analytic in $\mathbb{C}$, being the product of analytic functions. The antiderivative may be obtained via

$$F(z) = \int_{z_0}^{z} \zeta\, e^{\zeta^2+2}\, d\zeta,$$

where z_0 is any point in the plane. To satisfy the condition $F(i\sqrt{2}) = 0$, we take $z_0 = i\sqrt{2}$. Using the substitution $u = \zeta^2 + 2$, we have $du = 2\zeta\, d\zeta$; $u = 0$ at $\zeta = i\sqrt{2}$; $u = z^2 + 2$ at $\zeta = z$. Hence,

$$F(z) = \int_{i\sqrt{2}}^{z} \zeta\, e^{\zeta^2+2}\, d\zeta = \frac{1}{2}\int_0^{z^2+2} e^u\, du = \frac{1}{2}e^u \Big|_0^{z^2+1} = \frac{1}{2}(e^{z^2+2} - 1).$$

2. d) *Answer.* $f(z) = \frac{1}{2}z\sin 2z + \frac{1}{4}\cos 2z - \frac{1}{4}$.

Section 5.4

1. *Solution.* The equation $z^2 + \pi^2 = 0$ has two roots, $\pm i\pi$. Therefore the integrand has singularities at these two points, of which $-i\pi$ lies within Γ, a circle centered at $-i\pi$ of radius 2, and the other singularity $i\pi$ lies outside Γ (Fig. 108). We factor

$$z^2 + \pi^2 = (z + i\pi)(z - i\pi),$$

and rewrite the integral in the form

$$\int_\Gamma \frac{\left(\frac{\cos 2z}{z-i\pi}\right)}{z + i\pi}\, dz.$$

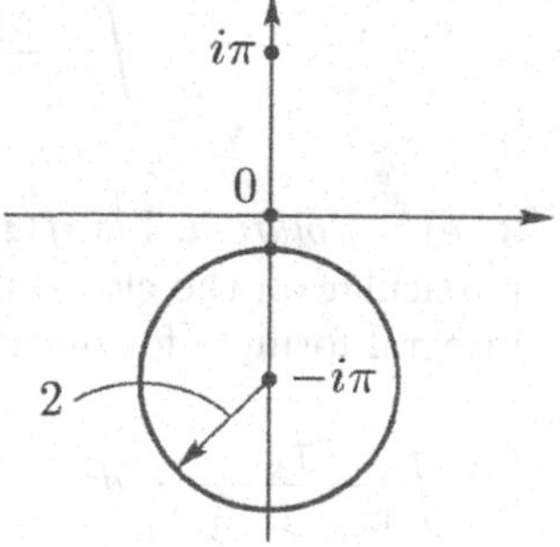

Fig. 108

The function

$$f(z) = \frac{\cos 2z}{z - i\pi}$$

is analytic in the closed disk bounded by Γ. Therefore we can apply the Cauchy integral formula (5.21) with $z = -i\pi$ and ζ replaced by z:

$$\int_\Gamma \frac{\left(\frac{\cos 2z}{z-i\pi}\right)}{z + i\pi}\, dz = \int_\Gamma \frac{f(z)}{z - (-i\pi)}\, dz = 2\pi i f(-i\pi) = 2\pi i \frac{\cos(-2\pi i)}{-\pi i - \pi i}$$

$$= -\cos 2\pi i = -\frac{1}{2}\left(e^{2\pi i^2} + e^{-2\pi i^2}\right) = -\frac{1}{2}\left(e^{-2\pi} + c^{2\pi}\right).$$

2. *Solution.* We will denote $f(z) = z^5 - 2z^4 + 3$. This function is analytic everwhere in $\mathbb{C}$, in particular on the closed disk $|z| \le 5$, and the point $z = i$ lies within this disk. Therefore we can apply Cauchy's integral formula (5.25) for derivatives, with $z = i$ and $n = 3$:

$$\int_\Gamma \frac{z^5 - 2z^4 + 3}{(z-i)^4}\,dz = \int_\Gamma \frac{f(z)}{(z-i)^4}\,dz = \frac{2\pi i}{3!} f'''(i)$$

$$= \frac{2\pi i}{6}(60i^2 - 48i) = 2\pi i(-10 - 8i) = 4\pi(4 - 5i),$$

where we have used

$$f'(z) = 5z^4 - 8z^3, \quad f''(z) = 20z^3 - 24z^2, \quad f'''(z) = 60z^2 - 48z.$$

3. a) *Solution.* We may rewrite the integrand as

$$\frac{\cos z}{z^2 - \pi^2} = \frac{\frac{\cos z}{z+\pi}}{z - \pi}.$$

Let

$$f(z) = \frac{\cos z}{z + \pi}.$$

Then f is analytic on the closed domain bounded by Γ, so we may apply the Cauchy integral formula (5.21) to get

$$\int_\Gamma \frac{f(z)}{z - \pi}\,dz = 2\pi i f(\pi) = 2\pi i \cdot \frac{-1}{2\pi} = -i.$$

3. e) *Solution.* Let $f(z) = z \sin z$. Then f is analytic everywhere in $\mathbb{C}$, in particular on the closed domain bounded by Γ, so we may apply the Cauchy integral formula for derivatives (5.25):

$$\int_\Gamma \frac{f(z)}{(z - (-\frac{\pi}{2}))^2}\,dz = \frac{2\pi i}{1!} f'(-\tfrac{\pi}{2}) = 2\pi i\,(\sin z + z\cos z)\Big|_{z=-\frac{\pi}{2}} = -2\pi i.$$

4. *Hint.* For $n = 1$ the formula (5.25) is already proved (in fact we may consider the Cauchy integral formula (5.21) as a special case of (5.25) with $n = 0$, $0! = 1$, $f^{(0)}(z) = f(z)$, and start induction with $n = 0$). Now assume that (5.25) is valid for some $n > 1$; we must show it holds also for $n + 1$, that is

$$\lim_{\Delta z \to 0} \left[\frac{f^{(n)}(z + \Delta z) - f^{(n)}(z)}{\Delta z} - \frac{(n+1)!}{2\pi i} \int_\Gamma \frac{f(\zeta)}{(\zeta - z)^{n+2}}\,d\zeta \right] = 0.$$

Since (5.25) is valid for n, the expression under the limit is equal to

$$\frac{n!}{2\pi i}\int_{\Gamma} A(z,\Delta z,\zeta)f(\zeta)\,d\zeta,$$

where $$A(z,\Delta z,\zeta)=\frac{(\zeta-z)^{n+1}-(\zeta-z-\Delta z)^{n+1}}{\Delta z(\zeta-z-\Delta z)^{n+1}(\zeta-z)^{n+1}}-\frac{n+1}{(\zeta-z)^{n+2}}.$$

According to the algebraic identity

$$a^k-b^k=(a-b)(a^{k-1}+a^{k-2}b+a^{k-3}b^2\cdots+b^{k-1}) \tag{9.51}$$

we have

$$\begin{aligned}&(\zeta-z)^{n+1}-(\zeta-z-\Delta z)^{n+1}\\ &\quad=\Delta z((\zeta-z)^n+(\zeta-z)^{n-1}(\zeta-z-\Delta z)+\cdots+(\zeta-z-\Delta z)^n).\end{aligned}$$

Cancel Δz and write $A(z,\Delta z,\zeta)$ as one fraction. Show that the numerator of this fraction is

$$(\zeta-z)^{n+1}+(\zeta-z)^n(\zeta-z-\Delta z)+\cdots+(\zeta-z)(\zeta-z-\Delta z)^n-(n+1)(\zeta-z-\Delta z)^{n+1}.$$

This expression can be written as a sum of $n+1$ differences:

$$\begin{aligned}&[(\zeta-z)^{n+1}-(\zeta-z-\Delta z)^{n+1}]+[(\zeta-z)^n(\zeta-z-\Delta z)-(\zeta-z-\Delta z)^{n+1}]\\ &\qquad+\ldots+[(\zeta-z)(\zeta-z-\Delta z)^n-(\zeta-z-\Delta z)^{n+1}].\end{aligned}$$

Use (9.51) to prove that each of these differences has the same factor Δz, and therefore the numerator of $A(z,\Delta z,\zeta)$ is equal to Δz multiplied by a bounded function.

Estimate the denominator of $A(z,\Delta z,\zeta)$ from below in the same way as in the proof of Theorem 5.18. Conclude that

$$\lim_{\Delta z\to 0}\frac{n!}{2\pi i}\int_{\Gamma} A(z,\Delta z,\zeta)f(\zeta)\,d\zeta=0.$$

5. *Hint.* The case $n=0$ is obvious. Suppose that $n>0$.

(a) Prove that for every $z_0\in\mathbb{C}$ the polynomial can be written in the form $P(z)=(z-z_0)Q(z)+r$, where $Q(z)$ is a polynomial of degree $n-1$, and r is a complex number.

(b) Prove that if $P(z_0)=0$, then $r=0$, that is $P(z)=(z-z_0)Q(z)$.

(c) Use this representation and Theorem 5.25 to prove the desired assertion.

Section 6.2

2. a) *Answer.* $\operatorname{Re} z < 0$.

3. *Answer.* a) The real axis $\operatorname{Im} z = 0$; b) Yes.

Section 6.3

2. c) *Answer.* $R = 1/e$.

5. *Hint.* Prove the proposition by contradiction. Assume that there are a positive number M and a point z_1 such that $|z_1| > R$ and $|c_n z_1^n| \le M$, $n = 1, 2, \ldots$. Use the same arguments as in the proof of Theorem 6.10 to show that the series $\sum_{n=0}^{\infty} c_n z^n$ converges on $|z| < |z_1|$. Hence, the radius of convergence is greater than R. We have arrived to contradiction with the assumption that the radius of convergence is R.

6. a), b) *Hint.* Prove that the series converges at every z with $|z| < \min(R_1, R_2)$ and with $|z| < R_1 R_2$ respectively.

6. c) *Hint.* We have to prove that if $|z| > \dfrac{R_1}{R_2}$, then the series $\sum_{n=0}^{\infty} \dfrac{a_n}{b_n} z^n$ diverges.

(i) Prove that there is $\delta > 0$ such that $|z| > \dfrac{(R_1 + \delta)}{(R_2 - \delta)}$;

(ii) by the previous problem 5, the sequence $\{|a_n|(R_1+\delta)^n\}$ is unbounded. Moreover, $\lim_{n\to\infty} |b_n|(R_2 - \delta)^n = 0$ (why?), and

$$\left|\frac{a_n}{b_n} z^n\right| > \frac{|a_n|(R_1 + \delta)^n}{|b_n|(R_2 - \delta)^n}.$$

(iii) Conclude that the series $\sum_{n=0}^{\infty} \dfrac{a_n}{b_n} z^n$ diverges.

Section 6.4

1. *Solution.* 1. First we expand the function

$$f(z) = \frac{e^{z^2} - 1}{z}$$

in a Taylor series about the same point. For this, we use the known series expansion for the function e^z at $z_0 = 0$ (formula (6.27)):

$$e^z = 1 + z + \frac{z^2}{2!} + \cdots + \frac{z^n}{n!} + \cdots = \sum_{n=0}^{\infty} \frac{z^n}{n!}, \quad \text{for} \quad z \in \mathbb{C}. \tag{9.52}$$

Substituting z^2 in for z, and subtracting 1, we get

$$e^{z^2} - 1 = z^2 + \frac{z^4}{2!} + \cdots + \frac{z^{2n}}{n!} + \cdots = \sum_{n=1}^{\infty} \frac{z^{2n}}{n!}.$$

This resulting series, like the original, converges everywhere in $\mathbb{C}$. We divide both sides by z:

$$\frac{e^{z^2} - 1}{z} = z + \frac{z^3}{2!} + \cdots + \frac{z^{2n-1}}{n!} + \cdots = \sum_{n=1}^{\infty} \frac{z^{2n-1}}{n!}. \tag{9.53}$$

This series also converges everywhere in $\mathbb{C}$, including at the point $z_0 = 0$, and consequently defines a function which is analytic in $\mathbb{C}$. By Property 6.17, it can be differentiated termwise.

2. We differentiate equation (9.53) with respect to z:

$$\left(\frac{e^{z^2} - 1}{z}\right)' = 1 + \frac{3z^2}{2!} + \cdots + \frac{(2n-1)z^{2n-2}}{n!} + \cdots = \sum_{n=1}^{\infty} \frac{(2n-1)z^{2n-2}}{n!},$$

so

$$f(z) = \sum_{n=1}^{\infty} \frac{(2n-1)z^{2n-2}}{n!}.$$

By Property 6.17, the series we have obtained has the same disk of convergence as the one in equation (9.53), namely all of $\mathbb{C}$.

2. *Solution.* 1. We make a change of variable, $w = z - z_0 = z - 3$, so that $z = w + 3$. Then

$$\frac{z-6}{(z+2)(z-5)} = \frac{w-3}{(w+5)(w-2)}.$$

2. Next we break up the resulting fraction into a sum of simpler ones:

$$\frac{w-3}{(w+5)(w-2)} = \frac{A}{w+5} + \frac{B}{w-2} = \frac{A(w-2) + B(w+5)}{(w+5)(w-2)},$$

$$\text{so that} \quad w - 3 = A(w-2) + B(w+5).$$

Substituting $w = 2$ and $w = -5$, we get $B = -\frac{1}{7}$ and $A = \frac{8}{7}$, respectively. In this way,

$$\frac{w-3}{(w+5)(w-2)} = \frac{8}{7} \cdot \frac{1}{w+5} - \frac{1}{7} \cdot \frac{1}{w-2}.$$

3. We wish to use the formula (6.30)

$$\frac{1}{1+z} = 1 - z + z^2 - \cdots + (-1)^n z^n + \cdots = \sum_{n=0}^{\infty} (-1)^n z^n, \quad |z| < 1, \quad (9.54)$$

to expand our two fractions into power series. In order to apply the formula, we shift some factors in the denominators:

$$\frac{8}{7} \cdot \frac{1}{w+5} = \frac{8}{7 \cdot 5} \cdot \frac{1}{1+\frac{w}{5}} \quad \text{and} \quad -\frac{1}{7} \cdot \frac{1}{w-2} = \frac{1}{7 \cdot 2} \cdot \frac{1}{1-\frac{w}{2}}.$$

Then with $z = \frac{w}{5}$ in (9.54) we get

$$\frac{8}{7} \cdot \frac{1}{w+5} = \frac{8}{7 \cdot 5} \left(1 - \frac{w}{5} + \frac{w^2}{5^2} - \cdots + (-1)^n \frac{w^n}{5^n} + \ldots \right) = \sum_{n=0}^{\infty} \frac{(-1)^n 8}{7 \cdot 5^{n+1}} w^n,$$

which converges when $|z| = |\frac{w}{5}| < 1$, or $|w| < 5$. With $z = -\frac{w}{2}$ we get

$$-\frac{1}{7} \cdot \frac{1}{w-2} = \frac{1}{7 \cdot 2} \left(1 + \frac{w}{2} + \frac{w^2}{2^2} + \cdots + \frac{w^n}{2^n} + \ldots \right) = \sum_{n=0}^{\infty} \frac{1}{7 \cdot 2^{n+1}} w^n,$$

which converges when $|z| = |-\frac{w}{2}| < 1$, or $|w| < 2$.

Combining the two expansions, we have

$$\frac{w-3}{(w+5)(w-2)} = \sum_{n=0}^{\infty} \frac{(-1)^n 8}{7 \cdot 5^{n+1}} w^n + \sum_{n=0}^{\infty} \frac{1}{7 \cdot 2^{n+1}} w^n$$
$$= \sum_{n=0}^{\infty} \frac{1}{7} \left(\frac{(-1)^n 8}{5^{n+1}} + \frac{1}{2^{n+1}} \right) w^n,$$

and this series converges when $|w| < 5$ and $|w| < 2$, i.e. on the disk $|w| < 2$.

4. Returning to the original variable z, we substitute $w = z - 3$ to obtain the desired expansion:

$$f(z) = \frac{z-6}{(z+2)(z-5)} = \sum_{n=0}^{\infty} \frac{1}{7} \left(\frac{(-1)^n 8}{5^{n+1}} + \frac{1}{2^{n+1}} \right) (z-3)^n, \quad |z-3| < 2.$$

Note that the disk of convergence of this Taylor series can also be determined just from the fact that the radius of convergence R is equal to the distance from z_0 to the closest singularity of the function $f(z)$. Our function has two singularities, at $z_1 = -2$ and $z_2 = 5$; the closest to $z_0 = 3$ is z_2, so $R = 5 - 3 = 2$.

3. a) *Solution.* Using the known Taylor series expansion of sin, we get

$$f(z) = \frac{d}{dz} \left(\frac{1}{z} \sum_{n=0}^{\infty} (-1)^n \frac{z^{2n+1}}{(2n+1)!} \right)$$
$$= \frac{d}{dz} \left(\sum_{n=0}^{\infty} (-1)^n \frac{z^{2n}}{(2n+1)!} \right) = \sum_{n=1}^{\infty} (-1)^n \frac{2n z^{2n-1}}{(2n+1)!}.$$

Notice that the function

$$F(z) = \frac{\sin z}{z}$$

is not defined at $z_0 = 0$, and is analytic in all other points. But the power series "automatically" defines $F(z)$ at z_0 in such a way that $F(z)$ and its derivative f are analytic in the entire complex plane. Therefore the radius of convergence for the series obtained is ∞.

3. d) *Solution.* Using partial fractions we may rewrite $f(z)$ as

$$\frac{z}{z^2 - 4} = \frac{1}{2} \cdot \frac{1}{z-2} + \frac{1}{2} \cdot \frac{1}{z+2}.$$

Now we use the geometric series to expand each term in powers of $(z-1)$.

$$\frac{1}{z-2} = \frac{1}{-1+(z-1)} = \frac{-1}{1-(z-1)} = -\sum_{n=0}^{\infty} (z-1)^n;$$

$$\frac{1}{z+2} = \frac{1}{3+(z-1)} = \frac{1}{3} \cdot \frac{1}{1+\frac{z-1}{3}} = \frac{1}{3} \sum_{n=0}^{\infty} (-1)^n \left(\frac{z-1}{3}\right)^n.$$

Putting these two parts together we get

$$f(z) = -\frac{1}{2} \sum_{n=0}^{\infty} (z-1)^n + \frac{1}{6} \sum_{n=0}^{\infty} (-1)^n \left(\frac{z-1}{3}\right)^n$$
$$= \sum_{n=0}^{\infty} \frac{1}{2} \left(\frac{(-1)^n}{3^{n+1}} - 1\right) (z-1)^n.$$

Since f is analytic everywhere except $z = \pm 2$, the closest singularity to the center of the expansion is at $z = 2$, and the radius of convergence is therefore $|2-1| = 1$.

4. *Hint.* Use Cauchy inequality for derivatives (5.28) to prove that $f^{(n)}(0) = 0$ for $n = k+1, k+2, \ldots$ Then apply (6.20) with $z_0 = 0$.

Section 6.5

2. a) *Hint.* Use the Taylor series for sin about $z_0 = 0$; *answer:* n; b) $2n$.

5. *Hint.* a) Consider $D = \{|z| < 1\}$ and $f(z) = z$.
b) Apply Theorem 6.29 to the function $g(z) = \frac{1}{f(z)}$ which is analytic in D.

Section 6.7

1. *Solution.* 1. Make the change of variable $w = z - 3$, or $z = w + 3$. Then

$$\frac{z-2}{(z+1)(z-3)} = \frac{w+1}{(w+4)w} = g(w).$$

2. Break the new fraction up into the sum of simpler ones:

$$\frac{w+1}{(w+4)w} = \frac{A}{w} + \frac{B}{w+4} = \frac{A(w+4)+Bw}{(w+4)w},$$
$$\text{so that} \quad w+1 = A(w+4) + Bw.$$

Then when w is 0 and -4, we get $A = \frac{1}{4}$ and $B = \frac{3}{4}$, respectively. So,

$$g(w) = \frac{1}{4} \cdot \frac{1}{w} + \frac{3}{4} \cdot \frac{1}{w+4}.$$

3. The function $g(z)$ has singularities at $w = 0$ and $w = -4$. Therefore it is analytic on the annuli

$$V_1 = \{ w : 0 < |w| < 4 \} \quad \text{and} \quad V_2 = \{ w : 4 < |w| < \infty \}.$$

We will find the Laurent expansion on each of them. For this we will have to expand the fraction $\dfrac{1}{w+4}$ into a series in powers of w, using the geometric series formula (9.54).

In the case of V_1, we shift a factor of 4 in the denominator:

$$\begin{aligned}\frac{3}{4} \cdot \frac{1}{w+4} &= \frac{3}{4 \cdot 4} \cdot \frac{1}{1 + \frac{w}{4}} \\ &= \frac{3}{4^2}\left(1 - \frac{w}{4} + \frac{w^2}{4^2} - \dots + (-1)^n \frac{w^n}{4^n} + \dots\right) = \sum_{n=0}^{\infty} \frac{(-1)^n 3}{4^{n+2}} w^n,\end{aligned}$$

setting $z = \frac{w}{4}$ in (9.54). Series (9.54) converges when $|z| < 1$; so, the obtained series converges when $|w| < 4$. Thus, for $w \in V_1$,

$$g(w) = \frac{1}{4} \cdot \frac{1}{w} + \sum_{n=0}^{\infty} \frac{(-1)^n 3}{4^{n+2}} w^n.$$

When $|w| > 4$ this series diverges, so we rewrite $\dfrac{1}{w+4}$ in a slightly different way, shifting a factor of w:

$$\frac{3}{4} \cdot \frac{1}{w+4} = \frac{3}{4w} \cdot \frac{1}{1 + \frac{4}{w}}.$$

Now if $|w| > 4$ then $\left|\frac{4}{w}\right| < 1$, so we substitute $z = \frac{4}{w}$ in (9.54):

$$\begin{aligned}\frac{3}{4}\cdot\frac{1}{w+4} &= \frac{3}{4w}\left(1-\frac{4}{w}+\frac{4^2}{w^2}-\cdots+(-1)^n\frac{4^n}{w^n}+\ldots\right)\\ &= \frac{3}{4w}-\frac{3}{w^2}+\frac{3\cdot 4}{w^3}-\cdots+(-1)^n\frac{3\cdot 4^{n-1}}{w^{n+1}}+\ldots\\ &= \frac{3}{4w}+\sum_{n=1}^{\infty}(-1)^n\frac{3\cdot 4^{n-1}}{w^{n+1}} = \frac{3}{4w}+\sum_{k=-2}^{-\infty}(-1)^{k+1}\frac{3}{4^{k+2}}w^k,\end{aligned}$$

where we have changed the index of summation to $k = -n-1$, so that $n = -k-1$ and $(-1)^n = (-1)^{-k-1} = (-1)^{k+1}$. Now we have, when $w \in V_2$,

$$g(w) = \frac{1}{4w}+\frac{3}{4w}+\sum_{k=-2}^{-\infty}(-1)^{k+1}\frac{3}{4^{k+2}}w^k = \frac{1}{w}+\sum_{k=-2}^{-\infty}(-1)^{k+1}\frac{3}{4^{k+2}}w^k.$$

4. To return to the original variable z, we substitute $w = z-3$. The desired expansions of the function $f(z)$ are then

$$f(z) = \frac{1}{4(z-3)}+\sum_{n=0}^{\infty}\frac{(-1)^n 3}{4^{n+2}}(z-3)^n, \qquad 0 <|z-3| < 4;$$

$$f(z) = \frac{1}{z-3}+\sum_{k=-2}^{-\infty}\frac{(-1)^{k+1}3}{4^{k+2}}(z-3)^k, \qquad 4 <|z-3| < \infty.$$

Had we been asked to find the Laurent expansion only on the punctured neighborhood of a singularity (for example $z_0 = 3$), then we would have needed to carry out this procedure only for the smaller of the annuli (in this case V_1).

2. d) *Solution.* Using partial fractions, we rewrite $f(z)$ as

$$\frac{z+1}{(z-1)(z-3)} = \frac{-1}{z-1}+\frac{2}{z-3}.$$

There will be two Laurent series expansions, one for $0 < |z-3| < 2$, and one for $|z-3| > 2$. For the first annulus we use the geometric series to expand the first fraction on the right in powers of $(z-3)$:

$$\frac{1}{z-1} = \frac{1}{2+(z-3)} = \frac{1}{2}\cdot\frac{1}{1+\frac{z-3}{2}} = \frac{1}{2}\sum_{n=0}^{\infty}(-1)^n\left(\frac{z-3}{2}\right)^n.$$

This series converges when $|z-3| < 2$. Therefore on the annulus $0 < |z-3| < 2$,

$$f(z) = \frac{2}{z-3}-\frac{1}{2}\sum_{n=0}^{\infty}\frac{(-1)^n}{2^n}(z-3)^n = \frac{2}{z-3}+\sum_{n=0}^{\infty}\frac{(-1)^{n+1}}{2^{n+1}}(z-3)^n.$$

For the other annulus $|z-3|>2$, we rework the geometric series so that it converges there:

$$\frac{1}{z-1}=\frac{1}{(z-3)+2}=\frac{1}{z-3}\cdot\frac{1}{1+\frac{2}{z-3}}$$
$$=\frac{1}{z-3}\sum_{n=0}^{\infty}(-1)^n\left(\frac{2}{z-3}\right)^n=\sum_{n=0}^{\infty}(-1)^n 2^n\frac{1}{(z-3)^{n+1}}.$$

Therefore when $|z-3|>2$,

$$f(z)=\frac{2}{z-3}-\sum_{n=0}^{\infty}(-1)^n2^n\frac{1}{(z-3)^{n+1}}=\frac{1}{z-3}+\sum_{n=1}^{\infty}(-1)^{n+1}2^n\frac{1}{(z-3)^{n+1}}.$$

3. *Hint.* Use the same idea as in the proof of the first part of Theorem 6.36.

Section 7.1

1. *Solution.* 1. Making the change of variable $w=z+2$, so that $z=w-2$, we get

$$f(z)=(z+2)^2e^{z/(z+2)}=w^2e^{(w-2)/w}=w^2e^{1-2/w}=ew^2e^{-2/w}.$$

2. Using the known expansion of the function e^z from formula (6.27), and replacing z with $-\frac{2}{w}$, we get

$$ew^2e^{-2/w}=ew^2\sum_{n=0}^{\infty}\frac{1}{n!}\left(-\frac{2}{w}\right)^n=\sum_{n=0}^{\infty}\frac{(-1)^n2^ne}{n!w^{n-2}}=\sum_{k=-2}^{+\infty}\frac{(-1)^k2^{k+2}e}{(k+2)!w^k}$$
$$=ew^2-2ew+2e+\sum_{k=1}^{+\infty}\frac{(-1)^k2^{k+2}e}{(k+2)!}w^{-k},$$

where we have changed the index of summation to $k=n-2$, or $n=k+2$.

Since the power series for e^z converges for all z in $\mathbb{C}$, the series we have obtained converges whenever $|z|=|-\frac{2}{w}|<\infty$, i.e. for all $w\neq 0$.

3. To return the original variable, we substitute $w=z+2$ to get the desired expansion of $f(z)$:

$$f(z)=e(z+2)^2-2e(z+2)+2e+\sum_{k=1}^{+\infty}\frac{(-1)^k2^{k+2}e}{(k+2)!}(z+2)^{-k},\quad z\neq -2.$$

The regular part of the expansion is the sum

$$e(z+2)^2-2e(z+2)+2e,$$

while the principal part is the series

$$\sum_{k=1}^{+\infty} \frac{(-1)^k 2^{k+2} e}{(k+2)!}(z+2)^{-k}.$$

Because the principal part has infinitely many coefficients distinct from zero, the point $z_0 = -2$ is an essential singularity.

2. a) *Solution.* We can either make the change of variable $w = z - 2$, or directly represent the fraction as

$$\frac{z}{z-2} = 1 + \frac{2}{z-2},$$

so that

$$f(z) = ee^{\frac{2}{z-2}}.$$

The Taylor series expansion of e^z is valid for all $z \in \mathbb{C}$, so we may replace z by $\frac{2}{z-2}$ to get a Laurent series valid when $z \neq 2$:

$$f(z) = e\sum_{n=0}^{\infty} \frac{1}{n!}\left(\frac{2}{z-2}\right)^n = \sum_{n=0}^{\infty} \frac{e2^n}{n!}\frac{1}{(z-2)^n} = e + \sum_{n=1}^{\infty} \frac{e2^n}{n!}\frac{1}{(z-2)^n}.$$

Here the regular part equals e, and the principal part makes up the rest of the series. Because the principal part has infinitely many terms, the singularity at $z_0 = 2$ is essential.

2. d) *Solution.* Using partial fractions we rewrite $f(z)$ as

$$f(z) = \frac{2z}{(z+i)(z-i)} = \frac{1}{z-i} + \frac{1}{z+i}.$$

Now we rewrite the second fraction on the right as a geometric series:

$$\frac{1}{z+i} = \frac{1}{2i + (z-i)} = \frac{1}{2i} \cdot \frac{1}{1 + \frac{z-i}{2i}} = \frac{1}{2i}\sum_{n=0}^{\infty}(-1)^n \left(\frac{z-i}{2i}\right)^n.$$

This gives us

$$f(z) = \frac{1}{z-i} + \sum_{n=0}^{\infty} \frac{(-1)^n}{(2i)^{n+1}}(z-i)^n, \quad 0 < |z-i| < 2.$$

The first fraction on the right is the principal part, and the series on the right is the regular part. Because the principal part has only finitely many terms, the singularity at $z_0 = i$ is a pole, in this case of first order.

3. *Solution.* We expand the numerator and denominator in Taylor series in powers of z. For this we use the power series for cos from (6.29):

$$\cos z = 1 - \frac{z^2}{2!} + \frac{z^4}{4!} - \cdots + (-1)^n \frac{z^{2n}}{(2n)!} + \cdots = \sum_{n=0}^{\infty} (-1)^n \frac{z^{2n}}{(2n)!}, \quad z \in \mathbb{C}. \quad (9.55)$$

From this we get the equality

$$\cos z - 1 + z^2 = z^2 - \frac{z^2}{2!} + \frac{z^4}{4!} - \dots + (-1)^n \frac{z^{2n}}{(2n)!} + \dots = z^2 \left(\frac{1}{2} + \frac{z^2}{4!} - \dots \right),$$

and this series converges in the entire complex plane.

To expand the denominator $e^{z^3} - 1 - z^3$, we use the expansion of e^z in formula (9.52), replacing z with z^3, and subtract $1 + z^3$:

$$e^{z^3} - 1 - z^3 = -1 - z^3 + 1 + z^3 + \frac{(z^3)^2}{2!} + \frac{(z^3)^3}{3!} + \dots = z^6 \left(\frac{1}{2} + \frac{z^3}{3!} + \dots \right).$$

This series, like that in (9.52), converges for all $z \in \mathbb{C}$. Therefore

$$f(z) = \frac{\cos z - 1 + z^2}{e^{z^3} - 1 - z^3} = \frac{z^2 \left(\frac{1}{2} + \frac{z^2}{4!} - \dots \right)}{z^6 \left(\frac{1}{2} + \frac{z^3}{3!} + \dots \right)} = \frac{1}{z^4} h(z),$$

where

$$h(z) = \frac{\frac{1}{2} + \frac{z^2}{4!} - \dots}{\frac{1}{2} + \frac{z^3}{3!} + \dots}.$$

The function h is the ratio of functions f_1 and f_2, both of which are analytic in $\mathbb{C}$, and $f_2(0) = \frac{1}{2} \neq 0$. Therefore $h(z)$ is also analytic in some neighborhood of the point $z_0 = 0$, and $h(0) = 1 \neq 0$. As $f(z) = \frac{1}{z^4} h(z)$, by Corollary 7.5 the point $z_0 = 0$ is a pole of order $N = 4$.

4. a) Solution. Using the Taylor series for the exponential and cosine functions, we have

$$e^{2z} = 1 + 2z + \frac{(2z)^2}{2} + \frac{(2z)^3}{3!} + \dots + \frac{(2z)^n}{n!} + \dots;$$
$$\cos z = 1 - \frac{z^2}{2} + \frac{z^4}{4!} - \dots + (-1)^n \frac{z^{2n}}{(2n)!} + \dots.$$

Therefore

$$f(z) = \frac{e^{2z} - 1}{\cos z - 1 + \frac{z^2}{2}} = \frac{2z + \frac{(2z)^2}{2} + \frac{(2z)^3}{3!} + \dots + \frac{(2z)^n}{n!} + \dots}{\frac{z^4}{4!} + \dots + (-1)^n \frac{z^{2n}}{(2n)!} + \dots}$$
$$= \frac{z}{z^4} \cdot \frac{2 + \frac{2^2 z}{2} + \frac{2^3 z^2}{3!} + \dots + \frac{2^n z^{n-1}}{n!} \dots}{\frac{1}{4!} - \dots + (-1)^n \frac{z^{2n-4}}{(2n)!} + \dots} = \frac{1}{z^3} h(z),$$

where h is a function which is analytic near $z_0 = 0$, and nonzero there. Therefore, the function f has a pole of order 3 at $z_0 = 0$.

5. *Hint.* Show that

$$G(w) = f\left(\frac{1}{w}\right) = \frac{(1+w^2)(1+3w)^2}{(1-w)w^3}.$$

Then show that $G(w)$ has a pole of order 3 at $w_0 = 0$.

6. *Solution.* Finite singular points are $z_1 = 0$, and also the points at which $z^2+9 = 0$, namely $z_2 = 3i$ and $z_3 = -3i$. Therefore the singularity at infinity is also isolated because its neighborhood $|z| > 3$ contains no other singularities.

Consider the point $z_1 = 0$. Define a sequence $\{z_n'\}$, $n = 1, 2, \ldots$ by

$$z_n' = \frac{1}{\pi n} \quad \text{or} \quad \frac{1}{z_n'} = \pi n.$$

Then $f(z_n') = 0$ for all n. Define another sequence $\{z_n''\}$, $n = 1, 2, \ldots$ by

$$z_n'' = \frac{2}{\pi + 4\pi n} \quad \text{or} \quad \frac{1}{z_n''} = \frac{\pi}{2} + 2\pi n,$$

so that $\sin \frac{1}{z_n''} = 1$ for all n. It is easy to see that both $z_n' \to 0$ and $z_n'' \to 0$ as $n \to \infty$. At the same time, the limits

$$\lim_{n\to\infty} f(z_n') = 0 \quad \text{and} \quad \lim_{n\to\infty} f(z_n'') = \lim_{n\to\infty} \frac{1}{((z_n'')^2 + 9)^2} = \frac{1}{9^2},$$

take different values. Therefore the function $f(z)$ does not have a limit as $z \to 0$, and so $z_1 = 0$ is an essential singularity.

To investigate the points z_2 and z_3, we factor the denominator. Since

$$z^2 + 9 = (z - 3i)(z + 3i),$$

we have

$$f(z) = \frac{\sin \frac{1}{z}}{(z - 3i)^2(z + 3i)^2}.$$

For $z_2 = 3i$, we will rewrite this as

$$f(z) = \frac{h(z)}{(z - 3i)^2}, \quad \text{where} \quad h(z) = \frac{\sin \frac{1}{z}}{(z + 3i)^2}.$$

Because the function h is analytic in a neighborhood of $3i$, and

$$h(3i) = \frac{\sin \frac{1}{3i}}{(6i)^2} \neq 0,$$

by Corollary 7.5 the point $z_2 = 3i$ is a pole of second order. Analogously, we can show that the point $z_3 = -3i$ is also a pole of second order.

There remains to consider the point $z_4 = \infty$. There are two possible methods here: either switch to the variable $w = \frac{1}{z}$ and examine the singularity

$w_0 = 0$ of the function $G(w) = f(\frac{1}{z})$ (see Example 7.12), or directly calculate the limit of $f(z)$ as $z \to \infty$ (or show it does not exist). In this case the second method is easier, since the limit is easy to calculate. For, as $z \to \infty$, $\frac{1}{z} \to 0$ and so $\sin \frac{1}{z} \to 0$; and clearly $(z^2+9)^2 \to \infty$. So we have

$$\lim_{z\to\infty} f(z) = \lim_{z\to\infty} \frac{\sin \frac{1}{z}}{(z^2+9)^2} = 0.$$

Since $f(z)$ has a finite limit as $z \to \infty$, then $z_4 = \infty$ is a removable singularity (that the limit is 0 is unimportant, what matters is only that it is finite). If we set $f(\infty) = 0$, then f will be analytic at $z_4 = \infty$.

7. b) *Hint and answer.* The function

$$f_1(z) = \cos \frac{1}{z-1}$$

has an essential singularity at $z = 1$, and is analytic elsewhere in $\mathbb{C}$. The function

$$f_2(z) = \frac{1}{(z^2+1)(z^2+4)}$$

has simple poles at $z = \pm i, \pm 2i$, and is analytic elsewhere. Therefore their product $f(z) = f_1(z)f_2(z)$ has those five finite singular points; the isolated singularity at ∞ is removable.

Section 7.2

1. *Solution.* This function has three singularities, $z_1 = -1$, $z_2 = 2i$, and $z_3 = -2i$. From Corollary 7.5 it is easy to see that z_1 is a pole of second order, and z_2 and z_3 are poles of first order.

At $z_2 = 2i$ we use the formula (7.15):

$$\begin{aligned}\operatorname{res}_{2i} f &= \lim_{z\to 2i} (z-2i)f(z) = \lim_{z\to 2i} \frac{(z-2i)z^4}{(z-2i)(z+2i)(z+1)^2}\\ &= \frac{(2i)^4}{(2i+2i)(2i+1)^2} = \frac{16}{4i(-3+4i)} = \frac{4}{-4-3i} = \frac{4(-4+3i)}{25}.\end{aligned}$$

Analogously,

$$\begin{aligned}\operatorname{res}_{-2i} f &= \lim_{z\to -2i} \frac{(z+2i)z^4}{(z-2i)(z+2i)(z+1)^2}\\ &= \frac{(-2i)^4}{(-2i-2i)(-2i+1)^2} = \frac{16}{-4i(-3-4i)} = \frac{4}{-4+3i} = \frac{-4(4+3i)}{25}.\end{aligned}$$

To calculate the residue at the point $z_1 = -1$, we use the formula (7.18) with $n = 2$:

$$\operatorname{res}_a f = \lim_{z \to a} ((z-a)^2 f(z))';$$

$$((z-(-1))^2 f(z))' = \left(\frac{(z+1)^2 z^4}{(z^2+4)(z+1)^2} \right)' = \left(\frac{z^4}{z^2+4} \right)'$$
$$= \frac{4z^3(z^2+4) - z^4 \cdot 2z}{(z^2+4)^2} = \frac{2z^3(z^2+8)}{(z^2+4)^2};$$
$$\operatorname{res}_{-1} f = \lim_{z \to -1} \frac{2z^3(z^2+8)}{(z^2+4)^2} = \frac{-2 \cdot 9}{5^2} = -\frac{18}{25}.$$

2. *Answer.* $\operatorname{res}_i f = 1$. *Hint.* The Laurent expansion of f was obtained in Section 7.1, problem 2 d).

3. a) *Answer and hint.* $\operatorname{res}_2 f = 2e$ —see Section 7.1, problem 2 a).

3. d) *Answer and hint.* $\operatorname{res}_{-2} f = -\frac{4e}{3}$ —see Section 7.1, problem 1.

3. e), f) *Hint.* Apply Theorem 7.17, formula (7.16).

4. *Solution.* Let us check that the singularity $z_0 = \infty$ is isolated, which means that there is some neighborhood of ∞—the exterior of some circle $|z| = R$—in which there are no other singularities beside $z_0 = \infty$. The function f has only one finite singularity, at $z_1 = 0$. Therefore the exterior of any circle $|z| = R$, for example, $|z| = 1$, has no singularities beside $z_0 = \infty$.

We can find the desired value $\operatorname{res}_\infty f$ in either of two ways.

1. Using formula (7.25). Here

$$g(w) = \frac{1}{w^2} f\left(\frac{1}{w}\right) = \frac{1}{w^2}\left(\frac{1}{w} + 2\right) e^{-w} = \frac{(1+2w)e^{-w}}{w^3} = \frac{h(w)}{w^3},$$

where $h(w) = (1+2w)e^{-w}$. Since the function h is analytic in the neighborhood of $w_0 = 0$, and $h(0) \neq 0$, according to Corollary 7.5 the point w_0 is a pole of order three. We can now find the value of $\operatorname{res}_0 g$ using the formula (7.19) with $n = 3$:

$$\operatorname{res}_0 g = \frac{h''(0)}{2!};$$
$$h'(w) = 2e^{-w} - (1+2w)e^{-w} = (1-2w)e^{-w};$$
$$h''(w) = -2e^{-w} - (1-2w)e^{-w} = (-3+2w)e^{-w};$$
$$\operatorname{res}_0 g = \frac{(-3+0)e^0}{2!} = -\frac{3}{2}.$$

Then according to formula (7.25), $\operatorname{res}_\infty f = -\operatorname{res}_0 g = \dfrac{3}{2}$.

2. Using formula (7.22). To find the coefficient c_{-1} we make the change of variable $w = \frac{1}{z}$, and substitute $z = \frac{1}{w}$ into $f(z)$:

$$G(w) = f\left(\frac{1}{w}\right) = \left(\frac{1}{w} + 2\right) e^{-w}.$$

Now we find the coefficient of w^1 in the Laurent expansion of $G(w)$ in powers of w; this will also be the coefficient c_{-1} of z^{-1} in the Laurent expansion of $f(z)$ in powers of z. To obtain the expansion of G, we will use formula (6.27), replacing z with $-w$:

$$\begin{aligned} G(w) &= \left(\frac{1}{w} + 2\right)\left(1 + (-w) + \frac{(-w)^2}{2} + \dots\right) \\ &= \frac{1}{w} + 2 - 1 - 2w + \frac{w}{2} + w^2 - \dots = \frac{1}{w} + 1 - \frac{3}{2}w + w^2 - \dots, \end{aligned}$$

where the remaining terms contain powers of w greater than or equal to two. So the coefficient of w^1 is $c_{-1} = -\frac{3}{2}$.

Then by formula (7.22), $\operatorname{res}_\infty f = -c_{-1} = \frac{3}{2}$.

Of course, both methods give the same result!

5. a) *Solution.* Introducing $w = \frac{1}{z}$ we have

$$\begin{aligned} f\left(\frac{1}{w}\right) &= \left(\frac{1}{w} + 3\right)\sin w = \left(\frac{1}{w} + 3\right)\left(w - \frac{w^3}{3!} + \frac{w^5}{5!} - \dots\right) \\ &= 1 + 3w - \frac{w^2}{3!} + \frac{3w^3}{3!} - \dots \end{aligned}$$

The coefficient of w^1 is $c_{-1} = 3$, and by formula (7.22), $\operatorname{res}_\infty f = -c_{-1} = -3$.

Section 7.3

1. *Solution.* The path of integration is the circle centered at $z_0 = i$ with radius 2. We denote this by Γ —see Fig. 109.

Denote the integrand by $f(z)$:

$$f(z) = \frac{z^4}{(z^2+4)(z+1)^2}.$$

This function has three singularities, $z_1 = -1$, $z_2 = 2i$, and $z_3 = -2i$, plotted in Fig. 109; of these, z_1 and z_2 lie within the contour, and z_3 is outside it.

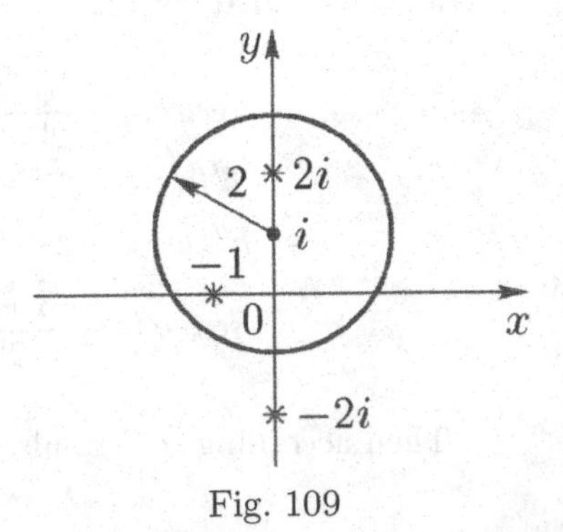

Fig. 109

The residues of the function f at these two interior points were found in problem 1, Section 7.2:

$$\operatorname{res}_{2i} f = \frac{4(-4+3i)}{25}; \quad \operatorname{res}_{-1} f = -\frac{18}{25}.$$

Using the Residue Theorem 7.15 and formula (7.12)

$$\int_{\Gamma} f(z)\, dz = 2\pi i \sum_{k=1}^{n} \operatorname{res}_{z_k} f,$$

we find the desired value of the integral:

$$\int_{|z-i|=2} \frac{z^4}{(z^2+4)(z+1)^2}\, dz = 2\pi i \left(\frac{4(-4+3i)}{25} - \frac{18}{25} \right) = -\frac{4\pi}{25}(6+17i).$$

2. c) *Solution.* The integrand

$$f(z) = \frac{z^4}{(z^2-9)(z-i)}$$

has simple poles at $z = \pm 3, i$. Of these, -3 lies outside of the path, while 3 and i lie in the interior. Therefore we can sum the residues at these latter two points to calculate the integral:

$$\operatorname{res}_i f = \lim_{z \to i} (z-i) f(z) = \frac{1}{-1-9} = -\frac{1}{10};$$

$$\operatorname{res}_3 f = \lim_{z \to 3} (z-3) f(z) = \frac{81}{6(3-i)} = \frac{27(3+i)}{2 \cdot 10} = \frac{81+27i}{20};$$

$$\int_{|z-1|=3} f(z)\, dz = 2\pi i \left(-\frac{1}{10} + \frac{81+27i}{20} \right) = \pi \frac{-27+79i}{10}.$$

2. d) *Solution.* The integrand

$$f(z) = \frac{z^8}{(z^2-1)^2}$$

has two second order poles at 1 and -1. Both of these singularities lie in the interior of the path of integration. We can either find these residues directly, or only calculate one residue at infinity and apply Theorem 7.24. Using Property 7.25, we get

$$g(w) = \frac{1}{w^2} f\left(\frac{1}{w}\right) = \frac{1}{w^2} \cdot \frac{w^{-8}}{(w^{-2}-1)^2} = \frac{1}{w^6(1-w^2)^2}.$$

The Laurent expansion of $g(w)$ at 0 contains only even powers of w. Hence, the coefficient of w^1 is equal to 0. Thus, the integral equals 0.

2. e) *Solution.* The integrand

$$f(z) = \frac{e^z}{z^2(z - i\pi)}$$

has a simple pole at $i\pi$ and a second order pole at 0. Both of these singularities lie in the interior of the path of integration. We can only calculate one residue at infinity, but in this case it's easier to find the residues at 0 and $i\pi$ directly.

$$\operatorname{res}_{i\pi} f = \lim_{z \to i\pi} (z - i\pi) f(z) = \frac{e^{i\pi}}{(i\pi)^2} = \frac{1}{\pi^2};$$

$$\operatorname{res}_0 f = \frac{1}{1!} \lim_{z \to 0} \frac{d}{dz} (z^2 f(z)) = \lim_{z \to 0} e^z \frac{z - i\pi - 1}{(z - i\pi)^2} = \frac{-i\pi - 1}{(-i\pi)^2} = \frac{1}{\pi^2} + \frac{i}{\pi};$$

$$\int\limits_{|z|=5} f(z)\, dz = 2\pi i \left(\frac{1}{\pi^2} + \frac{1}{\pi^2} + \frac{i}{\pi} \right) = -2 + i\frac{4}{\pi}.$$

3. *Solution.* We make a change of variable, setting $z = e^{it}$. Then

$$dz = e^{it} i\, dt = zi\, dt,$$

and so

$$dt = \frac{dz}{zi} = -\frac{i\, dz}{z}, \quad \sin t = \frac{1}{2i}(e^{it} - e^{-it}) = \frac{1}{2i}\left(z - \frac{1}{z}\right).$$

As t goes from 0 to 2π, the point $z = e^{it}$ describes the circle $|z| = 1$. Writing the integral in terms of the new variable z we get

$$\int_0^{2\pi} \frac{1}{5 - \sqrt{21} \sin t}\, dt = -\int\limits_{|z|=1} \frac{i}{z\left(5 - \sqrt{21}\frac{1}{2i}\left(z - \frac{1}{z}\right)\right)}\, dz$$

$$= \int\limits_{|z|=1} \frac{2}{-\sqrt{21} z^2 + 10iz + \sqrt{21}}\, dz.$$

This integral can be evaluated with the help of residues. First we find the singularities of the function

$$f(z) = \frac{2}{-\sqrt{21} z^2 + 10iz + \sqrt{21}},$$

by setting the denominator to zero:

$$-\sqrt{21} z^2 + 10iz + \sqrt{21} = 0.$$

Solving the quadratic equation gives us

$$z_{1,2} = \frac{-10i \pm \sqrt{(10i)^2 - 4(-\sqrt{21})\sqrt{21}}}{2(-\sqrt{21})} = \frac{5i \mp 2i}{\sqrt{21}},$$

so that $z_1 = \dfrac{3i}{\sqrt{21}}$ and $z_2 = \dfrac{7i}{\sqrt{21}}$.

Of these two points, only z_1 lies inside the circle $|z| = 1$, since $|z_2| = \frac{7}{\sqrt{21}} > 1$. Since $f(z)$ can be written as

$$f(z) = \frac{2}{-\sqrt{21}\left(z - \frac{3i}{\sqrt{21}}\right)\left(z - \frac{7i}{\sqrt{21}}\right)},$$

both singularities are simple poles.

Now we find the residue at z_1 using formula (9.57):

$$\begin{aligned}\operatorname{res}_{z_1} f &= \lim_{z \to \frac{3i}{\sqrt{21}}} \frac{2\left(z - \frac{3i}{\sqrt{21}}\right)}{-\sqrt{21}\left(z - \frac{3i}{\sqrt{21}}\right)\left(z - \frac{7i}{\sqrt{21}}\right)} \\ &= \lim_{z \to \frac{3i}{\sqrt{21}}} \frac{2}{-\sqrt{21}\left(z - \frac{7i}{\sqrt{21}}\right)} = \frac{2}{7i - 3i} = \frac{1}{2i}.\end{aligned}$$

Finally, from the Residue Theorem 7.15, we get

$$\int_{|z|=1} f(z)\, dz = 2\pi i \cdot \operatorname{res}_{z_1} f = 2\pi i \frac{1}{2i} = \pi.$$

So the original integral evaluates to π.

4. b) *Solution.* As we saw in Section 7.3.2, when $z = e^{it}$,

$$\cos t = \frac{1}{2}\left(z + \frac{1}{z}\right), \quad dt = \frac{dz}{iz}.$$

Therefore the integral is equivalent to

$$\int_{|z|=1} \frac{1}{\frac{3}{2}(z + \frac{1}{z}) + 5} \cdot \frac{1}{iz}\, dz = \int_{|z|=1} \frac{1}{i} \cdot \frac{2}{3z^2 + 10z + 3}\, dz.$$

Using the quadratic formula to get the roots of the denominator,

$$z_{1,2} = \frac{-10 \pm \sqrt{100 - 36}}{6} = \frac{-10 \pm 8}{6} = -\frac{1}{3}, -3.$$

These points are simple poles of the integrand, and $-\frac{1}{3}$ is within the unit circle, while -3 is without. Therefore, denoting the integrand by $f(z)$,

$$\operatorname{res}_{-\frac{1}{3}} f = \lim_{z\to-\frac{1}{3}} (z+\tfrac{1}{3})f(z) = \frac{1}{i}\lim_{z\to-\frac{1}{3}} \frac{2}{3}\cdot\frac{1}{z+3} = \frac{1}{i}\cdot\frac{2}{-1+9} = \frac{1}{4i};$$

$$\frac{1}{i}\int_{|z|=1} \frac{2}{3z^2+10z+3}\,dz = 2\pi i\cdot\frac{1}{4i} = \frac{\pi}{2}.$$

5. *Solution.* As in Theorem 7.30, we will consider the closed contour consisting of the segment $[-R,R]$ on the real axis, and the half circle $\gamma(R)$ of radius R in the upper half-plane (Fig. 110).

We introduce the function

$$f(z) = \frac{z^2-4}{(z^2+6z+13)(z^2+16)},$$

obtained by replacing the real variable x with the complex one z. We will show that

$$\lim_{R\to\infty}\int_{\gamma(R)} f(z)\,dz = 0, \qquad (9.56)$$

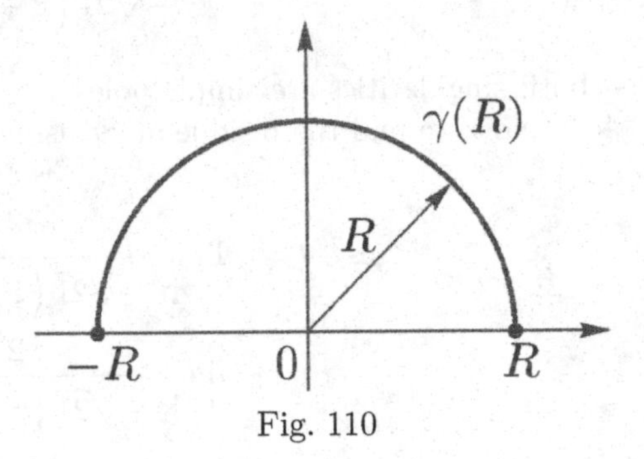

Fig. 110

i.e. that the second condition of Theorem 7.30 is satisfied. In fact,

$$f(z) = \frac{z^2\left(1-\frac{4}{z^2}\right)}{z^2\left(1+\frac{6}{z}+\frac{13}{z^2}\right)z^2\left(1+\frac{16}{z^2}\right)} = \frac{1}{z^2}\cdot\frac{1-\frac{4}{z^2}}{\left(1+\frac{6}{z}+\frac{13}{z^2}\right)\left(1+\frac{16}{z^2}\right)}$$
$$= \frac{1}{z^2}h(z), \quad \text{where } h(z) = \frac{1-\frac{4}{z^2}}{\left(1+\frac{6}{z}+\frac{13}{z^2}\right)\left(1+\frac{16}{z^2}\right)}.$$

Since $h(z)\to 1$ as $z\to\infty$, for all sufficiently large values of z we will have $|h(z)|<2$. Therefore

$$|f(z)| = \frac{|h(z)|}{|z|^2} < \frac{2}{|z|^2}.$$

Applying (5.9) with $M = 2/|z|^2$ and $\ell = \pi R$, for sufficiently large R we obtain the estimate

$$\left|\int_{\gamma(R)} f(z)\,dz\right| \le \frac{2}{R^2}\pi R = \frac{2\pi}{R}.$$

Taking the limit as $R\to\infty$, we get (9.56).

Now we can apply Theorem 7.30, which says that the desired real integral is equal to the sum of the residues of the function f at its singularities in the upper half-plane, multipled by $2\pi i$. In our case the singularities will be points at which the denominator becomes zero: either where $z^2 + 6z + 13 = 0$, i.e. $z_{1,2} = -3 \pm 2i$, or $z^2 + 16 = 0$, i.e. $z_{3,4} = \pm 4i$. Of these, only $z_1 = -3 + 2i$ and $z_3 = 4i$ lie in the upper half-plane. Since

$$f(z) = \frac{z^2 - 4}{(z + 3 - 2i)(z + 3 + 2i)(z - 4i)(z + 4i)},$$

by Corollary 7.5 each singularity is a first order pole. The residues at z_1 and z_3 can easily be found using either formula (7.17), or formula (7.15):

$$\operatorname{res}_{z_0} f = \lim_{z \to z_0} (z - z_0) f(z). \tag{9.57}$$

For comparison we will do it in both ways.

First we apply formula (7.17). We put $f(z)$ in the form

$$f(z) = \frac{z^2 - 4}{(z + 3 - 2i)(z + 3 + 2i)(z^2 + 16)} = \frac{h_1(z)}{z + 3 - 2i},$$

where

$$h_1(z) = \frac{z^2 - 4}{(z + 3 + 2i)(z^2 + 16)}.$$

Because h_1 is analytic at the point $z_1 = -3 + 2i$ and $h_1(-3 + 2i) \neq 0$, by formula (7.17) we have

$$\begin{aligned} \operatorname{res}_{-3+2i} f &= h_1(-3 + 2i) \\ &= \frac{(-3 + 2i)^2 - 4}{4i((-3 + 2i)^2 + 16)} = \frac{1 - 12i}{12i(7 - 4i)} = \frac{11 - 16i}{12 \cdot 13i}. \end{aligned}$$

Next we use formula (9.57):

$$\begin{aligned} \operatorname{res}_{-3+2i} f &= \lim_{z \to -3+2i} \frac{(z + 3 - 2i)(z^2 - 4)}{(z + 3 - 2i)(z + 3 + 2i)(z^2 + 16)} \\ &= \lim_{z \to -3+2i} \frac{z^2 - 4}{(z + 3 + 2i)(z^2 + 16)} \\ &= \frac{(-3 + 2i)^2 - 4}{4i((-3 + 2i)^2 + 16)} = \frac{11 - 16i}{12 \cdot 13i}. \end{aligned}$$

Note that the numerical computation is essentially the same for both methods; the choice is a matter of taste. Our taste is for formula (9.57), so we use it to calculate the residue at $z_3 = 4i$:

$$\begin{aligned} \operatorname{res}_{4i} f &= \lim_{z \to 4i} \frac{(z - 4i)(z^2 - 4)}{(z^2 + 6z + 13)(z - 4i)(z + 4i)} \\ &= \lim_{z \to 4i} \frac{z^2 - 4}{(z^2 + 6z + 13)(z + 4i)} \\ &= \frac{(4i)^2 - 4}{((4i)^2 + 24i + 13)8i} = \frac{-5}{6i(-1 + 8i)} = \frac{1 + 8i}{6 \cdot 13i}. \end{aligned}$$

So finally

$$\int_{-\infty}^{\infty} \frac{x^2-4}{(x^2+6x+13)(x^2+16)}\,dx = 2\pi i(\text{res}_{-3+2i} f + \text{res}_{4i} f)$$

$$= 2\pi i\left(\frac{11-16i}{12\cdot 13i} + \frac{1+8i}{6\cdot 13i}\right) = \frac{\pi}{6\cdot 13}(11-16i+2+16i) = \frac{\pi}{6}.$$

6. *Hint.* Consider the closed contour Γ consisting of the segment $[-R, R]$ on the real axis, and the half circle $\gamma(R)$ of radius R in the upper half-plane (Fig. 110). Prove that

$$\int_{\Gamma} F(z)\,dz = i\pi, \quad \int_{-R}^{R} F(x)\,dx = 0.$$

7. c) *Solution.* The function

$$f(z) = \frac{z^2+4}{(z^2+9)^2}$$

has second order poles at $z = \pm 3i$. Since this is a rational function whose denominator has degree 2 higher than the numerator, we may use Remark 7.32 to justify applying Theorem 7.30, and calculate the improper real integral by using the residues of f in the upper half-plane.

$$\text{res}_{3i} f = \frac{1}{1!}\lim_{z\to 3i}\frac{d}{dz}((z-3i)^2 f(z)) = \lim_{z\to 3i}\frac{d}{dz}\frac{z^2+4}{(z+3i)^2}$$

$$= \lim_{z\to 3i}\frac{2z(z+3i)^2-(z^2+4)\cdot 2(z+3i)}{(z+3i)^4} = \frac{-36+10}{(6i)^3} = -i\frac{13}{108};$$

$$\int_{-\infty}^{\infty} f(z)\,dz = 2\pi i\cdot -i\frac{13}{108} = \frac{13}{54}\pi.$$

7. d) *Hint.* Apply Jordan's lemma as in Example 7.34.

8. *Hint.* 1. Note that

$$\int_{0}^{\infty}\frac{\sin x}{x}\,dx = \text{Im}\int_{0}^{\infty}\frac{e^{ix}}{x}\,dx.$$

2. For positive numbers R and ε, $0 < \varepsilon < R$, consider the closed contour Γ consisting of the segments $\gamma_1 = [-R, -\varepsilon]$, $\gamma_2 = [\varepsilon, R]$ on the real axis, and the half circles $\gamma(\varepsilon)$, $\gamma(R)$ of radii ε, R respectively in the upper half-plane. Prove that

$$\int_{\Gamma}\frac{e^{iz}}{z}\,dz = 0, \quad \lim_{R\to\infty}\int_{\gamma(R)}\frac{e^{iz}}{z}\,dz = 0, \quad \lim_{\varepsilon\to 0}\int_{\gamma(\varepsilon)}\frac{e^{iz}}{z}\,dz = -i\pi.$$

To prove the last equality, note that $\frac{e^{iz}}{z} = \frac{1}{z} + \alpha(z)$, where $\alpha(z)$ is analytic in $\mathbb{C}$. Since the contour $\gamma(\varepsilon)$ is traversed clockwise, $\int\limits_{\gamma(\varepsilon)} \frac{1}{z}\,dz = -i\pi$.

Conclude that

$$\int\limits_{\gamma_1} \frac{\sin x}{x}\,dx = \int\limits_{\gamma_2} \frac{\sin x}{x}\,dx \to \int\limits_0^\infty \frac{\sin x}{x}\,dx = \frac{\pi}{2} \quad \text{as } \varepsilon \to 0, \quad R \to \infty.$$

Another method for calculating this integral is given in Example 9.22.

Section 7.4

1. *Hint.* Use the same arguments as in the proof of Lemma 7.41.

2. *Hint.* Use the result of problem 1 with $q = 0$. As U, choose any neighborhood of z_0 such that $|f(z) - w_0| < \mu$.

3. *Hint.* Let D' be the image of D under the mapping $f(z)$.

To prove that D' is connected, fix any two points $w_1, w_2 \in D'$. Let z_1 and z_2 be some preimages of w_1 and w_2, respectively. Since D is connected, there is a continuous curve $z(t)$ in D connecting z_1 and z_2. Then $f(z(t))$ will be a continuous curve in D' connecting w_1 and w_2. Hence, D' is connected.

To prove that D' is open, use the result of problem 1.

Section 8.1

1. (a) $f(z) = \log z$, where the branch of the logarithm is such that $-\frac{\pi}{2} < \arg z \le \frac{3\pi}{2}$; the principal branch is appropriate as well.

(b) $f(z) = \log(z+1) - \log(z-1) + i\pi$, where the branch of the logarithm the same as in (a).

2. $A = \dfrac{i2h}{\pi}, \quad f(z) = \dfrac{2h}{\pi} \sin^{-1} z.$

3. (i) $\alpha_1 = -\frac{1}{2}$, $\alpha_2 = 1$, $\alpha_3 = \frac{1}{2}$, $\alpha_4 = 1$.

(ii)

$$f(z) = A \int_1^z \frac{\sqrt{\zeta + a^2}}{\zeta\sqrt{\zeta - 1}}\,d\zeta, \quad B = 0$$

(iii) Integrate over the path consisting of two intervals and a small semicircle centered at 0 (Fig. 74), and get $h_1 = Ai\pi\frac{a}{\sqrt{-1}} = A\pi a$. Then integrate over the path consisting of the interval $(1, |x|)$ and a big semicircle Γ of radius $|x|$ (Fig. 74).

The imaginary part of the integral over $(1, |x|)$ is zero. When $|x| \to \infty$, the ratio $\sqrt{\zeta + a^2}/\sqrt{\zeta - 1}$ approaches 1 uniformly. Since $\int_\Gamma \frac{1}{\zeta}\, d\zeta = i\pi$, we have $h_2 = A\pi$. Therefore, $A = h_2/\pi$. From the equalities $h_1 = A\pi a = h_2 a$, we get $a = h_1/h_2$.

$$\text{(iv)} \quad f(z) = \frac{h_2}{\pi}\left(ia \log \frac{t - ia}{t + ia} + \log \frac{t+1}{t-1} \right), \quad t = \left(\frac{z + a^2}{z - 1} \right)^{1/2}. \tag{9.58}$$

6. (i) $f(z) = \frac{1}{2}\left(\sqrt{z} - \frac{1}{\sqrt{z}} \right)$; (ii) $f(z) = -\frac{1}{\sqrt{z^2 - 1}}$, where $\arg(z^2 - 1) = \mathrm{Arg}(z - 1) + \mathrm{Arg}(z + 1)$.

Section 8.2

4. (i) $f(z) = \frac{2H}{\pi}\mathrm{Log}(z + \sqrt{z^2 + 1})$, where $\arg(z^2 + 1) = \arg(z - i) + \mathrm{Arg}(z + i)$, and $-\frac{3\pi}{2} < \arg(z - i) < \frac{\pi}{2}$; (ii) $\mathbf{V}(iy) = \frac{2H}{\pi\sqrt{1-y^2}}$, $0 < y < 1$.

Section 8.3

2. (i) Equipotentials: $r = C_1 e^{-\kappa\phi/N}$, streamlines: $r = C_2 e^{N\phi/\kappa}$, where $z = re^{i\phi}$ and C_1, C_2 are positive constants.

(ii) The streamlines spiral into the origin clockwise if N and κ have the same sign, and counterclockwise if the signs are opposite.

4. *Hint.* Use the same operations with curves as in the proof of Theorem 5.8.

6. *Hint.* There is a solution based on conformal mappings. But here we sketch a very elementary direct method which requires only routine, if rather cumbersome, computations.

(i) Assuming for a moment that the image of γ_1 under the mapping $\phi(z)$ is an arc of a circle passing through $\pm a$, find R_1 and H from the triangle with vertices iH, 0, and a —Fig. 91.

(ii) Parametrize γ_1 by the equality $z = ih + re^{i\varphi}$, $-\pi < \varphi \le \pi$, and prove that

$$w = \phi(ih + re^{i\varphi}) = i\frac{2h^2 - r^2}{2h} + \frac{r^2}{2h}\frac{ir + he^{i\varphi}}{ih + re^{i\varphi}}e^{i\varphi}.$$

(iii) Prove that

$$\left|\frac{ir + he^{i\varphi}}{ih + re^{i\varphi}}\, e^{i\varphi}\right| = 1 \text{ for all } \varphi, \text{ and } \operatorname{Im}\left(\frac{ir + he^{i\varphi}}{ih + re^{i\varphi}}\, e^{i\varphi}\right) \geq 0.$$

Notice that $\phi(\pm a) = \pm a$ and $\phi(ih \pm r) = ih$.

7. *Hint.* (i) Prove that $\phi(z_1) = \phi(z_2)$, $z_1 \neq z_2$ if and only if $z_1 z_2 = a^2$. Use the same method as in the proof of (4.35).

(ii) The relations $z_1 \in G_e$, $z_2 \in G_i$ are equivalent to the inequalities $|z_1 - ih| > r$ and $|z_2 - ih| < r$ respectively. Square the inequalities and consider the expression $|z - ih|^2 - r^2$. For $z = x + iy$, we have

$$\begin{aligned} |z - ih|^2 - r^2 &= x^2 + (y - h)^2 - r^2 \\ &= x^2 + y^2 - 2hy + h^2 - r^2 = |z|^2 - 2h \operatorname{Im} z - a^2 \end{aligned}$$

(see the triangle $0(ih)a$ in Fig. 90). Therefore, $z_1 \in G_e$ or $z_2 \in G_i$ if and only if $|z|^2 - 2h \operatorname{Im} z - a^2$ is positive or negative correspondingly. Suppose that $\phi(z_1) = \phi(z_2)$, $z_1 \neq z_2$. Then $z_2 = \frac{a^2}{z_1}$, $\operatorname{Im} z_2 = -\frac{a^2}{|z_1|^2} \operatorname{Im} z_1$, and we have

$$\begin{aligned} |z_2|^2 - 2h \operatorname{Im} z_2 - a^2 &= \frac{a^4}{|z_1|^2} + 2h\frac{a^2}{|z_1|^2} \operatorname{Im} z_1 - a^2 \\ &= -\frac{a^2}{|z_1|^2}(|z_1|^2 - 2h \operatorname{Im} z_1 - a^2). \end{aligned}$$

Hence the expression $|z|^2 - 2h \operatorname{Im} z - a^2$ for $z = z_1$ and for $z = z_2$ has opposite signs. Therefore, z_1, z_2 are in different sets G_e, G_i.

8. Let A_e be the exterior of A. Since A is the image of γ_1, for every $z \notin \gamma_1$ we have $\phi(z) \in A_e$. On the other hand, let w be any point in A_e. The equation $\phi(z) = w$ is equivalent to the quadratic equation $z^2 - 2zw + a^2 = 0$ for z, which has two different solutions z_1, z_2. Since $\phi(z_1) = \phi(z_2) = w$, the points z_1 and z_2 lie on opposite sides of γ_1—see problem 7. Therefore, each point $w \in A_e$ has a unique preimage in A_e and in A_i, as desired.

Section 8.4

1. $f(z) = Vz/h$, $\mathbf{E} = -iV/h$.

2. *Hint.* The complex potential $w = f(z)$ maps D onto the strip $0 < \operatorname{Im} w < V$ in such a way that the lower and upper plates of the condenser go to the lower and upper boundaries of the strip, respectively. The function

$$w_1 = -i\frac{z-1}{z+1}$$

(see problem 13 in Section 4.1, formula (9.50) with $h = 1$) maps D onto the upper half-plane, moreover the upper semi-circle (where the potential equals V) goes to the positive real semi-axis. Next apply the mapping $w = iV - \frac{V}{\pi}\operatorname{Log} w_1$.

Answer. $f(z) = iV - \frac{V}{\pi}\operatorname{Log}\left(i\frac{1-z}{z+1}\right)$.

3. *Hint:* revise the arguments in (8.29)–(8.31).

4. $f(z) = T_1 - i\frac{T_2-T_1}{b-a}(z - ia)$, $u(x,y) = T_1 + \frac{T_2-T_1}{b-a}(y-a)$.

6. *Hint.* The complex potential $f(z)$ maps D onto the strip $-T < \operatorname{Re} w < T$, where D is the complement of two rays as in Example 8.16. Map this strip onto the strip $-\pi < \operatorname{Im} w_1 < \pi$ by the function $w_1 = i\frac{\pi}{T}w$. In the same way as in Example 8.16, we get

$$z = f^{-1}(w) = \frac{h}{\pi}\left(e^{i\pi w/T} + \frac{i\pi w}{T} + 1\right), \quad -T < \operatorname{Re} w < T.$$

The diagram for the isotherms and streamlines is the same as shown in Fig. 95: the solid lines are isotherms, and the dotted lines are heat streamlines.

Section 9.1

1. *Answers.* a) $\sigma_0 = 3$; c) conditions (2) and (3) are not satisfied; d) $\sigma_0 = 0$; f) condition (3) is not satisfied.

2. a) *Hint and answer.* We apply (9.3) for the intervals $[0,3)$ and $[3,4)$; according to our convention, we write 1 instead of $h(t)$:

$$\begin{aligned} f(t) &= [1 - h(t-3)](3t-1) + [h(t-3) - h(t-4)](t+2) \\ &= 3t - 1 + h(t-3)(3-2t) - h(t-4)(t+2). \end{aligned}$$

3. a) *Hint.* Write the integrand in the form $\dfrac{x^{10}}{e^x}e^{-x}$, and prove that $\dfrac{x^{10}}{e^x} < M$ for some constant M.

4. e) *Answer.* $F(p) = \dfrac{1 + e^{-\pi p}}{p^2 + 1}$.

Section 9.2

1. *Solution.* Let $f_1(t)$, $f_2(t)$, $f_3(t)$ be the "pieces" of $f(t)$ on the intervals $[0,1)$, $[1,2)$, and $[2,3)$ respectively. Then $f_1(t)$ is a "piece" of the function

$f(t) = t$, $f_2(t)$ is a "piece" of $f(t) = 1$, and $f_3(t)$ is a "piece" of $f(t) = 3 - t$. Applying (9.3) we have

$$f(t) = [1 - h(t-1)]t + [h(t-1) - h(t-2)] \cdot 1 + [h(t-2) - h(t-3)](3-t)$$
$$= t - h(t-1)(t-1) - h(t-2)(t-2) + h(t-3)(t-3).$$

Now we apply the time delay theorem (Theorem 9.11) and the linearity of the Laplace transform. For all four terms we use (9.16) with $f(t - \tau) = t - \tau$. So, $f(t) = t$, and according to formula (9.27) with $n = 1$, $F(p) = \frac{1}{p^2}$. We get

$$f(t) \overset{\mathcal{L}}{\rightarrow} (1 - e^{-p} - e^{-2p} + e^{-3p})\frac{1}{p^2}.$$

We could also have obtained this result by directly applying the definition of the Laplace transform to the original function $f(t)$. In that case we would have to break the integral into three separate pieces. This is rather more cumbersome than the method used above.

2. a) *Solution.* We write $f(t)$ in the form

$$f(t) = 3t - 1 + h(t-3)(3-2t) - h(t-4)(t+2)$$

— see the answer to problem 2 in Section 9.1. According to formulas 1 and 9 of the table,

$$3t - 1 \overset{\mathcal{L}}{\rightarrow} \frac{3}{p^2} - \frac{1}{p}.$$

To find the Laplace transform of other terms we apply Theorem 9.11 and the linearity of the Laplace transform.

Consider the term $h(t-3)(3-2t)$ and use formula (9.16). Here $f(t-\tau) = h(t-3)(3-2t)$ and $\tau = 3$. As in Example 9.12, to find $f(t)$ we introduce a new variable $x = t - 3$. Then $t = x + 3$, and

$$f(x) = 3 - 2(x+3) = -2x - 3.$$

We may re-denote the variable and write $f(t) = -2t - 3$. Hence, $F(p) = -\frac{2}{p^2} - \frac{3}{p}$, and therefore

$$h(t-3)(3-2t) \overset{\mathcal{L}}{\rightarrow} -e^{-3p}\left(\frac{2}{p^2} + \frac{3}{p}\right).$$

Analogously we consider the last term $h(t-4)(t+2)$. Here $\tau = 4$, $x = t - 4$, $t = x + 4$, and $f(x) = x + 6$. Hence, $F(p) = \frac{1}{p^2} + \frac{6}{p}$, and

$$h(t-4)(t+2) \overset{\mathcal{L}}{\rightarrow} e^{-4p}\left(\frac{1}{p^2} + \frac{6}{p}\right).$$

Combining all these results, we obtain the Laplace transform of the original function; we denote this transform by $F(p)$:

$$F(p) = \frac{3}{p^2} - \frac{1}{p} - e^{-3p}\left(\frac{2}{p^2} + \frac{3}{p}\right) - e^{-4p}\left(\frac{1}{p^2} + \frac{6}{p}\right).$$

3. a) *Answer.*

$$f(t) = h(t-1)(t-1) - 2h(t-2)(t-2) + h(t-3)(t-3);$$

$$F(p) = \frac{e^{-p}}{p^2} - \frac{2e^{-2p}}{p^2} + \frac{e^{-3p}}{p^2} = \frac{1}{p^2}(e^{-p} - 2e^{-2p} + e^{-3p}).$$

3. e) *Answer.*

$$f(t) = [1 - h(t-1)](1-t) + [h(t-2) - h(t-3)](t-2);$$

$$F(p) = \frac{1}{p} - \frac{1}{p^2} + \frac{e^{-p}}{p^2} + \frac{e^{-2p}}{p^2} - \frac{e^{-3p}}{p^2} - \frac{e^{-3p}}{p}.$$

6. a) *Hint and answer.* Apply (9.29) with $F(s) = \frac{1}{s-2} - \frac{1}{s}$.

$$\frac{e^{2t} - 1}{t} \xrightarrow{\mathcal{L}} \ln\frac{p}{p-2}, \quad \operatorname{Re} p > 2.$$

6. b) *Hint and answer.* Apply (9.29) with $F(s) = \frac{1}{s} - \frac{s}{s^2+\omega^2}$.

$$\frac{1 - \cos\omega t}{t} \xrightarrow{\mathcal{L}} \frac{1}{2}\ln\frac{p^2+\omega^2}{p^2}, \quad \operatorname{Re} p > 0.$$

Note that here we cannot set $p = 0$ unlike Example 9.22. This was to be expected because the integral $\int_0^\infty \frac{1-\cos\omega t}{t}\,dt$ diverges.

7. *Solution.* We break $F(p)$ up into factors for which the original functions are known:

$$F(p) = \frac{1}{((p+2)^2+3^2)^2} = \frac{1}{(p+2)^2+3^2}\cdot\frac{1}{(p+2)^2+3^2}.$$

From formula 7 of the table, with $a = -2$ and $\omega = 3$, we see that

$$e^{-2t}\sin 3t \xrightarrow{\mathcal{L}} \frac{3}{(p+2)^2+3^2}; \quad \text{hence} \quad \frac{1}{(p+2)^2+3^2} \xleftarrow{\mathcal{L}} \frac{1}{3}e^{-2t}\sin 3t.$$

According to the convolution Theorem 9.25 (formula 25 in the table), the product of the transforms corresponds to the convolution of the original functions. Therefore

$$F(p) = \frac{1}{(p+2)^2+3^2}\cdot\frac{1}{(p+2)^2+3^2}$$

$$\xleftarrow{\mathcal{L}} \int_0^t \frac{1}{3}e^{-2\tau}\sin 3\tau \cdot \frac{1}{3}e^{-2(t-\tau)}\sin 3(t-\tau)\,d\tau.$$

We first move the constant factor $\frac{1}{9}e^{-2t}$ out from the integral sign, and then use of the formula

$$\sin\alpha\sin\beta = \frac{1}{2}(\cos(\alpha-\beta) - \cos(\alpha+\beta)).$$

We get

$$\begin{aligned}
F(p) &\fallingdotseq \frac{1}{9}e^{-2t}\int_0^t \sin 3\tau \cdot \sin(3t-3\tau)\,d\tau \\
&= \frac{1}{18}e^{-2t}\int_0^t (\cos(6\tau-3t) - \cos 3t)\,d\tau \\
&= \frac{1}{18}e^{-2t}\left(\frac{1}{6}\sin(6\tau-3t)\Big|_0^t - \tau\cos 3t\Big|_0^t\right) \\
&= \frac{1}{18}e^{-2t}\left(\frac{1}{6}\sin 3t + \frac{1}{6}\sin 3t - t\cos 3t\right) = \frac{1}{18}e^{-2t}\left(\frac{1}{3}\sin 3t - t\cos 3t\right).
\end{aligned}$$

So

$$F(p) \fallingdotseq f(t) = \frac{1}{18}e^{-2t}\left(\frac{1}{3}\sin 3t - t\cos 3t\right).$$

This result is easy to check by finding the Laplace transform of f. In fact, using formulas 7 and 12 from the table,

$$e^{-2t}\sin 3t \xrightarrow{\mathcal{L}} \frac{3}{(p+2)^2+3^2}, \quad \text{and} \quad t\cos 3t \xrightarrow{\mathcal{L}} \frac{p^2-3^2}{(p^2+3^2)^2}.$$

Then applying formula 19 of the table to the latter, we see that

$$e^{-2t}t\cos 3t \xrightarrow{\mathcal{L}} \frac{(p+2)^2-3^2}{((p+2)^2+3^2)^2}.$$

Lastly, using the linearity of the Laplace transformation, we have

$$\begin{aligned}
f(t) &= \frac{1}{18\cdot 3}e^{-2t}\sin 3t - \frac{1}{18}e^{-2t}t\cos 3t \\
&\xrightarrow{\mathcal{L}} \frac{1}{18\cdot 3}\frac{3}{(p+2)^2+9} - \frac{1}{18}\frac{(p+2)^2-9}{((p+2)^2+9)^2} \\
&= \frac{1}{18}\frac{(p+2)^2+9-((p+2)^2-9)}{((p+2)^2+9)^2} = \frac{1}{18}\frac{18}{((p+2)^2+9)^2} \\
&= \frac{1}{((p+2)^2+9)^2} = \frac{1}{(p^2+4p+13)^2},
\end{aligned}$$

which agrees with the original transform $F(p)$.

8. b) *Answer.*

$$f(t) = \frac{t}{2}\cos 2t + \frac{1}{4}\sin 2t.$$

9. *Solution.* First the hard way, using the convolution formula 25 in the table. We rewrite the transform as

$$\frac{p}{(p^2+16)^2} = \frac{1}{4} \cdot \frac{4}{p^2+4^2} \cdot \frac{p}{p^2+4^2}.$$

Letting

$$F(p) = \frac{4}{p^2+4^2} \quad \text{and} \quad G(p) = \frac{p}{p^2+4^2},$$

we see that the given transform is $\frac{1}{4}FG$. Moreover, we can get the original functions f and g from formulas 3 and 4 in the table:

$$f(t) = \sin 4t \quad \text{and} \quad g(t) = \cos 4t.$$

By the convolution formula, the original function for FG is $f * g$:

$$f * g = \int_0^t \sin 4\tau \cos(4t - 4\tau)\, d\tau.$$

Using the trig identity

$$\sin\alpha\cos\beta = \frac{1}{2}(\sin(\alpha+\beta) + \sin(\alpha-\beta)),$$

we get

$$f * g = \int_0^t (\sin 4t + \sin(8\tau - 4t))\, d\tau = \tau \sin 4t \Big|_0^t - \frac{1}{8}\cos(8\tau - 4t)\Big|_0^t$$

$$= t\sin 4t - \frac{1}{8}\cos 4t + \frac{1}{8}\cos 4t = t \sin 4t.$$

Therefore the desired original function is

$$\frac{1}{4} f * g = \frac{1}{4} \cdot \frac{t}{2}\sin 4t = \frac{t}{8}\sin 4t.$$

Now we find the original function using formula 11 from the table. We rewrite the given transform as

$$\frac{p}{(p^2+16)^2} = \frac{1}{8} \cdot \frac{2p \cdot 4}{(p^2+4^2)^2},$$

and see that we can apply the formula with $\omega = 4$ to get

$$\frac{1}{8} \cdot t \sin 4t$$

for the desired function.

Section 9.3

1. *Solution.* **A.** *Method of partial fraction decomposition.*

(i) First we break $F(p)$ down into a sum of partial fractions. Since the equation $p^2 + 4p + 5 = 0$ has no real roots, the partial fraction decomposition takes the form

$$\frac{2p+1}{(p+1)(p^2+4p+5)} = \frac{A}{p+1} + \frac{Mp+N}{p^2+4p+5}.$$

(It would be possible to factor $p^2 + 4p + 5$ into $(p - p_1)(p - p_2)$ using complex p_1 and p_2, but that is less convenient.) Combining the fractions on the right over a common denominator, and then setting the numerators of the left and right sides of the equation equal, we get

$$2p + 1 = A(p^2 + 4p + 5) + (Mp + N)(p + 1).$$

From this we see that

$$\begin{aligned} &\text{when} \quad p = -1 \quad \text{then} \quad -1 = A \cdot 2 \quad \text{and} \quad A = -\tfrac{1}{2}; \\ &\text{when} \quad p = 0 \quad \text{then} \quad 1 = 5A + N \quad \text{and} \quad N = \tfrac{7}{2}; \\ &\text{and} \quad 0 \cdot p^2 = (A + M)p^2, \quad \text{so} \quad 0 = A + M \quad \text{and} \quad M = \tfrac{1}{2}. \end{aligned}$$

Thus

$$\frac{2p+1}{(p+1)(p^2+4p+5)} = -\frac{1}{2}\frac{1}{p+1} + \frac{\frac{1}{2}p + \frac{7}{2}}{p^2+4p+5}.$$

(ii) For each of these fractions we find the corresponding inverse transform, using the table. From formula 2, with $a = -1$, we see that

$$\frac{1}{p+1} \risingdotseq e^{-t}.$$

In the other fraction we complete the square to get

$$p^2 + 4p + 5 = p^2 + 4p + 4 + 1 = (p + 2)^2 + 1,$$

and so

$$\begin{aligned} \frac{\frac{1}{2}p + \frac{7}{2}}{p^2+4p+5} &= \frac{\frac{1}{2}(p+2-2) + \frac{7}{2}}{(p+2)^2+1} = \frac{\frac{1}{2}(p+2) + \frac{5}{2}}{(p+2)^2+1} \\ &= \frac{1}{2}\frac{p+2}{(p+2)^2+1} + \frac{5}{2}\frac{1}{(p+2)^2+1}. \end{aligned}$$

From formulas 8 and 7 from the table we see that

$$\frac{p+2}{(p+2)^2+1} \risingdotseq e^{-2t}\cos t \quad \text{and} \quad \frac{1}{(p+2)^2+1} \risingdotseq e^{-2t}\sin t.$$

(iii) Using the linearity of the Laplace transform, we now obtain the original function $f(t)$ those transform is $F(p)$. Since

$$F(p) = -\frac{1}{2}\frac{1}{p+1} + \frac{1}{2}\frac{p+2}{(p+2)^2+1} + \frac{5}{2}\frac{1}{(p+2)^2+1},$$

we have

$$f(t) = -\frac{1}{2}e^{-t} + \frac{1}{2}e^{-2t}\cos t + \frac{5}{2}e^{-2t}\sin t.$$

B. *Method using residues.*

(i) We find the zeros of the denominator of $F(p)$, which will be the poles of the function. Solving the equation $p^2 + 4p + 5 = 0$, we get

$$p_1 = \frac{-4+2i}{2} = -2+i, \quad p_2 = \frac{-4-2i}{2} = -2-i.$$

To determine the orders of the poles we factor the denominator into linear terms:

$$p^2 + 4p + 5 = (p+2-i)(p+2+i);$$

$$F(p) = \frac{2p+1}{(p+1)(p+2-i)(p+2+i)}.$$

From this it follows that F has three singularities,

$$p_1 = -1, \quad p_2 = -2+i, \quad \text{and} \quad p_3 = -2-i,$$

and each is a simple pole.

(ii) We find the residues of the function $F(p)e^{pt}$ at each of the poles, using formula (7.15) (we could also use formula (7.17)):

$$\begin{aligned}
\text{res}_{-1}(F(p)e^{pt}) &= \lim_{p\to-1}\frac{(p+1)(2p+1)e^{pt}}{(p+1)(p+2-i)(p+2+i)} \\
&= \lim_{p\to-1}\frac{(2p+1)e^{pt}}{(p+2-i)(p+2+i)} = \frac{-1e^{-t}}{(1-i)(1+i)} = -\frac{1}{2}e^{-t}; \\
\text{res}_{-2+i}(F(p)e^{pt}) &= \lim_{p\to-1}\frac{(p+2-i)(2p+1)e^{pt}}{(p+1)(p+2-i)(p+2+i)} \\
&= \frac{(-3+2i)e^{(-2+i)t}}{(-1+i)2i} = \frac{(1-5i)e^{(-2+i)t}}{4}; \\
\text{res}_{-2-i}(F(p)e^{pt}) &= \lim_{p\to-1}\frac{(2p+1)e^{pt}}{(p+1)(p+2-i)} \\
&= \frac{(-3-2i)e^{(-2-i)t}}{(-1-i)(-2i)} = \frac{(1+5i)e^{(-2-i)t}}{4}.
\end{aligned}$$

(iii) By Theorem 9.29,

$$f(t) = \sum_k \text{res}_{p_k}(F(p)e^{pt}).$$

In our case,

$$f(t) = -\frac{1}{2}e^{-t} + \frac{(1-5i)e^{(-2+i)t}}{4} + \frac{(1+5i)e^{(-2-i)t}}{4}.$$

This is equivalent to the function we found using the first method. Indeed,

$$\begin{aligned} f(t) &= -\frac{1}{2}e^{-t} + \frac{1}{4}e^{-2t}((1-5i)e^{it} + (1+5i)e^{-it}) \\ &= -\frac{1}{2}e^{-t} + \frac{1}{4}e^{-2t}(e^{it} + e^{-it} - 5i(e^{it} - e^{-it})) \\ &= -\frac{1}{2}e^{-t} + \frac{1}{4}e^{-2t}(2\cos t - 5i \cdot 2i \sin t) \\ &= -\frac{1}{2}e^{-t} + \frac{1}{2}e^{-2t}\cos t + \frac{5}{2}e^{-2t}\sin t. \end{aligned}$$

2. e) *Solution.* We use partial fractions to rewrite the transform in the form

$$\frac{2p+3}{(p-1)(p^2+4)} = \frac{A}{p-1} + \frac{Bp+C}{p^2+4}.$$

We get

$$2p + 3 = A(p^2+4) + (p-1)(Bp+C),$$

and when $p = 1$ we see that $A = 1$.
Since there is no p^2 term on the left, we see that $A + B = 0$, and so $B = -1$.
When $p = 0$, we have $3 = 4 - C$, so that $C = 1$.
Thus

$$\frac{2p+3}{(p-1)(p^2+4)} = \frac{1}{p-1} + \frac{-p+1}{p^2+4} = \frac{1}{p-1} + \frac{1}{p^2+4} - \frac{p}{p^2+4}.$$

Applying formulas 2, 3, and 4 from the table, the latter two with $\omega = 2$, we get the original function

$$f(t) = e^t + \frac{1}{2}\sin 2t - \cos 2t.$$

Now we use the method of residues, denoting the given transform by $F(p)$ and setting $G(p) = e^{pt}F(p)$. Calculating the residues, we see that G has singularities at $p = 1$ and $\pm 2i$, which are all simple poles.

$$\begin{aligned} \operatorname{res}_1 G &= \lim_{p\to 1}(p-1)G(p) = \lim_{p\to 1} e^t \frac{2p+3}{p^2+4} = \frac{e^t \cdot 5}{5} = e^t; \\ \operatorname{res}_{2i} G &= \lim_{p\to 2i}(p-2i)G(p) = \frac{e^{2it}(4i+3)}{(2i-1)(4i)} = -e^{2it}\left(\frac{1}{2} + \frac{i}{4}\right). \end{aligned}$$

The calculation for the residue at $-2i$ is the same as the previous one, except all values are replaced by their conjugates, so

$$\operatorname{res}_{-2i} G = -e^{-2it}\left(\frac{1}{2} - \frac{i}{4}\right).$$

By the residue theorem for Laplace transforms, Theorem 9.29, we get

$$\begin{aligned} f(t) &= e^t - e^{2it}\left(\frac{1}{2} + \frac{i}{4}\right) - e^{-2it}\left(\frac{1}{2} - \frac{i}{4}\right) \\ &= e^t - \frac{i}{4}(e^{2it} - e^{-2it}) - \frac{1}{2}(e^{2it} + e^{-2it}) = e^t + \frac{1}{2}\sin 2t - \cos 2t. \end{aligned}$$

3. *Solution.* (i) We will transform the original equation relating the functions $y = y(t)$ and $f(t) = e^t$ to an equation relating their transforms $Y = Y(p)$ and $F = F(p)$. To do this we will use Theorem 9.17 about Laplace transform of derivatives, as well as the table of Laplace transforms. From formula 21 in the table, we get that

$$y'' \xrightarrow{\mathcal{L}} p^2Y - py(0) - y'(0) = p^2Y + p,$$

from formula 20 we get

$$y' \xrightarrow{\mathcal{L}} pY - y(0) = pY + 1,$$

and formula 2 gives us

$$e^t \xrightarrow{\mathcal{L}} \frac{1}{p-1}.$$

Therefore the transformed version of the original differential equation is

$$p^2Y + p + pY + 1 - 2Y = \frac{1}{p-1}, \quad \text{or} \quad (p^2 + p - 2)Y + p + 1 = \frac{1}{p-1}.$$

(ii) This last equation is a linear algebraic equation in the unknown Y. Rearranging it in the form

$$(p^2 + p - 2)Y = \frac{1}{p-1} - p - 1 = \frac{1 - p^2 + p - p + 1}{p-1} = \frac{-p^2 + 2}{p-1},$$

we get

$$Y = \frac{-p^2 + 2}{(p-1)(p^2 + p - 2)}.$$

(iii) Now we wish to recover the function $y(t)$ from its Laplace transform $Y(p)$. That will be the solution to the original differential equation.

The equation $p^2 + p - 2 = 0$ has the roots 1 and -2, so $p^2 + p - 2$ factors into $(p-1)(p+2)$, and

$$Y = \frac{-p^2 + 2}{(p-1)^2(p+2)}.$$

To find the desired function $y(t)$ we may employ any of our now familiar methods (see problem 1 in this section). For example, using partial fractions, we have

$$\frac{-p^2+2}{(p-1)^2(p+2)} = \frac{A}{p-1} + \frac{B}{(p-1)^2} + \frac{C}{p+2} = \frac{A(p-1)(p+2)+B(p+2)+C(p-1)^2}{(p-1)^2(p+2)},$$

from which we get

$$-p^2+2 = A(p-1)(p+2)+B(p+2)+C(p-1)^2.$$

Therefore

$$\begin{aligned} &\text{when}\quad p=1 \quad\text{then}\quad 1=3B \quad\text{and}\quad B=\tfrac{1}{3};\\ &\text{when}\quad p=-2 \quad\text{then}\quad -2=9C \quad\text{and}\quad C=-\tfrac{2}{9};\\ \text{and}\quad &-p^2=(A+C)p^2 \quad\text{so}\quad -1=A+C \quad\text{and}\quad A=-\tfrac{7}{9}. \end{aligned}$$

Thus

$$Y = -\frac{7}{9}\frac{1}{p-1} + \frac{1}{3}\frac{1}{(p-1)^2} - \frac{2}{9}\frac{1}{p+2}.$$

Using formula 2 from the table, and formula 10 with $a=1$ and $n=1$, we get

$$y(t) = -\frac{7}{9}e^t + \frac{1}{3}te^t - \frac{2}{9}e^{-2t} \quad\text{or}\quad y(t) = -\frac{7}{9}e^t - \frac{2}{9}e^{-2t} + \frac{1}{3}te^t.$$

4. e) *Solution.* Taking the transform of the given differential equation we get

$$p^2Y(p) - py(0) - y'(0) - 9Y(p) = \frac{1}{p^2+1} - \frac{p}{p^2+1},$$

and substituting the initial conditions gives

$$p^2Y(p) + 3p - 2 - 9Y(p) = \frac{1-p}{p^2+1}.$$

Solving algebraically for Y,

$$(p^2-9)Y = \frac{1-p}{p^2+1} + 2 - 3p,$$

$$Y = \frac{1-p}{(p^2+1)(p^2-9)} + \frac{2-3p}{p^2-9}.$$

Now using partial fractions we have

$$\frac{1-p}{(p^2+1)(p^2-9)} + \frac{2-3p}{p^2-9} = \frac{Ap+B}{p^2+1} + \frac{C}{p-3} + \frac{D}{p+3},$$

$$A = \frac{1}{10},\quad B = -\frac{1}{10},\quad C = -\frac{6}{5},\quad D = -\frac{19}{10}.$$

Applying formulas 2–4 from the table, we obtain the desired solution:

$$y(t) = \frac{1}{10}\cos t - \frac{1}{10}\sin t - \frac{6}{5}e^{3t} - \frac{19}{10}e^{-3t}.$$

5. a) *Hint and answer.* First, we find the inverse of

$$F_1(p) = \frac{(p^2+6)}{(p^2+4)(p^2-3p+2)}.$$

It was done in Examples 9.28 and 9.30—see (9.41):

$$f_1(t) = -\frac{7}{5}e^t + \frac{5}{4}e^{2t} + \frac{3}{20}\cos 2t - \frac{1}{20}\sin 2t.$$

Secondary, we replace t by $t - \pi$ and multiply the obtained function by $h(t-\pi)$. We get

$$\begin{aligned} F(p) &= h(t-\pi)\left[-\frac{7}{5}e^{t-\pi} + \frac{5}{4}e^{2(t-\pi)} + \frac{3}{20}\cos 2(t-\pi) - \frac{1}{20}\sin 2(t-\pi)\right] \\ &= h(t-\pi)\left[-\frac{7}{5}e^{t-\pi} + \frac{5}{4}e^{2(t-\pi)} + \frac{3}{20}\cos 2t - \frac{1}{20}\sin 2t\right]. \end{aligned}$$

8. b) *Sketch of solution.* (i) Write $f(t)$ in the form

$$f(t) = \sin t - h(t-\pi)\sin t.$$

(ii) Find the Laplace transform of $f(t)$:

$$f(t) \xrightarrow{\mathcal{L}} \frac{1}{p^2+1} + \frac{e^{-\pi p}}{p^2+1}.$$

Note that the sign before $h(t-\pi)\sin t$ changes when we apply formula 18 of the table with $f(t-\pi) = \sin t$, so that $f(t) = \sin(t+\pi) = -\sin t$.

(iii) Take the Laplace transform of the given equation and find Y:

$$(p^2Y - p\cdot 0 - 1) - 3(pY - 0) + 2Y = \frac{1}{p^2+1} + \frac{e^{-\pi p}}{p^2+1},$$

$$Y = \frac{1}{p^2-3p+2} + \frac{1}{(p^2+1)(p^2-3p+2)} + \frac{e^{-\pi p}}{(p^2+1)(p^2-3p+2)}.$$

(iv) Find the inverse Laplace transform of Y which is the desired solution $y(t)$ of the initial value problem. We may consider the sum of first two fractions as one case. But we will need the inverse transform of the second fraction for the third one. Therefore we consider all three fractions separately. Using any of two methods (partial fractions or residues), we get

$$\frac{1}{p^2-3p+2} \risingdotseq -e^t + e^{2t},$$

$$\frac{1}{(p^2+1)(p^2-3p+2)} \risingdotseq -\frac{1}{2}e^t + \frac{1}{5}e^{2t} + \frac{3}{10}\cos t + \frac{1}{10}\sin t,$$

$$\frac{e^{-\pi p}}{(p^2+1)(p^2-3p+2)} \risingdotseq h(t-\pi)\Big[-\frac{1}{2}e^{t-\pi} + \frac{1}{5}e^{2(t-\pi)} + \frac{3}{10}\cos(t-\pi) + \frac{1}{10}\sin(t-\pi)\Big].$$

Adding the obtained functions and simplifying the last two terms, we have

$$y(t) = -\frac{3}{2}e^t + \frac{6}{5}e^{2t} + \frac{3}{10}\cos t + \frac{1}{10}\sin t + h(t-\pi)\Big[-\frac{1}{2}e^{t-\pi} + \frac{1}{5}e^{2(t-\pi)} - \frac{3}{10}\cos t - \frac{1}{10}\sin t\Big]$$

$$= \begin{cases} -\frac{3}{2}e^t + \frac{6}{5}e^{2t} + \frac{3}{10}\cos t + \frac{1}{10}\sin t, & 0 \le t < \pi, \\ -\frac{1}{2}(3+e^{-\pi})e^t + \frac{1}{5}(6+e^{-2\pi})e^{2t}, & t \ge \pi. \end{cases}$$

9. b) (i) *Sketch of solution.* The transform of the given differential equation is

$$p^2Y - 2p + 1 - 6(pY - 2) + 9Y = F(p).$$

Solving for Y, we get

$$Y = \frac{2p-13}{p^2-6p+9} + \frac{F(p)}{p^2-6p+9} = \frac{2p-13}{(p-3)^2} + \frac{F(p)}{(p-3)^2}.$$

The inverse transforms of the obtained rational functions are

$$\frac{2p-13}{(p-3)^2} \risingdotseq 2e^{3t} - 7te^{3t} = y_h(t),$$

$$\frac{1}{(p-3)^2} \risingdotseq te^{3t} = g(t).$$

Applying formula (9.45) we have

$$y(t) = 2e^{3t} - 7te^{3t} + \int_0^t f(\tau)(t-\tau)e^{3(t-\tau)}\, d\tau.$$

(ii) If $f(t) = \frac{e^{3t}}{t^2+1}$, then

$$y(t) = 2e^{3t} - 7tc^{3t} + \int_0^t \frac{e^{3\tau}}{\tau^2+1}(t-\tau)e^{3(t-\tau)}\, d\tau.$$

The integral equals

$$e^{3t}\int_0^t \frac{t-\tau}{\tau^2+1}\,d\tau = te^{3t}\int_0^t \frac{1}{\tau^2+1}\,d\tau - e^{3t}\int_0^t \frac{\tau}{\tau^2+1}\,d\tau$$

$$= te^{3t}\tan^{-1}t - \frac{e^{3t}}{2}\ln(\tau^2+1)\Big|_0^t = e^{3t}(t\tan^{-1}t - \tfrac{1}{2}\ln(t^2+1)).$$

So,

$$y(t) = 2e^{3t} - 7te^{3t} + e^{3t}(t\tan^{-1}t - \tfrac{1}{2}\ln(t^2+1)).$$

9. c) (i) *Answer.* $y(t) = \cos 3t + \frac{1}{3}\int_0^t f(\tau)\sin 3(t-\tau)\,d\tau$.
(ii) *Hint.* To evaluate the integral use the trig identity (9.34).

9. d) *Answer.* (i) $y(t) = \int_0^t f(\tau)e^{2(t-\tau)}\sin(t-\tau)\,d\tau$. (ii) $y(t) = e^{2t}(1-\cos t)$.

10. *Answer.* $y(t) = e^{3t} + \frac{1}{2}e^{3t}\tan^{-1}\frac{t}{2}$.

11. *Sketch of solution.* The transform of the given differential equation is

$$pY - y_0 + aY = F(p), \quad \text{so } Y = \frac{y_0}{p+a} + \frac{F(p)}{p+a}.$$

The inverse transform is

$$y(t) = y_0e^{-at} + \int_0^t f(\tau)e^{-a(t-\tau)}\,d\tau = y_0e^{-at} + e^{-at}\int_0^t f(\tau)e^{a\tau}\,d\tau.$$

12. a) *Solution.* We denote by X and Y the Laplace transforms of the unknown functions $x = x(t)$ and $y = y(t)$. Using the linearity of the transform, and formulas 20 and 1 from the table, we move to the transformed system

$$\begin{cases} pX - (-1) = X + 3Y + \frac{2}{p}, \\ pY - 2 = X - Y + \frac{1}{p}, \end{cases} \quad \text{or} \quad \begin{cases} (p-1)X - 3Y = \frac{2}{p} - 1, \\ -X + (p+1)Y = \frac{1}{p} + 2. \end{cases}$$

We solve this system of algebraic equations in X and Y. For this, we solve for X in the second equation and substitute the result into the first:

$$X = (p+1)Y - \frac{1}{p} - 2,$$

$$(p-1)\left((p+1)Y - \frac{1}{p} - 2\right) - 3Y = \frac{2}{p} - 1.$$

Then we simplify and solve for Y,

$$(p^2-4)Y = \frac{1}{p} + 2p - 2 = \frac{2p^2-2p+1}{p}, \quad Y = \frac{2p^2-2p+1}{p(p^2-4)},$$

and then solve for X,

$$X = (p+1)Y - \frac{1}{p} - 2 = \frac{(p+1)(2p^2 - 2p + 1)}{p(p^2-4)} - \frac{1}{p} - 2$$
$$= \frac{(p+1)(2p^2 - 2p + 1) - (1+2p)(p^2-4)}{p(p^2-4)} = \frac{-p^2 + 7p + 5}{p(p^2-4)}.$$

The reader may find it easier to apply Cramer's rule—see the next example.

Now we must find the functions x and y corresponding to these transforms X and Y. There are two methods available, both of which require about the same amount of work. To remind the reader of both methods, we will find x using partial fractions, and y using residues.

Breaking X down with partial fractions, we get

$$\frac{-p^2 + 7p + 5}{p(p-2)(p+2)} = \frac{A}{p} + \frac{B}{p-2} + \frac{C}{p+2}$$
$$= \frac{A(p-2)(p+2) + Bp(p+2) + Cp(p-2)}{p(p-2)(p+2)},$$

and

$$-p^2 + 7p + 5 = A(p-2)(p+2) + Bp(p+2) + Cp(p-2).$$

When $p = 0$ then $5 = -4A$ and $A = -\frac{5}{4}$;

when $p = 2$ then $-4 + 14 + 5 = B \cdot 2 \cdot 4$ and $B = \frac{15}{8}$;

when $p = -2$ then $-4 - 14 + 5 = C(-2)(-4)$ and $C = -\frac{13}{8}$.

Therefore

$$X = -\frac{5}{4} \cdot \frac{1}{p} + \frac{15}{8} \cdot \frac{1}{p-2} - \frac{13}{8} \cdot \frac{1}{p+2}.$$

From formulas 1 and 2 of the table, and using linearity, we find $x(t)$:

$$x(t) = -\frac{5}{4} + \frac{15}{8} e^{2t} - \frac{13}{8} e^{-2t}.$$

Now we find $y(t)$ using the residue formula (9.42)

$$y(t) = \sum_k \operatorname{res}_{p_k}(Y(p)e^{pt}).$$

The function

$$Y(p)e^{pt} = \frac{(2p^2 - 2p + 1)e^{pt}}{p(p^2-4)} = \frac{(2p^2 - 2p + 1)e^{pt}}{p(p-2)(p+2)}$$

has singularities at the points 0, 2, and -2, each of which is a simple pole. We find the residues at these points with the formula

$$\operatorname{res}_{p_k}(Y(p)e^{pt}) = \lim_{p \to p_k} (p - p_k)(Y(p)e^{pt}$$

(formula (7.15); formula (7.17) could also be used). We have

$$\operatorname{res}_0(Y(p)e^{pt}) = \lim_{p\to 0}\frac{p(2p^2-2p+1)e^{pt}}{p(p-2)(p+2)} = \lim_{p\to 0}\frac{(2p^2-2p+1)e^{pt}}{(p-2)(p+2)} = -\frac{1}{4};$$

$$\operatorname{res}_2(Y(p)e^{pt}) = \lim_{p\to 2}\frac{(2p^2-2p+1)e^{pt}}{p(p+2)} = \frac{5}{8}e^{2t};$$

$$\operatorname{res}_{-2}(Y(p)e^{pt}) = \lim_{p\to -2}\frac{(2p^2-2p+1)e^{pt}}{p(p-2)} = \frac{13}{8}e^{-2t}.$$

Consequently,

$$y(t) = -\frac{1}{4} + \frac{5}{8}e^{2t} + \frac{13}{8}e^{-2t}.$$

12. f) *Sketch of solution.* Transforming the given system of differential equations we get

$$\begin{cases} pX - 2 = 3X + Y, \\ pY - 0 = -5X - 3Y + \frac{2}{p}, \end{cases} \quad \text{or} \quad \begin{cases} (p-3)X - Y = 2, \\ 5X + (p+3)Y = \frac{2}{p}. \end{cases}$$

To remind the reader of Cramer's rule, we apply it to this system. From the coefficients at X and Y we compose the determinant

$$\Delta = \begin{vmatrix} p-3 & -1 \\ 5 & p+3 \end{vmatrix} = (p-3)(p+3) + 5 = p^2 - 4.$$

Replacing the first and second columns of this determinant with the column from the right-hand sides of the equations, we obtain the determinants Δ_X and Δ_Y respectively:

$$\Delta_X = \begin{vmatrix} 2 & -1 \\ \frac{2}{p} & p+3 \end{vmatrix} = 2p + 6 + \frac{2}{p} = 2\frac{p^2+3p+1}{p},$$

$$\Delta_Y = \begin{vmatrix} p-3 & 2 \\ 5 & \frac{2}{p} \end{vmatrix} = \frac{2p-6}{p} - 10 = -2\frac{4p+3}{p}.$$

Now we find X and Y by Cramer's formulas:

$$X = \frac{\Delta_X}{\Delta} = 2\frac{p^2+3p+1}{p(p^2-4)};$$

$$Y = \frac{\Delta_Y}{\Delta} = -2\frac{4p+3}{p(p^2-4)}.$$

Possibly this method requires less computations than the substitution method we applied in the previous example.

Using partial fractions we may write X and Y as

$$X = 2\left(-\frac{1}{4}\cdot\frac{1}{p} - \frac{1}{8}\cdot\frac{1}{p+2} + \frac{11}{8}\cdot\frac{1}{p-2}\right),$$

$$Y = -2\left(-\frac{3}{4}\cdot\frac{1}{p} - \frac{5}{8}\cdot\frac{1}{p+2} + \frac{11}{8}\cdot\frac{1}{p-2}\right).$$

Finally, using formulas 1, 2, and 3 from the table, we obtain the solution

$$x(t) = -\frac{1}{2} - \frac{1}{4}e^{-2t} + \frac{11}{4}e^{2t},$$
$$y(t) = \frac{3}{2} + \frac{5}{4}e^{-2t} - \frac{11}{4}e^{2t}.$$

Appendix

Proof of the Cauchy-Goursat Theorem

Our goal is to prove the following theorem.

Theorem (Cauchy-Goursat theorem). *Let $f(z)$ be an analytic function on the simply-connected domain D. Then for any piecewise smooth closed path Γ lying within D,*

$$\int_{\Gamma} f(z)\, dz = 0.$$

If Γ intersects itself, we can split it into parts forming closed Jordan curves. Thus, it's sufficient to prove the theorem for such curves, and we may assume that Γ does not intersect itself. We split the proof into several stages.

Lemma 1 (Nested squares). *Suppose that $Q_1, Q_2, \ldots$ is a sequence of closed squares in the complex plane with sides parallel to the coordinate axis and such that $Q_1 \supset Q_2 \supset \ldots$ Then the intersection $\bigcap_{n=1}^{\infty} Q_n$ is not empty, that is there is a point belonging to all these squares.*[2]

Proof. Let $[a_n, b_n]$ and $[c_n, d_n]$ be the projections of Q_n onto the x- and the y- coordinate axis, respectively. Let us prove first that the intersection $\bigcap_{n=1}^{\infty}[a_n, b_n]$ of the intervals is not empty.[3] Obviously, $a_k < b_n$ for every k and n (in particular, $a_k < b_1$). Hence, the sequence $\{a_k\}$ is bounded from above. Since $a_k \le a_{k+1}$, the sequence $\{a_k\}$ is non-decreasing. Hence, there is a finite limit a of a_k as $k \to \infty$, and $a \le b_n$ for every n. Therefore, $a_n \le a \le b_n$, that is a belongs to every interval $[a_n, b_n]$.

In the same way we can show that there is a c which belongs to every interval $[c_n, d_n]$. Hence, $(a, c) \in \bigcap_{n=1}^{\infty} Q_n$, and the lemma is proved. □

Let $\overline{G}$ be a closed domain consisting of Γ and its interior G. Obviously, $f(z)$ is analytic at every point of $\overline{G}$. We split the complex plane into closed squares $K_{1,k}$ with side-length $\ell_1 = \frac{1}{2}$ and with sides parallel to the coordinate axes. Denote by $S_{1,k}$ the intersection of a square $K_{1,k}$ with $\overline{G}$ containing at

[2]In fact a stronger assertion is valid: every nested sequence of nonempty closed sets has nonempty intersection.

[3]This is the so-called Nested Intervals Theorem. But the proof is short, and we give it here.

least two points. Every $S_{1,k}$ is a closed set, which is either a square or a subset of a square (perhaps not even connected). We will call $S_{1,k}$ a *set of generation one.* Note that, because Γ and its interior are bounded, there are only finitely many sets $S_{1,k}$.

Now we divide each square $K_{1,k}$ into four equal closed sub-squares $K_{2,k}$ with side-length $\ell_2 = \frac{1}{4}$. The intersections of $K_{2,k}$ with $\overline{G}$ containing at least two points we denote by $S_{2,k}$, and call them the sets of generation two. If $S_{1,k}$ is a complete square, it has four *children*, that is the sets of generation two contained in $S_{1,k}$. If $S_{1,k}$ is an incomplete square, the number of children may be less than four. Dividing the squares $K_{2,k}$ onto four equal parts and intersecting these smaller squares with $\overline{G}$, we get sets $S_{3,k}$ of generation three, and so on. Again, for each m there are only finitely many $S_{m,k}$.

Given $\varepsilon > 0$, we say that a set $S_{m,k}$ is ε-*good*, or just good, if there is a point $z_{m,k}$ in $S_{m,k}$ such that

$$\left| \frac{f(z) - f(z_{m,k})}{z - z_{m,k}} - f'(z_{m,k}) \right| < \varepsilon \tag{A.1}$$

for every $z \in S_{m,k}$, $z \neq z_{m,k}$. If a set is not good, it is called bad. If $S_{m,k}$ is bad, then for every point $z_{m,k} \in S_{m,k}$ there is $z \in S_{m,k}$ such that the left hand side in (A.1) is greater than or equals to ε.

Lemma 2. *For every $\epsilon > 0$, there exists $M > 0$ such that all generations starting with the M-th one, consist only of ϵ-good sets $S_{m,k}$.*

Proof. Suppose that Lemma 2 is incorrect, that is, for some $\epsilon > 0$ there are infinitely many generations containing bad sets. Hence, the amount of bad sets is also infinite. Since the number of sets $S_{1,k}$ is finite, there is a set among these sets which has infinitely many bad descendants (that is bad subsets $S_{m,k}$). We denote such a set $S_{1,k}$ by B_1 (if there are several such sets $S_{1,k}$, we choose any of them). Let Q_1 be the corresponding square $K_{1,k}$ containing $S_{1,k}$. The set B_1 has at most four children $S_{2,k}$. Hence at least one of them has infinitely many bad descendants. We select one such set $S_{2,k}$ and denote it and the corresponding square $K_{2,k}$ by B_2 and Q_2, respectively. Continuing in this way, we get an infinite sequence of nested closed sets B_m, having infinitely many bad descendants, and a sequence of nested closed squares Q_m. By Lemma 1 there is a point z_0 which belongs to all sets Q_m.

We prove that z_0 also belongs to all sets B_n. Fix n. Since side-lengths of squares Q_m tend to 0 as $m \to \infty$, for every $\delta > 0$ there is a square Q_m with $m > n$, contained in the δ-neighborhood of z_0. Since $B_m \subset Q_m$ and $B_m \subset B_n$, the δ-neighborhood of z_0 contains points of B_n. Thus, every neighborhood of z_0 contains points of B_n. Since B_n is closed, $z_0 \in B_n$ (otherwise z_0 is a boundary point of B_n not contained in B_n, and we come to a contradiction).

Since $z_0 \in \overline{G}$, the function $f(z)$ is differentiable at z_0, that is there exists the limit

$$\lim_{z \to z_0} \frac{f(z) - f(z_0)}{z - z_0} = f'(z_0). \tag{A.2}$$

Again, fix $\delta > 0$ and consider the square Q_m containing in the δ-neighborhood of z_0. Since Q_m is bad and $z_0 \in B_m$, there is a point z in B_m such that $z \neq z_0$ and

$$\left| \frac{f(z) - f(z_0)}{z - z_0} - f'(z_0) \right| \geq \varepsilon. \tag{A.3}$$

Therefore, there is $\varepsilon > 0$ such that for every $\delta > 0$ the inequality (A.3) holds for some z with $0 < |z - z_0| < \delta$. This means that

$$\lim_{z \to z_0} \frac{f(z) - f(z_0)}{z - z_0} \neq f'(z_0),$$

which contradicts (A.2). This contradiction arises from the assumption that there are infinitely many generations containing bad sets. Hence, this assumption is wrong, and Lemma 2 is proved. □

Proof of the Cauchy-Goursat Theorem. Fix $\varepsilon > 0$. According to Lemma 2, there is a generation consisting only of good sets $S_{m,k}$, that is for every set $S_{m,k}$ there is a point $z_{m,k} \in S_{m,k}$ satisfying (A.1). To simplify notations, we denote these sets $S_{m,k}$ and points $z_{m,k}$ by S_k and z_k, respectively, because the index m is fixed now.

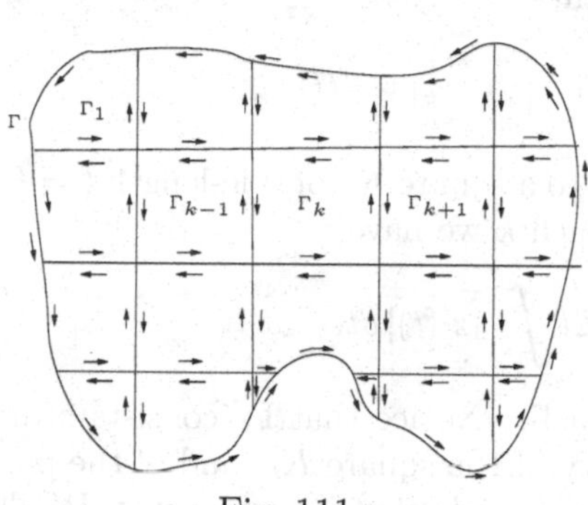

Fig. 111

In Fig. 111 we see the boundary Γ of the domain $\overline{G}$, oriented counterclockwise, together with the union of n smaller paths Γ_k $(1 \leq k \leq n)$, which are boundaries of the sets S_k. Notice that the boundaries Γ_k of incomplete squares S_k may consists of several closed contours. We will show that the integral along Γ is equal to that along $\bigcup_{k=1}^{n} \Gamma_k$, or equivalently

$$\sum_{k=1}^{n} \int_{\Gamma_k} f(z)\, dz = \int_{\Gamma} f(z)\, dz. \tag{A.4}$$

Note that each of the interior edges is integrated over twice, once in each orientation; therefore the integrals over these edges are cancelled. Hence, the only remaining contribution to the sum on the left is the integrals over the outer edges, which combine to form Γ. Thus, we get (A.4).

Let

$$\varphi_k(z) = \begin{cases} \frac{f(z)-f(z_k)}{z-z_k} - f'(z_k), & z \neq z_k; \\ 0, & z = z_k. \end{cases} \tag{A.5}$$

Since $\lim_{z \to z_k} \varphi_k(z) = f'(z_k) - f'(z_k) = 0$, the functions $\varphi_k(z)$ are continuous in $\overline{G}$. According to (A.1),

$$|\varphi_k(z)| < \varepsilon, \quad z \in S_k. \tag{A.6}$$

By (A.5) we have the equality

$$f(z) = f(z_k) + f'(z_k)(z - z_k) + \varphi_k(z)(z - z_k).$$

Obviously, the functions 1 and $z - z_k$ have continuous derivatives. According to the "weak" version of the Cauchy-Goursat Theorem which was proved in Section 5.2 under the additional assumption about continuity of $f'(z)$ (see Theorem 5.7),

$$\int_{\Gamma_k} 1\,dz = 0, \quad \int_{\Gamma_k} (z - z_k)\,dz = 0.$$

These equalities are also valid when S_k is not connected. In this case Γ_k consists of several closed contours. Notice that these integrals can be evaluated directly, without Theorem 5.7. The first equality follows from Example 5.2, because for a closed contour $a = b$. One can obtain the second equality in the similar way. Therefore,

$$\begin{aligned}\int_{\Gamma_k} f(z)\,dz &= f(z_k)\int_{\Gamma_k} dz + f'(z_k)\int_{\Gamma_k} (z - z_k)\,dz + \int_{\Gamma_k} \varphi_k(z)(z - z_k)\,dz \\ &= \int_{\Gamma_k} \varphi_k(z)(z - z_k)\,dz.\end{aligned} \tag{A.7}$$

Property 5.6 and (A.7) imply the inequality

$$\left|\int_{\Gamma_k} f(z)\,dz\right| \le \int_{\Gamma_k} |\varphi_k(z)| \cdot |z - z_k| \cdot |z'(t)|\,dt.$$

Points z and z_k are in S_k, and S_k belongs to a square K_k of side-length $\ell = \ell_m$; hence, $|z - z_k| \le \sqrt{2}\ell$. Using (A.6), for each k we have

$$\left|\int_{\Gamma_k} f(z)\,dz\right| \le \varepsilon\sqrt{2}\ell \int_{\Gamma_k} |z'(t)|\,dt.$$

The last integral represents the length of Γ_k. Notice that Γ_k consists of line segments which are parts of the boundary of the square K_k, and of the parts of the contour Γ which are inside K_k and which also form a part of Γ_k.The total length of the line segments does not exceed 4ℓ; the length of $\Gamma \cap K_k$ we denote by L_k. Thus,

$$\left|\int_{\Gamma_k} f(z)\,dz\right| \le \varepsilon\sqrt{2}\ell(4\ell + L_k) \le \varepsilon\sqrt{2}(4\ell^2 + L_k),$$

since $\ell \le \frac{1}{2} < 1$. Notice that ℓ^2 is the area of the square K_k, and the sum of these areas does not exceed the total area of the squares $K_{1,k}$ intersecting $\overline{G}$. This area is a fixed number independent of ε which we denote by A. The sum of L_k is the length of Γ which we denote by L. Therefore, (A.4) implies the estimate

$$\left|\int_{\Gamma} f(z)\,dz\right| \le \sum_{k=1}^{n}\left|\int_{\Gamma_k} f(z)\,dz\right| \le \varepsilon\sqrt{2}(4A + L).$$

Since $\varepsilon > 0$ can be arbitrarily small, the value of the original integral must be zero. □

Notice that these arguments work for multiply connected domains as well. In this case Γ is a union of several Jordan piece-wise smooth contours. Therefore, we have proved Theorem 5.8 as well.

Inversion theorem for the Laplace transform

Suppose that $F(p)$ is the Laplace transform of some function and $F(p)$ is analytic in a half-plane $\operatorname{Re} p > \sigma_0$. We assume that $F(p) \to 0$ as $|p| \to \infty$ in any half-plane $\operatorname{Re} p > \sigma \geq \sigma_0$. Let $f(t)$ be the function defined by the equality (9.11). We will give an informal argument explaining why the Laplace transform of this function $f(t)$ is the given function $F(p)$. Note that we don't prove the uniqueness of the function those Laplace transform is $F(p)$.

Fix p_1 such that $\operatorname{Re} p_1 > \sigma_0$, and choose $\sigma \in (\sigma_0, \operatorname{Re} p_1)$. Then

$$\int_0^\infty e^{-p_1 t} f(t)\, dt = \frac{1}{2\pi i} \int_0^\infty e^{-p_1 t} \left[\int_{\sigma - i\infty}^{\sigma + i\infty} e^{pt} F(p)\, dp \right] dt.$$

Now we change the order of integration in the last double integral. We will not justify the legality of this operation! For this reason, our argument is not rigorous. We get

$$\frac{1}{2\pi i} \int_{\sigma - i\infty}^{\sigma + i\infty} \int_0^\infty e^{-p_1 t} e^{pt} F(p)\, dt\, dp = \frac{1}{2\pi i} \int_{\sigma - i\infty}^{\sigma + i\infty} F(p) \int_0^\infty e^{(p - p_1)t}\, dt\, dp.$$

Since $\operatorname{Re} p_1 > \sigma$,

$$|e^{(p-p_1)t}| = e^{(\sigma - \operatorname{Re} p_1)t} \to 0 \quad \text{as} \quad t \to \infty.$$

Hence,

$$\int_0^\infty e^{(p-p_1)t}\, dt = \lim_{y \to \infty} \int_0^y e^{(p-p_1)t}\, dt = \lim_{y \to \infty} \frac{1}{p - p_1} e^{(p-p_1)t} \Big|_0^y = -\frac{1}{p - p_1}.$$

Therefore,

$$\int_0^\infty e^{-p_1 t} f(t)\, dt = -\frac{1}{2\pi i} \int_{\sigma - i\infty}^{\sigma + i\infty} \frac{F(p)}{p - p_1}\, dp = \frac{1}{2\pi i} \int_{\sigma + i\infty}^{\sigma - i\infty} \frac{F(p)}{p - p_1}\, dp.$$

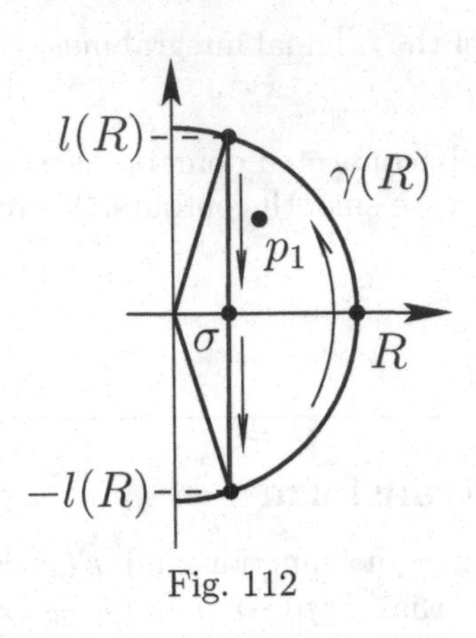

Fig. 112

Notice that in the last integral we are going through the vertical interval from top to bottom.

Let Γ_R be a closed contour consisting of the vertical interval with endpoints

$$\sigma + il(R) \quad \text{and} \quad \sigma - il(R),$$

where $l(R) = \sqrt{R^2 - \sigma^2}$, and the arc

$$\gamma(R) = \{p : |p| = R, \ \operatorname{Re} p \geq \sigma\};$$

see Fig. 112. Let M_R be the maximal value of $|F(p)|$ for $p \in \gamma(R)$. By assumption, $M_R \to 0$ as $R \to \infty$.

Hence,

$$\left| \int\limits_{\gamma(R)} \frac{F(p)}{p - p_1} \, dp \right| \leq \frac{M_R}{R - |p_1|} \cdot \pi R \to 0 \quad \text{as} \quad R \to \infty.$$

We have

$$\int\limits_{\gamma(R)} \frac{F(p)}{p - p_1} \, dp + \int\limits_{\sigma+il(R)}^{\sigma-il(R)} \frac{F(p)}{p - p_1} \, dp = \int\limits_{\Gamma_R} \frac{F(p)}{p - p_1} \, dp = 2\pi i \operatorname{res}_{p_1} \frac{F(p)}{p - p_1},$$

because the function $\frac{F(p)}{p-p_1}$ has singularity only at p_1 inside Γ_R. This point is either a removable singularity or a pole. Hence, $\operatorname{res}_{p_1} \frac{F(p)}{p-p_1} = F(p_1)$. Passing to the limit as $R \to \infty$ we get

$$\lim_{R\to\infty} \int\limits_{\gamma(R)} \frac{F(p)}{p - p_1} \, dp + \lim_{R\to\infty} \int\limits_{\sigma+il(R)}^{\sigma-il(R)} \frac{F(p)}{p - p_1} \, dp = 0 + \int\limits_{\sigma+i\infty}^{\sigma-i\infty} \frac{F(p)}{p - p_1} \, dp = 2\pi i F(p_1).$$

Finally we obtain the desired equality:

$$\int\limits_0^\infty e^{-p_1 t} f(t) \, dt = \frac{1}{2\pi i} \int\limits_{\sigma+i\infty}^{\sigma-i\infty} \frac{F(p)}{p - p_1} \, dp = \frac{1}{2\pi i} \cdot 2\pi i F(p_1) = F(p_1), \operatorname{Re} p_1 > \sigma_0.$$

Thus, the Laplace transform of $f(t)$ is indeed the given function $F(p)$.

Bibliography

[1] L. V. Ahlfors, *Complex Analysis*, McGraw-Hill Book Company, New York, 1953.

[2] R. V. Churchill, *Complex Variables and Applications*, McGraw-Hill Book Company, New York, Toronto, London, 1960.

[3] V. Ya. Eiderman, *Theory of Functions of a Complex Variable and Operational Calculus*, 2th Edition, Urait Publishing House, Moscow, 2018 (in Russian).

[4] S. D. Fisher, *Complex Variables*, Wadsworth & Brooks, 1986.

[5] M. A. Lavrent'ev and B. V. Shabat, *Methods of Theory of Functions of Complex Variables*, 2th Edition, Fizmatgiz, Moscow, 1958 (in Russian).

[6] D. E. Marshall, *Complex Analysis*, Cambridge University Press, 2019.

[7] R. A. Silverman, *Complex Analysis with Applications*, Prentice-Hall, Inc., 1974

[8] J. Stewart, *Single Variable Calculus: Early Transcendentals*, 8th Edition, Cengage Learning, 2014.

[9] J. Stewart, *Multivariable Calculus*, 8th Edition, Cengage Learning, 2015.

Index

Printed in the United States
by Baker & Taylor Publisher Services

Printed in the United States
by Baker & Taylor Publisher Services